# *Honey in Traditional and Modern Medicine*

# Traditional Herbal Medicines for Modern Times

Each volume in this series provides academia, health sciences, and the herbal medicines industry with in-depth coverage of the herbal remedies for infectious diseases, certain medical conditions, or the plant medicines of a particular country.

Series Editor: Dr. Roland Hardman

Traditional Herbal Medicines for Modern Times

# Honey in Traditional and Modern Medicine

Edited by
Laïd Boukraâ

CRC Press
Taylor & Francis Group
Boca Raton London New York

CRC Press is an imprint of the
Taylor & Francis Group, an **informa** business

**Cover plant image:** From Tann, John. 2011. *"Leptospermum polygalifolium"* Prince Edward Park, Woronora, NSW Australia. Available from The Encyclopedia of Life at http://eol.org/data_objects/21925804.

CRC Press
Taylor & Francis Group
6000 Broken Sound Parkway NW, Suite 300
Boca Raton, FL 33487-2742

First issued in paperback 2016

© 2014 by Taylor & Francis Group, LLC
CRC Press is an imprint of Taylor & Francis Group, an Informa business

No claim to original U.S. Government works

Version Date: 20130918

ISBN 13: 978-1-138-19927-9 (pbk)
ISBN 13: 978-1-4398-4016-0 (hbk)

**Visit the Taylor & Francis Web site at**
**http://www.taylorandfrancis.com**

**and the CRC Press Web site at**
**http://www.crcpress.com**

# Contents

# Series Preface

Global warming and global travel are contributing factors in the spread of infectious diseases such as malaria, hepatitis B, and HIV. These are not well controlled by the present drug regimens. Antibiotics also are failing because of bacterial resistance. Formerly less well-known tropical diseases are reaching new shores. A whole range of illnesses, such as cancer, for example, occurs worldwide. Advances in molecular biology, including methods of in vitro testing for a required activity, give new opportunities to draw judiciously on the use and research of traditional herbal remedies from around the world. The reexamining of the herbal medicines must be done in a multidisciplinary manner.

Since 1997, there have been 49 volumes published in the book series Medicinal and Aromatic Plants—Industrial Profiles (volumes 476–479 have been on vanilla, sesame, and citrus oils, respectively). The series continues.

The same series editor is also covering Traditional Herbal Medicines for Modern Times. Each volume reports on the latest developments and discusses key topics relevant to interdisciplinary health sciences research by ethnobiologists, taxonomists, conservationists, agronomists, chemists, clinicians, and toxicologists. The series is relevant to all these scientists and will enable them to guide business, government agencies, and commerce in the complexities of these matters. The background of the subject is outlined next.

Over many centuries, the safety and limitations of herbal medicines have been established by their empirical use by the "healers" who also took a holistic approach. The healers are aware of the infrequent adverse effects and know how to correct them when they occur. Consequently and ideally, the preclinical and clinical studies of an herbal medicine need to be carried out with the full cooperation of the traditional healer. The plant composition of the medicine, the stage of the development of the plant material, when it is to be collected from the wild or when from its cultivation, its postharvest treatment, the preparation of the medicine, the dosage and frequency, and many other essential pieces of information are required. A consideration of the intellectual property rights and appropriate models of benefit sharing may also be necessary.

However the medicine is being prepared, the first requirement is a well-documented reference collection of dried plant material. Such collections are encouraged by organizations such as the World Health Organization and the United Nations Industrial Development Organization. The Royal Botanic Gardens at Kew (United Kingdom) is building its collection of traditional Chinese dried plant material relevant to its purchase and use and sells or prescribes traditional Chinese medicine in the United Kingdom.

In any country, control of the quality of plant raw material, its efficacy, and its safety in use are essential. The work requires sophisticated laboratory equipment and highly trained personnel. This kind of "control" cannot be applied to the locally produced herbal medicines in the rural areas of many countries, on which millions of people depend. Local traditional knowledge of healers has to suffice.

Conservation and protection of plant habitats are required, and breeding for biological diversity is important. Gene systems are being studied for medicinal exploitation. There can never be too many seed conservation "banks" to conserve genetic diversity. Unfortunately, such banks are usually dominated by agricultural and horticultural crops, with little space for medicinal plants. Developments such as random amplified polymorphic DNA enable the genetic variability of species to be checked. This can be helpful in deciding whether specimens of close genetic similarity warrant storage.

From ancient times, a great deal of information concerning diagnosis and the use of traditional herbal medicines has been documented in the scripts of China, India, and elsewhere. Today, modern formulations of these medicines exist in the form of powders, granules, capsules, and tablets. They are prepared in various institutions, such as government hospitals in China and North and South Korea, and by companies such as the Tsumura Company of Japan, with good quality control. Similarly, products are produced by many other companies in India, the United States, and elsewhere with varying degrees of quality control. In the United States, the Dietary Supplement and Health Education Act of 1994 recognized the class of physiotherapeutic agents derived from medicinal and aromatic plants. Furthermore, under public pressure, the US Congress set up an Office of Alternative Medicine. In 1994, this office assisted in the filing of several Investigational New Drug (IND) applications required for clinical trials of some Chinese herbal preparations. The significance of these applications was that each Chinese preparation involved several plants and yet was handled with a single IND. A demonstration of the contribution to efficacy of each ingredient of each plant was not required. This was a major step forward toward more sensible regulations with regard to phytomedicines.

The subject of western herbal medicines is now being taught to medical students in Germany and Canada. Throughout Europe, the United States, Australia, and other countries, pharmacy and health-related schools are increasingly offering training in phytotherapy. Traditional Chinese medicine clinics are now common outside China. An Ayurvedic hospital now exists in London with a BSc Honors degree course in Ayurveda available (Professor Shrikala Warrier, Registrar/Dean, MAYUR, Ayurvedic University of Europe, 81 Wimpole Street, London, WIG 9RF, United Kingdom. E-mail: sw@unifiedherbal.com). This is a joint venture with a university in Manipal, India.

The term "integrated medicine" is now being used, which selectively combines traditional herbal medicine with "modern medicine." In Germany, there is now a hospital in which traditional Chinese medicine is integrated with western medicine. Such co-medication has become common in China, Japan, India, and North America through those educated in both systems. Benefits claimed include improved efficacy, reduction in toxicity and the period of medication, as well as a reduction in the cost of the treatment. New terms such as "adjunct therapy," "supportive therapy," and "supplementary medicine" now appear as a consequence of such co-medication. Either medicine may be described as an adjunct to the other depending on the communicator's view. Great caution is necessary when traditional herbal medicines are used by doctors not trained in their use and likewise when modern medicines are used by traditional herbal doctors. Possible dangers from drug interactions need to be stressed.

The world's population is said to be growing at the rate of 73 million per year. This equates to 139 more mouths to feed every minute. People are living longer. Others can now afford to eat meat, so more grain is in demand for animal feed.

Farmers say that without the use of, for example, neonicotinoid to treat their crops of wheat, oil seed rape, and maize, their yields will be down by 30%, and the food shortage will get worse. The bees are trying to overcome this chemical affront to their Darwinian epigenes and also attack by, for example, the latest generation of mites. Fortunately, there are some saintly beekeepers who strive always to have healthy flourishing hives and succeed in dispatching these by overnight plane to farmers and fruit growers all over the world. Thus, in addition to the wild bees, we continue to have honey available for a wide range of medicinal and dietetic purposes with the attendant analytical procedures to ensure they fulfill their function.

For all his hard work, I thank the editor, Laïd Boukraâ, and all the other chapter contributors for their enthusiasm and specialized knowledge. My thanks are also due to the staff of CRC Press, Barbara Norwitz and John Sulzycki, for their ready support and to the project coordinator, Joselyn Banks-Kyle, for her help in the important final stage.

**Roland Hardman, BPharm, BSc (Chemistry), PhD**
**(London University), FRPharmS**
*Reader and Head of Pharmacognosy (retired),*
*School of Pharmacy and Pharmacology,*
*University of Bath, United Kingdom*

# Preface

*Thy Lord has inspired the Bees,*
*to build their hives in hills,*
*on trees and in man's habitations,*
*From within their bodies comes*
*a drink of varying colors,*
*wherein is healing for mankind,*
*Verily in this is a Sign,*
*for those who give thought.*
*(Koran, Surah XVI: 68–69.)*

With accumulated experience, natural medicinal products have been used for millennia to treat many ailments. The resurgence of this interest by both the general public and physicians continues.

Honey has a long history of traditional use in medicine in a large number of societies. Both the Bible and the Koran recommend its use, and in this volume, evidence is given from Ayurvedic medicine and ethnomedicine.

Honey is in constant high demand. China is the leading producer on the world market. Inevitably, the Chinese honey is a blend of honeys from the temperate and tropical regions of the country from many thousands of diligent beekeepers.

Local people know their own honeys and are aware of how they change in color, odor, flavor, and composition with the different species of flowers as these become available.

The flowers and the bees were on this Earth long before man. The bees' epigenes protect them from the toxic principles of, for example, the alkaloids in the flowers of the *Datura* species, but the honey from this source is toxic to man. Local knowledge is an advantage, as in the case of the "mad honey" from *Rhododendron* plants of Turkey, which is rarely fatal.

The importance of the botanical origin of the honey in relation to its medicinal use cannot be overstressed. However, an extensive study of the composition of unifloral honeys of named countries by independent researchers affords much useful data (compounds present and those absent, etc.) but sadly also reveals the overwhelming complexities of the subject. More straightforward are the important "no peroxide" manuka honeys (from *Leptospermum scoparium* of New Zealand and *L. polygalifolium* of Australia). These are relatively rare with their methylglyoxal content, along with various glycosides, polysaccharides, peptides, and proteins. The composition gives these honeys a broad antibacterial activity range including that against antibiotic-resistant strains of bacteria—at a time of severe criticism of the pharmaceutical industry for its failure to produce new antibiotics.

Honey has been proven to be beneficial in burn and wound management due to its physical and biochemical properties but especially because of the absence of adverse effects. Indeed, honey has a role to play in many medical conditions, such as diabetic ulcers, gastrointestinal disorders, cardiovascular diseases, and cancers, and in pediatrics and in animal health. Bee products (propolis, royal jelly, and bee venom) all find use in human and animal medicine.

As a nutrient, honey has a very long history as a widely consumed safe food—its low water content protecting it from microbial spoilage. Hence, there is a long tradition of the use of honey in culinary practice, and today, honey has new roles in the food science industry.

Chemical analysis of honey has never been more important and the modern methods for achieving this are described with examples.

I wish to thank Dr. Roland Hardman, Series Editor, whom I have not personally met, but who was always involved and guided this project from the first day when the book was just a dream.

My thanks go also to the CRC Press senior editors Barbara Norwitz and John Sulzycki and to their project coordinator Joselyn Banks-Kyle.

I would like to acknowledge the immense contribution that all authors have made. Without them, the book would not know the light of day.

Finally, I thank my wife Djoher for her patience and the obvious sacrifices she has made during the period of the project.

# Editor

**Laïd Boukraâ, DVM, MSc, PhD**, associate professor, is a veterinary surgeon who graduated from the Ibn-Khaldoun University of Tiaret (Algeria). He did his postgraduate studies in microbiology at the Faculty of Veterinary Medicine at Montreal University in Quebec (Canada). He obtained a postdoctoral position at the Department of Pharmacology at Universiti Sains Malaysia (Malaysia). Besides teaching, he also chairs the Laboratory of Research on Local Animal Products at Ibn-Khaldoun University of Tiaret, and is leading a national research project on the use of medicinal plants and api-products in dairy herds. He is also a researcher at "Mohammed Hussein Al Amoudi Chair for Diabetic Foot Research" at King Abdulaziz University, Jeddah, Saudi Arabia. He has published many papers and book chapters related to the bioactive properties of api-products and attended many international conferences as well.

# Contributors

**Fatiha Abdellah**
Laboratory of Research on Local
 Animal Products
and
Institute of Veterinary Sciences
Ibn-Khaldoun University of Tiaret
Tiaret, Algeria

**Leila Ait Abderrahim**
Laboratory of Research on Local
 Animal Products
and
Faculty of Natural and Life Sciences
Ibn-Khaldoun University of Tiaret
Tiaret, Algeria

**Saâd Aissat**
Laboratory of Research on Local
 Animal Products
Ibn-Khaldoun University of Tiaret
Tiaret, Algeria

**Nadia Alam**
Department of Botany
Rajshahi University
Rajshahi, Bangladesh

**Rezzan Aliyazicioglu**
Faculty of Pharmacy
Karadeniz Technical University
Trabzon, Turkey

**Hasan Ali Alzahrani**
Department of Surgery
Medical College
King Abdulaziz University
Jeddah, Saudi Arabia

**Faik Ahmet Ayaz**
Department of Biology
Faculty of Science
Karadeniz Technical University
Trabzon, Turkey

**Yuva Bellik**
Faculty of Nature and Life Sciences
Abderrahmane Mira University
Bejaia, Algeria

**Hama Benbarek**
Faculty of Sciences of Nature and Life
Mascara University
Mascara, Algeria

**Laïd Boukraâ**
Laboratory of Research on Local
 Animal Products
Ibn-Khaldoun University of Tiaret
Tiaret, Algeria

**Akhlaq A. Farooqui**
Department of Molecular and Cellular
 Biochemistry
The Ohio State University
Columbus, Ohio

**Tahira Farooqui**
Department of Entomology
Center for Molecular Neurobiology
The Ohio State University
Columbus, Ohio

**Abdülkadir Gunduz**
Department of Emergency Medicine
Faculty of Medicine
Karadeniz Technical University
Trabzon, Turkey

**Mokrane Iguerouada**
Faculty of Nature and Life Sciences
Abderrahmane Mira University
Bejaia, Algeria

**Aldina Kesic**
University in Tuzla
Faculty of Science
Department of Chemistry
Tuzla, Bosnia and Hercegovina

**Md. Ibrahim Khalil**
Department of Pharmacology
School of Medical Sciences
University Sains Malaysia
Kelantan, Malaysia

and

Department of Biochemistry and
    Molecular Biology
Jahangirnagar University
Dhaka, Bangladesh

**Sevgi Kolayli**
Department of Chemistry
Faculty of Sciences
Karadeniz Technical University
Trabzon, Turkey

**Juraj Majtan**
Institute of Zoology
Slovak Academy of Sciences
and
Department of Microbiology
Faculty of Medicine
Slovak Medical University
Bratislava, Slovakia

**Abdelmalek Meslem**
Veterinary Sciences Institute
Ibn Khaldoun University
Tiaret, Algeria

**Lutfun Nahar**
Leicester School of Pharmacy
De Montfort University
Leicester, United Kingdom

**Nor Hayati Othman**
School of Medical Sciences
Universiti Sains Malaysia
Kelantan, Malaysia

**Surya Prakash**
Post Graduate Department of Salya
Dr. B.R.K.R. Government Ayurvedic
    College
Hyderabad, India

**Ratna Rao**
Apollo Hospital
Hyderabad, India

**Hüseyin Sahin**
Department of Chemistry
Faculty of Sciences
Karadeniz Technical University
Trabzon, Turkey

**Satyajit D. Sarker**
Department of Pharmacy
School of Applied Sciences
University of Wolverhampton
Wolverhampton, United Kingdom

**Yasmina Sultanbawa**
Queensland Alliance for Agriculture
    and Food Innovation (QAAFI)
Centre for Nutrition and Food Sciences
The University of Queensland
Queensland, Australia

**Oktay Yildiz**
Department of Chemistry
Faculty of Sciences
Karadeniz Technical University
and
Maçka Vocational School
Trabzon, Turkey

# 1 Honey
## *An Ethnomedicine*

*Yuva Bellik and Laïd Boukraâ*

## CONTENTS

## INTRODUCTION

Many products based on traditional knowledge are important sources of income, food, and health care for the large parts of the populations throughout the world. By the time ancient civilizations began to evolve, a large number of treatments had been discovered by prehistoric and primitive peoples alongside a body of magical or mythological therapies and won widespread use (Forrest 1982). Honey has been used for centuries for its nutritional as well as medicinal properties (Ajibola et al. 2012). The human use of honey is traced to some 8000 years ago as depicted by Stone Age paintings. Different traditional systems of medicine have elaborated the role of honey as a medicinal product. Sumerian clay tablets (6200 BC), Egyptian papyri (1900–1250 BC), Vedas (5000 years), Holy Koran, the Talmud, both the Old and New Testaments of the Bible, sacred books of India, China, Persia, and Egypt (Mcintosh 1995; Beck and Smedley 1997), and Hippocrates (460–357 BC) (Jones 2001) have described the uses of honey. The latter described its use for baldness, contraception, wound healing, laxative action, cough and sore throat, eye diseases, topical antisepsis, and the prevention and treatment of scars (Bansal et al. 2005).

Undoubtedly, honey is in no way a new discovery since its use goes back to ancient times and it has been used as an ethnoremedy in many parts of the world, both internally and externally. Nevertheless, its use continues into modern folk medicine. It is still one of the most frequently used remedies.

## HONEY IN THE PREHISTORY

The history of the use of honey is parallel to the history of man. A reappraisal of the evidence from the Stone Age, antiquity, the Middle Ages, and early modern times suggests that ordinary people ate much larger quantities of honey than has previously been acknowledged (Allsop and Miller 1996).

There are many cave drawings or paintings of people collecting honey. The most important history cave drawing is at Bicorp in eastern Spain, known as the Arana or Spider Cave (Figure 1.1). Its walls bear a painting figure of a human on a rope ladder ascending a cliff face (McGee 1984). The painting is one of the earliest firm evidences on honey hunting by humans.

There are several other such paintings of raids on bees' nests, dating from the remotest antiquity (Figure 1.2). Their large numbers may indicate that honey was highly valued and considered as an important food source by early humans (Allsop and Miller 1996).

Studies of primitive tribal diets indicate that honey is highly nutritious (Zucoloto 2011). It is well known that, for the Hadza of Tanzania, meat and honey combined constitutes 20% by weight of food eaten (Woodburn 1963). The Mbuti pygmies of the Congo obtain as much as 80% of their dietary energy from honey during the honey season. The Veddas or Wild Men of Sri Lanka esteem honey so highly that they regularly risk their lives to obtain it (Crane 1983).

Many Australian aboriginal tribes regard the honey of the native bee as the supreme delicacy (Low 1989). Meehan lived for a year (1972–1973) with the native Anbarra people of northern Australia. Over four 1-month periods, chosen to be representative of the different tropical seasons, she recorded the weights of foods consumed. She reported an average intake of 2 kg of honey per person per year (Meehan 1982).

In the New World, the Guayaki Indians of Paraguay have honey as the very basis of their diet and culture (Crane 1975). In the rest of the world, it is usually the males of a tribe who hunt for honey, but among some Australian aborigines this task falls to the women. One method they employ involves capturing a bee and attaching a small feather to its body, so that, on release, it can be more easily seen and followed

**FIGURE 1.1**  Prehistoric man gathering honey. A rock painting made in 6000 BC, Cueva de la Arana, near Valencia, Spain.

**FIGURE 1.2** Beekeeping in ancient Egypt (Tomb of Pabasa). 26th Dynasty, 760–656 BC.

all the way back to the nest. On removing the contents, Australian aborigines eat everything: honey, wax, dead bees, and brood (which provides protein), with relish.

## HONEY IN THE CIVILIZATIONS

Every culture evidence can be found of the use of honey as a food source and as a symbol employed in religious, magic, and therapeutic ceremonies (Cartland 1970; Crane 1980; Zwaenepoel 1984).

The ancient Egyptian civilization existed for several centuries. Egyptian temples kept bees to satisfy the demand for honey as offerings to the gods as well as for other domestic uses such as mummifying, boat and ship building, and a binding agent for paints and in metal casting (Harris Papyrus). Egyptians offered their gods honeycombs overflowing with honey as a valuable gift to show devotion and worship. In the 12th century BC, Ramses III offered 15 tons of honey to Hapi, the god of the Nile (Crane 1975). Jars of honey were buried with the dead as sustenance for the afterlife. Archaeologists have found clay pots filled with honey in a pharaoh's tomb in the city of Thebes. Large quantities of honey in jars were also found in the tomb of Tutankhamen (Hymns of the Atharva-Veda). Burying the dead in or with honey was common practice in Egypt, Mesopotamia [Iraq], and other regions. It is rumored that Alexander the Great was buried in honey (Hajar 2002).

Some writers believe that, for the ancient Egyptians, honey was a luxury item, sold at prices only the wealthiest could afford (Darby et al. 1977). Egyptian hieroglyphics dating back at least 3000 years indicate that honey was used as a sweetener—mixing it with various fruits, nuts, herbs, and spices in breads, cakes, and pastries (Tannahill 1975). The pharaohs also used honey in their wedding celebrations. This custom was passed on to Greco-Roman culture and handed down to medieval Europe. Newlyweds drank honey wine (mead) for a month after the wedding ceremony for good luck and happiness. The ritual gave rise to the word honeymoon, a custom still practiced today (Hajar 2002).

The ancient Greeks no doubt acquired knowledge of beekeeping from Egypt. Pythagoras is said to have lived largely on honey and bread, and the bodies of his countrymen who died some distance from home were sometimes preserved in honey (Free 1982).

In ancient Rome, honey is used in a wide range of dishes and it was a necessary ingredient of many sauces. Wines drunk at the beginning and end of meals were sweetened with honey; meat, fruit, and vegetables were sometimes preserved by immersion in it (Free 1982). Half of the 468-odd recipes in a late Roman cookery book, credited to Apicius, call for honey as an ingredient (Style 1992).

China is known to have imported honey. One official was granted on his retirement in AD 500 a quart of white honey each month (Crane 1975). This amounts to just more than 19 kg of honey per year.

In late Bronze Age Britain, beeswax was an important commodity because it was required for bronze casting (Wilson 1973). Large volumes of wax had the pleasant concomitance of large volumes of honey, and the Britons were pleased to make wide use of it in their food.

Mead was an alcoholic drink made by fermenting the final washings of honey from the comb in a solution with water. Although it is almost unknown today, it was very widely enjoyed from the very early Middle Ages until as late as the 17th century (McGee 1984), especially in those areas where grapes were not available to produce grape wines (Tannahill 1975). Large quantities of honey must have been available just to meet the demand for this beverage.

The banquets held by the nobility of the Middle Ages were spectacular affairs. Huge numbers of guests in fine apparel would partake of a truly vast array of exotic dishes served on silver platters (Cosman 1976). However, such feasts are hardly representative of the everyday food habits of that time. Historians and others hold the view that honey was not widely used in medieval times. Crane (1975) speculates that a typical per capita consumption would have been approximately 2 kg/year, and Free (1982) believes that honey was not extensively used in cooking in medieval Europe. Rogers (1866) expresses his surprise that bees were not more commonly kept because an occupied hive was the only agricultural commodity that could double in value over a year. In fact, the medieval manor frequently employed a beekeeper to tend to hives on-site (Wilson 1973). The lower classes too must often have kept bees for their family's use, but no record of this could be expected to remain. It is well known that, in Ireland in about AD 440, honey was stored in different types of jars (Free 1982). A garden near the border of Flanders in AD 800 is recorded to have produced three muids of honey, that is, a little less than a tonne, in 1 year (Crane 1980). Another writer holds the view that a manor near Rheims, France, possessed 21 beehives. In England in about AD 1350, honey cost approximately 7 pence/gallon (equivalent to 1.3 pence/kg) (Rogers 1866).

Most European monasteries and abbeys kept bees, although not primarily for the honey. The bee was seen as an industrious, selfless worker for the greater good, which, in addition, was chaste. As such, its pure white and relatively odorless wax was considered far more suitable for the manufacture of church candles than animal fats (Free 1982). In the 1520s, the dissolution of the monasteries reduced the demand for beeswax for church candles and brought about a small decrease in the production of honey. Almost simultaneous with this came an increase in the supply of sugar, imported from the new European colonies. Sugar was still considerably more expensive than honey, but this combination of events gained it a more complete following among the wealthy. Many it seems were even indulging to excess, for in 1598 a foreign visitor remarked of Queen Elizabeth that her teeth were black, a defect that the English seemed subject to from their too great use of sugar (Best 1986). Cookery books were used exclusively by the well-to-do at this time and clearly illustrate that, for this section of society, sugar had, by the 1550s, usurped honey's place in the diet. It was not until the early 1700s, however, when the supply of sugar boomed, its price fell, and coffee, tea, and chocolate entered the British diet, that ordinary people finally began to buy significant amounts, so that the per capita consumption reached 1.8 kg/year (McGee 1984). The change over from honey to sugar occurred more gradually in rural areas than in the cities. Just 80 years later, sugar consumption had trebled to reach 5.4 kg per capita per year. Honey was no longer the standard sweetening agent. From this point, sugar consumption rose inexorably, whereas honey consumption declined. Beekeeping ceased to be the general custom that it had been in former years; there was no longer a hive in every garden.

## HONEY AND RELIGIONS

The Bible (both the Old and New Testaments), the Talmud, the Koran, and the sacred books of India, China, Persia, and Egypt, all speak of honey in laudatory terms as a food, beverage, and medicine.

Honey received religious endorsement by both Christianity and Islam (Zumla and Lulat 1989). When the Children of Israel were in Egypt or were journeying through the desert, the promise was made that their destination was to be "a land flowing with milk and honey" (Taylor 1971). Honey was very important and has been mentioned 54 times in the Old Testament. The most famous is the saying of King Solomon in his Proverbs 24:13 advised, "My son, eat honey; it is good. And just as honey from the comb is sweet on your tongue, you may be sure that wisdom is good for the soul. Get wisdom and have a bright future." The Jews believed honey made a person "mentally keen" (Hajar 2002). In the New Testament, it plays a role in the resurrection of Christ: "The day Christ rose from the dead and appeared before His Disciples, He asked for food. They gave Him broiled fish and a honeycomb (Luke 24:42). Christ ate the food to prove to the Apostles that He was truly resurrected and not merely a Spirit or a Thought."

In early Christianity, honey had a deep mysterious meaning and it was given in christening ceremonies as a symbol of renovation and spiritual perfection. When St. John the Baptist was in the desert, he lived on honey and locusts. Bees were revered because of their ability to produce wax and therefore provide light, in many cases for religious practices. Before the advent of electricity, candles were important for lighting. In many religions, candles are still used for religious ceremonies. The Roman Catholic Church regarded the bee as an example of godliness and believed beeswax to be "pure" because virgins—that is, nonmating worker bees—produce it. Monks kept bees to provide beeswax to make candles for the church. The Catholic Church still requires that their candles contain a certain amount of beeswax (Hajar 2002).

The holy Koran vividly illustrates the potential therapeutic value of honey (Irving et al. 1987).

> Thy Lord has inspired the Bees,
> to build their hives in hills,
> on trees and in man's habitations,
> From within their bodies comes
> a drink of varying colours,
> wherein is healing for mankind,
> Verily in this is a Sign,
> for those who give thought. (Koran, Surah XVI:68–69)

The Prophet Mohammed (peace be upon him) pronounced: "Honey is a remedy for all diseases." The Prophet ordered the eating of honey not only because it was an exquisite food and an important healing substance but also because it brought good luck. The followers of Islam looked upon honey as a talisman (Bogdanov 2009).

In many obvious cases, many companions of the Prophet (peace be upon Him) could treat a patient suffering from certain diseases during that time successfully without any knowledge of medicine as practiced today, but they merely followed the instruction of the Prophet (peace be upon Him) relating to the treatment of disease, that is, a gulp of honey, cupping (hijamah), and cauterization ("kayy": to burn a wound with hot metal or a chemical to stop the blood or stop it from becoming infected) (Deuraseh 2006). In the Islamic medical system, as in most other medical systems, honey is considered as a healthy drink. It is not surprising that Imam Bukhari entitled chapter four of his Kitab al-Tibb (book of medicine) as "al-Dawa'bi al-Asal wa Qawlihi Ta'ala 'Fihi Shifa li al-Nas (treatment with honey and the statement of Allah: where is healing for men)." Accordingly, the Prophet (peace be upon Him) said, "A man came to the Prophet and said: 'My brother has some abdominal trouble.'" The Prophet (peace be upon Him) replied to him, "Let him drink honey." The man came for the second time and the Prophet replied to him, "Let him drink honey." He came for the third time and the Prophet replied, "Let him drink honey." He returned again and said, "I have done that." The Prophet (peace be upon Him) then said, "The statement of God is true and the stomach of your brother lies (sadaqa Allah wa kadhiba Batn Akhika). Let him drink honey. So he made him drink honey and he was cured."

With this illness, it is honey that expels the excess moisture, because the moisture is driven out and expelled downward when honey is eaten. In "Umdah al-Qari," Ibn Ahmad al-Ayni expressed and recognized that drinking honey may open up the obstructions of the blood vessels, dissolve the excessive food by evacuating the stomach and intestines, and clear the chest and liver. Furthermore, al-Baghdadi was of the opinion that honey, which contains a variety of sugar and mineral, is good to purify what is in the veins and stomach. Consequently, it has the potential to make the blood to circulate better and provide more air to areas of the body such as the brain (Deuraseh 2006).

## HONEY AS AN ETHNOREMEDY

The use of honey as a medicine is referred to in the most ancient written records, it being prescribed by the physicians of many ancient races of people for a wide variety of ailments (Ransome 1937). It has continued to be used in folk medicine ever since. There are abundant references to honey as medicine in ancient scrolls, tablets, and books. It was prescribed for a variety of illnesses. Excavated medical tablets from Mesopotamia indicate that honey was a common ingredient in many prescriptions (Hajar 2002).

In ancient Egyptian medicine, honey was the most frequent ingredient in all the drug recipes for both internal and external use listed in the Ebers and Edwin Smith Papyri. According to the Ebers papyrus (1550 BC), it is included in 147 prescriptions in external applications. Also, according to the Smith papyrus (1700 BC), it was used in wound healing: "Thou shouldst bind [the wound] with fresh meat the first day [and] treat afterwards with grease, honey [and] lint every day until he recovers." Honey was used for treatment of stomach pain and urinary retention and as ointment for dry skin. It was used as ointment for wounds and burns, skin irritation, and eye diseases. The Ebers Papyrus contains a description on how to make ointment from honey and how to apply it, with a note: "Notice that this is a very good therapy." The author of the Smith Papyrus directed that honey be applied topically, with few if any other possibly active ingredients, to wounds.

In old Egypt, honey was the only active ingredient in an ointment described in the Ebers Papyrus for application to the surgical wound of circumcision. Ebers also specifies that an ointment for the ear be made of one-third honey and two-thirds oil. The concentration of honey in seven oral remedies in the Chester Beatty VI Papyrus ranges from 10% to 50%, whereas its proportion in other remedies ranges from 20% to 84%. Honey could very well have provided some kind of protection from the kinds of bacteria most likely to infect wounds, at least enough protection to permit wounds to begin healing on their own.

The ancient Egyptians were not the only people who used honey as medicine. The Chinese, Indians, ancient Greeks, Romans, and Arabs used honey in combination with other herbs and on its own to treat wounds and various other diseases.

In old Greece, the honeybee, a sacred symbol of Artemis, was an important design on Ephesian coins for almost six centuries. Aristotle (384–322 BC) described for the first time the production of honey. Aristotle believed that eating honey prolonged life. Hippocrates (460–377 BC) speaks about the healing virtues of honey: "cleans sores and ulcers, softens hard ulcers of the lips, heals cabuncles and running sores." Hippocrates is quoted as saying, "I eat honey and use it in the treatment of many diseases because honey offers good food and good health." Dioscorides (AD 40–90), a Greek physician who traveled as a surgeon with the armies of the Roman emperor Nero, compiled *De Materia Medica* around AD 77, which was the foremost classic source of modern botanical terminology and the leading pharmacologic text until the 15th century. In addition to excellent descriptions of nearly 600 plants and 1000 simple drugs, Dioscorides described the medicinal and dietetic value of animal derivatives such as milk and honey. Dioscorides stated that honey could be used as a treatment for stomach disease, for a wound that has pus, for hemorrhoids, and to stop coughing. "Honey opens the blood vessels and attracts moisture. If cooked and applied to fresh wounds, it seals them. It is good for deep dirty wounds. Honey mixed with salt could be dropped inside a painful ear. It will reduce the pain and swelling of the ear. It will kill lice if infested children skin is painted with it. It may also improve vision. Gargle with honey to reduce tonsil swelling. For coughing, drink warm honey and mix with rose oil." Galen recommended warming up the honey or cooking it, then using it to treat hemorrhoids and deep wounds.

In ancient Rome, honey was mentioned many times by the writers Vergil, Varro, and Plinius. Especially Virgil's Georgics is a classic where he describes in detail how honey is made. During the time of Julius Caesar, honey was used as a substitute for gold to pay taxes. In the first century AD, Apicus, a wealthy Roman gourmet, wrote a series of books in which more than half the recipes included honey (Bogdanov 2009). A Roman Catholic saint (St. Ambrose) stated, "The fruit of the bees is desired of all and is equally sweet to kings and beggars and is not only pleasing but profitable and healthful, it sweetens their mouths, cures their wounds, and conveys remedies to inward ulcers." The Roman, Pliny the Elder, said that mixing fish oil with honey was an excellent treatment for ulcers.

In medieval high cultures of the Arabs, the Byzantines, and medieval Europe, honey was important too, and in these cultures, most sweet meals contained honey.

The *Compendium of Medicine* by Gilbertus Anglicus is one of the largest sources of pharmaceutical and medical information from medieval Europe. Translated in the early 15th century from Latin to Middle English, the text consists of medicinal recipes with guides to diagnosis, medicinal preparation, and prognosis. The text names more than 400 ingredients. Treatments are presented roughly from "head to tail," so to speak, beginning with headache and ending with hemorrhoids. Honey was a frequent ingredient to many of the remedies and it was combined with other medicinal herbs commonly used at that time. Excerpts appear below:

> Headache … let him use oxymel … made of honey and vinegar; two parts of vinegar and the third part of honey, mixed together and simmered. Pimples … anoint it with clean honey, or with the powder of burnt beans and honey, or with the powder of purslane and honey mixed together. Pennyroyal … taken with honey, cleanse the lungs and clear the chest of all gross and thick humors. (Fay Marie Getz 1991)

Germans used honey and cod liver oil for ulcerations, burns, fistulas, and boils in addition to a honey salve, which was mixed with egg yolk and flour for boils and sores (Newman 1983).

AlBasri (Ali Bin Hamzah AlBasri), a 10th century Arab philosopher, mentioned uncooked honey for swollen intestine, whereas cooked honey was good for inducing vomiting when a poisonous drug was ingested. For that purpose, he recommended mixing one pound of sesame oil with one-third pound of cooked honey. Al Razi (Rhazes, AD 864–932), a renowned Muslim physician famous for writing a treatise distinguishing measles from smallpox, claimed that honey ointment made of flour and honey vinegar was good for skin disease and sports nerve injuries and recommended the use of honey water for bladder wounds. His book, *Al Hawi* (*Encyclopedia of Medicine*), a comprehensive medical textbook of medicine, which was translated from Arabic to Latin in the 13th century and became a standard textbook of medicine up to the 1700s stated: "Honey is the best treatment for the gums. To keep the teeth healthy mix honey with vinegar and use as mouth wash daily. If you rub the teeth with such a preparation it will whiten the teeth. Honey does not spoil and could also be used to preserve cadavers." Likewise, Ibn Sina (Avicenna), another famous Muslim physician whose great medical treatise, the *Canon*, was the standard

**TABLE 1.1**
**Selected Historical Uses of Honey**

| Source | Use |
|---|---|
| Hippocrates (460–357 BC) | Wound healing |
| Aristotle (384–322 BC) | Refers to pale honey as a "good salve for sore eyes and wounds" |
| Celsius (circa 25 AD) | Laxative; cure for diarrhea and upset stomach; for coughs and sore throats; to agglutinate wounds; eye diseases |
| Dioscorides (circa 50 AD) | States that pale yellow honey from Attica was the best honey, being "good for all rotten and hollow ulcers" as well as for sunburn, inflammation of throat and tonsils, and as a cure for coughs |
| Ancient Greeks, Romans and Chinese | Topical antiseptic for sores, wounds, and skin ulcers |
| Chinese | Prevent scars; remove discoloration and freckles and improve the general appearance of the skin; abscesses and hard swelling or callosity of skin; for cancer applied to inflamed wounds to reduce the pain and lessen the drawing; diluted honey used for inflammation of the eye; treat small pox; remove worms and for diseases of the mouth and throat |
| Culpepper's *Complete Herbal* (17th Century) | Many remedies, for example, uses of garden rue for "worms of the belly" |
| Sir John Hill (1759) | Book on "*The Virtues of Honey in Preventing Many of the Worst Disorders, and in the Cure of Several Others*" <br> Antibacterial activity reported |
| American Pharmaceutical Association (1916, 1926, 1935) | For general coughs—mixture of barley water, honey, and the juice of one lemon; honey of rose with borax for treating sore throats and skin ulcers |
| U.S. Medical Archives | Many applications and prescriptions; prevention of infection in wounds |

*Sources:* American Pharmaceutical Association, *The National Formulary IV*, 1916, 1926, 1935; Kelly, E.C., *Medical Classics*, Williams and Wilkins Company, Baltimore, Maryland, 1940–1941, V 1–5, 10; Newman, T.G., *Honey Almanac*, Newman, Chicago, Illinois, 1983; Zumla, A. and Lulat, A., *Journal of the Royal Society of Medicine*, *82*, 384–385, 1989; Mcintosh, E.N., *American Food Habits in Historical Perspective*, Praeger Publishers, Westport, Connecticut, 1995; Beck, D.F., and Smedley, D., *Honey and Your Health: A Nutrimental, Medicinal and Historical Commentary* (originally published in 1938), Health Resources, Inc., Silver Springs, Maryland, 1997; Jones, R., Honey and healing through the ages, in *Honey and Healing*, eds. Munn, P., and Jones, R., International Bee Research Association, Cardiff, U.K., 2001; Molan, P.C., Why honey is effective as a medicine. 1. Its use in modern medicine, in *Honey and Healing*, eds. Munn, P. and Jones, R., International Bee Research Association, Cardiff, U.K., 2001.

textbook on medicine in the Arab world and Europe until the 17th century, wrote: "Honey is good for prolonging life, preserve activity in old age. If you want to keep your youth, take honey. If you are above the age of 45, eat honey regularly, especially mixed with chestnut powder. Honey and flour could be used as dressing for wounds. For lung disease, early stage of tuberculosis, use a combination of honey and shredded rose petals. Honey can be used for insomnia on occasions."

The Hindu Scripture, *Veda*, which was composed about 1500 BC and written down about 600 BC, speak of "this herb, born of honey, dripping honey, sweet honey, honied, is the remedy for injuries. Lotus honey is used for eye diseases. It is used as topical eye ointment in measles to prevent corneal scarring" (Imperato and Traore 1969), "moreover it crushes insects." In the section on *Hymn to All Magic and Medicinal Plants*, honey is used as a universal remedy: "The plants ... which removes disease, are full of blossoms, and rich in honey ... do I call to exempt him from injury" (Bogdanov 2009).

In ancient China, honey has been mentioned in the book of songs *Shi Jing*, written in the 6th century BC. According to Chinese medicine, honey acts according to the principles of the Earth element, acting mainly on the stomach and on the spleen. It has Yang character, acting on the Triple Heater Meridian (Shaoyang) (Bogdanov 2009).

In Central and South America, honey from stingless bees was used for ages, long before Columbus. Honey of the native stingless bees was used and regarded as a gift of the gods; it was also a sign of fertility and was given as an offering to the gods (Bogdanov 2009).

Africa has also a long tradition of a bee use for honey, both in the high cultures of Mediterranean Africa and in the more primitive cultures in regions to the south. Honey is used to treat infected leg ulcers in Ghana (Ankra-Badu 1992) and earaches in Nigeria (Obi et al. 1994). Other uses include treatment of gastric ulcers and constipation (Molan 1999).

Table 1.1 summarizes some of the ways honey has been used through the ages.

It is clear that, throughout the ages, honey was prescribed for a variety of uses, frequently mixed with herbs, grains, and other botanicals. Obviously, remedies were passed down through the millennia simply because they seemed to be effective. No one knew why the remedies worked. The ancient remedies survive today, lumped together by modern medicine under the term "folk medicine" because their effectiveness has not been scientifically proven through clinical trials.

## REFERENCES

Ajibola, A., Chamunorwa, J.P. and Erlwanger, K.H. 2012. Nutraceutical values of natural honey and its contribution to human health and wealth. *Nutrition and Metabolism* 9: 61.

Allsop, K.A. and Miller J.B. 1996. Honey revisited: A reappraisal of honey in pre-industrial diets. *British Journal of Nutrition* 75: 513–520.

American Pharmaceutical Association. 1916, 1926, 1935. *The National Formulary IV.*

Ankra-Badu, G.A. 1992. Sickle cell leg ulcers in Ghana. *East African Medical Journal* 69: 366–369.

Bansal, V., Medhi, B. and Pandhi, P. 2005. Honey—A remedy rediscovered and its therapeutic utility. *Kathmandu University Medical Journal* 3: 305–309.

Beck, D.F. and Smedley, D. 1997. *Honey and Your Health: A Nutrimental, Medicinal and Historical Commentary*. (Originally published in 1938). Silver Springs, MD: Health Resources, Inc.

Best, M.R. 1986. *Gervase Markham: The English Housewife*. (First published c. 1600). Kingston, ON, Canada: McGill-Queen's University Press.

Bogdanov, S. 2009. *A Short History of Honey*. Bee Product Science.

Cartland, B. 1970. *The Magic of Honey*. London: Corgi Books.

Crane, E. 1975. *Honey: A Comprehensive Survey*. London: Heinemann.

Crane, E. 1980. *A Book of Honey*. Oxford: Oxford University Press.

Crane, E. 1983. *The Archaeology of Beekeeping*. Ithaca, NY: Cornell University Press.

Cosman, M.P. 1976. *Fabulous Feasts—Medieval Cookery and Ceremony*. New York: George Braziller.

Darby, W.J., Ghalioungui, P. and Grivetti, L. 1977. *Food. The Gift of Osiris*. London: Academic Press.

Deuraseh, N. 2006. Health and medicine in the Islamic tradition based on the book of medicine (Kitab Al-Tibb) of Sahih Al-Bukhari. *Journal of the International Society for the History of Islamic Medicine* 5: 1–14.

Forrest, R.D. 1982. Early history of wound treatment. *Journal of the Royal Society of Medicine* 75: 198–205.

Free, J.B. 1982. *Bees and Mankind*. Boston, MA: George Allen and Unwin.

Getz, F.M. 1991. *Healing and Society in Medieval England, A Middle English Translation of the Pharmaceutical Writings of Gilbertus Angelicus*. Madison, WI: University of Wisconsin Press.

Hajar, R. 2002. History of medicine. *Heart Views* 3:10.

Hill, J. 1759. *The virtues of honey in preventing many of the worst disorders: And in the certain cure of several others: Particularly the gravel, asthmas, coughs, hoarseness, and a tough morning phlegm*. London: J. Davis and M. Cooper.

Imperato, P.J. and Traore. 1969. Traditional beliefs about measles and its treatment among the Bambara of Mall. *Tropical and Geographical Medicine* 21: 62–67.

Irving, T.B., Ahmad, K. and Ahsan, M.M. 1987. The story of creation. In *The Qur'an-Basic Teachings*, ch 5. Bath: Pitman Press.

Jones, R. 2001. Honey and healing through the ages. In *Honey and Healing*, ed. P. Munn and R. Jones. Cardiff, U.K.: International Bee Research Association.

Kelly, E.C. 1940–1941. *Medical Classics*. Baltimore, MD: Williams and Wilkins Company, V 1–5: 10.

Low, T. 1989. *Bush Tucker—Australia's Wild Food Harvest*. Sydney, Australia: Angus and Robertson.

Meehan, B. 1982. *Shell Bed to Shell Midden*. Canberra, Australia: Australian Institute of Aboriginal Studies.

McGee, H. 1984. *On Food and Cooking: The Science and Lore of the Kitchen*. London: Harper Collins.

Mcintosh, E.N. 1995. *American Food Habits in Historical Perspective*. Westport, CT: Praeger Publishers.

Molan, P.C. 1999. Why honey is effective as medicine. 1. Its use in modern medicine. *Bee World* 80: 80–92.

Molan, P.C. 2001. Why honey is effective as a medicine. 1. Its use in modern medicine. In *Honey and Healing*, ed. P. Munn and R. Jones. Cardiff, U.K.: International Bee Research Association.

Nahl (The Bee), surah XVI, verse 69. In *The Holy Qur'an* (text, translation and commentary by Abdullah Yusuf Ali, 1987), 2nd ed. American Trust Publications.

Newman, T.G. 1983. *Honey Almanac*. Chicago: Newman.

Obi, C., Ugoji, E., Edun, S.A. Lawal, S.F. and Anyiwo, C.E. 1994. The antibacterial effect of honey on diarrhoea causing bacterial agents isolated in Lagos, Nigeria. *African Journal of Medical Sciences* 23: 257–260.

Ransome, H.M. 1937. *The sacred bee in ancient times and folklore*. London: George Allen and Unwin.

Rogers, J.E.T. 1866. *A History of Agriculture and Prices in England*. Oxford: Clarendon Press.

Style, S. 1992. *Honey: From Hive to Honeypot*. London: Pavilion.

Tannahill, R. 1975. *Food in History*. St Albans, U.K.: Paladin.

Taylor, K. 1971. *The Living Bible*. London: Hodder and Stoughton.

Wilson, C.A. 1973. *Food and Drink in Britain: From the Stone Age to Recent Times*. London: Constable.

Woodburn, J. 1963. An introduction to Hazda ecology. In *Man the Hunter*, ed. R.B. Lee and I. DeVore. New York, NY: Aldine de Gruyter, pp. 49–55.

Zucoloto, F.S. 2011. Evolution of the human feeding behavior. *Psychology & Neuroscience* 4: 131–141.

Zumla, A. and Lulat, A. 1989. Honey—A remedy rediscovered. *Journal of the Royal Society of Medicine* 82: 384–385.

Zwaenepoel, C. 1984. *Honey: Facts and Folklore*. Edmonton, Canada: Alberta Beekeepers' Association Edmonton.

# 2 Honey in Ayurvedic Medicine

*Surya Prakash and Ratna Rao*

## CONTENTS

Ayurveda is the complete knowledge for long life. In Sanskrit, the word *āyus* means "longevity or life" and the word *veda* means "related to knowledge" or "science."[1] Ayurveda, an ancient system, addresses information or knowledge of or for a healthy living. It is afar a system of medicine but also describes a lifestyle for happy and healthy living. Ayurveda defines health as a state of complete physical, mental, and social well-being (*Satva, atma, sareeram*) and not merely the absence of disease or infirmity.[2] It views illness as caused by an imbalance in a person's bodily or mental constitution. Ayurveda describes both preventive and curative measures for illness. It recommends/specifies curative medicine, diet, and changes in lifestyle as therapeutic measures in treating an illness (*Oushadam, aaharam*, and *viharam*).[3] Here, medicine involves both medical and surgical management. The ancient gurus of Ayurveda have recited that every physical material on Mother Earth is a resource/ raw material for medicine[4] (*Nanoushadam*).

## PRINCIPLES OF AYURVEDA

Ayurveda adopted the physics of the "five elements" (Pancha maha booth)—*Prithvi*-(Earth), *Jala* (water), *Agni* (fire), *Vāyu* (air), and *Ākāśa* (space/sky)—that compose the universe, including the human body.[5] Chyle or plasma (called *rasa*), blood (*rakta dhātu*), flesh (*māṃsa dhātu*), fat/adipose tissue (*medha dhātu*), bone (*asthi dhātu*), marrow (*majja dhātu*), and semen or female reproductive tissue (*śukra dhātu*) are held to be the seven primary constituent elements (*saptadhātu*) of the body.[6] Ayurvedic literature deals elaborately with measures of healthy living during the entire span of life and its various phases. Ayurveda asserts a balance of three elemental energies or humors: *vāta, pitta* (fire), and *kapha* (material). According to Ayurvedic medical theory, these three substances—*doṣas* (literally that which deteriorates)—are important for health because the body is healthy when they exist in equilibrium and unhealthy when they exist in inequilibrium.

The earliest literature on Indian medical practice appeared during the Vedic period in India, that is, in the mid-second millennium BCE. The *Suśruta* and the *Caraka* are great encyclopedias of medicine compiled from various sources from the mid-first millennium BCE to about 500 CE.[7] They are among the foundational works of Ayurveda.

### HONEY IN AYURVEDA

In Ayurveda, honey is called "Madhu." Its qualities are explained as follows. Honey is said to be a substance that is to be licked (Lehyam). Lehyam is a jam-like substance of herbal origin/herbal infused jam. Usage of honey as food and medicine has been advocated since the Vedic period, that is, mid-second millennium or mid-first millennium. This is because the types, properties, actions, and indications of honey were mentioned in the Ayurvedic great encyclopedias, such as *Charaka samhita* and *Susrutha samhitha*. Thus, the uses of honey in health care stretch back into antiquity.

### PROPERTIES OF HONEY

> Vaatalam guru sheetam cha raktapittakaphapaham |
> Sandhatru cchedanam ruksham kashayam madhuram madhu ||
> It has sweetness (*madhura rasa*) with added astringent as end taste (*Kashaya anu rasa*). It is heavy (*guru guna*), dry (*ruksha*) and cold (*sheeta*). Its effect on *doshas* is as follows. It aggravates *vāta*, scrapes *kapha* and pacifies *pitta* and *rakta*. It promotes healing process.

Honey in Ayurved or Sanskrit is known as *Madhu*. *Madhu*/honey is commonly used as *Anupana* (i.e., given along with primary active medication to enhance activity, or enhances faster absorption and assimilation of medicament) and *Sahapana* (along with primary/active medicament), sometimes, for its primary medical property systemically as well as locally, either alone or in combination with other drugs. It is also used as a vehicle along with some medicines to improve its efficacy or to mitigate the side effects of the other medicines it is mixed with. It has been described to have properties such as *Lekhana* (scraping), *Sandhana* (union), *Shodhana* (purification), *Ropana* (healing), and *Tridoshaghna* (pacifying all three doshas: *vāta, pitta,* and *kapha*).[8] It is used as an external application in *Vrana* (wound), either alone or in combination with *Sarpi* (*Goghrita*, i.e., ghee made from cow's milk).

*Madhu* has Vranaropak properties as per the principles of the 60 Upakramas of Vrana management described in the *Sushruta Samhita. Madhu*[9] is believed to act by "pacifying" the three vitiated doshas (i.e., *vāta, pitta,* and *kapha*) by multiple actions attributable to its *Madhura* (sweet) *Rasa, Kashaya* (astringent) *Uparasa* or *anurasa, Ruksha* (dry) *Guna, Sheeta* (cold) *Virya, Madhura Vipaka,* and *Sukshma Marga Anusari* (ability to permeate in microchannels) *Prabhava*.

Its value, beyond a sweetener, is being rediscovered. The ancients of Vedic civilization considered honey as one of nature's most remarkable gifts to mankind.

## HONEY AND RASAYAN AND REJUVENATION THERAPY

The Sanskrit word *rasayan* means *rasa* (juice or elixir vitae) + *ayan* (Path). *Rasa* is concerned with the conservation, transformation, and revitalization of energy. *Rasa* nourishes our body, boosts immunity, helps to keep the body and mind in the best of health, and promotes, vitality, longevity, intelligence, and complexion. Honey is said to be a potent *rasayana* and hence conserves, transforms, and revitalizes energy in the body.

## HONEY AS YOGAVAHI AND ASUKARI

It is called "Yogavahi," which means it strengthens and enhances all other herbs and processes in the body. The process of penetrating deep into the tissues and cells through porous channels is rare and is known as "Yogavahi" or bioavailability enhancer. Honey through this property is known as *yogavahi. Asukari* is the property of honey in penetrating faster to the cell level.

## TYPES OF HONEY[10]

Eight types of honey are described in Ayurveda depending on the type of bee that collects it. They are *Pouttikam, Bhramaram, Kshoudram, Makshikam, Chatram, Arghyam, Oudalakam,* and *Dalam*.

1. *Pouttikam*: This honey is collected by very large black honey bees from the nectar of poisonous flowers. It increases *vāta* and causes gout and burning sensation in the chest. It is also sedative/hypnotic, promotes wound healing, and reduces fat. It is indicated in urinary tract infections, tumors, ulcers or wounds, and diabetes (*Prameha*).
2. *Bhramaram*: This honey is collected by large bees and sticky in nature. It is used in cases of hematemesis and bleeding disorders.

3. *Kshoudram*: This is honey procured by honey bees of moderate size. It is yellowish brown colored, light, and cold in nature. It dissolves *kapha* and used in the treatment of diabetes (*Prameha* or *madhumeha*).

4. *Makshikam*: This is honey collected by small red colored honey bees. It is very light and dry natured. It is indicated in *VataKapha* diseases and *kapha* diseases. It is used in the treatment of eye diseases, hepatitis, piles (*Arsas*), asthma (*Swasa*), cough (*Kasa*), and tuberculosis (*Kshaya*).

5. *Chatram*: This is honey collected by bees of Himalayan regions. Beehive is of umbrella shape (*chata*). It is heavy and cool in nature and useful in gout, Leucoderma (*Shwitra*), worm infestations, and diabetes.

6. *Arghyam*: It is good for eyes and treatment of cough and anemia and causes arthritis.

7. *Oudalakam*: It is useful in skin diseases and leprosy. It is also used as a detoxicating agent and helps in modulation of voice.

8. Dalam: It is dry and reduces vomiting. It increases digestion and is used in diabetes.

The *Makshikam* type of honey is considered as the best type with immense medicinal properties.

## THERAPEUTICS OF HONEY

- Honey is an appetizer and also promotes digestion/metabolism.
- It is very good for the eyes, vision, and heart.[11]
- It quenches thirst and stops hiccups.
- It dissolves or mitigates *kapha*.
- It is a natural detoxifying agent; the effects of toxin are nullified.
- It is very useful in urinary tract disorders, worm infestations, bronchial asthma, cough, diarrhea and nausea, vomiting.
- It cleanses and promotes wound healing by facilitating formation of granulation tissue.
- Honey that is newly collected from the beehive increases body weight and is a mild laxative.
- Honey that is stored and is old helps in fat metabolism and is indicated in obesity and scrapes *kapha* and adipose tissue (*medas*).
- It promotes nourishment and stimulates union of tissues.
- It acts as a sedative or hypnotic and is useful in bed wetting disease.
- Honey, being a very good antioxidant, restores the damaged skin and gives soft, young looks.

---

Chakshushayam Chedi tritshleshmavishahidmaasrapittanut ।
Mehakushtakrimicchardishwaasakaasaatisaarajit ॥
Vranashodhana sandhaanaropanam vaatalam madhu ॥

**Sushruta Samhita, Nibandha sangraha, Yadavji Trikamji**

---

## Precautions to Be Taken before Using Honey

- Honey is not to be heated/boiled and is not to be mixed with foods that are hot.
- Honey is not to be consumed warm and when working in hot atmosphere or where you are exposed to more heat.
- Honey should never be mixed with rainwater, hot and spicy foods, fermented beverages such as whisky, rum, and brandy, ghee, and mustard.
- Honey is to be used only after storing for 1 year.
- If taken in excess, honey causes *madhvajirna*.

## Dosage

- Adults: One teaspoonful three times a day
- Child: Half-teaspoonful three times a day

## Honey in Wound Healing or *Vrana Ropana*

*Madhura Rasa* gives nutrition to the tissue, which helps in granulation tissue formation, whereas *Kashaya Rasa* provides *Lekhana* (scraping), which helps in desloughing, preparing the wound for healing. Thus, *Madhu* has excellent properties to heal the wound by virtue of its *Sodhana* (purification), *Ropana* (healing), and *Sandhana* (union) actions.

- Honey possesses antimicrobial properties.
- It helps in promoting autolytic debridement.
- It deodorizes malodorous wounds.
- It speeds up the healing process by stimulating wound tissues.
- It helps in initiating the healing process in dormant wounds.
- It also helps in promoting moist wound healing.

## Honey and Heart

Ayurveda recommends it in cases of arteriosclerosis and weak hearts: a glass of water with honey and lemon juice in it at bedtime.

## Honey in Pulmonary Diseases/Respiratory Diseases

Honey is highly useful for treatment of cough, cold, and chronic respiratory diseases such as bronchial asthma and bronchitis. It usually brings relief whether the air flowing over the honey is inhaled or whether it is eaten or taken either in milk or water. Ayurveda recommends that honey be preserved for 1 or more years for respiratory diseases.

Honey is highly beneficial in the treatment of irritating cough. As a demulcent or soothing agent, it produces a soothing effect on the inflamed mucus membrane of the upper respiratory tract and retrieves irritating cough and symptoms such as dysphagia (difficulty in swallowing).

Mix half a gram of black pepper powder with honey and ginger juice mixture (equal quantity). Take this mixture at least three times a day. It helps to cure asthma.

## HONEY AND ANEMIA

Honey due to its hemopoietic property is indicated in anemia. Take one glass of warm water and mix one to two teaspoonfuls of honey and one teaspoonful of lemon juice. Take this preparation daily before evacuation. It helps to purify the blood. It also helps to reduce fat and to clean the bowels.

## HONEY IN SKIN OR DERMATOLOGY

The texts of Ayurveda also describe honey as one that is applied externally and is considered useful in the treatment of wounds and sores. Honey soothes pain, acts as an antiseptic, hastens healing, and is especially effective in healing burns and carbuncles.

Honey, being a very good antioxidant, restores the damaged skin and gives soft, young looks. Honey's natural antioxidant and antimicrobial properties and ability to absorb and retain moisture have been recognized and used extensively in skin care treatments because they help to protect the skin from the damage of the sun's rays and rejuvenate depleted skin.

## HONEY AND INSOMNIA

Honey, due to its hypnotic property, has been used for centuries for the treatment of insomnia because it produces sound sleep or induces sleep. It should be taken with water, before going to bed, in doses of two teaspoonfuls in a big cupful of water. Children fall asleep after consuming honey.

## HONEY AND EYE DISEASES

Honey has a long history of Ayurvedic use for various eye ailments. Applied daily in the eyes, it improves the vision or eyesight. It is also indicated in the treatment of itching of the eyes, trachoma, conjunctivitis, and other similar diseases. Its regular internal as well as external application will prevent glaucoma in the initial stage of the disease.

Ayurveda asserts that honey is valuable in the prevention of cataract formation. Two grams of onion juice and honey each, mixed together, should be kept safe in a clean bottle. It is to be applied locally to the eyes with a glass rod. This is a very effective remedy for immature cataract. It dissolves the already coagulated protein fibers as well as prevents further coagulation.

## HONEY AND THE STOMACH

Ayurvedic experts have long regarded honey as useful in maintaining the health of the stomach. It tones up the stomach, helps in proper digestion, and prevents stomach diseases. It also decreases the overproduction of hydrochloric acid, thereby preventing symptoms such as nausea, vomiting, and heartburn. Honey, due to its laxative nature and being an emetic, clears the digestive canal of waste matter, when in constipation or indigestion.

## HONEY AND AGE

The texts also state that honey is especially useful in providing energy and heat to the body in old age. It dries up the phlegm and clears the system of mucus. One or two teaspoonfuls of honey in a cupful of boiled and cooled water, taken in lukewarm condition, is a refreshing and strengthening drink.

## HONEY AND ORAL DISEASES

Honey is valuable in keeping the mouth or oral cavity healthy. Applied daily over the teeth and gums, it cleans and shines the teeth. It prevents tarter deposition at the base of the teeth and prevents decay and early falling of the teeth. Being a mild antiseptic, it prevents the growth of harmful microorganisms on teeth. It also keeps the gums in the healthy state by increasing the vascularity. In case of ulcers in the oral cavity, honey helps in their early healing and prevents further sepsis and pyogenic infection-related bad odor and pus formation. Gargling with honey water is very useful in gingivitis due to inflammation of the gums.

## HONEY AND PEDIATRICS

Honey and ghee, when mixed in equal quantities and applied on the tongue of the newborn baby (day 1) or neonate, help as an immune booster to the child (*Suvarna prasam*). Honey acts as a sedative or hypnotic and is useful in bed-wetting disease.

## TESTING OF ADULTERATIONS/IMPURITIES

If honey is taken and exposed to flame and it produces sound, it is impure; if it is soundless, it is pure honey.

## CONCLUSION

Honey, due to its *Yogavahi*, *sookshmamarganusari* or *asukari*, and *rasayan* properties, is used as *sahapana* and *anupana* (vehicle) for various Ayurvedic medicines along with several preparations. Honey due to its varied therapeutic properties has also been used individually for ages and has a unique place in Ayurvedic pharmacology.

## FURTHER READINGS

1. *Hitahitam*...Charaka Sootra stanam-Deergamjeevatheeyamadyayam-1/41, Acharya Chakrapani–Ayurveda deepika commentary.
2. *Sareerendriya*...Charaka Sootra stanam-Deergamjeevatheeyamadyayam-1/41 and 42.

3. *Hetuvyadi*…Madava nidanam-Panchalakshna nidanam-1/7.
4. *Anenopadesena*…Charaka Sootra stanam-Atreya Badrakapyeeyam-26/29.
5. *Panchaboutikam*…Charaka Sareera stanam-4/13, Charaka Chikitsa stanam-Grahani-15/11.
6. *Rasasrik*…Ashtanga Hridaya Sootra stanam-1/12.
7. Ayurveda Itihasa by Acharya Priyavat Sharma-Chowkamba Publications, Varanasi.
8. *Chakchushyam*…Reprint edition. Varanasi: Chaukhambha orientalia; 2009. Sushruta Chikitsa Sthan 1/8; Sushruta Samhita, Nibandha sangraha, Yadavji Trikamji; p. 397.
9. *Vranasodhana*…Reprint edition. Varanasi: Chaukhambha orientalia; 2009. Sushruta Chikitsa Sthan 1/8; Sushruta Samhita, Nibandha sangraha, Yadavji Trikamji; p. 397.
10. *Makshikam*…Susrutha Sootra stanam or Reprint edition. Varanasi: Chaukhambha orientalia; 2009. Sushruta Sutra Sthan 45/132; Sushruta Samhita, Nibandha sangraha, Yadavji Trikamji; p. 207.
11. (i) *Vatalam*…Charaka Sootrastanam-Annapanavidimadyayam-27/243-246, vol. 1, Varanasi, India: Chowkamba Orientalia, reprinted 2011.
    (ii) Natakashtatamam…Charaka Sootrastanam-Annapanavidimadyayam-27/247.
    (iii) Chakchusyam…Reprint edition. Varanasi: Chaukhambha orientalia; 2009. Sushruta Chikitsa Sthan 1/8; Sushruta Samhita, Nibandha sangraha, Yadavji Trikamji; p. 397.

# 3 Biochemistry and Physicochemical Properties of Honey

*Sevgi Kolayli, Oktay Yildiz, Hüseyin Sahin, and Rezzan Aliyazicioglu*

## CONTENTS

## INTRODUCTION

Honey is a natural sweet substance produced by honeybees from nectar of blossoms or sweet deposits from plants, modified and stored in honeycombs (NHB 1996). Honey is also a very important energy food and is used as an ingredient in hundreds of manufactured foods, mainly in cereal-based products, for its sweetness, color, flavor, caramelization, and viscosity (Karaman et al. 2010), and is considered to be important in traditional medicine (White et al. 1975). Its composition is variable and depends on many factors, such as botanical/floral origins and the bee's species, but consists of approximately 75%–80% carbohydrates, 17%–20% water, and 1%–2% minerals and organic matters. In general, honey contains 65%–70% monosaccharides and disaccharides, trisaccharides, and oligosaccharides. The organic matter has at least 200 substances, including amino acids, enzymes, proteins, vitamins, organic acids, pigments, phenolics, vitamins, Maillard reaction products (MRP), and volatile compounds (Küçük et al. 2007; Karaman et al. 2010). Honey is also used as a source of dietary antioxidant. The protein content of honey is normally less than 0.5%. The mineral honey content is about 0.04%–0.2% and depends on the type of soil in which the original nectar-bearing plant was located. The major minerals present are K, Na, Ca, Mg, and Mn followed by lower concentrations of Fe, Zn, Cu, Se, and Rb. In general, dark honey types contain higher levels of minerals and phenolic substances, which are related to their biological activities.

The chemical composition determines the physicochemical characteristics of honey, such as color, rheology, electrical conductivity (EC), pH, and water activity. The physical properties of honey are substantial factors determining its price in the world market and also its acceptability by consumers.

## PHYSICAL PROPERTIES

### RHEOLOGICAL PROPERTIES

Honey, a viscous and aromatic product, is prepared by bees mainly from the nectar of flowers or honeydew (Dustmann 1993). Honey contains high amounts of total sugar, which are composed of approximately 95% of honey dry weight (Bogdanov et al. 2004).

In the food industry, reliable rheological data are essential for the design, process and quality control, sensory assessment, stability, and consumer acceptance of a product (Chhinnan et al. 1985; Steffe 1996). The viscosity of honey is the most remarkable of its rheological properties (Kayacier and Karaman 2008a). Viscosity is a measure of the resistance of a fluid that is being deformed by either shear stress or tensile stress. Thus, water is "thin," having a lower viscosity, whereas honey is "thick," having a higher viscosity. Put simply, the less viscous the fluid, the greater its ease of movement or fluidity (Symon 1971). In honey, the viscosity depends on both the physicochemical composition (moisture, floral source, sugar composition, floral origin, crystallization, and colloidal substances) and the temperatures of honey. However, knowledge of the flow behavior is useful in quality control, calculating energy usage, process control, and equipment selection (Kaya and Belibağlı 2002).

Fluid foods are divided into Newtonian and non-Newtonian fluids according to the viscosity–shear rate relationship.

A Newtonian fluid is a fluid whose stress versus strain rate curve is linear and passes through the origin. The constant of proportionality is known as the viscosity. As points of reference, the following represent typical Newtonian viscosities at room temperature: honey, 10,000 cP; water, 1 cP; mercury, 1.5 cP; coffee cream, 10 cP; vegetable oil, 100 cP; and glycerol, 1000 cP (Weast et al. 1985; Steffe 1996). The viscosity of non-Newtonian fluids is not independent of shear rate or shear rate history. However, there are some non-Newtonian fluids with shear-independent viscosity

**TABLE 3.1**
**Viscosity Change in Honeys with Temperature, Moisture Content, and Floral Source**

| Water Content | Viscosity (Poise) at 25°C |
|---|---|
| 15.5% | 138.0 |
| 17.1% | 69.0 |
| 18.2% | 48.1 |
| 19.1% | 34.9 |
| 20.2% | 20.4 |

| Temperature (°C) | Viscosity (Poise) at 25°C |
|---|---|
| 13.7 | 600.0 |
| 29.0 | 68.4 |
| 39.4 | 21.4 |
| 48.1 | 10.7 |
| 71.1 | 2.6 |

| Floral Source | Viscosity (Poise) at 25°C (16.5% $H_2O$) |
|---|---|
| Sage | 115.0 |
| Clover | 87.5 |
| White clover | 94.0 |

that nonetheless exhibit normal stress differences or other non-Newtonian behavior. Many salt solutions and molten polymers are non-Newtonian fluids, as are many commonly found substances such as ketchup, custard, and starch suspensions (Steffe 1996; Penna et al. 2001; Kayacier and Karaman 2008b; Kirby 2010).

Rheological properties of honey are measured using viscosimeters, rheometers, cone plates, controlled strained method, controlled stress method, double-gap cylinder, dynamic tests, and frequency sweep tests (Bhandari et al. 1999; Mossel et al. 2000; Abu-Jdayil et al. 2002; DaCosta and Pereira 2002; Anupama et al. 2003; Kayacier and Karaman 2008b).

The viscosity of honey generally decreases as temperature increases. The effect caused by temperature is more important in the low range of temperatures, whereas, at high temperatures, the viscosity shows less variation (Gómez-Díaz et al. 2009).

Table 3.1 shows how the viscosity changes as temperature, moisture content, and floral source change. Thus, 1% moisture is equivalent to about 3.5°C in its effect on viscosity (National Honey Board's Guide 2005).

In the literature, the viscosity values of some honey were measured at different temperatures and found to be between 0.5 and 600 Pa s (Steffe 1996; Juszczak and Fortuna 2006).

The Arrhenius equality or Williams–Landel–Ferry (WLF) equation is used to express the change in viscosity with temperature. The WLF equation is generally used in low temperatures (below 0°C). The consistency index can be used to describe the variation in viscosity with temperature using the Arrhenius equation (Briggs and Steffe 1997; Mossel et al. 2000; Sengul et al. 2005):

$$\ln k = \ln k_0 + Ea/RgTa, \tag{3.1}$$

where $k_0$ is the Arrhenius constant (Pa $s^n$), $Ea$ is the activation energy (J $mol^{-1}$), $Rg$ is the universal gas constant (J $mol^{-1}$), and $Ta$ is the absolute temperature (K). $k_0$ (Pa $s^n$) and $Ea$ (J $mol^{-1}$) parameters were obtained from the Arrhenius-type equation with the linear regression analysis (Equation 3.1).

It was previously reported that $Ea$ values of some different honeys were between 63.4 and 93.75 kJ $mol^{-1}$ (Bhandari et al. 1999; Kayacier and Karaman 2008b). High-activation energy shows that viscosity of honey is more affected by temperature. $k_0$ values were found to be between 2.93 and 8.88 (Steffe 1996).

There are a number of researches about the rheological properties of honeys. However, there is little information about the rheological properties for use as markers.

## COLOR

The color index is one of the most important factors in the quality of honeys. Honey's floral origin, mineral composition, chemical content, and heating process can affect color. Honey coloring components are plant pigments such as chlorophyll, carotene, xanthophylls, and yellow-green color pigments.

Honey is classified by the U.S. Department of Agriculture into seven color categories: water white, extra white, white, extra light amber, light amber, amber, and dark amber. The most commonly used methods are based on optical comparison using simple color grading after Pfund (Fell 1978) or Lovibond (Aubert and Gonnet 1983), and Table 3.2 shows the Pfund scale for the color index that measured at 560 nm.

Other more objective methods have also been tested as the determination of all color parameters through the CIE $L*a*b*$ tristimulus method (Aubert and Gonnet 1983; Ortiz Valbuena and Silva Losada 1990; Bogdanov et al. 2004). Color measurement of the honeys is carried out by using a colorimeter and the measurements were recorded as Hunter $L$, $a$, and $b$ color values. Color values for each sample are computed by using three measurements from different positions. The color values are expressed as $L$ (darkness/lightness; 0, black; 100, white), $a$ ($-a$, greenness; $+a$,

**TABLE 3.2**
**Pfund Scale for the Color Index of Honey (560 nm)**

| Color Name | Pfund Scale (mm) |
| --- | --- |
| Water white | <8 |
| Extra white | 9–17 |
| White | 18–34 |
| Extra light amber | 35–50 |
| Light amber | 51–85 |
| Amber | 86–114 |
| Dark amber | >114 |

redness), and $b$ ($-b$, blueness; $+b$, yellowness) (Yildiz and Alpaslan 2012). In general, darker honeys have been shown to be higher in antioxidant content than lighter honeys (Gheldof and Engeseth 2002).

The color of honey is related to the content of pollen, total phenolics, mineral composition, and hydroxymethylfurfural (HMF) and is characteristic of floral origin (Gonzalez-Miret et al. 2005; Bertoncelj et al. 2007). However, most of the honey studies showed that dark-colored honeys exhibited distinct antioxidant quenching and free radical scavenging potential against different radical species. It has been reported by several researchers that light-colored honeys usually have low total phenolic content, whereas dark-colored honeys generally have higher phenolic content. Chestnut, thyme, pine, rhododendron, and linen vine of honeys are dark-colored honeys with high phenolic contents (Krop et al. 2010; Tezcan et al. 2010).

Among most honey samples, darker honeys showed higher antioxidant capacity related to the total phenolic compound contents. Because of their high phenolic constituents of the honeys, they may also possess anticancer activities; also, the honeys are called medicinal honeys, such as chestnut honey (Küçük et al. 2007). Because of the high phenolic natural substances, honey can be suggested for regular consumption and use in food industries that might serve to protect health and fight against several diseases.

## Water Activity

Water activity is defined as the ratio of the vapor pressure of water in a material ($p$) to the vapor pressure of pure water ($p_0$) at the same temperature. Water activity ($a_w$) is the amount of water that is available to microorganisms. Many microorganisms prefer an $a_w$ of 0.99 and most need an $a_w$ higher than 0.91 to grow. Relative humidity (RH) and $a_w$ are related. Water activity refers to the availability of water in a food or beverage. RH refers to the availability of water in the atmosphere around the food or beverage (Bell and Labuza 2000).

$$a_w = p/p_0; \text{RH} = a_w \times 100,$$

where $a_w$ is the water activity, $p$ is the vapor pressure of a solution, and $p_0$ is the vapor pressure of pure water.

Honey's water activity varies between 0.562 and 0.620 in the 40°F–100°F (4°C–37°C) temperature range (McCarthy 1995).

## Electrical Conductivity

EC is the ability of a material to carry the flow of an electric current (a flow of electrons). EC is applicable to all food, compost, manure, soil, organic fertilizer, growth media, and water samples. In honey, EC depends predominantly on the mineral content of honey (Andrade et al. 1997; Ruoff 2006).

The conductivity is the main quality parameter for honey, which is specified in *Codex Alimentarius* Draft revised standard for honey (2001), EU Directive relating to honey (2002), and Turkish Food Honey Directive. According to those documents, the values of EC should be not more than 0.8 mS cm$^{-1}$ for nectar honey and mixture of blossom and honeydew honeys and not less than 0.8 mS cm$^{-1}$ for honeydew and chestnut honeys (Szczesna and Chmielewska 2004; Ruoff 2006).

The method for the determination of EC is described in Bogdanov and Baumann (1997). Honey EC values are expressed in mS cm$^{-1}$ at 20°C; nowadays, the international reference measurements should be carried out at 25°C (Bogdanov et al. 2004).

## SPECIFIC GRAVITY

Specific gravity of honey is a measure of the density of honey compared to water and dependent on water content. The specific gravity of honey is 1.40 to 1.45; that is, it is heavier than water (Table 3.3).

Other factors such as floral source slightly affect the specific gravity of honey. Honeys from different origins or batches should be thoroughly mixed to avoid layering (National Honey Board's Guide 2005).

## VISCOSITY

Honey is a viscous liquid. Its viscosity depends on a large variety of substances (Table 3.4) and therefore varies with its composition and particularly with its water content and temperature (Abu-Jdayil et al. 2002; Yanniotis et al. 2006). The viscosity of honey is essential to its processing and it has an important link to its technological applications, extraction, pumping, setting, filtration, mixing, and bottling. Honey of high quality is usually thick and viscous. If the concentration of water is increased, honey becomes less viscous. Proteins and other colloidal substances increase honey viscosity, but their amount in honey may be insignificant. The percentage of fructose content in honey has also been found to affect its viscosity and rheological properties. Honeys become less viscous with an increase in fructose content (James et al. 2009).

## HYGROSCOPICITY

Hygroscopicity is another property of honey and describes the ability of honey to absorb and hold moisture from the environment. In the case of honey, it results largely from the high concentration of fructose. It is a desirable property in baking because it assists in keeping baked goods from drying out. It is a very useful property

**TABLE 3.3**
**Relation between Specific Gravity and Water Content of Honey**

| Water Content (%) | Specific Gravity (20°C) |
| --- | --- |
| 15 | 1.4350 |
| 18 | 1.4171 |

**TABLE 3.4**

**Honey Viscosity Depending on Its Water Content, Temperature, and Floral Source**

| Type | Water Content (%) | Temperature (°C) | Viscosity (Poise) |
|---|---|---|---|
| Sweet clover (*Melilotuss*) | 16.1 | 13.7 | 600 |
| | | 20.6 | 189.6 |
| | | 29 | 68.4 |
| | | 39.4 | 21.4 |
| | | 48.1 | 10.7 |
| | | 71.1 | 2.6 |
| | | 25.0 | |
| White clover (*Trifolium repens*) | 13.7 | | 420 |
| | 14.2 | | 269 |
| | 15.5 | | 138 |
| | 17.1 | | 69.0 |
| | 18.2 | | 48.1 |
| | 19.1 | | 34.9 |
| | 20.2 | | 20.4 |
| | | | 13.6 |
| | | 11.7 | |
| Sage (*Salvia*) | 18.6 | 20.2 | 729.6 |
| | | 30.7 | 184.8 |
| | | 40.9 | 55.2 |
| | | 50.7 | 19.2 |
| Sage | | | 9.5 |
| Sweet clover | | 25 | |
| White clover | 16.5 | 25 | 115 |
| | 16.5 | 25 | 94 |
| | 16.5 | | 87.5 |

*Source:* Munro, J.A., *Journal of Economic Entomology*, 36, 769–777, 1943.

of honey in food manufacturing. A dehydrated honey will revert to thick syrup in a short time when exposed to the air (White 2000).

Normal honey with water content of 18.8% or less will absorb moisture from air of a RH of more than 60%. During processing or storage, however, the same hygroscopicity can become problematic, causing difficulties in preservation and storage due to excess water content (Olaitan et al. 2007).

## OPTICAL ROTATION

Optical rotation is a parameter that is discussed in relation to the determination of botanical origin and adulteration of honey (Piazza et al. 1991; Bogdanov et al. 1999). In some countries, the rotation is applied to differentiation of honey groups—blossom, honeydew, and compound honeys, but the limit values have not been harmonized so far.

Honey has the property to rotate the polarization plane of polarized light. This depends largely on types and relative proportions of sugars in honey (Dinkov 2003). Each sugar has a specific effect, and the total optical rotation is dependent on concentration. Some sugars (fructose) exhibit a negative optical rotation, whereas others (glucose) a positive one. Blossom honey has negative values and honeydew honeys have mostly positive values (Oddo and Piro 2004).

## CHEMICAL COMPOSITION

### Sugars

Honey is one of the most complex natural mixtures and mainly consists of carbohydrates. A great number of factors influencing honey composition and properties make the product an interesting topic for research. In general, honey is a supersaturated sugar solution, and sugars are the main constituents of honey accounting for about 95 g/100 g dry matter. Fructose and glucose are the major sugars, the former being a dominant component almost in all honey types, except for some of honeys of rape (*Brassica napus*), dandelion (*Tarazacum officinale*), and blues curls (*Trichostema lanceolatumi*) origin, when glucose is present in higher amounts (Cavia et al. 2002; Kaškonienè et al. 2010). The concentration of fructose and glucose and their ratio are useful indicators for the classification of unifloral honeys (Kaškonienè et al. 2010). Fructose is approximately two times sweeter than sucrose, glucose is less sweeter, and maltose is even less sweeter; therefore, fructose is responsible for honey sweetness and commonly used as a bulk sweetener (Miller and Adeli 2008; Rizkalla 2010). Maltose, isomaltose, kojibiose, turanose, trehalose, nigerose, melibiose, maltulose, gentiobiose, palatinose, nigerose, and laminaribiose are the major disaccharides and relatively found in lower concentration in honey. Although several works have been focused on the study of monosaccharides and disaccharides, the knowledge about trisaccharides and tetrasaccharides is still limited. However, melesitose, isomaltotriose, theanderose, isopanose, erlose, panose, maltotriose, kestose, and cellobiose are the oligopolysaccharides and are found especially in honeydew honeys (Özcan et al. 2006). In addition, the presence of some sugars and the ratio of fructose/glucose may be used to determine the adulteration of honey (Tezcan et al. 2010).

### Phenolic Compounds

Honey has been used in old medicine since the early ages of human beings; in recent times, its application in the treatment of burns, gastrointestinal disorders, infected wounds, and skin lesions has been investigated (Tezcan et al. 2010). Most of the biological properties of honey are based on phenolic substances. Honey contains about 0.1%–0.5% phenolic compounds that are responsible for its antioxidant, antimicrobial, antiviral, anti-inflammatory, anticarcinogenic, and many biological activities. Phenolic compounds are considered to be one of the most important nutritional substances and these compounds play a major role in human nutrition.

Phenolic compounds include different subclasses that are flavonoids, phenolic acids, anthocyanins, stilbenes, lignans, tannins, and oxidized polyphenols (Ferreira et al. 2009). The amount and diversity of phenolic compounds were investigated by several scientists in honeys and the studies have shown that phenolic substances were also changeable (Kassim et al. 2010). Some researchers classified phenolic compounds into three groups: flavonoids, cinnamic acids, and benzoic acids, and some of them were classified into two groups: phenolic acids including phenolic esters and flavonoids (Amiot et al. 1989; Yaoa et al. 2005). Until the present time, about 25–30 phenolic compounds (phenolic acids and flavonoids) have been characterized in different honeys, many of them being unifloral (Table 3.2). The phenolic fraction was determined by advanced chromatographic techniques: high-performance liquid chromatography, liquid chromatography–mass spectrometry, and gas chromatography–mass spectrometry. Phenolic acids and flavonoids may also be used as biomarkers for the origin of a honey. Traditionally, the geographical origin of honey was previously determined by the melissopalynologic technique, which is achieved by pollen analysis. Although it gives satisfactory results, it cannot stand as a reliable method on its own mainly because it is tedious and very dependent on the ability and judgment of the expert (Alissandrakis et al. 2003). For this reason, in recent years, the identification of phenolic compounds appears to be one of the most promising techniques for the determination of botanic origin and floral source, and tricetin, myricetin, quercetin, luteolin, and kaempferol for *Eucalyptus* honey, hesperetin for citrus honey, kaempferol for rosemary honey, quercetin for sunflower honey, and abscisic acid for heather honey are suitable markers (Bertoncelj et al. 2011).

## ESSENTIAL OILS

Another important substances in honeys are volatile compounds that play an important role in the aroma and nutritional value of honey. Honey volatile compounds, similar to many natural plants, consist of many monoterpenes, diterpenes, and sesquiterpenes and terpenoids, fatty acids, alcohols, ketones, and aldehydes. Essential oils have been used for thousands of years for food preservation, pharmaceuticals, alternative medicine, and natural therapies (Okoh et al. 2010). Taste and flavor are two of the most significant attributes of honey, and aroma is produced by complex mixtures of volatile compounds, which vary depending on nectar origin, processing, and storage conditions (De la Fuente et al. 2007). The volatile compounds of honey are responsible for its characteristic taste, smell, and flavor, playing a considerable role to protect honey from many harmful agents. Essential oils have been of great interest for their potential uses as alternative remedies for the treatment of many infectious diseases, pharmaceutical alternative medicine, and natural therapies, and it is therefore the essential oils of honey that are responsible for honey's biological values. In addition, the volatile compounds of honey are potentially useful to prove the authenticity of honey samples (De la Fuente et al. 2007). For example, germacrene-D proved to be most suitable for the identification of goldenrod honey origin (Amtmann 2010). The combination of (+)-8-hydroxylinalool with methyl anthranilate and caffeine could be proposed as a fingerprint marker reported for the description of citrus honey (Melliou and Chinou 2011).

## MINERALS

The total mineral content of honey is approximately 0.04%–0.2%. Many factors affect mineral composition of honey, including soil type, floral source, climatic conditions, and fertilization, and a great variability has been reported in honey mineral contents (White 1978; Anklam 1998). K, Na, Ca, and Mg are the major mineral contents of all the honeys, and Fe, Mn, Cu, Zn, Al, B, Sr, and Na are present in lower concentrations. Potassium is present in the highest content and varied from 500 to 5000 mg $kg^{-1}$ depending on honey's floral sources. It was reported that K levels were 1290 mg $kg^{-1}$ in Manuka and 3640 mg $kg^{-1}$ in honeydew honey in the New Zealand region (Vanhanen et al. 2011). It was reported that, up until now, 2011 mineral elements have been detected in all honey samples, and this may increase next time; however, none has been shown to contain all 27 elements (Vanhanen et al. 2011). Many studies have showed that honeys with dark color have a higher total mineral content and consequently higher essential elements. It was reported that dark honey samples, such as chestnut and honeydew (from *Pinus*), always contain higher levels of minerals (Golob et al. 2005; Pisani et al. 2008).

It was reported that honey may also be useful as an environmental marker for assessing the presence of environmental contaminants and pharmaceutical toxic agents, such as heavy metal elements and pesticides. Some of the toxic trace element levels (As, Cd, Pb, Hg, Ag, Ni, Cr, Co, Zn, and Cu) may also reflect the quality of honey (Anklam 1998; Przyblowksi and Wilczyńska 2011). This kind of contamination may be caused by external sources or by incorrect procedures during honey processing (Pisani et al. 2008). Minerals, such as phenolics and essential oils, also can be highly indicative of the geographical origin of honey and can be useful markers to determine floral authenticity (Tuzen et al. 2007; Przyblowksi and Wilczyńska 2011). Some researchers have reported that chestnut honeys are richest in Mn and K (Golob et al. 2005; Kolayli et al. 2008) and proposed manganese as chestnut honey markers.

## ORGANIC ACIDS AND OTHER ORGANIC MATERIALS

Organic acids comprise a small proportion of honey and honey acidity is mainly due to organic acids whose quantity is lower than 0.5% (w/w). Acidity contributes to honey flavor, stability against microorganisms, enhancement of chemical reactions, antibacterial and antioxidant activities, and granulation (Cavia et al. 2002). Organic acids can be used as indicators of deterioration on account of storage and aging or even to measure purity and authenticity (White 1978). There is a limited amount of literature on the individual organic acid content of honey samples. However, apart from phenolic acids and free amino acids, a few organic acids, such as formic, oxalic, malic, maleic, succinic, citric, gluconic, glutaric, and fumaric acids, have been detected in many different honey types in different concentrations (Mato et al. 2006; Cavia et al. 2002; Tezcan et al. 2011).

## PROTEINS AND AMINO ACIDS

Honey contains about 0.2% protein, which is of bee and plant pollen. Proteins and amino acids in honey are attributable both to bee and floral sources, the majority of these being pollen (Anklam 1998; Hermosín et al. 2003). Amino acids account for 1% (w/w), and prolin is the major contributor with 50%–85% of the total amino acids. All essential and some nonessential amino acids, such as γ-aminobutyric acid, amino isobutyric acid, and ornithine, were detected in many honey samples, with their relative proportions depending on the honey origin (nectar or honeydew). Besides proline, the amino acids found in honey in the order of decreasing concentration are phenylalanine, tyrosine, lysine, arginine, glutamic acid, histidine, and valine. The concentration of prolin serves as an additional determinant of quality and in some cases as a criterion for estimating the maturity of honey as well as an indicator for detecting sugar adulteration (Meda et al. 2005). The proline contents of qualified honey must be above 180 mg $kg^{-1}$ honey, the minimum value allowed by the Food Codex and Council of the European Union (Bogdanov and Baumann 1997). Honey contains also α-amylase, invertase, catalase, glucose oxidase, and phosphatase, which are related to plant origin, pollens, and nectars (Anklam 1998).

## MAILLARD REACTION PRODUCTS

Honey also contains a small quantitation of HMF, furfural, melanoidins, and acrylamide, which are the best known MRPs. The products are produced in honey between free amino groups of proteins and reducing sugars or lipid oxidation products and may depend on the types and quantity of reactants and the pH, temperature, and water activity of the medium. The products have also produced many foods that are partly responsible for the taste and color of bread, cookies, cakes, meat, beer, chocolates, popcorn, and rice. However, it is not always desired for this reaction to take place in foods because of the cytotoxic, genotoxic, and carcinogenic effects of the advanced reaction products. HMF is the best known MRP and its concentration reflects freshness and the heating treatment of honey; in *Codex Alimentarius* (2001), the limit for HMF content is 40 mg $kg^{-1}$ honey. The HMF content is lower in fresh honeys than older honeys (Ajlouni and Sujirapinyokul 2010).

## REFERENCES

Abu-Jdayil, B., Ghzawi, A.A.M., Al-Malah, K.I.M., and Zaitoun, S. 2002. Heat effect on rheology of light- and dark-colored honey. *Journal of Food Engineering* 51: 33–38.

Alissandrakis, E., Daferera, D., Tarantilis, P.A., Polissiou, M., and Harizanis, P.C. 2003. Ultrasound-assisted extraction of volatile compounds from citrus flowers and citrus honey. *Food Chemistry* 82: 575–582.

Ajlouni, S. and Sujirapinyokul, P. 2010. Hydroxmethylfurfuraldehyde and amylase contents in Australian honey. *Food Chemistry* 119: 1000–1005.

Amiot, M.J., Aubert, S., Gonnet, M., and Tacchini, M. 1989. Honey phenolic compounds: A preliminary study on their identification and quantitation by families. *Apidologie* 20: 115–125.

Amtmann, M. 2010. The chemical relationship between the scent features of goldenrod (*Solidago canadensis* L.) flower and its unifloral honey. *Journal of Food Composition and Analysis* 23: 122–129.

Andrade, P., Ferreres, F., and Amaral, M.T. 1997. Analysis of honey phenolic acids by HPLC, its application to honey botanical characterization. *Journal of Liqued Chromatography and Related Technologies* 20: 2281–2288.

Anklam, E. 1998. A review of the analytical methods to determine the geographical and botanical origin of honey. *Food Chemistry* 63: 549–562.

Anupama, D., Bhat, K.K., and Sapna, V.K. 2003. Sensory and physico-chemical properties of commercial samples of honey. *Food Research International* 36: 183–191.

Aubert, S. and Gonnet, M. 1983. Mesure de la couleur des miels. *Apidologie* 14: 105–118.

Bell, L.N. and Labuza, T.P. 2000. *Practical Aspects of Moisture Sorption Isotherm Measurement and Use.* 2nd Edition. Egan, MN: AACC Egan Press.

Bertoncelj, J., Doberśek, U., Jamnik, M., and Golob, T. 2007. Evaluation of the phenolic content, antioxidant activity and colour of Slovenian honey. *Food Chemistry* 105: 822–828.

Bertoncelj, J., Polak, T., Krop, U., Korošec, M., and Golob, T. 2011. LC-DAD-ESI/MS analysis of flavonoids and abscisic acid with chemometric approach for the classification of Slovenian honey. *Food Chemistry* 127: 296–302.

Bhandari, B., D'Arcy, B., and Chow, S. 1999. A research note: Rheology of selected Australian honeys. *Journal of Food Engineering* 41: 65–68.

Briggs, J.L. and Steffe, I.F. 1997. Using brookfield data and the mitschka method to evaluate power law foods. *Journal of Texture Studies* 28: 517–522.

Bogdanov, S. and Baumann, S.E. 1997. Harmonised methods of the European Honey Commission. Determination of sugars by HPLC. *Apidologie* (Extra Issue), 42–44.

Bogdanov, S., and 21 other members of the International Commission. 1999. Honey quality methods of analysis and international regulatory standards: Review of the work of the International Honey Commission. *Mitteilungen aus Lebensmitteluntersuchung und Hygiene* 90: 108–125.

Bogdanov, S., Ruoff, K., and Persano Oddo, L. 2004. Physico-chemical methods for the characterisation of unifloral honeys: A review. *Apidologie* 35: 4–17.

Cavia, M.M., Fernàndez-Muin, M.A., Gömez-Alonso, E.G., Montes-Pèrez, M.J., Huidobro, J.F., and Sancho, M.T. 2002. Evolution of fructose and glucose in honey over one year influence of induced granulation. *Food Chemistry* 78: 157–161.

Chhinnan, M.S., McWatters, K.H., and Rao, V.N.M. 1985. Rheological characterization of garin legume pastes and effect of hydration time and water level on apparent viscosity. *Journal of Food Science* 50: 1167–1171.

CODEX STAN 12-1981 (Rev. 2 - 2001). 2001. *Revised Codex Standard for Honey.* (Formerly Codex Stan-12-1987) Rome: FAO, WHO, 7.

De la Fuente, E., Sanz, M.L., Martínez-Castro, I., Sanz, J., and Ruiz-Matute, A.I. 2007. Volatile and carbohydrate composition of rare unifloral honeys from Spain. *Food Chemistry* 105: 84–93.

DaCosta, C.C. and Pereira, R.G. 2002. The influence of propolis on the rheological behavior of pure honey. *Food Chemistry* 76: 417–421.

Dinkov, D. 2003. A scientific note on the specific optical rotation of three honey types from Bulgaria. *Apidologie* 34: 319–320.

Ferreira, I.C.F.R., Edmur, A., Barreira, J.C.M., and Estevinho, L.M. 2009. Antioxidant activity of Portuguese honey samples: Different contributions of the entire honey and phenolic extract. *Food Chemistry* 114: 1438–1443.

Fell, R.D. 1978. The color grading of honey. *American Bee Journal* 18: 782–789.

Gheldof, N. and Engeseth, N.J. 2002. Antioxidant capacity of honeys from various floral sources based on the determination of oxygen radical absorbance capacity and inhibition of in vitro lipoprotein oxidation in human serum samples. *Journal of Agricultural and Food Chemistry* 50: 3050–3055.

Golob, T., Doberšek, U., Kump, P., and Nečemer, M. 2005. Determination of trace and minor elements in Slovenian honey by total reflection X-ray fluorescence spectroscopy. *Food Chemistry* 91: 593–600.

Gómez-Díaza, D., Navazaa, J.M., and Quintáns-Riveiro, L.C. 2009. Effect of temperature on the viscosity of honey. *International Journal of Food Properties* 12 (2): 396–404.

Gonzalez-Miret, M.L., Terrab, A., Hernanz, D., Fernandez-Recamales, M.A., and Heredia, F.J. 2005. Multivariate correlation between color and mineral composition of honey and their botanical origin. *Journal of Agricultural and Food Chemistry* 53: 2574–2580.

Hermosin, I., Chicon, R.M., and Cabezudo, M.D. 2003. Free amino acid composition and botanical origin of honey. *Food Chemistry* 83: 263–268.

James, O.O., Mesubi, M.A., Usman, L.A., Yeye, S.O., Ajanaku, K.O., Ogunniran, K.O., Ajani, O.O., and Siyanbola, T.O. 2009. Physical characterisation of some honey samples from North-Central Nigeria. *International Journal of Physical Sciences* 4: 464–470.

Juszczak, L. and Fortuna, T. 2006. Rheology of selected polish honeys. *Journal of Food Engineering* 75: 43–49.

Karaman, T., Buyukunal, S.K., Vurali, A., and Altunatmaz, S.S. 2010. Physico-chemical properties in honey from different regions of Turkey. *Food Chemistry* 123: 41–44.

Kayacier, A. and Karaman, S. 2008a. Balların Reolojik Karakterizasyonu, Türkiye 10. Gıda Kongresi; 21–23 Mayıs 2008, Erzurum.

Kayacier, A. and Karaman, S. 2008b. Rheological and some physicochemical characteristics of selected Turkish honeys. *Journal Texture Studies* 39: 17–27.

Kaya, A. and Belibağlı, K.B. 2002. Rheology of solid Gaziantep pekmez. *Journal of Food Engineering* 54: 221–226.

Kassim, M., Achoui, M., Mustafa, M.R., Mohd, M.A., and Yusoff, K.M. 2010. Ellagic acid, phenolic acids, and flavonoids in Malaysian honey extracts demonstrate in vitro anti-inflammatory activity. *Nutrition Research* 30: 650–659.

Kaškonienè, V., Venskutonis, P.R., and Ceksterytè, V. 2010. Carbohydrate composition and electrical conductivity of different origin honeys from Lithuania. *LWT-Food Science and Technology* 43: 801–807.

Kirby, B.J. 2010. *Micro- and Nanoscale Fluid Mechanics: Transport in Microfluidic Devices.* Cambridge: Cambridge University Press. http://www.kirbyresearch.com/textbook.

Kolayli, S., Kongur, N., Gündoğdu, A., Kemer, B., Duran, C., Aliyazıcıoğlu, R., and Küçük, M. 2008. Mineral composition of selected honeys from Turkey. *Asian Journal of Chemistry* 20: 2421–2424.

Küçük, M., Kolayli, S., Karaoglu, Ş., Ulusoy, E., Baltacı, C., and Candan, F. 2007. Biological activities and chemical composition of three honeys of different types from Anatolia. *Food Chemistry* 100: 526–534.

McCarthy, J. 1995. The antibacterial effects of honey. *American Bee Journal* May: 171–172.

Mato, I., Huidobro, J., Simal-Lozano, J., and Sancho, M.T. 2006. Rapid determination of non-aromatic organic acids in honey by capillary zone electrophoresis with direct ultraviolet detection. *Journal of Agricultural and Food Chemistry* 54: 1541–1550.

Melliou, E. and Chinou, I. 2011. Chemical constituents of selected unifloral Greek bee-honeys with antimicrobial activity. *Food Chemistry* 129: 284–290.

Miller, A. and Adeli, K. 2008. Dietary fructose and the metabolic syndrome. *Current Opinion Gastroenterology* 24: 204–209.

Meda, A., Lamien, C.E., Romito, M., Millogo, J., and Nacoulma, O.G. 2005. Determination of the total phenolic, flavonoid and proline contents in Burkina Fasan honey, as well as their radical scavenging activity. *Food Chemistry* 91: 571–577.

Munro, J.A. 1943. The viscosity and thixotropy of honey. *Journal of Economic Entomology* 36: 769–777.

Mossel, B., Bhandari, B., D'Arcy, B., and Caffin, N. 2000. Use of Arrhenius model to predict rheological behaviour in some Australian honeys. *Lebensmittel-Wissenschaft und-Technologie* 545–552.

National Honey Board's Guide. Honey: A Reference Guide to Nature's Sweetener. Available at http://www.honey.com/images/downloads/refguide.pdf, accessed July 24, 2013.

NHB 1996. Definition of honey and honey products approved by the national honey board, June 15, 1996. Updated September 27, 2003.

Oddo, L.P. and Piro, R. 2004. Main European unifloral honeys: Descriptive sheets. *Apidologie* 35: S38–S81.

Okoh, O.O., Sadimenko, A.P., and Afolayan, A.J. 2010. Comparative evaluation of the antibacterial activities of the essential oils of *Rosmarinus officinalis* L. obtained by hydrodistillation and solvent free microwave extraction methods. *Food Chemistry* 120: 308–312.

Olaitan, P.B., Adeleke, O.E., and Ola, I.O. 2007. Honey: A reservoir for microorganisms and an inhibitory agent for microbes. *African Health Sciences* 7: 159–165.

Ortiz Valbuena, A. and Silva Losada, M.C. 1990. Caracterizacion cromatica (CIE L10, a10, b10) de las mieles de la alcarria y zonas adyacentes. *Cuadernos de Apicultura* 3: 8–11.

Özcan, M., Arslan, D., and Durmuş, A.C. 2006. Effect of inverted saccharose on some properties of honey. *Food Chemistry* 99: 24–29.

Penna, A.L.B., Sivieri, K., and Oliveira, M.N. 2001. Relation between quality and rheological properties of lactic beverages. *Journal of Food Engineering* 49: 89–93.

Piazza, M.G., Accorti, M., and Oddo, L.P. 1991. Electrical conductivity, ash, colour and specific rotatory power in Italian unifloral honeys. *Apicoltura* 7: 51–63.

Pisani, A., Protano, G., and Riccobono, F. 2008. Minor and trace elements in different honey types produced in Siena County (Italy). *Food Chemistry* 107: 1553–1560.

Przyblowksi, P. and Wilczyńska, A. 2001. Honey as an environmental marker. *Food Chemistry* 74: 289–291.

Rizkalla, S.W. 2010. Health implications of fructose consumption: A review of recent data. *Nutrition and Metabolism* 7: 82.

Ruoff, K. 2006. *Doctor Tesis/Authentication of the Botanical Origin of Honey*. Helsinki University, Diss. Eth. no. 16850.

Sengul, M., Ertugay, F.M., and Sengul, M. 2005. Rheological, physical and chemical characteristics of mulberry pekmez. *Food Control* 16: 73–76.

Steffe, J.F. 1996. *Rheological Methods in Food Process Engineering*. East Lansing, MI: Freeman Press.

Symon, K. 1971. *Mechanics*. 3rd Edition. Reading, MA: Addison-Wesley.

Szczesna, T. and Chmielewska, R. 2004. The temperature correction factor for electrical conductivity of honey. *Journal of Apicultural Science* 48 (2): 97–103.

Tezcan, F., Kolaylı, S., Sahin, H., Ulusoy, E., and Erim, B.F. 2010. Evaluation of organic acid, sacchraride composition and antioxidant properties of some authentic Turkish honeys. *Journal of Food Nutrition Research* 50: 33–40.

Terrab, A., Recamales, A.F., Hernanz, D., and Heredia, F.J. 2004. Characterisation of Spanish thyme honeys by their physicochemical characteristics and mineral contents. *Food Chemistry* 88: 537–542.

Tuzen, M., Silici, S., Mendil, D., and Soylak, M. 2007. Trace element levels in honeys from different regions of Turkey. *Food Chemistry* 103: 325–330.

Vanhanen, L.P., Emmertz, A., and Savage, G.P. 2011. Savage mineral analysis of mono-floral New Zealand honey. *Food Chemistry* 128: 236–240.

Weast, R.C., Astle, W.H., and Beyer, W.H. 1985. *CRC Handbook of Chemistry and Physics*. Boca Raton, FL: CRC Press, Inc.

White, J.W., Willson, R.B., Maurizio, A., and Smith, F.G. 1975. *Honey. A Comprehensive Survey*. London: Heinemann, 608 pp.

White J.W. 1978. *Journal of Apicultural Science* 17: 234–238.

White, J.W. 1979. Composition of honey, In Crane, E. (ed.), *Honey. A Comprehensive Survey*. London: Heinemann, pp. 157–206.

White, J.W. 2000. Ask the honey expert. *American Bee Journal* 365–367.

Yanniotis, S., Skaltsi, S., and Karaburnioti, S. 2006. Effect of moisture content on the viscosity of honey at different temperatures. *Journal of Food Engineering* 72: 4372–4377.

Yaoa, L., Jiang, Y., Singanusong, R., Datta, N., and Raymont, K. 2005. Phenolic acids in Australian *Melaleuca, Guioa, Lophostemon, Banksia* and *Helianthus* honeys and their potential for floral authentication. *Food Research International* 38: 651–658.

Yildiz, O. and Alpaslan, M. 2012. Properties of rosehip marmalades. *Food Technology and Biotechnology* 50 (1): 98–106.

# 4 Healing Properties of Honey

*Laïd Boukraâ*

## CONTENTS

## INTRODUCTION

Natural products have been used for thousands of years in folk medicine for several purposes. Among them, honey has attracted increased interest in recent years due to its antimicrobial activity against a wide range of pathogenic microorganisms (Table 4.1). Various studies attribute the antibacterial, antifungal, anti-inflammatory, antiproliferative, and anticancer potentiating properties to honey (Skiadas and Lascaratos 2001). Approximately 70% of bacteria that cause infections in hospitals are resistant to at least one of the antibiotics most commonly used to treat infections. This antibiotic resistance is driving up health care costs, increasing the severity of disease and the fatality of certain infections. Sepsis is another serious medical condition resulting from severe inflammatory response to systemic bacterial infections (Martin et al. 2003). More desirably, honey has the capacity to bind bacterial endotoxin and neutralize bacterium-induced inflammatory response. Because of the dual capability to kill bacteria and neutralize endotoxins, this antimicrobial natural product holds great promise as a new class of antimicrobial and antisepsis agent (Finaly and Hancock 2004). The major antimicrobial properties are correlated to

## TABLE 4.1
## Infection Caused by Bacterial Pathogens That Are Sensitive to the Antibacterial Activity of Honey

| Infection | Pathogen |
|---|---|
| Anthrax | *Bacillus anthracis* |
| Diphtheria | *Corynebacterium diphtheriae* |
| Diarrhea, septicemia, urinary infections, wound infections | *Escherichia coli* |
| Ear infections, meningitis, respiratory infections, sinusitis | *Haemophilus influenzae* |
| Pneumonia | *Klebsiella pneumoniae* |
| Meningitis | *Listeria monocytogenes* |
| Tuberculosis | *Mycobacterium tuberculosis* |
| Infected animal bites | *Pasteurella multocida* |
| Septicemia, urinary infections, wound infections | *Proteus* species |
| Urinary infections, wound infections | *Pseudomonas aeruginosa* |
| Diarrhea | *Salmonella* species |
| Septicemia | *Salmonella choleraesuis* |
| Typhoid | *Salmonella typhi* |
| Wound infections | *Salmonella typhimurium* |
| Septicemia, wound infections | *Serratia marcescens* |
| Dysentery | *Shigella* species |
| Abscesses, boils, carbuncles, impetigo, wound infections | *Staphylococcus aureus* |
| Urinary infections | *Streptococcus faecalis* |
| Dental caries | *Streptococcus mutans* |
| Ear infections, meningitis, pneumonia, sinusitis | *Streptococcus pneumoniae* |
| Ear infections, impetigo, puerperal fever, rheumatic fever, scarlet fever, sore throat, wound infections | *Streptococcus pyogenes* |
| Cholera | *Vibrio cholerae* |

*Source:* Molan, P.C., *Bee World, 73*, 5–28, 1992.

the hydrogen peroxide ($H_2O_2$) level, which is determined by relative levels of glucose oxidase and catalase (Weston et al. 2000), whereas the nonperoxide factors that contribute to honey's antibacterial and antioxidant activity are lysozyme, phenolic acids, and flavonoids (Snowdon and Cliver 1996). Apart from antibacterial properties, honey also plays a therapeutic role in wound healing and the treatment of eye and gastric ailments. This is partly due to its antioxidant activity (Gheldof et al. 2002), because some of these diseases have been recognized as being a consequence of free radical damage (Aljadi and Kamaruddin 2004). Topical application of growth factors to wound has shown good capability to speed up tissue repair in animal models (Mustoe et al. 1991; Pierce et al. 1992). Moreover, human recombinant platelet-derived growth factor (PDGF), which directly interferes and favors the repair process, has given good results in the healing of diabetic patient ulcers (Pierce et al. 1992; Steed 1998). However, they are high-cost dressings that are not affordable to most patients who have chronic ulcers (Mendonça and Coutinho-Netto 2009). It is also to be mentioned that infection is important for the treatment of an infected wound. Therefore, different types of medicated collagen dressings with antibiotics have been developed (Lee et al. 2002). It is well established that honey has antibacterial activity in vitro, and clinical case studies have shown that application of honey to severely infected cutaneous wounds is capable of improving the healing process (Moore et al. 2001). Health care professionals are aware that wound dressings should be judged on effectiveness, safety, and cost. As such, honey that is to be used for medicinal purposes has to be free of residual herbicides, pesticides, heavy metals, and radioactivity. It also has to be sterilized by γ-radiation to prevent wound infection (wound botulism). Furthermore, glucose oxidase in honey has to be controlled during processing to maintain the potency for infection prevention without doing harm to the wound tissues. Besides these primary conditions, the application of honey should be easy (Bogdanov 1996; Molan and Allen 1996; Emsen 2007). Understanding the scientific basis of the anti-inflammatory properties of honey could potentially lead to the development of novel therapeutic agents with a view to rationalizing and optimizing its use for wound therapy. Topical application of cytokines such as interleukin (IL)-8, although not very effective, or other agents such as emu oil (Politis and Dmytrowich 1998; Li et al. 2004), which inhibits proinflammatory cytokine production, has been shown experimentally to promote wound healing. Ointments that contain enzymatic agents such as DNase and collagenase are used to promote wound debridement (Hebda et al. 1990). However, these drugs used in clinical practice have low efficacy in healing of chronic wounds (Mendonça and Coutinho-Netto 2009). Modern hydrocolloid wound dressings are presently favored as moist dressing, although such wound dressings are expensive. Foams, gels, and alginates are also available for treating chronic wounds. Although moist wound care enhances the healing process through tissue regrowth, such moist conditions favor the growth of infecting bacteria.

Honey has a wide range of phytochemicals including polyphenols that act as antioxidants. Polyphenols and phenolic acids found in the honey vary according to the geographical and climatic conditions. Some of them were reported as a specific marker for the botanical origin of the honey. The antioxidant activity of phenolic compounds might significantly contribute to the human health benefits of plant foods (Hertog et al. 1993; Bravo 1998) and beverages such as tea (Bravo 1998).

## ANTIMICROBIAL PROPERTIES OF HONEY

The antibacterial effectiveness of honey is attributed to its physicochemical characteristics and phytochemical compounds.

### PHYSICOCHEMICAL PROPERTIES OF HONEY

It has been demonstrated in many studies that honey has antibacterial effects, attributed to its high osmolarity, low pH, $H_2O_2$ content, and content of other, uncharacterized compounds (Molan 1995). The low water activity of honey is inhibitory to the growth of the majority of bacteria, but this is not the only explanation for its antimicrobial activity. Molan (1992) has studied sugar syrups of the same water activity as honey and found them to be less effective than honey at inhibiting microbial growth in vitro. Honey is mildly acidic, with a pH between 3.2 and 4.5. The low pH alone is inhibitory to many pathogenic bacteria and, in topical applications at least, could be sufficient to exert an inhibitory effect. When consumed orally, the honey would be so diluted by body fluids that any effect of low pH is likely to be lost (Molan 1995). $H_2O_2$ was identified as the major source of antibacterial activity in honey (White and Subers 1964). It is produced by the action of glucose oxidase on glucose, producing gluconic acid. This is inhibited by excessive heat and low water activity (White and Subers 1964). The fact that the antibacterial properties of honey are increased when diluted was clearly observed and reported in 1919 (Sackett 1919). The explanation for this apparent paradox came from the finding that honey contains an enzyme that produces $H_2O_2$ when diluted (White et al. 1963). This agent was referred to as "inhibine" before its identification as $H_2O_2$ (Dold et al. 1937); $H_2O_2$ is a well-known antimicrobial agent, initially hailed for its antibacterial and cleansing properties when it was first introduced into clinical practice (White et al. 1963). In more recent times, it has lost favor because of inflammation and damage to tissue (Salahudeen et al. 1991; Halliwell and Cross 1994; Saissy et al. 1995). However, the $H_2O_2$ concentration produced in honey activated by dilution is typically about 1000 times less than in the 3% solution commonly used as an antiseptic (Molan 1992). Although the level of $H_2O_2$ in honey is very low, it is still effective as an antimicrobial agent. It has been reported that $H_2O_2$ is more effective when supplied by continuous generation with glucose oxidase than when added in isolation (Pruitt and Reiter 1985). Additional nonperoxide antibacterial factors have been identified (Allen et al. 1991). Moreover, the antibacterial components of medical-grade honey have been completely characterized by Kwakman et al. (2010). Besides the presence of $H_2O_2$, some minerals, particularly copper and iron, present in honey may lead to the generation of highly reactive hydroxyl radicals as part of the antibacterial system (Molan 1992; Mccarthy 1995). Therefore, there must be mechanisms involved in honey to control the formation and removal of these reactive oxygen species (ROS). Contributions of free radicals and quenching properties of honeys in wound healing have been demonstrated by Henriques et al. (2006).

## PHYTOCHEMICAL PROPERTIES OF HONEY

Recently, methylglyoxal has been successfully identified as the dominant bioactive component in manuka honey (*Leptospermum scoparium*) and its concentration was correlated to the nonperoxide activity of the honey (Adams et al. 2008; Mavric et al. 2008). Phenolic compounds are among the most important groups of compounds occurring in plants, which are found to exhibit anticarcinogenic, anti-inflammatory, antiatherogenic, antithrombotic, immunomodulating, and analgesic activities and may exert these functions as antioxidants (Vinson et al. 1998). They are also present in honey and have been reported to have some chemoprotective effects in humans (Arreaz-Roman et al. 2006). The phenolic acids are generally divided into two subclasses: the substituted benzoic acids and cinnamic acids, whereas the flavonoids present in honey are categorized into three classes with similar structure: flavonols, flavones, and flavanones. These contribute significantly to honey color, taste, and flavor and have beneficial effects on health (Estevinho et al. 2008). The composition of honey, including its phenolic compounds, is variable depending mainly on the floral

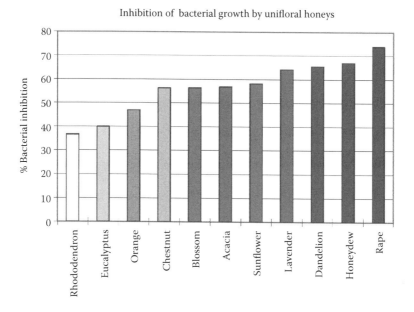

**FIGURE 4.1** Nonperoxide activity of different honeys against *Staphylococcus aureus*. (From Bogdanov, S., in *Bee Products: Properties Application and Apitherapy*, Eds. Mizrahi, A. & Lensky, Y., Plenum Publishing Corporation, New York, 1996, 39–47.)

source and also other external factors including seasonal and environmental factors as well as processing (Arreaz-Roman et al. 2006). Thus, with different compositions of active compounds in honey collected from different locations, differences in honey properties are to be expected (Figure 4.1). Davidson et al. (2005) have shown that individual phenolic compounds have growth inhibition on a wide range of gram-positive and gram-negative bacteria. Other honey therapeutic action studies carried out had been mainly on screening the raw honey samples on antimicrobial activity (Taormina et al. 2001; Basualdo et al. 2007) and antioxidant capacity (Frankel et al. 1998; Rauha et al. 2003). Because not all honeys are created equal in terms of antimicrobial activity because of differences in levels of nonperoxide factors, which vary by floral source and processing, a comparative study has been conducted to set out the antibacterial potency of Sahara honey and those of northern Algerian origin. Results have shown that Sahara honey is more potent against bacteria compared with other Algerian honeys. This is most likely due to its phenolic and flavonoid components (Boukraâ and Niar 2007).

## VARIATION IN HONEY POTENCY AND BACTERIAL SUSCEPTIBILITY

In almost all studies in which more than one type of honey has been used, differences in the antibacterial activity of the honeys have been observed (Table 4.2). The degree of difference observed has in some cases been very large, and in many others where it has been smaller, this possibly is the result of a more limited range of testing rather than of less variance in the activity of the honeys. In many studies, the antibacterial activity of different honeys has been compared by way of the "inhibine

**TABLE 4.2**

**Minimum Concentration of Honey (%, v/v) in the Growth Medium Needed to Completely Inhibit the Growth of Various Species of Wound-Infecting Bacteria**

| Species | Manuka Honey + Catalase | Other Honey |
|---|---|---|
| *Escherichia coli* | 3.7 | 7.1 |
| *Proteus mirabilis* | 7.3 | 3.3 |
| *Pseudomonas aeruginosa* | 10.8 | 6.6 |
| *Salmonella typhimurium* | 6.0 | 4.1 |
| *Serratia marcescens* | 6.3 | 4.7 |
| *Staphylococcus aureus* | 1.8 | 4.9 |
| *Streptococcus pyogenes* | 3.6 | 2.6 |

*Source:* Willix, D.J. et al., *Journal of Applied Bacteriology, 73,* 388–394, 1992.

*Note:* The manuka honey had catalase added to remove hydrogen peroxide, so that only the unique *Leptospermum* antibacterial was being tested.

number" determined by the method devised by Dold and Witzenhausen (1955) for such comparisons, who coined the term "inhibine number" in 1955 to describe the degree of dilution to which a honey will retain its antibacterial activity. This is a term that has been widely used since as a measure of the antibacterial activity of honey. The "inhibine number" involves a scale of 1 to 5 representing sequential dilutions of honey in steps of 5% from 25% to 5%. There have since been various minor modifications to this method, so that the actual concentration corresponding to the "inhibine number" reported may vary. One modification has been to estimate fractional "inhibine numbers" by visual assessment of partial inhibition on the agar plate with the concentration of honey that just allows growth (Duisberg and Warnecke 1959; Stomfay-Stitz and Kominos 1960; Adcock 1962). In three other studies (Stomfay-Stitz and Kominos 1960; Chwastek 1966; Molan et al. 1988), activity was found to range over a fourfold difference in concentration in the dilution series. With some honeys not active at the highest concentration tested in some of the studies and others still active at the greatest dilutions, it is possible that if greater and lesser degrees of dilution had been included in the testing, then a wider range of activities would have been detected. One study using a wider range of dilutions (honey from 50% to 0.25%) found the minimum inhibitory concentrations of the honeys tested to range from 25% to 0.25% (Agostino et al. 1961). Another testing from 50% to 0.4% found the minimum inhibitory concentrations to range from >50% (i.e., not active at 50%) to 1.5% (Dustmann 1979). Other studies with wide ranges tested also found some honeys without activity at the highest concentration tested, and other honey with activity at the lowest concentration tested: the ranges were from 20% to 0.6% (Buchner 1966) and from 50% to 1.5% (Christov and Mladenov 1961). When the data are examined, activities are seen to be fairly well spread over these ranges. Duisberg and Warnecke (1959) plotted the distribution of the activity of 131 samples of honey tested and found that it deviated from a normal Gaussian distribution because of the large number of samples with low activity. (In 7% of the samples, the activity was below the level of detection.) They attributed this to destruction of activity by exposure to heat and light and estimated that 50% of the samples had lost more than half of their original activity and 22% had lost more than three quarters.

## ANTIVIRAL PROPERTIES OF HONEY

A large amount of research has established the potent antibacterial activity of honey, but its activity against viral species has been the subject of only a small number of studies. These were with viruses that cause localized infections in which honey could be used topically. Recent studies demonstrating the safety of intrapulmonary administration of honey in sheep and humans raised the possibility of using honey to treat respiratory infections. Attempts at isolating the antiviral component in honey demonstrated that the sugar was not responsible for the inhibition of respiratory syncytial virus (RSV) but that methylglyoxal may play a part in the greater potency of manuka honeys against RSV. Addition of honey to moderately infected cells was observed to halt the progression of infection, thus showing the potential benefit of using honey as a treatment in individuals already experiencing symptoms of infection. It was shown that sugar, in similar levels found in 2% natural honey, is not an effective antiviral agent. In contrast, further investigations determined that methylglyoxal had potent antiviral activity even at very low concentrations and may therefore be the component responsible for the greater inhibition seen in the high-NPA honeys. It is concluded from the findings in this study that honey may possibly be an effective antiviral treatment for the therapy of respiratory viral infections and provides justification for future in vitro studies and clinical trials. Previous studies suggest other roles that honey may play if used as a treatment for viral infections. As well as having antibacterial activity that would protect patients from secondary infections, honey has also been found to have some immunostimulatory effects (Tonks et al. 2003, 2007), which would also augment the direct antiviral action and therefore contribute to the clearing of virus and healing of infection by the body's own defenses.

## ANTIFUNGAL PROPERTIES OF HONEY

Although an earlier brief review (Haydak et al. 1975) of the biological effects of honey expressed the opinion that honey had no effect on fungi beyond its osmotic action, many recent studies show that some honeys, at least, must have antifungal factors present, because some fungi are inhibited under conditions where the sugar content of the honey is clearly not responsible (Boukraâ and Bouchougrane 2007; Boukraâ et al. 2008).

# ANTI-INFLAMMATORY PROPERTIES OF HONEY

## INFLAMMATION: PHYSIOLOGIC OVERVIEW

During the inflammatory phase, neutrophils and macrophages appear in the wounded area to phagocytize bacteria and debris (Molnar 2007). The release of bradykinin, histamines, and free radicals from leukocytes leads to increased vascular permeability. It has become clear that ROS produced by neutrophils and macrophages not only kill microbes but are also responsible for the tissue injury in acute inflammation. Oxidative stress likely influences the wound healing and local metabolism throughout the healing process. The vasoconstriction/vasodilatation that occurs at the time of wound creation can produce ischemia-reperfusion injury in the tissues due to the generation of ROS (e.g., superoxide, $H_2O_2$, hydroxyl, and peroxyl radicals). Likewise, neutrophils that infiltrate the wound during the early stages of inflammation and monocytes/macrophages that appear later in the healing process are an abundant source of ROS (Singer and Clark 1999).

It has been shown that superoxide (but not $H_2O_2$) is able to stimulate the production of IL-1–like factors from human peripheral blood monocytes and poly-morphonuclears (PMNs) (Kasama et al. 1989). These findings suggest that there is a feedback mechanism of inflammation, in which ROS stimulate the enhanced IL-1–like factor production that, in turn, increases the formation of oxygen radicals (Denisov and Afanas'ev 2005). On the contrary, infiltration of cells into the wound can induce production of tumor necrosis factor (TNF), which is considered to be a proinflammatory cytokine, along with IL-1, IL-17, and other cytokines (Feldmann and Steinman 2005). At low concentrations in tissues, TNF is thought to have beneficial effects, such as increasing the host defense mechanisms against infections, but at high concentrations TNF can lead to excess inflammation and organ injury (Tracey et al. 2008). After the PMN leukocyte influx, the monocytes are next to arrive. Monocytes become phagocytic macrophages. Macrophages are central to the regulation of cutaneous wound healing (Riches 1996). In its molecular form, oxygen is required for oxidative metabolism-derived energy synthesis, protein synthesis, and the maturation (hydroxylation) of extracellular matrices such as collagen. Molecular oxygen is also required for nitric oxide synthesis, which in turn plays a key role in the regulation of vascular tone as well as in angiogenesis. In a wound setting, large amounts of molecular oxygen are partially reduced to form ROS. ROS includes oxygen free radicals such as superoxide anion as well as its nonradical derivative $H_2O_2$.

Superoxide anion radical is the one-electron reduction product of oxygen. NADPH oxidases represent one major source of superoxide anion radicals at the wound site. NADPH oxidases in phagocytic cells help fight infection. Superoxide anion also drives endothelial cell signaling such as required during angiogenesis. Endogenous $H_2O_2$ drives redox signaling, a molecular network of signal propagation that supports key aspects of wound healing such as cell migration, proliferation, and angiogenesis. Neutrophil-derived $H_2O_2$ may be utilized by myeloperoxidase to mediate peroxidation of chloride ions resulting in the formation of hypochlorous acid (HOCl), a potent disinfectant that removes debris as well as bacteria from the wound (Chandan 2009). These macrophages secrete proteases, producing interferon and prostaglandins as well as cytokines. These cytokines, among other factors, are chemoattractants for mesenchymal cells, which will differentiate into fibroblasts, one of the major cell types involved in the proliferative phase and connective tissue formation (Barbul 1992; Monaco and Lawrence 2003). Excessive protease activity in a wound can slow or prevent healing by destroying growth factors, protein fibers, and fibronectin in the wound matrix. Consequently, a functioning immune system and adequate supply of growth factors are necessary in this phase of wound healing (Fonder et al. 2008).

During the proliferative phase, fibroblasts produce extracellular matrix required for cell growth (collagen matrix), new blood vessel tissues invade the forming granulation (carry oxygen and nutrients required for local cellular metabolism) (Singer and Clark 1999), and epidermal cells migrate across the wound surface to close the breach. During the healing process, the formation of new blood vessels (angiogenesis) becomes necessary to form new granulation tissues, in which the blood vessel cells correspond to about 60% of the repair tissue (Arnold and West 1991). The remodeling phase is marked by maturation of elements that are released to the extracellular matrix, leading to proteoglycan and collagen deposits (Mendonça and Coutinho-Netto 2009). Later, fibroblasts reorganize the collagen matrix and upon transformation to myofibroblasts behave as contractile tissue to effect connective tissue compaction and wound contraction (Fonder et al. 2008). As a result of maturation and remodeling processes, most vessels, fibroblasts, and inflammatory cells disappear from the wound site through a process of migration, apoptosis, or other unknown cell death mechanisms, which results in a scar with fewer cells (Mendonça and Coutinho-Netto 2009). Wounds gain about 20% of their final strength in the first 3 weeks of normal wound healing through collagen deposition, remodeling, and wound contraction (Singer and Clark 1999). The main cytokines involved in this phase are TNF-$\alpha$, IL-1, PDGF, and transforming growth factor (TGF)-$\alpha$ produced by fibroblasts in addition to those produced by epithelial cells such as epidermal growth factor and TGF-$\alpha$ (Karukonda et al. 2000).

## ROLE OF HONEY IN PREVENTING AND CURING INFLAMMATION

One way in which honey may work is through its stimulation of an inflammatory response in leukocytes (Abuharfeil et al. 1999; Tonks et al. 2001, 2003). There are also suggestions that a small amount of bacteria may be beneficial to the wound healing process (Edwards and Harding 2004), and many reports suggest that honey could contain osmophilic bacteria that stimulate an inflammatory response (Molan 2002), as inflammation triggers the cascade of cellular events that give rise to the production of growth factors that control angiogenesis and proliferation of fibroblasts and epithelial cells (Denisov and Afanas'ev 2005). Honey may influence the activation of various cellular and extracellular matrix components and cells. Fukuda et al. (2011) described the properties of a protein (MRJP1) from jungle honey as an effective component that could have potential therapeutic effect for the treatment of wounds. It was reported that incubation of human fibroblasts from normal skin or chronic wounds in the presence of the immunostimulatory component of honey enhances cellular proliferation. Other authors have found that concentrations of honey as low as 0.1% stimulate the proliferation of lymphocytes in cell culture and activate phagocytes from blood (Abuharfeil et al. 1999). Immunomodulatory effects were demonstrated in vitro by cytokine release from the monocytic cell line Mono Mac 6 (MM6) and human peripheral monocytes after incubation with honey (Tonks et al. 2003). Honey could also induce IL-6, IL-1$\beta$, and TNF-$\alpha$ release (Tonks et al. 2001). Recently, Tonks et al. (2007) discovered a 5.8-kDa component of manuka honey, which stimulates the production of TNF-$\alpha$ in macrophages via Toll-like receptor 4. These molecules also induced the production of IL-1$\alpha$ and IL-6. Honey can also stimulate angiogenesis, granulation, and epithelialization in animal models (Bergman et al. 1983; Gupta et al. 1992). In wound and burn management, hydration is an important factor for optimal wound healing (Atiyeh et al. 2005a,b). Moisture retaining dressings provide a protective barrier, prevent eschar formation, reduce dermal necrosis seen in wounds that have been allowed to dry, and significantly accelerate wound reepithelialization (Jurjus et al. 2007). Healing under both wet and moist environments is significantly faster than under dry conditions (Eaglstein et al. 1987). Thus, it is suggested that honey dressings do not stick to the surface of wounds as they sit on a layer of diluted honey without any growth of new tissue into dressing (Tonks et al. 2007). It seems that the optimum environment would be an intermediate gelatinous environment between moist and dry such as seen under highly vapor-permeable dressings (Bernabei et al. 1999; Braddock et al. 1999). Formulation of honey-based gel compositions containing protein growth factors and/or debriding enzymes is provided to achieve rapid and optimal wound healing. Moreover, honey has been combined with fatty ester, wax, and a wax-like compound to form an ointment that could be applied on wounds. Several aspects of healing such as cell proliferation and migration are supported by redox signaling where low-level ROS produced by nonphagocytic oxidases serve as messenger molecules (Sen and Roy 2008) and induces Nrf2, a transcription factor implicated in the transactivation of genes encoding antioxidant enzymes. However, an intermediate amount of ROS triggers an inflammatory response through the activation of nuclear factor-$\kappa$B (NF-$\kappa$B) and AP-1. The transcription factor NF-$\kappa$B is crucial in a series of cellular processes, such as inflammation, immunity, cell proliferation, and apoptosis (Gloire et al. 2006). Wound fluid from healing tissues contains the highest concentration of $H_2O_2$ compared with all other body fluids (Roy et al. 2006; Ojha et al. 2007) and this serves to stimulate the growth of fibroblasts and epithelial cells to repair the damage (Ojha et al. 2007). The $H_2O_2$ produced in honey would also be a factor responsible for the rapid rate of healing observed when wounds are dressed with honey. It has been shown that the optimal dilution at which the honey will produce maximal amounts of $H_2O_2$ is between 40% and 60% (Henriques et al. 2006). Honey provides such a controlled delivery of $H_2O_2$, the enzymatic production giving a slow release achieving equilibrium concentrations of 20 to 95 $\mu$mol $L^{-1}$. Prolonged exposure to elevated levels of ROS causes cell damage and may eventually inhibit healing of both acute and chronic wounds. Typically, burn injuries show excessive activity of free radicals (Wan and Evans 1999; Subrahmanyarn et al. 2003). It is likely that reducing levels of free radicals and other oxidants in the wound bed will aid wound management. Natural honey has been shown to reduce the production of ROS in endotoxin primed MM6 cells (Tonks et al. 2001) and serves as a source of natural antioxidants, which are effective in reducing the risk of different inflammatory processes (National Honey Board 2003). Honey has a wide range of phytochemicals including polyphenols (Saravana and Mahitosh 2009). Important groups of polyphenols found in honey are flavonoids (chrysin, pinocembrin, pinobanksin, quercetin, kaempferol, luteolin, galangin, apigenin, hesperetin, and myricetin), phenolic acids (caffeic, coumaric, ferrulic, ellagic, and chlorogenic), organic acids, enzymes, and vitamins (Gheldof et al. 2002). These components found in honey are responsible for its antioxidative effect. However, the proportion of polyphenols and phenolic acid in honey vary according to the geographical and climatic conditions (Saravana and Mahitosh 2009). Medical honey could also hasten the healing of wounds by reducing edema and the amount of exudates. Another effect of honey is to reduce pain, because the pain in wounds results from the nerve endings being sensitized by prostaglandins produced in the process of inflammation as well from the pressure on tissues resulting from edema (Simon et al. 2009).

## ANTIOXIDANT PROPERTIES OF HONEY

The enzyme catalase present in honey has an antioxidant property; thus, honey may have a role as an antioxidant in thermal injury (Subrahmanyam 1996a). The nutrient contents of honey, such as glucose and fructose, improve local substrate supply and may help promote epithelialization. Furthermore, the increase in vitamin C and other antioxidants caused by honey in the blood concentration is particularly important for granulation tissue development and wound healing (Al-Waili 2003; Schramm et al. 2003). The reduction in reactive oxygen intermediates (ROIs) seen in the presence of honey may serve to limit tissue damage by activated macrophages during the healing process. Natural honey has also been shown to reduce the production of ROS in endotoxin primed MM6 cells (Senyuva et al. 1997). The harmful effects of $H_2O_2$ are further reduced because honey sequesters and inactivates the free iron, which catalyzes the formation of oxygen free radicals produced by $H_2O_2$ (Bunting 2001) and its antioxidant components help to mop up oxygen free radicals (Frankel et al. 1998). Although free radicals of oxygen are a natural byproduct of metabolism within the organism, they cause cellular damage and break down the structure of DNA. Exactly these processes cause premature aging. Antioxidants bind these dangerous molecules, preventing their harmful effects (Drost et al. 1993). Unlike synthetic compounds, honey represents a natural product that does not carry side effects that can be harmful to health. Among the compounds found in honey, vitamin C, phenol compounds, catalase, peroxides, and glucose oxidase enzymes have antioxidant properties. Honey also contains flavonoids and carotenoids. High levels of these indicators ensure a high level of antioxidants in honey. Antioxidant properties of honey act as an antidepressant during high emotional, physical, and intellectual stress (Drost et al. 1993). Various polyphenols are reported in honey. Some of the polyphenols of honey such as caffeic acid, caffeic acid phenyl ester, chrysin, galangin, quercetin, acacetin, kaempferol, pinocembrin, pinobanksin, and apigenin have evolved as promising pharmacologic agents in the treatment of cancer (Drost et al. 1993). "Gram for gram, antioxidants in buckwheat honey equal those of fruits and vegetables," said Dr. May Berenbaum, head of the University of Illinois entomology department. "It packs the antioxidant power of vitamin C in a tomato." Researchers at the University of Illinois-Champaign/Urbana have identified the antioxidant values of 14 unifloral honeys. The antioxidative components of honey were compared to an ascorbic acid standard. The water-soluble antioxidant content of the honey samples varied more than 20-fold, from a high value of $4.32 \times 10^{-3}$ Eq for Illinois buckwheat honey to a low value of $21.3 \times 10^{-5}$ Eq for California button sage honey (Endo et al. 1993). Research showed a correlation between color and antioxidant capacity, with the darker honeys providing the highest levels of antioxidants.

With antioxidant levels reaching $4.32 \times 10^{-3}$ mEq, honey rivals those levels found in tomatoes ($2.83 \times 10^{-3}$ mEq) and sweet corn ($1.36 \times 10^{-3}$ mEq). Although honey by itself may not serve as a major source of dietary antioxidants, it demonstrates the potential for honey to play a role in providing antioxidants in a highly palatable form. Due to honey's pleasing taste, it may be more readily consumed by individuals reluctant to ingest plant-derived antioxidants. Certainly, compared to sucrose that has no antioxidant value, honey can be a flavorful, supplementary source of antioxidants (Drost et al. 1993).

## BOTANICAL ORIGIN AND ANTIOXIDANT ACTIVITIES OF HONEY

Honey has been found to contain significant antioxidant compounds including glucose oxidase, catalase, ascorbic acid, flavonoids, phenolic acids, carotenoid derivatives, organic acids, Maillard reaction products, amino acids, and proteins (Frankel et al. 1998; Fahey and Stephenson 2002; Aljadi and Kamaruddin 2004; Beretta et al. 2005; D'Arcy 2005; Inoue et al. 2005; Blasa et al. 2006). The antioxidative activity of honey polyphenols can be measured in vitro by comparing the oxygen radical absorbance capacity (ORAC) with the total phenolic concentration (Table 4.3). There is a significant correlation between the antioxidant activity, the phenolic content of honey, and the inhibition of the in vitro lipoprotein oxidation of human serum (Gheldof et al. 2003). Furthermore, in a lipid peroxidation model system, buckwheat honey showed a similar antioxidant activity as 1 mM α-tocopherol (Nagai et al. 2006). The influence of honey ingestion on the antioxidative capacity of plasma was tested in two studies (Al-Waili 2003; Schramm et al. 2003). In the first one, the trial persons were given maize syrup or buckwheat honeys with a different antioxidant capacity in a dose of $1.5$ g $kg^{-1}$ body weight. In comparison to the sugar control, honey caused an increase in both the antioxidant and the reducing serum capacities.

### TABLE 4.3
### Antioxidative Activity (ORAC) and Total Phenol Content of Different Unifloral Honeys

| Honey Type ORAC (μmol) | Total Phenolics (TE g⁻¹) | GAE (mg kg⁻¹) |
|---|---|---|
| Buckwheat Illinois | $16.95 \pm 0.76$ | $796 \pm 32$ |
| Buckwheat New York | $9.75 \pm 0.48$ | $456 \pm 55$ |
| Soy | $8.34 \pm 0.51$ | $269 \pm 22$ |
| Hawaiian Christmas berry | $8.87 \pm 0.33$ | $250 \pm 56$ |
| Clover | $6.05 \pm 1.00$ | $128 \pm 11$ |
| Tupelo | $6.48 \pm 0.37$ | $183 \pm 9$ |
| Fireweed | $3.09 \pm 0.27$ | $62 \pm 6$ |
| Acacia | $3.00 \pm 0.16$ | $46 \pm 2$ |

*Source:*   Gheldof, N. et al., *Journal of Agricultural and Food Chemistry, 50,* 5870–5877, 2002.

*Note:*   GAE, gallic acid equivalent; TE, Trolox equivalent.

In the second study, humans received a diet supplemented with a daily honey serving of 1.2 g kg$^{-1}$ body weight. Honey increased the body antioxidant agents: blood vitamin C concentration by 47%, β-carotene by 3%, uric acid by 12%, and glutathione reductase by 7% (Al-Waili 2003). It should be borne in mind that the antioxidant activity depends on the botanical origin of honey and varies to a great extent in honeys from different botanical sources (Baltrusaityte et al. 2007; Kücük et al. 2007).

## ANTIOXIDANT IN HONEY ADDS HEALTH BENEFITS

The Departments of Nutrition and Internal Medicine at the University of California and National Honey Board showed that free radicals and ROS have been implicated in contributing to the processes of aging and disease (Schramm et al. 2003). Humans protect themselves from these damaging compounds, in part, by absorbing antioxidants from high antioxidant foods. This report describes the effects of consuming 1.5 g kg$^{-1}$ body weight of corn syrup or buckwheat honey on the antioxidant and reducing capacities of plasma in healthy human adults. Following consumption of the two honey treatments, plasma total phenolic content increased ($P < 0.05$) as did plasma antioxidant and reducing capacities ($P < 0.05$). These data support the concept that phenolic antioxidants from processed honey are bioavailable and that they increase antioxidant activity of plasma. It can be speculated that these compounds may augment defenses against oxidative stress and that they might be able to protect humans from oxidative stress. Given that the average sweetener intake by humans is estimated to be in excess of 70 kg year$^{-1}$, the substitution of honey in some foods for traditional sweeteners could result in an enhanced antioxidant defense system in healthy adults (Schramm et al. 2003). Beretta et al. (2005) demonstrated the protective activity of a honey of multifloral origin, standardized for total antioxidant power and analytically profiled (high-performance liquid chromatography–mass spectrometry) in antioxidants, in a cultured endothelial cell line (EA.hy926) subjected to oxidative stress. Cumene hydroperoxide (CuOOH) was used as a free radical promoter. Native honey (1%, w/v [pH 7.4], 10$^6$ cells) showed strong quenching activity against lipophilic cumoxyl and cumoperoxyl radicals, with significant suppression/prevention of cell damage, complete inhibition of cell membrane oxidation, intracellular ROS production, and recovery of intracellular reduced glutathione. Experiments with endothelial cells fortified with the isolated fraction from native honey enriched in antioxidants, exposed to peroxyl radicals from 1,1-diphenyl-2-picrylhydrazyl (10 mM) and to H$_2$O$_2$ (50–100 μM), indicated that phenolic acids and flavonoids were the main causes of the protective effect. They suggested that, through the synergistic action of its antioxidants, honey by reducing and removing ROS may lower the risks and effects of acute and chronic free radical-induced pathologies in vivo (Beretta et al. 2007).

## IMMUNOMODULATORY PROPERTIES OF HONEY

During the inflammatory phase, there is activation of the immune system involving cytokines, growth factors, vascular endothelial cells, and other cells (Dehne et al. 2002). The roles of TNF-α, IL-1β, and IL-6 are also important in the inflammatory response. These three cytokines originate principally from macrophage activation and have been described as indicators of severity of injury (Ueyama et al. 1992; Drost et al. 1993; Endo et al. 1993; Yamada et al. 1996). Later, macrophages dominate and, as in phagocytosis, produce a number of specific mediators including ROIs, lipid metabolites, and cytokines (Fels and Cohn 1986). The importance of TNF-α in the healing process has been stressed by a number of authors (Schlenger et al. 1994; Groves et al. 1995; Moore et al. 1997). It stimulates angiogenesis (Leibovich et al. 1987), fibroblast proliferation (Sugarman et al. 1985), and the synthesis of prostaglandins and collagenase by fibroblasts (Dayer et al. 1985). It has also been suggested that the stimulation of TNF-α production may exert much of its beneficial effects by affecting the increasing levels of IL-6 (Grossman et al. 1989). IL-6 is a pleiotropic cytokine with a significant impact on healing. It is mitogenic for keratinocytes (Sato et al. 1999), and its contribution to epithelialization has been demonstrated in IL-6–deficient transgenic mice in which the wound healing rates were retarded threefold compared with wild-type control mice (Gallucci et al. 2000). Topical application of honey to burn wounds and other wounds has been found to be effective in controlling infection and producing a clean granulating bed (Bulmann 1955; Subrahmanyam 1991, 1996b,c). Tonks et al. (2003) suggested that the wound healing effect of honey might in part be related to the release of inflammatory cytokines from surrounding tissue cells, mainly monocytes and macrophages. The findings show that natural honeys can induce IL-6, IL-1β, and TNF-α release. Artificial honey only induces release of these cytokines to a negligible extent (Tonks et al. 2001). Honey has been shown to have mitogenic activity on human B and T lymphocytes (Abuharfeil et al. 1999) and a human myeloid cell line (Watanabe et al. 1998). Proteins present in honey will be highly glycosylated because of high sugar content. Glycosylated proteins have been shown to activate a number of cell types including monocytic cells (Brownlee 1995). It is possible that these may affect the monocytic cell line MM6 activation. It has been reported, in a recent study, that honey significantly stimulated the rate of burn wound healing as the rate of wound contraction at day 7 was greater compared with untreated control groups (Norimah et al. 2007). Similarly, Suguna et al. (1993) and Aljadi et al. (2000) reported that honey hastens wound healing by accelerating wound contractions.

Malignant tumors are linked with a decrease in the immune function. Stimulation of macrophages results in killing of cancer cells. Thus, immunomodulating activity is often linked to anticancer action. All five bee products have immunomodulating and antitumor effects of honey: stimulates T lymphocytes in cell culture to multiply and activates neutrophils (Abuharfeil et al. 1999). It was shown in a study with humans that honey causes an increase in monocytes, lymphocytes, and eosinophil serum percentages (Al-Waili 2003). Honey increases proliferation of B and T lymphocytes and neutrophils in vitro (Abuharfeil et al. 1999). In another study with rats, feeding of honey caused an increase in lymphocytes in comparison with the sucrose-fed controls (Chepulis 2007). Honey antitumor effects have been recently reviewed (Orsolic 2009). A significant antimetastatic effect of honey was demonstrated in methylcholanthrene-induced fibrosarcoma of CBA mouse and in anaplastic colon adenocarcinoma of Y59 rats (Orsolic and Basic 2004). A pronounced antimetastatic effect was observed when honey was applied before tumor cell inoculation (peroral

2 g kg$^{-1}$ for mice or 1 g kg$^{-1}$ for rats, once a day for 10 consecutive days) (Orsolic et al. 2003). Jaganathan and Mandal (2009) demonstrated an antiproliferative effect in colon cancer cells. Honey ingestion by rats induced antitumor and pronounced antimetastatic effects (Gribel and Pashinskii 1990). In another study, the antitumor effect of bee honey against bladder cancer was demonstrated in vitro and in vivo in mice (Swellam et al. 2003). Greek honey extracts (thyme, pine, and fir honey) had an anticancer effect and suggested that a honey-enriched diet may prevent cancer-related processes in breast, prostate, and endometrial cancer cells (Tsiapara et al. 2009). Jungle honey, collected from tree blossoms by wild honeybees that live in the tropical forest of Nigeria, enhanced immune functions and has antitumor activity in mice (Fukuda et al. 2011). The in vitro evaluation of two honeydew honey samples carried out using the blast transformation test indicated a possible stimulatory influence on the avian leukocytes' growth capacity (Niculae et al. 2008).

## REFERENCES

Abuharfeil N, Al-Oran R, and Abo-Shehada M. 1999. The effect of bee honey on proliferative activity of human B- and T-lymphocytes and the activity of phagocytes. *Food Agri Immunol*, 11:169–177.

Adams CJ, Boult CH, Deadman BJ et al. 2008. Isolation by HPLC and characterisation of the bioactive fraction of New Zealand manuka (*Leptospermum scoparium*) honey. *Carbohydr Res*, 343(4):651–659.

Adcock D. 1962. The effect of catalase on the inhibine and peroxide values of various honeys. *J Apicul Res*, 1:38–40.

Agostino BAD, Rosa CLA, and Zanelli C. 1961. Attività antibatterica di mieli Siciliani. *Quad Nutr*, 21(1/2):30–44.

Aljadi AM, Kamaruddin MY, Jamal AM, and Mohd Yassim MY. 2000. Biochemical study on the efficacy of Malaysian honey on inflicted wounds: An animal model. *Med J Islamic Acad Sci*, 13(3):125–132.

Aljadi AM and Kamaruddin MY. 2004. Evaluation of the phenolic contents and antioxidant capacities of two Malaysian floral honeys. *Food Chem*, 85:513–518.

Allen KL, Molan PC, and Reid GM. 1991. A survey of the antibacterial activity of some New Zealand honeys. *J Pharm Pharmacol*, 43(12):817–822.

Al-Waili NS. 2003. Effects of daily consumption of honey solution on haematological indices and blood levels of minerals and enzymes in normal individuals. *J Med Food*, 6: 135–140.

Arnold F and West DC. 1991. Angiogenesis in wound healing. *Pharmacol Ther*, 52:407–422.

Arreaz-Roman D, Gomez-Caravaca AM, Gomez-Romero M, Segura-Carretero A, and Fernandez-Gutierrez A. 2006. Identification of phenolic compounds in rosemary honey using solid-phase extraction by capillary electrophoresis–electrospray ionization-mass spectrometry. *J Pharm Biomed Anal*, 41:1648–1656.

Atiyeh BS, Gunn SW, and Hayek SN. 2005a. New technologies for burn wound closure and healing-review of the literature. *Burns*, 31:944–956.

Atiyeh BS, Hayek SN, Atiyeh RCB, Jurjus RA, Tohme R, Abdallah I et al. 2005b. Cicatrisation des plaies et onguent pour les brûlures exposées humides (MEBO). *J Plaies Cicatrisation*, 49:7–13.

Baltrusaityte V, Venskutonis PR, and Ceksteryte V. 2007. Radical scavenging activity of different floral origin honey and beebread phenolic extracts. *Food Chem*, 101:502–514.

Barbul A. 1992. Role of the immune system. In Cohen IK, Diegelmann RF, and Lindblad WB, Eds. *Wound Healing: Biochemical and Clinical Aspects*. WB Saunders, Philadelphia, PA.

Basualdo C, Sgroy V, Finola MS, and Marioli JM. 2007. Comparison of the antibacterial activity of honey from different provenance against bacteria usually isolated from skin wounds. *Vet Microbiol*, 124:375–381.

Beretta G, Granata P, Ferrero M, Orioli M, and Facino RM. 2005. Standardization of antioxidant properties of honey by a combination of spectrophotometric/fluorimetric assays and chemometrics. *Anal Chim Acta*, 533:185–191.

Beretta G, Orioli M, and Facino RM. 2007. Antioxidant and radical scavenging activity of honey in endothelial cell cultures (EA.hy926). *Planta Med*, 73(11):1182–1189.

Bergman A, Yanai J, Weiss J, Bell D, and David M. 1983. Acceleration of wound healing by topical application of honey. An animal model. *Am J Surg*, 145:374–376.

Bernabei R, Landi F, Bonini S, Onder G, Lambiase A, Pola R et al. 1999. Effect of topical application of nerve-growth factor on pressure ulcers. *Lancet*, 354(9175):307.

Blasa M, Candiracci M, Accorsi A, Piacentini M, Albertini M, and Piatti E. 2006. Raw *Millefiori* honey is packed full of antioxidants. *Food Chem*, 97:217–222.

Bogdanov S. 1996. Non-peroxide antibacterial activity of honey. In Mizrahi A and Lensky Y, Eds. *Bee Products: Properties Application and Apitherapy*. Plenum Publishing Corporation, New York, 39–47.

Boukraâ L and Niar A. 2007. Sahara honey shows higher potency against *Pseudomonas aeruginosa* compared to north Algerian types of honey. *J Med Food*, 10(4):712–714.

Boukraâ L, Benbarek H, and Ahmed M. 2008. Synergistic action of starch and honey against *Aspergillus niger* in correlation with Diastase Number. *Mycoses*, 51:520–522.

Boukraâ L and Bouchougrane S. 2007. Additive action of honey and starch against *Candida albicans* and *Aspergillus niger. Rev Iberoam Micol*, 24:309–313.

Braddock M, Campbell CJ, and Zuder D. 1999. Current therapies for wound healing: Electrical stimulation, biological therapeutics, and the potential for gene therapy. *Int J Dermatol*, 38:808–817.

Bravo L. 1998. Polyphenols: Chemistry, dietary sources, metabolism, and nutritional significance. *Nutr Rev*, 56:317–333.

Brownlee M. 1995. Advanced protein glycosolation diabetes and aging. *Annu Rev Med*, 46:223–234.

Buchner R. 1966. Vergleichende Untersuchungen über die antibakteriellen Wirkung von Blüten- und Honigtauhonigen. *Südwestdeutscher Imker*, 18:240–241.

Bulman MW. 1955. Honey as a surgical dressing. *Middlesex Hosp J*, 15:18–19.

Bunting CM. 2001. The production of hydrogen peroxide by honey and its relevance to wound healing. MSc thesis, University of Waikato, New Zealand.

Chandan KS. 2009. Wound healing essentials: Let there be oxygen. *Wound Repair Regen*, 17:1–18.

Chepulis LM. 2007. The effects of honey compared with sucrose and a sugar-free diet on neutrophil phagocytosis and lymphocyte numbers after long-term feeding in rats. *JCIM*, 4, DOI: 10.2202/1553-3840.1098.

Christov G and Mladenov S. 1961. Propriétés antimicrobiennes du miel. *C R Acad Bulg Sci*, 14(3):303–306.

Chwastek M. 1966. Jakosc miodówpszczelich handlowych na podstawie oznaczania ich skladników niecukrowcowych: Czesc II. Zawartosc inhibiny w miodach krajowych. *Rocziniki Panstwowego Zakladu Higieny*, 17(1):41–48.

D'Arcy BR. 2005. Antioxidants in Australian floral honeys—Identification of health enhancing nutrient components. RIRDC Publication, No 05/040, 1.

Davidson PM, Sofos JN, and Brenem AL. 2005. *Antimicrobials in Foods*. (third ed.), Marcel Dekker Inc., New York, pp. 291–306.

Dayer JM, Beutler B, and Cerami A. 1985. Cachectin/tumor necrosis factor stimulates collagenase and prostaglandin E2 production by human synovial cells and dermal fibroblasts. *J Exp Med*, 162:2163–2168.

Dehne MG, Sablotzki A, Hoffmann A, Muhling J, Dietrich FE, and Hempelmann G. 2002. Alterations of acute phase reaction and cytokine production in patients following severe burn injury. *Burns*, 28:535–542.

Denisov ET and Afanas'ev IB. 2005. *Oxidation and Antioxidants in Organic Chemistry and Biology*. Taylor & Francis Group, London, 932.

Dold H, Du DH, and Dziao ST. 1937. Detection of the antibacterial heat and light-sensitive substance in natural honey. *Z Hyg Infektionskr*, 120:155–167.

Dold H and Witzenhausen R. 1955. Ein Verfahren zur Beurteilung der örtlichen inhibitorischen (keimvermehrungshemmenden) Wirkung von Honigsorten verschiedener Herkunft. *Z Hyg Infektionskr*, 141:333–337.

Drost AC, Burleson DG, Cioffi WG Jr, Mason AD Jr, and Pruitt BA Jr. 1993. Plasma cytokines after thermal injury and their relationship to infection. *Ann Surg*, 218:74–78.

Duisberg H and Warnecke B. 1959. Erhitzungs- und Lichteinfluß auf Fermente und Inhibine des Honigs. *Z LebensmittUntersuch*, 111:111–119.

Dustmann JH. 1979. Antibacterial effect of honey. *Apiacta*, 14(1):7–11.

Eaglstein WH, Mertx PM, and Falanga V. 1987. Occlusive dressings. *Am Fam Physician*, 35:211–216.

Edwards R and Harding KG. 2004. Bacteria and wound healing. *Curr Opin Infect Dis*, 17:91–96.

Emsen IM. 2007. A different and safe method of split thickness skin graft fixation: Medical honey application. *Burns*, 33:782–787.

Endo S, Inada K, Yamada Y et al. 1993. Plasma tumour necrosis factor alpha (TNF-$\alpha$) levels in patients with burns. *Burns*, 19:124–127.

Estevinho L, Pereira AP, Moreira L, Dias LG, and Pereira E. 2008. Antioxidant and antimicrobial effects of phenolic compounds extracts of Northeast Portugal honey. *Food Chem Toxic*, 46(12):3774–3779.

Fahey JW and Stephenson KK. 2002. Pinostrobin from honey and Thai ginger (*Boesenbergia pandurata*): A potent flavonoid inducer of mammalian phase 2 chemoprotective and antioxidant enzymes. *J Agric Food Chem*, 50:7472–7476.

Feldmann M and Steinman L. 2005. Design of effective immunotherapy for human autoimmunity. *Nature*, 435, 612–619.

Fels A and Cohn Z. 1986. The alveolar macrophage. *J Appl Physiol*, 60:353–369.

Finaly BB and Hancock RE. 2004. Can innate immunity be enhanced to treat microbial infections? *Nat Rev Microbiol*, 2:497–504.

Fonder MA, Lazarus GS, Aronson-Cook B, Kohli AR, and Mamelak AJ. 2008. Treating the chronic wound: A practical approach to the care of nonhealing wounds and wound care dressings. *J Am Acad Dermatol*, 58(2):185–206.

Frankel S, Robinson GE, and Berenbaum MR. 1998. Antioxidant capacity and correlated characteristics of 14 unifloral honeys. *J Apic Res*, 37(1):27–31.

Fukuda M, Kobayashi K, Hirono Y, Miyagawa M, Ishida T, Ejiogu EC et al. 2011. Jungle honey enhances immune function and antitumor activity. *Evid Based Complement Alternat Med*, doi:10.1093/ecam/nen086.

Gallucci RM, Simeonova PP, Matheson JM et al. 2000. Impaired cutaneous wound healing in interleukin-6-deficient and immunosuppressed mice. *FASEB J*, 14:2525–2531.

Gheldof N, Wang XH, and Engeseth NJ. 2003. Buckwheat honey increases serum antioxidant capacity in humans. *J Agric Food Chem*, 51:1500–1505.

Gheldof N, Wang XH, and Engeseth NJ. 2002. Identification and quantification of antioxidant components of honeys from various floral sources. *J Agric Food Chem*, 50: 5870–5877.

Gloire G, Legrand-Poels S, and Piette J. 2006. NF-kB activation by reactive oxygen species: Fifteen years later. *Bioch Pharm*, 72:1493–1505.

Gribel' NV and Pashinskii VG. 1990. The antitumor properties of honey. *Vopr Onkol*, 36:704–709.

Grossman RM, Krueger J, Yourish D et al. 1989. Interleukin 6 is expressed in high levels in psoriatic skin and stimulates proliferation of cultured human keratinocytes. *Proc Natl Acad Sci*, 86:6367–6371.

Groves RW, Allen MH, Ross EL, Barker JNWN, and MacDonald DM. 1995. TNF-a is proinflammatory in normal human skin and modulates cutaneous adhesion molecule expression. *Br J Dermatol*, 132:345–352.

Gupta S, Singh H, Varshney A, and Prakash P. 1992. Therapeutic efficacy of honey in infected wounds in buffaloes. *Indian J Anim Sci*, 62:521–523.

Halliwell B and Cross CE. 1994. Oxygen-derived species: Their relation to human disease and environmental stress. *Environ Health Perspect*, 102(suppl 10):5–12.

Haydak MH, Crane E, Duisberg H, Gochnauer TA, Morse RA, White JW, and Wix P. 1975. Biological properties of honey. In Crane E, Ed. *Honey: A Comprehensive Survey*. Heinemann, London, pp. 258–266.

Hebda PA, Klingbeil CK, Abraham JA, and Fiddes JC. 1990. Basic fibroblast growth factor stimulation of epidermal wound healing in pigs. *J Invest Dermatol*, 95:626–631.

Henriques A, Jackson S, Cooper R, and Burton N. 2006. Free radical production and quenching in honeys with wound healing potential. *J Antimicrob Chemother*, 58(4):773–777.

Hertog MG, Feskens EJ, Hollman PC, Katan MB, and Kromhout D. 1993. Dietary antioxidant flavonoids and risk of coronary heart disease: The Zutphen Elderly Study. *Lancet*, 342:1007–1111.

Inoue K, Murayarna S, Seshimo F, Takeba K, Yoshimura Y, and Nakazawa H. 2005. Identification of phenolic compound in manuka honey as specific superoxide anion radical scavenger using electron spin resonance (ESR) and liquid chromatography with coulometric array detection. *J Sci Food Agric*, 85:872–878.

Jaganathan S and Mandal M. 2009. Honey constituents and their apoptotic effect in colon cancer cells. *JAAS*, 1:29–36.

Jurjus A, Atiyeh BS, Abdallah IM, Jurjus RA, Hayek SN, Abou Jaoude M et al. 2007. Pharmacological modulation of wound healing in experimental burns. *Burns*, 33:892–907.

Karukonda SR, Flynn TC, Boh EE, McBurney EI, Russo GG, and Millikan LE. 2000. The effects of drugs on wound healing—Part II. Specific classes of drugs and their effect on healing wounds. *Inter J Derm*, 39:321–333.

Kasama T, Kobayashi K, Fukushima T, Tabata M, Ohno I, Negishi M et al. 1989. Production of interleukin 1-like factor from human peripheral blood monocytes and polymorphonuclear leukocytes by superoxide anion: The role of interleukin 1 and reactive oxygen species in inflamed sites. *Clin Immunol Immunopathol*, 53:439–448.

Kücük M, Kolayli S, Karaoglu S, Ulusoy E, Baltaci C, and Candan F. 2007. Biological activities and chemical composition of three honeys of different types from Anatolia. *Food Chem*, 100:526–534.

Kwakman PH, Te Velde AA, De Boer L, Speijer D, Vandenbroucke-Grauls CM, and Zaat SA. 2010. How honey kills bacteria. *FASEB J*, 24(7):2576–2582.

Lee JE, Park JC, Lee HK, Oh SH, and Suh H. 2002. Laminin modified infection-preventing collagen membrane containing silver sulfadiazine-hyaluronan microparticles. *Artif Organs*, 26:521–528.

Leibovich SJ, Polverini PJ, Shepard HM, Wiseman DM, Shively V, and Nuseir N. 1987. Macrophage-induced angiogenesisis mediated by tumour necrosis factor-alpha. *Nature*, 329:630–632.

Li ZQ, Wang JH, Ren JL, and Yi ZH. 2004. Effects of topical emu oil on wound healing in scalded rats. *Di Yi Jun Yi Da Xue Xue Bao*, 24:1255–1256.

Martin GS, Mannin DM, Eaton S, and Moss M. 2003. The epidemiology of sepsis in the United States from 1979 through 2000. *N Engl J Med*, 348:1546–1554.

Mavric E, Wittman S, Barth G, and Henle T. 2008. Identification and quantification methyl-glyoxal as the dominant antibacterial constituent of Manuka (*Leptospermum scoparium*) honeys from New Zealand. *Mol Nutr Food Res*, 52(4):483–489.

Mccarthy J. 1995. The antibacterial effects of honey: Medical fact or fiction? *Am Bee J*, 135:341–342.

Mendonça RJ and Coutinho-Netto J. 2009. Cellular aspects of wound healing. *An Bras Dermatol*, 84:257–262.

Molan PC, Smith IM, and Reid GM. 1988. A comparison of the antibacterial activity of some New Zealand honeys. *J Apicult Res*, 27(4):252–256.

Molan PC. 1992. The antibacterial activity of honey. The nature of the antibacterial activity. *Bee World*, 73:5–28.

Molan PC. 1995. The antibacterial properties of honey. *Chem NZ*, 59(4):10–14.

Molan PC and Allen KL. 1996. The effect of gamma irradiation on the antibacterial activity of honey. *J Pharm Pharmacol*, 48:1206–1209.

Molan PC. 2002. Re-introducing honey in the management of wounds and ulcers—Theory and practice. *Ostomy Wound Manage*, 48:28–40.

Molnar J. 2007. *Nutrition and Wound Healing*. Taylor & Francis, Boca Raton, FL, pp. 18–24.

Monaco JL and Lawrence WT. 2003. Acute wound healing: An overview. *Clin Plast Surg*, 30(1):1–12.

Moore K, Thomas A, and Harding KG. 1997. Iodine released from the wound dressing Iodosorb modulates the secretion of cytokines by human macrophages responding to lipopolysaccharide. *Int J Biochem Cell Biol*, 29:163–171.

Moore OA, Smith LA, Campbell F, Seers K, McQuay HJ, and Moore RA. 2001. Systematic review of the use of honey as a wound dressing. *BMC Complement Alternat Med*, 1:2.

Mustoe TA, Pierce GF, Morishima C, and Deuel TF. 1991. Growth factorinduced acceleration of tissue repair through direct and inductive activities in a rabbit dermal ulcer model. *J Clin Invest*, 87:694–703.

Nagai T, Inoue R, Kanamori N, Suzuki N, and Nagashima T. 2006. Characterization of honey from different floral sources. Its functional properties and effects of honey species on storage of meat. *Food Chem*, 97:256–262.

National Honey Board. 2003. *Honey Health and Therapeutic Qualities*. http://www.honey.com, pp. 11–18.

Niculae M, Spinu M, Dana Sandru C, Brudască GhF et al. 2008. Preliminary study on the ability of honeydew honey to modulate the in vitro reactivity of avian leucocytes. *Bull UASVM, Vet Med*, 65(1):266–269.

Norimah Y, Ainul Hafiza AH, Rozaini MZ, and Bakar MdZA. 2007. Development of honey hydrogel dressing for enhanced wound healing. *Radiat Phys Chem*, 76:1767–1770.

Ojha N, Roy S, He G, Biswas S, Velayutham M, Khanna S et al. 2007. Assessment of wound-site redox environment and the significance of Rac2 in cutaneous healing. *Free Radic Biol Med*, 44(4):682–91.

Orsolic N, Knezevic H, Sver L, Terzic S, Heckenberger BK, and Basic I. 2003. Influence of honey bee products on transplantable murine tumours. *Vet Comp Oncology*, 1:216–226.

Orsolic N and Basic I. 2004. Honey as a cancer-preventive agent. *Period Biol*, 106:397–401.

Orsolic N. 2009. Honey and cancer. *JAAS*, 1:93–103.

Pierce GF, Tarpley JE, Yanagihara D, Mustoe TA, Fox GM, and Thomason A. 1992. Platelet-derived growth factor (BB homodimer), transforming growth factor-beta 1, and basic fibroblast growth factor in dermal wound healing. Neovessel and matrix formation and cessation of repair. *Am J Pathol*, 140:1375–1388.

Politis MJ and Dmytrowich A. 1998. Promotion of second intention wound healing by emu oil lotion: Comparative results with furasin, polysporin, and cortisone. *Plast Reconstr Surg*, 102:2404–2414.

Pruitt KM and Reiter B. 1985. Biochemistry of peroxidase system: Antimicrobial effects. In Pruitt KM and Tenovuo JO, Eds. *The Lactoperoxidase System: Chemistry and Biological Significance.* Marcel Dekker, New York, pp. 144–178.

Rauha JP, Remes S, Heinonen M et al. 2003. Antimicrobial effects of Finnish plant extracts containing flavonoids and other phenolic compounds. *Int J Food Microbiol,* 56:3–12.

Riches DW. 1996. Macrophage involvement in wound repair, remodelling and fibrosis. In Clarke R, Ed. *The Molecular and Cellular Biology of Wound Repair.* Plenum Press, New York, pp. 95–141.

Roy S, Khanna S, Nallu K, Hunt TK, and Sen CK. 2006. Dermal wound healing is subject to redox control. *Mol Ther* 13:211–220.

Sackett WG. 1919. Honey as a carrier of intestinal diseases. *Bull Colorado State Univ Agric Exp Stn,* 252:1–18.

Saissy JM, Guignard B, Pats B, Guiavarch M, and Rouvier B. 1995. Pulmonary oedema after hydrogen peroxide irrigation of a war wound. *Intensive Care Med,* 21(3):287–288.

Salahudeen AK, Clark EC, and Nath KA. 1991. Hydrogen peroxide-induced renal injury. A protective rolefor pyruvate in vitro and in vivo. *J Clin Invest,* 88(6):1886–1893.

Saravana KJ and Mahitosh M. 2009. Antiproliferative effects of honey and of its polyphenols: A review. *J Biomed Biotech,* 830616.

Sato M, Sawamura D, Ina S, Yaguchi T, Hanada K, and Hashimoto I. 1999. In vivo introduction of the interleukin 6 gene into human keratinocytes: Induction of epidermal proliferation by the fully spliced form of interleukin 6, but not by the alternatively spliced form. *Arch Dermatol Res,* 291:400–404.

Schlenger K, Hockel M, Schwab R, Frischmann-Berger R, and Vaupel P. 1994. How to improve the uterotomy healing: Effects of fibrin and tumour necrosis factor-a in the rat uterotomy model. *J Surg Res,* 56:235–241.

Schramm DD, Karim M, Schrader HR, Holt RR, Cardetti M, and Keen CL. 2003. Honey with high levels of antioxidants can provide protection to healthy human subjects. *J Agric Food Chem,* 51:1732–1735.

Sen CK and Roy S. 2008. Redox signals in wound healing. *Biochim Biophys Acta,* 1780:1348–1361.

Senyuva C, Yucel A, Erdamar S, Seradjmir M, and Ozdemir C. 1997. Fate of alloplastic materials placed under a burn scar: An experimental study. *Burns,* 23:484.

Simon A, Traynor K, Santos K, Blaser G, Bode U, and Molan PC. 2009. Medical honey for wound care-still the "Latest Resort"? *eCAM,* 6:165–173.

Singer AJ and Clark RA. 1999. Cutaneous wound healing. *N Engl J Med,* 341:738–746.

Skiadas PK and Lascaratos JG. 2001. Dietetics in ancient Greek philosophy: Plato's concepts of healthy diet. *Eur J Clin Nutr,* 55:532–537.

Snowdon JA and Cliver DO. 1996. Microorganisms in honey. *Inter J Food Microbiol,* 31:1–26.

Steed DL. 1998. Modifying the wound healing response with exogenous growth factors. *Clin Plast Surg,* 25:397–405.

Stomfay-Stitz J and Kominos SD. 1960. Über bakteriostatische Wirkung des honigs. *Z Lebensm Unters For,* 113:304–309.

Subrahmanyam M. 1991. Topical application of honey treatment of burns. *Br J Surg,* 78(4):497–498.

Subrahmanyam M. 1996a. Addition of antioxidants and polyethylene glycol 4000 enhances the healing property of honey in burns. *Ann Burns Fire Disast,* 9:93–95.

Subrahmanyam M. 1996b. Honey dressing for burns: An appraisal. *Ann Burns Fire Disast,* 9:33–35.

Subrahmanyam M. 1996c. Honey dressing versus boiled potato peel in the treatment of burns: A prospective randomised study. *Burns,* 22:491–493.

Subrahmanyarn M, Shahapure AG, Naganer NS, Bhagwat VR, and Ganu JV. 2003. Free radical control—The main mechanism of the action of honey in burns. *Ann Burns Fire Disasters*, 16:135–137.

Sugarman BJ, Aggarwal BB, Hass PE, Figari IS, Palladino MA Jr, and Shepard HM. 1985. Recombinant human tumor necrosis factor alpha: Effects on proliferation of normal and transformed cells in vitro. *Science*, 230:943–945.

Suguna L, Chandrakasan G, Ramamoorthy U, and Joseph KT. 1993. Influence of honey on biochemical and biophysical parameters of wounds in rats. *J Clin Biochem Nutr*, 14:91–99.

Swellam T, Miyanaga N, Onozawa M et al. 2003. Antineoplastic activity of honey in an experimental bladder cancer implantation model: In vivo and in vitro studies. *Inter J Urol*, 10:213–219.

Taormina PJ, Niemira BA, and Bauchat LR. 2001. Inhibitory activity of honey against foodborne pathogens as influenced by the presence of hydrogen peroxide and level of antioxidant power. *Inter J Food Microbiol*, 69:217–225.

Tonks A, Cooper RA, Price AJ, Molan PC, and Jones KP. 2001. Stimulation of TNF-alpha release in monocytes by honey. *Cytokine*, 14:240–242.

Tonks AJ, Cooper RA, Jones KP, Blair S, Parton J, and Tonks A. 2003. Honey stimulates inflammatory cytokine production from monocytes. *Cytokine*, 21:242–247.

Tonks AJ, Dudley E, Porter NG, Parton J, Brazier J, Smith EL et al. 2007. A 5.8-kDa component of manuka honey stimulates immune cells via TLR4. *J Leukoc Biol*, 82:1147–1155.

Tracey D, Klareskog L, Sasso EH, Salfeld JG, and Tak PP. 2008. Tumor necrosis factor antagonist mechanisms of action: A comprehensive review. *Pharmacol Ther*, 117:244–279.

Tsiapara A, Jaakkola M, Chinou I, Graikou K, Tolonen T, Virtanen V, and Moutsatsou P. 2009. Bioactivity of Greek honey extracts on breast cancer (MCF-7), prostate cancer (PC-3) and endometrial cancer (Ishikawa) cells: Profile analysis of extracts. *Food Chem*, 116:702–708.

Ueyama M, Maruyama I, Osame M, and Sawada Y. 1992. Marked increase in plasma interleukin-6 in burn patients. *J Lab Clin Med*, 120:693–698.

Vinson JA, Hao Y, Su X, and Zubik L. 1998. Phenol antioxidant quantity and quality in foods: Vegetables. *J Agric Food Chem*, 46:3630–3634.

Wan KC and Evans JH. 1999. Free radical involvement in hypertrophic scar formation. *Free Rad Biol Med*, 26:603–608.

Watanabe K, Shinmoto H, Kobori M et al. 1998. Stimulation of cell growth in the U-937 human myeloid cell line by honey royal jelly protein. *Cytotechnology*, 26:23–27.

Weston RJ, Broncklebank LK, and Lu Y. 2000. Identification and quantitative levels of antibacterial components of some New Zealand honeys. *Food Chem*, 70:427.

White JW, Subers MH, and Schepartz AI. 1963. The identification of inhibine, the antibacterial factor in honey, as hydrogen peroxide and its origin in a honey glucose-oxidase system. *Biochim Biophys Acta*, 73:57–70.

White JW and Subers MH. 1964. Studies on honey inhibine: Effect of heat. *J Apicult Res*, 3:45–50.

Yamada Y, Endo S, and Inada K. 1996. Plasma cytokine levels in patients with severe burn injury with reference to the relationship between infection and prognosis. *Burns*, 22:587–593.

Willix DJ, Molan PC, and Harfoot CJ. 1992. A comparison of the sensitivity of wound-infecting species of bacteria to the antibacterial activity of manuka honey and other honey. *J Appl Bacteriol*, 73:388–394.

# 5 Importance of Botanical Origin of Honeys

*Saâd Aissat and Hama Benbarek*

## CONTENTS

## INTRODUCTION

Honey is a very complex mixture and no honey is completely the same as another one. Its composition and also different biological effects vary to a great extent, but not completely, on its nectar and/or honeydew providing plant species (botanical origin). Caution must be taken in applying the ranges characteristic of unifloral honeys from country to country. Due to the variation of botanical origin, honey differs in appearance, sensory perception, and composition (Bogdanov et al. 2008). There are three types of honeys with regard to their origin: nectar honey (made from plant nectar), honeydew honey (made mostly from the secretion of insects feeding on plant juices or plant secretion), and mixed honey (honeydew and nectar honey). The same floral origin honey composition may be quite different depending on several factors, such as geographic area (soil characteristics and climate), season, bee species, harvest technology and condition, and mode of storage. This variability could be a handicap, given the market requirement for a consistent product, but when properly managed, it also could represent an opportunity for enhancing honey by offering to the consumer a number of typical products with special characteristics, according to the particular botanical origin (Bogdanov 2009). The price of honey is usually dictated by its botanical and/or geographic origin. In the case of botanical origin, the most expensive are unifloral honeys (Cajka et al. 2009); thus, the floral determination and certification of unifloral honey play an important role in quality control.

Most unifloral honeys are marketed in Europe. Up to now, the European Union (EU) has specified 18 protected denominations of origin regions for honey (1 Greek, 1 Italian, 1 Luxembourgian, 1 Polish, 2 French [including the island of Corsica], 3 Spanish, and 9 Portuguese) (Cajka et al. 2009). In non-European countries, with the exception of the manuka New Zealand honey, unifloral honeys have a smaller importance (Bogdanov 2009). Research indicates that honeys have functional properties in human health promotion, which depend largely on the floral source of the honey. Moreover, the increasing interest in the therapeutic or technological uses of certain honey varieties may also contribute to the demand of a reliable determination of their botanical origin. Therefore, different honey properties were expected since the composition of active compounds in honey from different locations should be different.

According to the *Codex Alimentarius* Standard for Honey (Codex Committee on Sugars 2001) and the EU Council Directive (EU Council 2002) relating to honey, the use of a botanical designation of honey is allowed if it originates predominantly from the indicated floral source. Although the composition of unifloral honeys has been described in various studies, internationally accepted criteria and the measures to be considered for their authentication have not been defined yet (Ruoff 2006).

The very first problem that arises when approaching the subject of unifloral honeys is when to define a honey as unifloral. A continuous series of intermediate possibilities exists between a multifloral honey and a unifloral honey. At what point and on what basis does the unifloral/multifloral discrimination take place (Persano Oddo et al. 1995)? It is difficult to find a unanimous opinion about the floral markers of honey collected from different floral origins; indicators that might distinguish one honey from another depend on many factors. Therefore, unambiguous and unique indicators for honey collected from the same plant source have not been identified so far; very often, honey from the same botanical origin analyzed in various studies is characterized by different compounds (Kaškonien and Venskutonis 2010).

## THERAPEUTIC PROPERTIES OF HONEY
## FOR SOME BOTANICAL ORIGINS

Over the centuries, honey has been used to treat everything from sore throats and coughs, conjunctivitis, and corneal damage to fungal infections and dry skin and used as a facial softener. Paradoxically, honey has also been recommended to treat constipation and diarrhea, sleepiness, and insomnia. Several studies on the honeybee products, including honey as the most familiar product, started a long time ago. Most of these studies have focused on their potential health benefits for humans. In light of modern science, several important therapeutic effects of honey have been elucidated (Table 5.1). But neither of these explanations seems likely to provide the whole answer. Whatever the reasons why honey works, and why different honeys have different medicinal actions, it is accepted that the type of flower on which the bees have been feeding is important and can affect the medicinal quality of honey. However, the fact that honey's different biological effects depend to a great extent

## TABLE 5.1
### Therapeutic Properties of Honeys for Some Botanical Origins

| Honey Source | Therapeutic Properties | Model and Country of Study | References |
|---|---|---|---|
| Acacia (*Acacia modesta*) | Antinematodal agent (*Caenorhabditis elegans* [strain N2]) | *In vitro* immersion tests (PAK) | Azim and Sajid 2009 |
| Alfalfa | Stimulate the growth of five intestinal bifidobacteria | Human intestinal *Bifidobacterium* (USA) | Shin and Ustunol 2005 |
| Anzer (high plateau in Turkey) | Hepatoprotective | Rats (TR) | Korkmaz and Kolankaya 2009 |
| | Protect stomach against ethanol-induced increased vascular permeability | Rats (TR) | Doğan and Kolankaya 2005 |
| | Could be used as scolicidal agent safely in the presence of biliary-cystic communication | Rats (TR) | Kilicoglu et al. 2008 |
| Astragale | Significant reduction on radiation-induced mucositis (patients with head-and-neck cancer) | Humans (IR) | Motallebnejad et al. 2008 |
| Basswood honey | Had less effect on serum glucose, C-peptide, and insulin values than the honey-comparable glucose-fructose solution | Humans (GER) | Münstedt et al. 2008 |
| Buckwheat | Induces mammalian phase 2 detoxication enzymes | Murine hepatoma cells (USA) | Fahey and Stephenson 2002 |
| | Increases serum antioxidant capacity | Human (USA) | Gheldof et al. 2003; Schramm et al. 2003 |
| | Treatment for the cough and sleep difficulty associated with childhood upper respiratory tract infection | Humans (USA) | Paul et al. 2007 |
| Blossom honeydew honey | May have a laxative effect on normal subjects | Humans (GR) | Ladas et al. 1995 |
| Chestnut | Reduces the ulcer index, microvascular permeability, and myeloperoxidase activity of the stomach | Rats (IT) | Nasuti et al. 2006 |
| | Believed to be a good ethno-remedy for asthma and respiratory diseases | Humans (TR) | Orhan et al. 2003 |
| | Increase the formation of granulation tissue, epithelization, angiogenesis, and fibroplasia levels | Rabbits (TR) | Nisbet et al. 2010 |

(*continued*)

**TABLE 5.1 (Continued)**

## Therapeutic Properties of Honeys for Some Botanical Origins

| Honey Source | Therapeutic Properties | Model and Country of Study | References |
|---|---|---|---|
| Citrus | Possibly as a complement in the management of alcohol intoxication | Humans (NIG) | Onyesom 2004 |
| Clover | Positive effect of honey on hepatitis by causing a decrease in the alanine aminotransferase activity and a decrease in bilirubin production | Humans (LIT) | Baltuskevicius et al. 2001 |
| Cotton | Addition of 10% cotton honey in processed food can inhibit the harmful and genotoxic effects of mycotoxins and improve the gut microflora | Mice (EG) | Ezz El-Arab et al. 2006 |
| Durian (*Duriozibethinus*) | Improves healing with regards to the tensile strength property | Rats (MAL) | Rozaini et al. 2004 |
| Gelam honey | Reduction in edema volume and the inhibition of pain | Rats (MAL) | Kassim et al. 2010 |
| Heather | Protect toward cytotoxicity of acrylamide in human hepatoma cells | Human hepatoma cells (HepG2) (ES) | Mademtzoglou et al. 2010 |
| Honeydew honey (HD19) | May therefore be an effective replacement for sucrose in individuals who suffer from poor glycemic control or who are at high risk for CHD | Rats (NZ) | Chepulis and Starkey 2008 |
| Jungle honey (from timber and blossom) | Enhances immune functions and antitumor activity in mice. Enhancement of chemotactic activity in neutrophils | Mice (JP) Neutrophils from guinea pigs (MAL) | Fukuda et al. 2009 Miyagawa et al. 2010 |
| Mad honey (*Rhododendron ponticum*) | Causes significant decreases in blood glucose and lipid (cholesterol, triglyceride, and VLDL). Thought to be helpful in healing some gastrointestinal diseases such as gastritis and peptic ulcer. Believed to reduce coronary heart diseases and increases life expectancy when used continuously | Rats (TR) Case report (TR) | Öztaşan et al. 2005 Dilber et al. 2002 |

| | | | |
|---|---|---|---|
| Manuka | Positive effect against dental plaque development and gingivitis | Human (NZ) | English et al. 2004 |
| | Inhibit the growth of *H. pylori*, the organism thought to be responsible for the development of dyspepsia and peptic ulcers | | Al Somai et al. 1994 |
| | Incorporated in a gel for topical use, suppress the itching and erythema associated with mosquito bites and other inflammatory skin reactions due to local anaphylactic reactions | Human (NZ) | Molan, personal communication |
| | Oral administration and combination with sulfadiazine significantly reduce colonic inflammation | Rat (IND) | Medhi et al. 2008 |
| Natural wild honey | May be useful in the treatment of some cardiac arrhythmias, especially those associated with hyperadrenergic activity | Isolated toads' hearts (EG) | Hussein et al. 2003 |
| | May exert cardioprotective and therapeutic effects against epinephrine-induced cardiac disorders and vasomotor dysfunction | Rats (EG) | Rakha et al. 2008 |
| Nenas (*Ananas comosus*) | Improves healing with regards to the tensile strength property | Rats (MAL) | Rozaini et al. 2004 |
| *Plectranthus* | Antinematodal agent (*Caenorhabditis elegans* [strain N2]) | *In vitro* immersion tests (PAK) | Azim and Sajid 2009 |
| Pine tree | Intraperitoneally decreased the formation of postoperative intraabdominal adhesions without compromising wound healing | Rats (TR) | Yuzbasioglu et al. 2009 |
| | Beneficial effect on trichlorfon induced some biochemical alterations | Mice (TR) | Eraslan et al. 2010 |
| Rape | Positive effect of honey on hepatitis by causing a decrease in the alanine aminotransferase activity and a decrease in bilirubin production | Humans (LIT) | Baltuskevicius et al. 2001 |
| Rosemary | Protect toward cytotoxicity of acrylamide in human hepatoma cells | Human hepatoma cells (HepG2) (ES) | Mademtzoglou et al. 2010 |
| Rhododendron | Increase the formation of granulation tissue, epithelization, angiogenesis, and fibroplasia levels | Rabbits (TR) | Nisbet et al. 2010 |
| Sage and sour-wood | Stimulate the growth of five intestinal bifidobacteria | Human (USA) | Shin and Ustunol 2005 |

*(continued)*

## TABLE 5.1 (Continued)
## Therapeutic Properties of Honeys for Some Botanical Origins

| Honey Source | Therapeutic Properties | Model and Country of Study | References |
|---|---|---|---|
| Thyme | Good anti-Rubella activity | Monkey kidney cell cultures (ENG) | Zeina et al. 1996 |
| | Oral honey administration following tonsillectomy in pediatric cases may reduce the need for analgesics via relieving postoperative pain | Humans (TR) | Ozlugedik et al. 2006 |
| | Significant reduction on radiation-induced mucositis (patients with head-and-neck cancer) | Humans (IR) | Motallebnejad et al. 2008 |
| | May prevent cancer-related processes in breast, prostate, and endometrial cancer cells | Cell culture (GR) | Tsiapara et al. 2009 |
| Tualang tree (*Koompassia excelsa*) (*Apis dorsata*) | Antiproliferative on effect oral squamous cell carcinoma and osteosarcoma cell | Cell culture (ML) | Ghashm et al. 2010 |
| | Protective effect against cigarette smoke–induced impaired testicular functions | Rats (ML) | Mahaneem et al. 2010 |
| | *In vivo* wound contraction of burn wounds dressed with Tualang honey was markedly greater than with the hydrofiber silver and hydrofiber dressing | Rats (ML) | Khoo et al. 2010 |
| | Beneficial effects on menopausal (ovariectomized) rats by preventing uterine atrophy, increased bone density, and suppression of increased body weight | Rats (ML) | Zaid et al. 2010 |
| | In combination with glibenclamide or metformin, improves glycemic control and provides additional metabolic benefits, not achieved with either glibenclamide or metformin alone | Rats (ML) | Erejuwa et al. 2001 |
| | Honey could be used as one of the prophylactic measures in reducing acute respiratory symptoms among Hajj pilgrims | Humans (ML) | Sulaiman et al. 2011 |

*Note:* Country of study: EG, Egypt; ENG, England; ES, Spain; GR, Greece; IND, India; IR, Iran; IT, Italy; GER, Germany; JP, Japan; LIT, Lithuania; ML, Malaysia; NIG, Nigeria; NZ, New Zealand; PAK, Pakistan; TR, Turkey; USA, United States of America.

on its botanical origin was often not considered in the reviewed studies (Bogdanov et al. 2008). Although the pharmacologic properties of all its many constituents are being explored, it will be years before they are understood. However, its only scientifically proven use is as an antibacterial wound dressing. Numerous reports of clinical studies, case reports, and randomized controlled trials (RCTs) show that it rates favorably alongside modern dressing materials in its effectiveness in managing wounds (Ahmed et al. 2003). The effects of different types of honey on the tensile strength of burn wound tissue healing were evaluated in 105 male Sprague-Dawley rats. Their floral sources were from *Melaleuca* spp. (Gelam) trees, *Cocos nucifera* spp. (Kelapa) trees, *Ananas comosus* spp. (Nenas) trees, and *Durio zibethinus* spp. (durian) trees manuka honey. Tensile strength of the animals treated with manuka honey demonstrated the highest value throughout the study, except at day 21 postburn, followed by *D. zibethinus* (durian) and *A. comosus*. Since wounds treated with *D. zibethinus* (durian) honey and *A. comosus* (Nenas) honey showed greater tensile strength compared to wounds treated with two other honeys (*Melaleuca* [Gelam] and *C. nucifera* [Kelapa]), it may be inferred that the former honey groups increased collagen concentration and stabilization of the fibers (Rozaini et al. 2004). Studies have shown good results using Medihoney dressings for patients with diabetic ulcers and in its advantage over topical antibiotics in preventing catheter-associated infections (Medical Journals Buzzing over Honey for Wound Care 2007). However, skepticism still exists among the medical and nursing fraternity in the use of honey in the treatment of legs ulcers, partially because there is a lack of level and evidence to support the fact that this type of dressing and topical agent will have a definitive bearing on ulcer healing (Moore et al. 2001).

The PubMed, MEDLINE, EMBASE, CINAHL database, and the Cochrane Library were searched for relevant publications on the efficacy of honey as an antibacterial agent and in the promotion of wound healing in chronic leg ulcers from 1980 to 2004. Thirteen publications had been found concerning the use of honey in chronic leg ulcers, but only two were clinical trials of relevance to our study. The studies analyzed were influenced by different sources of bias, especially lack of blinding, poor reporting quality, and poor sample size. None of those studies was an RCT. In order to elucidate the evidence for the use of honey as a first-line treatment in chronic leg ulcers, RCTs and laboratory studies on cellular effects are urgently needed (Mwipatayi et al. 2004). Jessica Beiler, M.P.H. (Penn State College of Medicine Pediatric Division, Hershey, Pennsylvania), delivered a presentation summarizing the findings of a rather elegant research study funded by the National Honey Board comparing honey as a cough suppressant to dextromethorphan or no treatment in pediatric patients. Citing numerous studies that showed no significant beneficial effect of over-the-counter cough medicines in treating children's cough due to cold symptoms, Beiler presented a strong rationale for using honey, a recommendation of the World Health Organization 2001 topic review, which concluded that honey was a cheap and safe demulcent treatment for cough and cold symptoms (Committee for the Promotion of Honey and Health 2008). On the contrary, Oduwole et al. (2010) presented an RCT of 108 children with upper respiratory tract infections comparing the effect of honey, dextromethorphan, and no treatment on cough and sleep quality for coughing children and their parents. Comparing symptoms and sleep quality scores of children that received honey with those that received no treatment showed that honey was more effective in reducing frequency of cough (mean difference, bothersome cough, and sleep quality of the child, but did not differ significantly between the honey versus no treatment groups in resolving severity of cough and sleep quality of the parents). Dextromethorphan and honey did not differ significantly on cough frequency, cough severity, bothersome cough, and sleep quality of the children or their parents. The authors' conclusion was that there was insufficient evidence to advise for or against the use of honey for acute cough in children. Manuka honey was shown to have bacteriostatic properties against *Helicobacter pylori* at concentrations of 5% (v/v), with most clinical strains showing complete inhibition of growth with honey concentrations of 20% (v/v). Importantly though, solutions of glucose, fructose, and their combinations have also been shown to inhibit the growth of *H. pylori* at concentrations equal to 15% (Osato et al. 1999), suggesting that the inhibition may be due to the osmotic action of the honey/sugar solutions rather than any inherent antibacterial activity. In addition, whereas manuka honey has been shown to be effective at retarding the growth of *H. pylori in vitro*, it has not been demonstrated to have any effect *in vivo*. In a small clinical study, McGovern et al. (1999) reported that all patients ($n = 6$) remained positive for *H. pylori* 4 weeks after a 2-week regimen of a tablespoon of manuka honey taken four times a day. However, honey may offer an effective therapeutic treatment for gastric ulcers both via its antibacterial effect on the *H. pylori* organisms directly and via prevention of ammonia-induced gastric lesions (Ali 2003). After a study designed to determine whether the anti-*H. pylori* activity of honey differed regionally (honey from Texas, Iowa, and New Zealand) and to determine whether this activity was due to the presence of hydrogen peroxide, Osato et al. (1999) concluded that regional differences in honey activity against *H. pylori* were not detected nor was the effect of killing related to the presence of hydrogen peroxide in the honey samples. Osmotic effects were shown to be the most important parameter for killing *H. pylori*.

The oligosaccharides have been reported to cause an increase in population of bifidobacteria and lactobacilli, which are responsible for maintaining a healthy intestinal microflora in humans (Kleerebezem and Vaughan 2009). A small number of researchers have demonstrated that honey can stimulate the growth of bifidobacteria *in vitro* (Chick et al. 2001; Ustunol and Gandhi 2001; Kajiwara et al. 2002; Sanz et al. 2005). In a large study funded by the US National Honey Board, 3% to 5% (w/v) of honey significantly enhanced the growth and survival of bifidobacteria BF-1 and BF-6 and *Bifidobacterium longum in vitro*. There is emerging interest in studying the potential health benefits of eating honey. The prebiotic effects of different honey samples (Leatherwood, Banksia, Bees Creek Woolybutt, Grey Ironbark, and Mugga Ironbark) from Australia were compared to inulin and sucrose using intestinal microcosms inoculated with human fecal material and hence containing the full complex microbial community of the human intestine. All tested honeys showed higher prebiotic index values than either inulin or sucrose. It is noted, however, that it is possible that this outcome reflects a synergistic effect of the simple and complex sugars present in the honey, which differs from the test conditions for sugar or inulin alone (Conway et al. 2010). In another study carried out in Malaysia, both wild and commercial Tualang honeys were found to support the growth of *B. longum* BB 536 (Jan Mei et al. 2010).

It was shown that consumption of honey has a favorable effect on diabetes patients, causing a significant decrease in plasma glucose (Peretti et al. 1994; Al-Waili 2003, 2004). The consumption of honey types with a low glycemic index (GI) (e.g., acacia honey) might have beneficial physiologic effects and could be used by type 2 diabetes patients. An intake of 50 g honey of unspecified type by healthy people and diabetes patients led to smaller increases in blood insulin and glucose than the consumption of the same amounts of glucose or of a sugar mixture resembling honey (Al-Khalidi et al. 1980; Jawad et al. 1981). According to different studies, long-term consumption of food with a high GI is a significant risk factor for type 2 diabetes patients (Liu et al. 2001). However, the GI concept for the general population is still an object of discussion (Pi-Sunyer 2002).

The antimutagenic activity of honeys from seven different floral sources (acacia, buckwheat, fireweed, soybean, tupelo, and Christmas berry) against Trp-p-1 was tested by the Ames assay and compared to a sugar analog as well as to individually tested simple sugars. All honeys exhibited a significant inhibition of Trp-p-1 mutagenicity. Glucose and fructose were found to have a similar antimutagenic activity as honey (Wang et al. 2002). Rady and Yahya (2011) showed that adiponectin combined with honey extracts exerted their antitumor effects on HepG2 cells through the inhibition of cell growth and the enhancement of alkaline phosphatase activity rather than through acceleration of apoptosis via Bcl-2 downregulation.

## Pollen Analysis for Characterization of Honey Botanical Origin

Traditionally, the botanical origin of honey has been determined in many laboratories by pollen analysis, a method that is called melissopalynology. The melissopalynologic method, which was elaborated and proposed by the International Commission for Bee Botany in 1970 (Louveaux et al. 1970) and later revised and updated in 1978 (Louveaux et al. 1978), is frequently used until the present time. Moreover, in the EU Council Directive (2002) related to honey, it is indicated that the product names may be supplemented by information referring to floral or vegetable origin, if the product comes wholly or mainly from the indicated source and possesses the organoleptic, physicochemical, and microscopic characteristics of the source (Kaškonien and Venskutonis 2010). The nectar in honey is characterized by the pollen grains, and so honeydew as a source of honey is characterized by algae, fungal spores, and molds. These microscopic particles can be found on the leaves and needles of conifers or can get in the honeydew by rain or wind (Karabournioti 2000). However, the microscopic analysis of the pollen in the honey is the most time-consuming part, which, in addition, can only be done by specialists (Louveaux et al. 1970; Von der Ohe et al. 2004). It has to be borne in mind that melissopalynologic methods will not meet the quality standards for a modern validated and quality-assured analytical method. The methods are based on experience and are thus subjective and have not been tested by modern proficiency test trials (Bogdanov and Martin 2002). Since the early 1980s, in a study carried out in Canada, Adams and Smith (1981) showed very clearly the unreliability of melissopalynology in determining the floral source of a honey. Nowadays, it is assumed that such procedure has severe drawbacks.

There are many reasons why the results from melissopalynology may not be showing the nectar sources correctly. Molan (1998) reported that melissopalynology is valid only for the determination of the geographical origin of honey, while it is less valid for determining its botanical origin by the fact that there may occur possible alterations of pollen content in honey by the action of bees or contamination by the actions of apiarists; in addition, the source of some honeys, for example, collected from cotton plants, castor-oil plants, and rubber trees, cannot be identified by melissopalynology. According to Bogdanov and Martin (2002), melissopalynology is very efficient only for the differentiation of honeys produced in distinctly different geographic and climatic areas. If the geographical differences are less pronounced, the determination of the pollen spectrum will generally not yield a confident authenticity proof. In recent years, honey plants typical of certain countries or regions are now grown in many other different areas or countries. For example, *Eucalyptus* species, which are endemic in Australia and New Zealand, have been used in recent years in Greece as boulevard trees and *Eucalyptus*-type pollen grains are found in Greek honeys (Karabournioti 2000).

It has also been regarded that several types of honey whose pollen is considered are "underrepresented" by the International Commission for Bee Botany (Louveaux et al. 1970). Their guidelines for floral origin state that honeys with only 10% to 20% citrus pollen may be considered as largely of that origin ("monofloral") rather than the 45% required for most other types. In *Citrus* species, anthers yield little pollen or are completely sterile (Maurizio 1975). However, there are other studies that report high percentages of *Citrus*-type pollen grains (Bonvehi et al. 1987; Munuena and Carrion 1994). In a study on the intraannual variations of the pollen spectrum of honey from Patagonia, obtained by successive harvests (between September and March), Forcone et al. (2003) showed that the pollen composition of honey was heterogeneous in all periods, and when comparing pollen composition noted of honey for the three periods, they found highly significant differences. The same authors noted that some species that frequently visited for nectar had very little or almost no presence in the honey pollen as occurred in *Salix* spp. and *Glycyrrhiza astragalina*. The low representation of these taxa was attributed to the predominance of the female foot in the most widespread *Salix* spp. and the small pollen production of *G. astragalina*. Furthermore, there are several plants poor or rich in pollen grains, and it is impossible to determine from pollen analysis their contribution to the honey. Fossel (1958) reported this finding in honey of a pollen substitute that is fed to bees.

Pollen analysis still cannot define which quantity of honey is represented by a certain number of pollen grains (Karabournioti 2000). Honey that has been filtered with diatomaceous earth has no pollen left in it to be identified. Also, honey produced from secretions of extrafloral nectaries (a major source of honey from cotton plants [*Gossypium hirsutum*], castor-oil plants [*Ricinus cornmunis*], and rubber trees [*Hevea brasiliensis*]) and the honeydew secreted by sap-sucking insects will contain no pollen other than airborne grains that become trapped and which will not necessarily be from the plant that was the source of the secretion. There are also honeys that come from flowers such as those of the rewarewa (*Knightia excelsa*) that are pollinated by nectar-eating birds: bees can collect nectar from these flowers without dislodging pollen from the anthers (Molan 1998).

In some cases, botanical origin of honey was based on the claims of local bee-keepers, when determination of honey origin is performed by sensory analysis or by considering the predominant flowers surrounding the hive. The latter approach may be quite precise, for example, allowing collecting honey with 72% to 75% of a predominant pollen (Jerković et al. 2009a,b).

However, regarding the growing interest accorded to controlling the quality and authenticity of honey, to preserve the production area, to develop particular standards of quality, and to protect consumers from commercial speculation without forgetting the relation of botanical origin of honey and its medicinal properties, many other studies were performed to establish new methodologies and different analytical methods, mostly combined with chemometric analysis, to assess the geographical and/or botanical origin of honey. A number of techniques and a search for faster methods, which would be suitable for routine analysis, particularly involving many samples in a short work period, to determine honey authenticity and botanical origin are now carried out, including the determination of aromatic compounds and flavonoids, amino acids, and sugars by high-performance liquid chromatography (HPLC), detection of aroma compounds by gas chromatography–mass spectrometry (GC-MS), determination of anions and cations by ion chromatography, and mineral content (Anklam 1998; Mateo and Bosch-Reig 1998; Anapuma et al. 2003; Serrano et al. 2004). Spectroscopic techniques, such as mid-infrared, near-infrared, and Raman spectroscopy, were also used to determine chemical characteristics (e.g., sugars) and contamination in honey samples from different origins (Cozzolino and Corbella 2005; Bertelli et al. 2007), electrospray MS (Beretta et al. 2007), inductively coupled plasma–optical emission spectrometry (Terrab et al. 2005), thin-layer chromatography (Rezić et al. 2005), atomic absorption spectroscopy (Hernández et al. 2005), high-performance anion-exchange chromatography with pulsed amperometric detection (Nozal et al. 2005), nuclear magnetic resonance (NMR) (Tuberoso et al. 2010), Fourier transform-Raman (FT-Raman) (Paradkar and Irudayaraj 2002), Fourier transform infrared (Etzold and Lichtenberg-Kraag 2008), rheology (Wei et al. 2010), and voltammetric electronic tongue with the chemometrics method (Wang 2011).

## POSSIBLE MARKERS OF HONEY

### VOLATILE COMPOUNDS

Unifloral honeys differ from each other, among other features, in volatile compound composition, which influences remarkably the individual sensory characteristics of each honey type. The composition of the honey's volatile fractions can be one of the distinguishable characteristics of honeys collected from different sources. Honey aroma has been studied since the early 1960s. Unifloral honeys possess highly characteristic aromas due to the presence of specific volatile organic components derived from the original nectar sources. The main volatile compounds in honey have their origins, in general terms, in different chemical families, such as alcohols, ketones, aldehydes, acids, esters, and terpenes (Zhou et al. 2002; Bastos and Alves 2003). The studies of a Belgian analyst using GC-MS found the presence of 400 volatile compounds in 11 floral honeys (Wolski et al. 2006). Radovic et al. (2001) identified

110 volatile compounds in 43 samples of honey with the Purge-and-Trap GC-MS method from various botanical and geographical origins. In four analyzed honeys, 86 compounds were identified as solid-phase microextraction (SPME) and GC-MS (Wolski et al. 2006). To date, 600 compounds have been identified by GC-MS (Montenegro et al. 2009) and the list is certainly far from being exhaustive. Some of these organic compounds are used as marker compounds for certain commercial honeys. However, honey volatiles may be also derived from the transformation of plant compounds by a honeybee, directly generated by honeybees, from heating or handling during honey processing or storage, from the isolation method, and from microbial or environmental contamination (Bastos and Alves 2003; Serra and Ventura 2003; Iglesias et al. 2004; Alissandrakis et al. 2005a,b; Baroni et al. 2006; Castro-Vázquez et al. 2006; Jerković et al. 2007). It is worth noting that some furan derivatives, considered for discrimination purposes, are known to arise from honey heat processing or from honey storage (Soria et al. 2004) or due to the extractive solvent used. Solvent extraction is the most common technique for volatiles isolation but not suitable for reliable honey fingerprinting due to intense promotion of artifact formation (Bonvehi and Coll 2003; Alissandrakis et al. 2005a,b). Bouseta and Collin (1995) optimized the Bicchi et al. (1983) method by preextraction with dichloromethane instead of acetone under an inert atmosphere followed by an optimized steam distillation extraction. They detected less furan derivatives in the dichloromethane extracts than in the acetone extracts, which are due to a nonenzymatic browning reaction (Cuevas-Glory et al. 2007). However, some aldehydes and alcohols reflect product quality and may also be a consequence of microbiological activity, heat exposure, and honey aging (Kaškonien and Venskutonis 2010). Some organic acids, ketones and benzenes, such as 2-hydroxy-2-propanone, butanoic acid, benzyl alcohol, or 2-phenyl-ethanol found in fresh honey, gradually increase their concentration with higher temperature and storage (Castro-Vázquez et al. 2008). Heating honey at temperatures as low as 50°C leads to the formation of new volatile flavor compounds, and the peak areas of many compounds vary significantly as a result of different heating conditions (Visser et al. 1988). Overton and Manura (1999) affirmed the presence of octane, hexanal, octanal, nonanal, and decanal forming in honey during its storage as a consequence of oxygenation of fatty acids (mainly of linoleic and linolenic acids). An example of such a situation is the presence of ethanol, particularly in the lime-honeydew honey, probably as a result of the fermentation process (Wolski et al. 2006). A research conducted by Rothe and Thomas (1963) on the aroma thresholds revealed that not all volatile compounds contributed to aroma. This implies a strategy change when searching for odor activity values or compounds of major sensorial activities. Using prepared synthetic samples, Grosch (2001) determined that no more than 5% of the sample's volatile compounds contributed to its aroma. However, a few compounds with low concentrations can still contribute significantly to honey aroma (Castro-Vázquez et al. 2007). Specific volatile compounds can be considered as aroma fingerprints because they provide information about the botanic origin of the honey (Alissandrakis et al. 2005a,b; Escriche et al. 2009) (Figure 5.1).

In some cases, certain honey types present a rich volatile chromatographic profile that can be used as a fingerprint of these honeys (Figures 5.2 through 5.4).

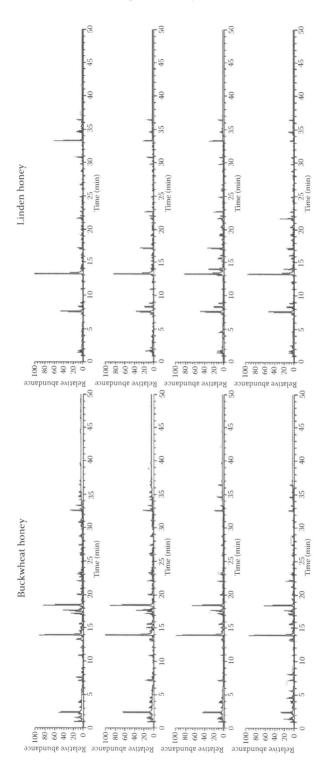

**FIGURE 5.1** Fingerprints of buckwheat and linden honey samples. HS-SPME as the sample preparation method and GC-MS as the method of final determination. In the result, the following conditions were applied: CAR/PDMS/DVB SPME fiber. (From Plutowska, B. et al., *Food Chemistry, 126,* 1288–1298, 2011.)

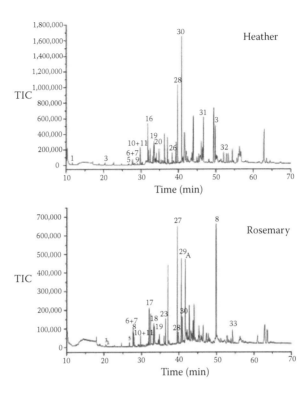

**FIGURE 5.2**  TIC profiles obtained by GC-MS from heather and rosemary honey volatiles fractionated by SPME using PA fiber. For identification of numbered peaks, see Soria et al. (2003).

Many studies have been carried out in order to specify volatile compounds that are most closely associated with a particular type of honey and consequently would be helpful for a fast and reliable identification of its botanical source (Kaškonien and Venskutonis 2010). It has been claimed that methyl anthranilate is a marker of *Citrus* honey (Bonvehi 1988; Ferreres et al. 1994b; White and Bryant 1996; Nozal et al. 2001; Piasenzotto et al. 2003; Soria et al. 2003; Alissandrakis et al. 2007). It was reported that heather (*Calluna vulgaris*) honey may be distinguished by benzoic acid, decanoic acid, and dehydrovomifoliol (Häusler and Montag 1989; Guyot et al. 1999) and chestnut honey by 2-aminoacetophenone (Guyot et al. 1998) and 3-aminoacetophenone (Radovic et al. 2001). However, only a few compounds seem really specific for certain unifloral honeys and many of them can be found in variable concentrations in various honey types. 3-Aminoacetophenone and dehydrovomifoliol have been detected in Tasmanian leatherwood (*Eucryphia lucida*) honey (Rowland et al. 1995). More recently, Sesta et al. (2008) reported that methyl anthranilate cannot be considered suitable as a chemical marker to assess the level of uniflorality of this honey type. It may be used only as a further descriptive element to complete the analytical picture of unifloral *Citrus* honey, together with other authenticity criteria. Later, Castro-Vazquez et al. (2009) distinguished sinensal isomers, linalool derivatives, lilac aldehydes, lilac alcohols, limonyl alcohol, and

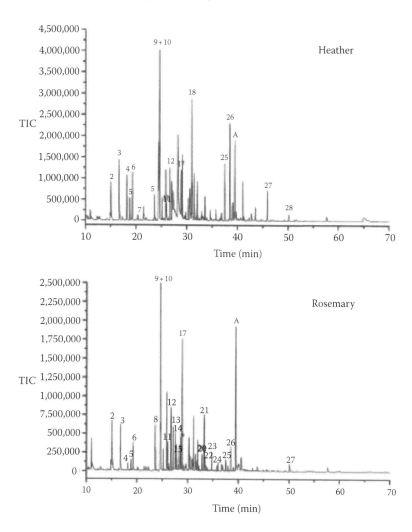

**FIGURE 5.3** TIC profiles obtained by GC-MS from heather and rosemary honey volatiles fractionated by SPME using C/PDMS fiber. For identification of numbered peaks, see Soria et al. (2003).

α-4-dimethyl-3-cyclohexene-1-acetaldehyde as other constituents involved in the formation of citrus honey aroma. D'Arcy et al. (1997) have reported the presence of 2,6-dimethylocta-3,7-diene-2,6-diol in Australian blue gum and yellow box honey extracts and Rowland et al. (1995) have identified it in leatherwood honey extracts.

Some volatile compounds have been used for the geographical discrimination of honey samples. It was suggested that English honey samples can be distinguished by the presence of 1-penten-3-ol and honey from Denmark by the absence of 3-methylbutanal. 2,2,6-Trimethylcyclohexanone, ethyl-2-hydroxypropanoate, 3-hexenyl-formate, and a few not as yet identified compounds with characteristic fragment ions have been suggested as possible marker compounds for honey from Portugal.

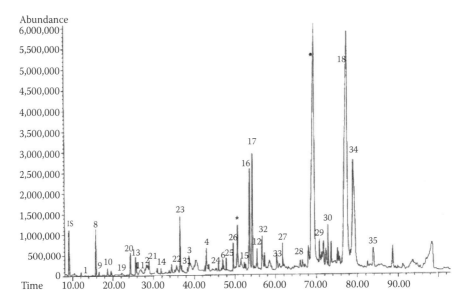

**FIGURE 5.4**   TIC profiles obtained of the SPE extract and GC-MS from *Eucalyptus* honey. For identification of numbered peaks, see Castro-Vázquez et al. (2006).

1-Octen-3-ol or 2,6,6-trimethyl-2,4-cycloheptadien-1-one has been suggested as possible marker compounds for honey from Spain, while pentanal and cis-linalool oxide were not found in the Spanish honeys (Radovic et al. 2001). However, it should be noted that Radovic et al. analyzed only three English samples, one from Denmark, two from Portugal, and two from Spain; therefore, their data may be considered as rather preliminary; considerably more samples from the same country should be evaluated in order to obtain more reliable conclusions (Kaškonien and Venskutonis 2010). Coumarin, previously proposed to characterize French lavender honeys, emerges here rather as an indicator of the freshness of lavender honey, being mainly released from glycosides during storage (Guyot-Declerck et al. 2002). Four volatile compounds in honey from the central valley of Ñuble Province, Chile, had never been reported before as honey compounds, these being 1,3-propanodiol, 2-methyl butanoic acid, 3,4-dimethyl-3-hexen-2-one, and 6-methyl-5-octen-2-one (Barra et al. 2010). The head-space SPME (HS-SPME)–based procedure, coupled to comprehensive 2D GC–time-of-flight (TOF) MS (GC×GC-TOF-MS), was employed for fast characterization of honey volatiles. In total, 374 samples were collected over two production seasons in Corsica ($n = 219$) and other European countries ($n = 155$), with the emphasis to confirm the authenticity of the honeys labeled as "Corsica" (protected denomination of origin region). For the chemometric analysis, artificial neural networks with multilayer perceptrons (ANN-MLP) were tested. The best prediction (94.5%) and classification (96.5%) abilities of the ANN-MLP model were obtained when the data from two honey harvests were aggregated in order to improve the model performance compared to separate year harvests (Cajka et al. 2009) (Figure 5.5).

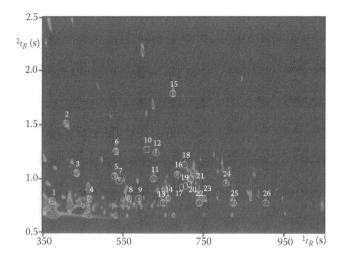

**FIGURE 5.5** (See color insert.) HS-SPME-GCxGC-TOF-MS chromatogram of honey volatiles (Corsica sample) with marked markers. For the description of analytes, see Cajka et al. (2009).

The main compounds suggested as possible markers of various honeys are listed in Table 5.2.

The absence of a particular compound in a particular honey could be a characteristic feature of its type (Table 5.3). The absence of 2-buten-1-ol, 2-methyl-; 2(3H)-furanone, dihydro-3-methyl-; linalol; and methyl salicylate in heather honey was affirmed, and the absence of oxime-, methoxy-phenyl-; benzenemethanol, α-methyl-; phenol, 3,4,5-trimethyl in buckwheat honey and butanoic acid, 3-methyl-; and butanoic acid, 2-methyl in lime honeydew honey (research conducted in Poland) (Wolski et al. 2006). Radovic et al. (2001) reported that the absence of 2-methyl-1-propanol and the simultaneous presence of dimethyl disulfide in the volatile fraction is characteristic for rape honey; they reported also lack of phenyl acetaldehyde as a characteristic headspace feature of acacia honeys. Unfortunately, Plutowska et al. (2011) by using GC-MS as the method of final determination and optimized HS-SPME as the sample preparation method, taking into account the type of fiber coating, effect of sample dilution, sample amount, salt addition, pH decrease, extraction temperature, and extraction time, on the enrichment of volatiles as well as influence of thermostating time and sample sonification before extraction on results, showed that the presence of disulfide and the simultaneous absence of isobutyl alcohol, except for honeydew honey, is not a peculiar feature for the examined rape honey and can also be found in other studied honeys (i.e., heather honey, linden honey, buckwheat honey, and acacia honey). These authors also showed that phenyl acetaldehyde occurred in greatest quantities in acacia and heather honeys. However, it must be pointed out that the Radovic et al. studies were carried out on honeys from Denmark, Germany, Italy, France, the Netherlands, Spain, Portugal, and England and those of Plutowska et al. on Polish honeys. Pérez et al. (2002), using HS-SPME/GC-MS for Spanish honeys,

**TABLE 5.2**

**Most Important Volatile Compounds in Unifloral Honeys**

| Honey Source | Compound | Extraction Method | GO | Reference |
|---|---|---|---|---|
| Acacia | Acetone, furfural, benzaldehyde | Dynamic HS GC-MS | EU | Radovic et al. 2001 |
| Alfalfa (*Medicago sativa*) | Benzene acetaldehyde, nonanal, and 2-methoxyphenol | SPME, GC-MS | AR | Baroni et al. 2006 |
| Almond tree (*Prunus dulcis*) | 2,6,6-Trimethyl-2,4-cycloheptadien-1-one (eucarvone) | SPME | ES | De la Fuente et al. 2007b |
| Amorpha | 2-Phenylethanol | HS-SPME, GC-MS | | Jerković et al. 2009b |
| Blue gum (*Eucalyptus leucoxylon*) | 8,9-Dehydrotheaspirone and 3-oxo-α-ionone | Solvent | AU | D'Arcy et al. 1997 |
| Buckwheat | 3-Methylbutanal, 3-hydroxy-4,5-dimethyl-2(5H)-furanone (sotolon), β-damascenone | Solvent | USA and ENG | Zhou et al. 2002 |
| | Aldehydes–furfural, 2-methylbutyraldehyde, 3-methylbutyraldehyde | HS-SPME, GC-MS | PO | Plutowska et al. 2011 |
| Calden (*Prosopis caldenia*) | 1-Octanol | SPME, GC-MS | AR | Baroni et al. 2006 |
| Carob (*Prosopis* spp.) | Nonanal and octanal | SPME, GC-MS | AR | Baroni et al. 2006 |
| Cambara | Benzaldehyde, 2,3-methylbutanoic acid, benzonitrile, 2-phenylethanol | Solvent | BR | Moreira and De Maria 2005 |
| Cashew (*Anacardium occidentale*) | Furfuryl mercaptan, benzyl alcohol, δ-decalactone, eugenol, benzoic acid, isovaleric acid, phenylethyl alcohol, 2-methoxyphenol | Purge-and-Trap (Tenax TA trap) | BR | Moreira et al. 2002 |
| Christ's thorn (*Paliurus spina-christi*) | 4-Hydroxy-3,5-dimethylbenzaldehyde, 4-hydroxybenzoic acid, and 4-methoxybenzoic acid (butanoic, hexanoic, octanoic, and nonanoic acid) | HS-SPME, GC, GC-MS | HR | Jerković et al. 2009a |

| | | | | |
|---|---|---|---|---|
| Citrus | (Z) (E)-linalool oxide, a-terpineol, terpineal, isomers of lilac aldehyde lilac alcohol; sinensal isomers | GC-MS | ES | Castro-Vázquez et al. 2006 |
| | Methyl anthranilate | Likens-Nickerson | (USA) | White and Bryant 1996 |
| | Lilac aldehydes, dill ether, methyl anthranilate, 1-p-menthen-9-al isomers | SPME | GR | Alissandrakis et al. 2007 |
| | (E)-linalool oxide, lilac aldehydes, lilac alcohols, limonyl alcohol, sinensal isomers, α-4-dimethyl-3-cyclohexene-1-acetaldehyde | Likens-Nickerson | ES | Castro-Vázquez et al. 2009 |
| Citrus | Limonene diol, methyl anthranilate | SPME | IT | Piasenzotto et al. 2003 |
| | Lilac aldehydes, methyl anthranilate | SPME | ES | De la Fuente et al. 2005 |
| | Isomers of lilac aldehydes | GC-MS | NS | Soria et al. 2008 |
| | Limonene diol, hotrienol | SPME | Sicilia, IT | Verzera et al. 2001 |
| | Lilac aldehyde, lilac alcohols | HS-SPME, GC-MS | NS | Soria et al. 2009 |
| | Methyl anthranilate, p-menth-1(7),8(10)-dien-9-ol | HS-SPME, GC-MS (PA) | NS | Soria et al. 2009 |
| Chestnut | 2-Aminoacetophenone, 1-phenylethanol, acetophenone | Likens-Nickerson | FR, IT | Guyot et al. 1998 |
| | 2-Methylcyclopentanol, diethylphenol | SPME | IT | Guidotti and Vitali 1998 |
| | Styrene1 | Dynamic HS GC-MS | EU | Radovic et al. 2001 |
| | 3-Aminoacetophenone, 2-methyldihydrofuranone, amethylbenzyl alcohol, 3-hexen-1-ol2, dimethylstyrene | Dynamic HS GC-MS | EU | Radovic et al. 2001 |
| | Acetophenone, 1-phenylethanol and 2-acetophenone | SPME | IT, Sicilia IT | Piasenzotto et al. 2003, Verzera et al. 2001 |
| | Acetophenone, 2-aminoacetophenone | Likens-Nickerson | ES | Bonvehi and Coll 2003 |
| | High percentage of phenyl acetic acid, low percentage of benzoic acid, also 4-aminoacetophenone | Solvent | HR | Jerković et al. 2007 |
| | Kynurenic acid | NMR spectroscopy | Dif | Donarski et al. 2008 |
| | n-Decane, phenyl acetaldehyde, phthalic acid, α-α-dimethylphenyl acetate, p-anisaldehyde | SPME | TR | Senyuva et al. 2009 |

*(continued)*

**TABLE 5.2 (Continued)**
**Most Important Volatile Compounds in Unifloral Honeys**

| Honey Source | Compound | Extraction Method | GO | Reference |
|---|---|---|---|---|
| Cotton (*Gossypium hirsutum*) | Cinnamaldehyde, cinnamyl alcohol, cinnamic acid, neryl and geranyl nitrile, benzene propanol, homovanillyl alcohol, (E)- and (Z)-p-methoxy-cinnamic acid, 2-methyl-p-phthalaldehyde, coniferaldehyde, p-coumaric acid, ferulic acid, scopoletin and scoparone | Solvent | GR | Alissandrakis et al. 2005b |
| Corontillo (*Escallonia pulverulenta*) | Safranal | SPME-GC-MS | AR | Montenegro et al. 2009 |
| Eucalyptus | Acetoin, aldimethyl disulfide, dimethyl trisulfide, alkane, nonane | Purge-and-Trap (cryogenic trap) | FR | Bouseta et al. 1992 |
| | 2-Hydroxy-5-methyl-3-hexanone and 3-hydroxy-5-methyl-3-hexanone, as well as exo-2-hydroxycineole, acetoin, nonanal, methyl nonanoate, and dehydrovomifoliol | USE, GC-MS | GR | Alissandrakis et al. 2011 |
| | Nonanol, nonanal, nonanoic acid | SPME | IT | Verzera et al. 2001 |
| | 1-Octene or 2,3-pentanedione | Dynamic HS GC-MS | EU | Radovic et al. 2001 |
| | 2,3-Pentanedione, acetoin, 1-hexyl alcohol, 2-acetyl-5-methylfuran, furfuryl propionate, 2-phenylacetaldehyde, nerolidol | Likens-Nickerson | ES | Bonvehi and Coll 2003 |
| | Acetoin, diacetyl, 2,3-pentanedione, dimethyl disulfide | SPME | ES | De la Fuente et al. 2005 |
| | 2-Hydroxy-5-methyl-3-hexanone, 3-hydroxy-5-methyl-2-hexanone | SPME | ES | De la Fuente et al. 2007a |
| | Nonanal, ethylphenyl acetate, phenethyl alcohol | SPME | TR | Senyuva et al. 2009 |
| Eucalyptus | Acetoin, 5-hydroxy-2,7-dimethyl-4-octanone, p-cymene derivatives, 3-caren-2-ol and spathulenol | Likens-Nickerson | ES | Castro-Vázquez et al. 2009 |

| | | | | |
|---|---|---|---|---|
| Fir honeydew | Acetonitrile, methyl-2-buten-1-ol, n-hexanol, 3-hexanol, 1-propyne, 2-furanmethanol, 5-methyl-2(5H)-furanone, 4-methylphenol, hexadecanoic acid, methylheptanoate | SPME | HR | Lušić et al. 2007 |
| Haze | Methyl-p-anisaldehyde, trimethoxybenzene, 5-hydroxy-2-methyl-4H-pyran-4-onen, lilac aldehyde isomers (A–D) | Dynamic HS GC-MS | JP | Shimoda et al. 1996 |
| Heather (Erica arborea) | Benzoic acid, decanoic acid, high levels of cinnamic acid, isophorone, 4-(3-oxobut-1-enylidene)-3,5,5-trimethylcyclohexen-2-en-1-one; shikimate-pathway derivatives: 4-methoxybenzaldehyde, 4-methoxybenzoic acid, methyl vanillate | Likens-Nickerson | EU | Guyot et al. 1999 |
| (Calluna vulgaris) | Phenyl acetic acid, dehydrovomifoliol, (4-(3-oxo-1-butynyl)-3,5,5-trimethylcyclohexen-2-en-1-one)m, higher levels of 3,5,5-trimethylcyclohexene derivatives | Likens-Nickerson | EU | Guyot et al. 1999 |
| Heather | 1-Penten-3-ol, 4-methylbicyclo[2,2,2]octan-1-ol,phenylacetaldehyde | Dynamic HS GC-MS | EU | Radovic et al. 2001 |
| | Isophorone | SPME / Solvent | NS / NZ | Soria et al. 2003 / Tan et al. 1989 |
| | High contents of benzene and guaiacol, p-anisaldehyde, propylanisole and p-cresol | Likens-Nickerson | ES | Castro-Vázquez et al. 2009 |
| | Isophorone, 2-hydroxy-3,5,5-trimethylcyclohexanone, furfuryl alcohol, benzyl alcohol, 2-phenylethanol | SPME | ES | De la Fuente et al. 2005 |
| | (S)-(+)-Dehydrovomifoliol (4-hydroxy-4-[3-oxo-1-butenyl]-3,5,5-trimethylcyclohex-2-en-1-one) | | EE | Häusler and Montag 1991 |

*(continued)*

**TABLE 5.2 (Continued)**
**Most Important Volatile Compounds in Unifloral Honeys**

| Honey Source | Compound | Extraction Method | GO | Reference |
|---|---|---|---|---|
| Honeydew | n-Decane, nonanal, α-α-dimethylphenyl acetate, nonanol, 2-methyl heptanoic acid | SPME | TR | Senyuva et al. 2009 |
| Lavender | n-Hexanal, n-heptanal, n-heptanol, phenyl acetaldehyde, coumarin | Dynamic headspace | FR | Bouseta et al. 1992, 1996 |
| | Ethanol, 2-methyl-1-propanol, 3-methyl-1-butanol, 3-methyl-3-buten-1-ol, 3,7-dimethyl-1,5,7-octatrien-3-ol (hotrienol), furfuryl alcohol, hexanal, heptanal, 1-hexanol, furfural, phenyl acetaldehyde, benzaldehyde | Dynamic HS GC-MS | EU | Radovic et al. 2001 |
| | n-Hexanal, n-heptanal, hexanol, phenyl acetaldehyde | Likens-Nickerson | PT and FR | Guyot-Declerck et al. 2002 |
| | Benzaldehyde, hexan-1-ol, phenyl acetaldehyde | HS-SPME, GC-MS | NS | Soria et al. 2009 |
| | Hexanal, nerolidol oxide, coumarin, high concentrations of hexanol and hotrienol | Likens-Nickerson | ES | Castro-Vázquez et al. 2009 |
| Kamahi | 2,6-dimethylocta-3,7-diene-2,6-diol, meliracemoic acid, and kamahines A-C | Solvent | NZ | Senanayake 2006 |
| Lime | Ethylmethylphenol, estragole, carvacrol | Likens-Nickerson | FR | Bouseta and Collin 1995; Guyot et al. 1998 |
| | 2-Pentanone, acetoin, furfural,4-methylacetophenone, methyl isopropyl benzene, dimethylstyrene | Dynamic HS GC-MS | EU | Radovic et al. 2001 |
| Linden | 4-tert-butylphenol, an unidentified isomer of ethylmethylphenol, estragol, menthol, methyl(1-methylethenyl)benzene, p-methyl-acetophenone, 8-p-menthen-1,2-diol, carvacrol, thymol | Likens-Nickerson | FR, IT | Guyot et al. 1998 |
| | p-Methylacetophenone, isomeric cymenols, methylstyrene | HS-SPME, GC-MS | PO | Plutowska et al. 2011 |

| | | | | |
|---|---|---|---|---|
| Maple (*Acer* spp.) | Syringaldehyde | USE, GC and GC-MS | | Jerković et al. 2010a, 2010b |
| Marmeleiro *Croton* species | Isovaleric acid, $\gamma$-decalactone, benzoic acid, vanillin | Dynamic HS GC-MS | BR | Moreira et al. 2002 |
| *Mentha* spp. | Methyl syringate, vomifoliol, 3,7-dimethylocta-1,5-dien-3,7-diol (terpendiol I), hotrienol | HS-SPME, USE, GC, GC-MS | HR | Jerković et al. 2010a, 2010b |
| Nodding thistle | (E)-2,6-Dimethyl-6-hydroxy-2,7-octadienoic acid, (E)-2,6-dimethyl-3,7-octadiene-2,6-diol, (Z)-2,6-dimethyl-2,7-octadiene-1,6-diol, (E)-2,6-dimethyl-6-hydroxy-2,7-octadienal, lilac aldehydes, lilac alcohols | Solvent | NZ | Wilkins et al. 1993 |
| Oak honeydew | trans-oak lactone, aminoacetophenone, propylanisol | Solvent | ES | Castro-Vázquez et al. 2006 |
| Quillay (*Quillaja saponaria*) | Megastigmatrienone, 2-p-hydroxyphenylalcohol, $\beta$-pinene, linalool oxide | SPME-GC-MS | AR | Montenegro et al. 2009 |
| Strawberry | $\alpha$-Isophorone, $\beta$-isophorone, 4-oxoisophorone | Dynamic HS GC-MS | Sardinia (IT) | Bianchi et al. 2005 |
| | a-Isophorone,2,5-dihydroxyphenylacetic acid | NMR spectroscopy | EU (majority from COR) | Donarski et al. 2008 |
| Rape | Dimethyl disulfide | Dynamic HS GC-MS | EU | Radovic et al. 2001 |
| Rhododendron | n-Decane, lilac aldehyde, 2-aminoaceto phenone, benzenedicarboxylic, nonanal, isobutylphthalate, damascenone | SPME | TR | Senyuva et al. 2009 |
| Rosemary | Acetone, 2-pentanone, benzaldehyde, 4-oxoisophorone | Dynamic HS GC-MS | EU | Radovic et al. 2001 |
| | Lilac aldehyde isomers | Likens-Nickerson | ES | Castro-Vázquez et al. 2009 |
| Sage | Benzoic acid, phenyl acetic acid, p-anisaldehyde, a-isophorone, 4-ketoisophorone, dehydrovomifoliol, 2,6,6-trimethyl-4-oxocyclohex-2-ene-1-carbaldehyde, 2,2,6-trimethylcyclohexane-1,4-dione, and coumaran | Solvent | HR | Jerković et al. 2006 |
| | Tetrahydro-2,2,5,5-tetramethylfuran, 3-hexenyl ester of butanoic acid, 2-methylbenzene, maltol, methyl ester of 3-furanocarboxylic acid, benzene acetic acid | SPME | HR | Lušić et al. 2007 |

(continued)

**TABLE 5.2 (Continued)**
**Most Important Volatile Compounds in Unifloral Honeys**

| Honey Source | Compound | Extraction Method | GO | Reference |
|---|---|---|---|---|
| Sulla (*Hedysarum coronarium* L.) | Vomifoliol, as α-isophorone, 4-ketoisophorone, 3-oxo-α-ionol or 3-oxo-α-ionone, 3-hydroxy-4-phenylbutan-2-one, methyl syringate | GC and GC-MS | Sardinia (IT) | Jerković et al. 2010a, 2010b |
| Sunflower | α-Pinene or 3-methyl-butan-2-ol | Dynamic HS GC-MS | EU | Radovic et al. 2001 |
| | 2-Methoxyphenol | SPME, GC-MS | AR | Baroni et al. 2006 |
| Strawberry tree | α-Isophorone, β-isophorone, 4-oxoisophorone | Purge-and-Trap (Carbopack trap) | Sardinia (IT) | Bianchi et al. 2005 |
| | Isophorone | SPME | ES | De la Fuente et al. 2007b |
| Thyme | 3,4,5-Trimethoxybenzaldehyde | SPME | TR | Mannas and Altug 2007 |
| | 1-Phenyl-2,3-butanedione, 3-hydroxy-4-phenyl-2-butanone, 3-hydroxy-1-phenyl-2-butanone, phenyl acetonitrile, and carvacrol | SPME | GR | Alissandrakis et al. 2007 |
| | 3-Hydroxy-4-phenyl-2-butanone, 3-hydroxy-1-phenyl-2-butanone, 2-methylpropionic acid, 4-(4-hydroxy-2,2,6-trimethyl-7-oxabicyclo[4.1.0]hept-1-yl)-3-buten-2-one | Solvent | GR | Alissandrakis et al. 2009 |
| *Thymus capitatus* | 1,3-Diphenyl-2-propanone, (3-methylbutyl)benzene, 3,4,5-trimethoxy benzaldehyde, 3,4-dimethoxy benzaldehyde, vanilline, thymol | SPME | PS | Odeh et al. 2007 |
| *Thymelaea hirsute* | Benzene propanol, benzyl alcohol, nonanol, hexanol, 4-methoxyphenol | SPME | PS | Odeh et al. 2007 |

| | | | | |
|---|---|---|---|---|
| *Tolpis virgata* | 3,5-Dihydroxytoluene, tridecane | SPME | PS | Odeh et al. 2007 |
| Turkish pine | 3-Carene and unidentified compound (*m/z*: 55, 79, 91, 107, 123, 162) | Dynamic HS GC-MS | TR | Tananaki et al. 2007 |
| Ulmo (*Eucryphia cordifolia*) | Isophorone and cetoisophorone | SPME-GC-MS | AR | Montenegro et al. 2009 |
| Willow (*Salix* spp.) | Methyl salicylate | SPME | ES | De la Fuente et al. 2007b |
| Willow (*Salix* spp.), honeydew honey | Benzoic acid, phenylacetic acid, 2-hydroxybenzoic acid, 4-hydroxyphenylacetic acid, 4-methoxybenzoic acid, 4-hydroxyphenylethanol, and 4-hydroxybenzoic acid, methyl salicylate | HS-SPME, USE, GC, GC-MS | HR | Jerković et al. 2010a, 2010b |
| White clover (*Melilotus albus*) | 2-H-1-benzopyran-2-one | SPME, GC-MS | AR | Baroni et al. 2006 |

*Note:* Geographic origin: AU, Australia; AR, Argentina; BR, Brazil; CN, China; ENG, England; ES, Spain; EU, Europe; GR, Greece; HU, Hungary; HR, Croatia; IR, Iran; IT, Italy; FR, France; JP, Japan; NZ, New Zealand; PO, Poland; PS, Palestine; PT, Portugal; RO, Romania; SK, Slovakia; TR, Turkey; USA, United States of America; NS, not précised.

**TABLE 5.3**

**Absence and Presence of Honey Marker Compounds Typical for Certain Floral Origins**

| Honey Source | Marker Compound | | Method | References |
|---|---|---|---|---|
| | Presence Of | Absence Of | | |
| Acacia | Cis-linalool oxide and heptanal | Both phenyl acetaldehyde and dimethyl disulfide | Dynamic HS GC-MS | Radovic et al. 2001 |
| Buckwheat | 2-Butenal, 2-methyl-; butanoic acid; nonane; pentanoic acid; 2(3H)-furanone, dihydro-4-methyl-; trans-linalol oxide; 2-acetylaniline; decanoic acid; 3,5-dimethoxybenzaldehyde; isopropyl myristate; dibutyl phthalate | Oxime-, methoxy-phenyl-; benzenemethanol, α-methyl-; phenol, 3,4,5-trimethyl- | SPME GC-MS | Wolski et al. 2006 |
| Heather | Acetone; formic acid; 2-propanone, 1-phenyl-; 2-hydroxy-3,5,5-trimethyl-cyclohex-2-enone; benzene propanol; benzene acetic acid; cinnamyl alcohol; phenol, 2,6-dimethoxy-; damascenone; cumene, 2,4,5-trimethyl- | 2-Buten-1-ol,2-methyl-; 2(3H)-furanone, dihydro-3-methyl-; linalol; methyl salicylate | SPME GC-MS | Wolski et al. 2006 |
| Lavender | Heptanal | 4-Oxoisophorone | Dynamic HS GC-MS | Radovic et al. 2001 |
| Lime honeydew | Ethyl acetate, 1-propanol, 2-methyl-; 3-buten-1-ol, 3-methyl-; 2-buten-1-ol, 3-methyl-; bicyclo[2.2.1]hept-2-ene, 2,3-dimethyl-; furfural, 5-methyl-; octanal, 1H-inden-1-one, 2,3-ihydro-; 2(3H)-furanone,dihydro-5-pentyl-. | Butanoic acid, 3-methyl-; butanoic acid, 2-methyl- | SPME, GC-MS | Wolski et al. 2006 |
| Rape | Dimethyl disulfide | 2-Methyl-proan-1-ol | Dynamic HS GC-MS | Radovic et al. 2001 |
| Sunflower | α-Pinene or 3-methyl-butan-2-ol | Both heptanal and 4-oxoiosphorone | Dynamic HS GC-MS | Radovic et al. 2001 |

orange, eucalyptus, jasmine, lavender, and thyme, determined unique compounds for each variety, but in the Plutowska et al. (2011) study it was stated that some of these compounds, such as 3-methyl-2-buten-1-ol, hotrienol (3,7-dimethyl-1,5,7-octatrien-3-ol), or lilac aldehydes, are typical elements of the volatile fraction in Polish honeys.

Sensory quality, especially aroma and taste, are often the most important factors from the consumer's point of view. Flavor qualities of honey are very much dependent on the volatile organic compounds; on the contrary, honey volatiles exhibit a potential role in distinguishing honeys as a function of botanical origin.

Volatile compound identification with the purpose of assessing the botanical origin of honey has the potential to be an extremely useful strategy, but these compounds occur most frequently in very low concentrations and are a complicated mixture of substances with various physicochemical properties and various levels of stability. The isolation of volatile components from a complex mixture such as honey in order to obtain representative extracts is very difficult. Methods of extracting the honey volatiles may display a varying degree of selectivity and effectiveness depending upon the compounds involved and the extraction conditions (Cuevas-Glory et al. 2007). Different methods can be used, but the use of one single technique is not adequate for reliable honey volatiles profiling (Jerković et al. 2009a,b). In addition, aroma profile analysis should be combined with other methods for the determination of other constituents.

## PHENOLIC COMPOUNDS

Phenolic compounds or polyphenols are one of the most important groups of compounds occurring in plants, where they are widely distributed, comprising at least 8000 different known structures (Bravo 1998). In general, phenolic compounds can be divided into at least 10 types depending on their basic structure: simple phenols, phenolic acids, coumarins, isocoumarins, naphthoquinones, xanthones, stilbenes, anthraquinones, flavonoids, and lignins. Flavonoids constitute the most important polyphenolic class, with more than 5000 compounds already described (Wollgast and Anklam 2000); they are divided into several groups according to their chemical structure, including flavonols (quercetin and kaempferol), flavanols (the catechins), flavones (apigenin), and isoflavones (genistein) (Young and Woodside 2001). Polyphenols are reported to exhibit anticarcinogenic, anti-inflammatory, antiatherogenic, antithrombotic, immunomodulating, and analgesic activities (Catapano 1997; Vinson et al. 1998); antiproliferative and vasodilatory actions (Lule and Xia 2005); antiallergic, antiplatelet, and anti-ischemic actions (Shi 2001); and antimicrobial action (Taguri et al. 2006). Epidemiologic studies point to the possible role of flavonoids in preventing cardiovascular diseases and cancer. Flavonoids behave as antioxidants in a variety of ways, including direct trapping of reactive oxygen species, inhibition of enzymes responsible for producing superoxide anions, chelation of transition metals involved in processes forming radicals, and prevention of the peroxidation process by reducing alkoxyl and peroxyl radicals (Rice-Evans et al. 1996). Dietary phenolic acids are also considered to be powerful antioxidants. Their antioxidant activity is much higher *in vitro* than of well-known vitamin antioxidants (Tsao and Deng 2004). The occurrence of flavonoids in honey has been reported a number of times over the past 30 years (Bogdanov 1984; Amiot et al. 1989; Ferreres et al. 1991; Sabatier et al. 1992); their content can vary between 60 and 460 μg/100 g of honey and was higher in samples produced during a dry season with high temperatures (Kenjerić et al. 2007). Bioactive substances may be transferred from the plant to the nectar and further to honey by imparting to the final product different properties that may reasonably depend on the floral origin of honey (Kaškonien and Venskutonis 2010).

Honey is rich in phenolic acids and flavonoids, which exhibit a wide range of biological effects and act as natural antioxidants (Pyrzynska and Biesaga 2009), but the levels of single phenolics or other compounds in honey are too low to have a major individual significance. Hence, the total antioxidant capacity of honey can be the result of the combined activity and interactions of a wide range of compounds, including phenolics, peptides, organic acids, enzymes, Maillard reaction products, and other minor components (Pérez et al. 2006). Phenolic compounds contribute to organoleptic properties such as color, taste, or flavor of honey. They also have antibacterial and antioxidant activities together with other substances (Gheldof et al. 2002).

The phenolic acids in honey are divided in two subclasses: the substituted benzoic acids and cinnamic acids and the flavonoids in three classes with similar structure: flavonols, flavones, and flavanones (Davidson et al. 2005). Moreover, honey phenolic compound composition depends on their floral sources used to collect honey, the predominance of which is dependent on seasonal and environmental factors (Al-Mamary et al. 2002; Yao et al. 2003). Considerable differences in both composition and content of phenolic compounds have been found in different unifloral honeys (Amiot et al. 1989); thus, phenolic acids and flavonoids are considered potential markers of the honey botanical origin (Davidson et al. 2005). Flavonoid analysis is a very promising technique in studies of the botanical (Amiot et al. 1989; Ferreres et al. 1992; Anthony et al. 2000; Merken and Beecher 2000; Tomás-Barberán et al. 2001; Gómez-Caravaca et al. 2006) and geographical (Ferreres et al. 1991, 1992; Tomás-Barberán et al. 1993; Yao et al. 2004a) origins of honey (Table 5.4). The characteristic polyphenols present in honeys that are able to perform the role of biomarkers are flavonoids such as hesperetin, kaempferol, quercetin and chrysin, and phenolic acids: abscisic, ellagic, p-coumaric, gallic, and ferulic (Ferreres et al. 1994a,b,c; Kenjerić et al. 2007; Soler et al. 1995; Tomás-Barberán et al. 2001; Yao et al. 2003, 2005).

Ferreres et al. (1993) found that hesperetin (hesperetin-7-rutinoside) was the major flavonoid detected in *Citrus* honey and suggested that this compound could be a useful floral marker for citrus honey. Naringenin (Andrade et al. 1997b) and luteolin (Ferreres et al. 1994a) were suggested as markers of lavender honey. Martos et al. (2000a,b) and Yao et al. (2004b) found tricetin (5,7,3′,4′,5′-pentahydroxyflavone), quercetin, luteolin (5,7,3′,4′-tetrahydroxyflavone), myricetin, and kaempferol to be present in Australian monofloral *Eucalyptus* honey and suggested that this flavonoid profile could be used to verify the botanical origin of *Eucalyptus* honey.

Quercetin was suggested as a marker for sunflower honey (Tomás-Barberán et al. 2001); 8-methoxykaempferol was the main compound in rosemary (Ferreres et al. 1994a); naringenin (Andrade et al. 1997b) and luteolin (Ferreres et al. 1994a) were suggested as markers of lavender honey. Myricetin, luteolin, and tricetin were found to be the main flavonoids in Australian jelly bush honey (*Leptospermum polygalifolium*) (Yao et al. 2003). Myricetin, tricetin, quercetin, luteolin, and an unknown flavonoid were found to be the main flavonoids in Australian crow ash honey, while quercetin, quercetin 3,3′-dimethyl ether, myricetin, and luteolin were characteristic only for the Australian sunflower honey (Yao et al. 2004c).

Several studies made on honey show that European honeys have a rich phenolic profile, consisting of benzoic, cinnamic acids, and flavonoid aglycones (Ferreres et al. 1992, 1994a,b; Andrade et al. 1997a; Martos et al. 2000a,b; Tomás-Barberán et al. 2001).

There are high differences between quantities of phenolic in the samples originating from different geographic areas and not between the profiles of these compounds. This could be due to the climatic conditions, purity of the sample, method of analysis, and sensitivity of the apparatus (Marghitas et al. 2010). Australian *Eucalyptus* honeys showed a common flavonoid profile comprising tricetin, quercetin, and luteolin, which, together with myricetin and kaempferol, were previously suggested as floral markers for European *Eucalyptus* honeys (Yao et al. 2004c). Ferreres et al. (1996) found that abscisic acid isomers were the main constituents in Portuguese "heather" honeys derived from *Erica* spp. and they suggested that these compounds were the main floral markers for the Portuguese heather (*Erica* spp.) honeys. High levels of abscisic acid isomers were also found in New Zealand *Erica* spp. honeys (Senanayake 2006).

Hesperetin, previously indicated as a possible marker for citrus honey (Ferreres et al. 1993, 1994b), was found, in different amounts, also in other honey types (Pulcini et al. 2006). Liang et al. (2009) using HPLC-electrochemical detection found that hesperetin content in Chinese citrus honey was lower than that reported by Ferreres et al. (1994). Homogentisic acid, indicated as a possible marker for *Arbutus* honey (Cabras et al. 1999), was found also in honey from *Erica* (which belongs to the same botanical family as *Arbutus* [Ericaceae]) (Pulcini et al. 2006). Gallic acid was found in Australian honeys from five botanical species (Yao et al. 2003, 2005) and it was suggested that gallic acid is the dominant component in the profile of phenolic acids in Australian honeys (Yao et al. 2005). However, it was determined only in linden honey by Michalkiewicz et al. (2008) after selection of suitable chromatographic and detection conditions.

## Amino Acids

The origin of amino acids in honey is attributable both to animal and vegetal sources, although the main source is the pollen, so the amino acid profile of a honey could be characteristic of its botanical origin. There are studies that appear to classify the geographical origin of the honey with the use of its amino acid content (Bosi and Battaglini 1978; Gilbert et al. 1981; Davies et al. 1982; Cometto et al. 2003; Iglesias et al. 2004; Seif Eldin and Elfadil 2010). The highest values were reported in honey of thymus origin, while the lowest were in honey samples from *Rhododendron*, *Brassica*, and *Robinia* (Persano Oddo and Piro 2004). In the 1970s, Davies (1975, 1976) attempted to correlate floral origin with the free amino acid composition but failed to identify any simple relationships between floral sources and the resulting

**TABLE 5.4**

**Phenolic and Some Other Compounds Characteristic to Unifloral Honeys**

| Honey Source | Compound | Extraction Method | GO | Reference |
|---|---|---|---|---|
| Acacia | Kaempferol rhamnosides | SPE | IT and SK | Truchado et al. 2008 |
| | Ferulic acid, acacetin | SPE | RO | Bobis et al. 2007 |
| | Cinnamic acid derivatives | SPE | EU | Dimitrova et al. 2007 |
| | cis,trans- and trans,trans-abscisic acids, ellagic acid | SPE | EU | Tomás-Barberán et al. 2001 |
| | Acacetin | SPE | RO | Marghitas et al. 2010 |
| Alder | 8-Methoxykaempferol | SPE | FR | Soler et al. 1995 |
| Arbutus | Homogentisic acid | | | Cabras et al. 1999 |
| Asphodel | Methyl syringate | Solvent | Sardinia (IT) | Tuberoso et al. 2009 |
| Brush box (*Lophostemon conferta*) | Gallic acid, ellagic acid, | SPE | AU | Lihu Yaoa et al. |
| Calluna | Ellagic acid | SPE | FR | Soler et al. 1995 |
| Citrus | Hesperetin | SPE | EU | Ferreres et al. 1993 |
| | | | FR | Soler et al. 1995 |
| | | | EU | Tomás-Barberán et al. 2001 |
| | Hesperetin, methyl anthranilate | SPE | ES | Ferreres et al. 1994b |
| | Quercetin, hesperetin, and chrysin | SPE | IR | Hadjmohammadi et al. 2009 |
| | Caffeic acid, p-coumaric acid, ferulic acid, hesperetin | Solvent | CN | Liang et al. 2009 |
| Chestnut | Caffeic acid, p-coumaric acid, ferulic acid | SPE | IT | Cherchi et al. 1994 |
| | 4-Hydroxybenzoic, DL-p-hydroxyphenyl lactic, ferulic, and phenylacetic acids | SPE | EU | Dimitrova et al. 2007 |
| Crow ash (*Guioa semiglauca*) | Myricetin, tricetin, quercetin, luteolin, and an unknown flavonoid | SPE | AU | Yao et al. 2004a, 2004b |
| | Gallic acid, abscisic acid | SPE | AU | |

| Botanical origin | Compounds | Method | Country | Reference |
|---|---|---|---|---|
| Eucalyptus | Myricetin, tricetin, quercetin, luteolin, and kaempferol | SPE | EU | Martos et al. 2000a, 2000b |
| | Quercetin and six unidentified compounds | SPE | EU | Tomás-Barberán et al. 2001 |
| | Gallic acid AU | SPE | AU | Yao et al. 2004a |
| | Benzoic acid derivatives | SPE | EU | Dimitrova et al. 2007 |
| Heather | Myricetin | SPE | FR | Soler et al. 1995 |
| | Nyricetin, myricetin 3-methyl ether, myricetin 3′-methyl ether, tricetin | SPE | PT | Ferreres et al. 1994c |
| | cis,trans- and trans,trans- abscisic acids | SPE | PT | Ferreres et al. 1996 |
| | Ellagic, p-hydroxybenzoic, syringic, and o-coumaric acids | SPE | PT | Andrade et al. 1997a |
| | Ellagic acid | SPE | EU | Andrade et al. 1997b |
| | DL-β-Phenyl lactic acid, phenylacetic, benzoic acids | SPE | EU | Dimitrova et al. 2007 |
| Heath (B. ericifolia) | Gallic acid, an unknown phenolic acid and coumaric acid | SPE | AU | Lihu Yaoa et al. |
| Jelly bush (Leptospermum polygalifolium) | Myricetin, luteolin, tricetin, gallic and coumaric acids, abscisic acid | SPE | AU | Yao et al. 2003 |
| Jelly bush (Leptospermum scoparium) | Quercetin, isorhamnetin, chrysin, luteolin and an unknown flavanone, gallic acid, abscisic acid | SPE | NZ | Yao et al. 2003 |
| Kanuka | Methoxyphenyl lactic acid | SPE | NZ | Stephens et al. 2010 |
| Lavender | Luteolin | SPE | ES | Ferreres et al. 1994a |
| | Gallic acid | SPE | PT | Andrade et al. 1997a |
| | Naringenin | SPE | EU | Andrade et al. 1997b |
| | Gallic and caffeic acids | SPE | EU | Dimitrova et al. 2007 |
| Lime tree | 3-Hydroxybenzoic acid | SPE | EU | Dimitrova et al. 2007 |
| Rosemary | Kaempferol | SPE | ES | Gil et al. 1995 |
| | 8-Methoxykaempferol | SPE | ES | Ferreres et al. 1994a |
| | Kaempferol, 8-methoxykaempferol | SPE | EU | Tomás-Barberán et al. 2001 |

(continued)

## TABLE 5.4 (Continued)
## Phenolic and Some Other Compounds Characteristic to Unifloral Honeys

| Honey Source | Compound | Extraction Method | GO | Reference |
|---|---|---|---|---|
| Strawberry tree | Gluconic acid | SPE | IT | Cherchi et al. 1994 |
| | Homogentisic acid | Solvent | Sardinia (IT) | Cabras et al. 1999 |
| | Homogentisic, (±)-2-cis,4-trans, (±)-2-trans,4-trans-abscisic acids, unedone | Solvent | Sardinia (IT) | Tuberoso et al. 2010 |
| | 2,5-Dihydroxyphenylacetic acid and α-isophorone | No extraction | EU | Donarski et al. 2010 |
| Sunflower | Quercetin | SPE | FR | Soler et al. 1995 |
| | Quercetin, quercetin 3,30-dimethyl ether, myricetin, and luteolin | SPE | AU | Yao et al. 2004a, 2004b |
| Thyme | Coumaric acid, gallic acid | SPE | AU | Lihu Yaoa et al. |
| | Rosmarinic acid | SPE | EU | Andrade et al. 1997b |

honey. Bonaga et al. (1986) attempted to determine if honey amino acid profiles could be correlated with biochemical modification(s) of amino acids carried out by honeybees; however, no such correlations were identified. Bosi and Battaglini (1978) were able to distinguish the geographical origin of some European honeys on the basis of free amino acid composition. Davies et al. (1982) found that amino acid levels could be used to distinguish British and foreign honeys. Gilbert et al. (1981) in a study of 45 samples of honey, by using GC, found that some groups of honey samples originating from different countries could be distinguished from each other on the basis of their amino acid profiles. On the basis of the results obtained for 56 honey samples from three Argentinian regions, Cometto et al. (2003) suggested that free amino acid profile could be used to verify the floral, or geographical, origin of a honey. However, the amino acid profile is characteristic of the flora in the immediate vicinity of the apiary rather than the overall floral characteristic of the geographic region from which the honey was collected. Using HPLC analysis and statistical processing by principal component analysis (PCA) on seven selected floral varieties, Cotte et al. (2004b) concluded that only lavender honeys were perfectly characterized, but complete satisfaction was not obtained with the six other varieties. However, the method enabled them to detect the addition of sugar syrup to rape and fir honeys.

After a study carried out on 48 Spanish honey samples from different botanical and geographical origin (eucalyptus, lavender, rosemary, orange blossom, thyme, heather, holm oak, forest, oak, and multifloral), Hermosin et al. (2003) concluded that amino acid composition was not absolutely able to distinguish the botanical origins of honey samples. However, according to a study by Iglesias et al. (2004) on free amino acid content of 46 honey samples (vipers bugloss, multiflower, heather, rosa bush, rosemary, and honeydew honey samples) collected from central Spain, glutamic acid and tryptophan were considered to be good floral source and/or honeydew marker substances. Pirini et al. (1992) reported that the content of proline (the main amino acid in honey) of chestnut honey is higher than that of the other honeys analyzed. Bouseta et al. (1996) found higher concentrations of proline and phenylalanine in lavender honeys than in *Eucalyptus* honeys. Sanchez et al. (2001), Sabatini and Grillenzoni (2002), and Oddo and Piro (2004) noted that if proline (the main amino acid in honey) shows characteristic values in different unifloral honeys, the variation of this parameter in different unifloral honeys is quite high and it is not possible to classify unifloral honeys on the basis of proline content only. Kanematsu et al. (1982) showed that free amino acid patterns and certain ratios (e.g., proline/phenylalanine) could be used to characterize the floral origins of honey. Bouseta et al. (1996) found higher concentrations of proline and phenylalanine in lavender honeys than in *Eucalyptus* honeys. Proline content of honey is roughly correlated with the enzyme activity (Sabatini and Grillenzoni 2002). Pirini et al. (1992) suggested that the presence of arginine is an important discriminating factor useful for the characterization of chestnut honey. Arginine tryptophan and cysteine contents were found similar for honeys produced from different botanical sources and this finding indicated that an amino acid or an amino acid group was not enough for discrimination of specific honey (Nisbet et al. 2009). A study with Turkish honeys showed that phenylalanine and tyrosine contents may contribute to distinguishing honeys, with valine, leucine, and isoleucine being other important amino acids for recognizing botanical and geographic origins (Senyuva et al. 2009). Furthermore, the authors suggested combining one or two amino acids with other phytochemicals such as volatile compounds or oligosaccharides to achieve improved discrimination. Fifteen Italian honey samples from 10 different floral types and 3 honeydew samples were analyzed. Amino acids were determined by an automatic precolumn derivatization with fluorenylmethyl-chloroformate and by reverse-phase liquid chromatography with fluorescence detector. The results obtained show that glutamic acid, proline, and tyrosine were major components in floral honeys. However, the overall amino acid profile did not enable the botanical origin of honey samples to be distinguished, as a rather high variation among honeys from the same botanical source and between honey types from different floral origin was found (Carratù et al. 2011); as a significant part of the free honey amino acids is added by the bees (Bergner and Hahn 1972), that fact leads to a high variability of the amino acid content within honeys from the same botanic source (Davies 1975; Gilbert et al. 1981).

With all these results, data on honey amino acids are too scarce to make sound assumptions on the use of these components in the characterization of different honey types. Baroni et al. (2002) applied immunoblot assays for the analysis of honey proteins and showed that it was possible to differentiate the pollen originated from sunflowers and *Eucalyptus* spp. and suggested this method as an alternative or complementary method to the melissopalynologic analysis. This method, however, will have all the shortcomings of pollen analysis (Bogdanov et al. 2004). There was no protein specificity in honeys produced from different botanical sources (rhododendron, chestnut, blossom, pine, etc.). Arginine, tryptophan, and cysteine contents were found similar for honeys produced from different botanical sources and this finding indicated that an amino acid or an amino acid group was not enough for discrimination of specific honey (Nisbet et al. 2009).

Cometto et al. (2003) found that, under some circumstances (see above), amino acid profiles can be used to verify the geographical origin of Argentinian honeys.

## Amino Acids (Biochemical Modification)

Bonaga et al. (1986) attempted to determine if honey amino acid profiles could be correlated with biochemical modification(s) of amino acids carried out by honeybees; however, no such correlations were identified.

## Carbohydrates

Honey is a natural syrup containing primarily fructose (38.5%) and glucose (31.3%). Other sugars in honey include maltose (7.2%), sucrose (1.5%), and a variety of oligosaccharides (4.2%) (National Honey Board [NHB] 1996).

Carbohydrate composition and content of honey are variable and are dependent on the floral source of the honey (Swallow and Low 1990; Weston and Brocklebank 1999). The characteristics of the honey sugar profile depend on two parameters: the first is the action of enzymes that transform disaccharides and trisaccharides into simple sugars (i.e., slow or rapid honey flow), which affects the final sugar contents, and the second is that the honey profile is correlated with the initial profile of its precursor. If a monosaccharide is present at high concentrations in one nectar compared to another, it will also be present in larger quantities in the honey produced (Cotte et al. 2004a). Trisaccharides and tetrasaccharides, which are formed mainly by the action of honey enzymes, were reported for Spanish and New Zealand honeys by Ruiz-Matute et al. (2010).

Available methods can be used to evaluate fructose and sucrose in honey (Palmer and Brandes 1974; Zürcher et al. 1975; Thean and Funderburk 1977; Deifel 1985; Bogdanov and Baumann 1988; Horváth and Molmir-Perl 1997; Mateo and Bosch-Reig 1997; Batsoulis et al. 2005). However, honey proved to be a difficult matrix because its sugar composition is one of the most complex mixtures of natural carbohydrates. Honey is a matrix of several saccharides with the same degree of polymerization (Horváth and Molmir-Perl 1997). The simultaneous mass percentage determination of fructose and glucose in honey has been achieved with the application of FT-Raman spectroscopy combined with the appropriate software. The proposed method further demonstrates the efficiency of transmittance spectroscopy methods for the quantitative analysis of complex mixtures. The returned results were statistically tested with those of the HPLC method. Both methods appear to score equally in terms of reproducibility (Batsoulis et al. 2005).

Use of sugar profiles such as fructose, glucose, sucrose, maltose contents, and glucose/water ratios has been suggested for characterization of unifloral honeys (Mateo and Bosch-Reig 1998; Da Costa Leite et al. 2000).

Acacia and chestnut honeys are characterized by high fructose/glucose ratios in contrast to rape and sunflower honeys (Persano Oddo et al. 1995; Cotte et al. 2004a). The fructose/glucose ratios could be used to classify groups of floral origins (Cotte et al. 2004a). Fructose and glucose are the major sugars, the former one being a major component almost in all honey types, except for some honeys of rape (*Brassica napus*), dandelion (*Taraxacum officinale*), and blue curls (*Trichostema lanceolatumi*) origin, when glucose is present in higher amounts (Cavia et al. 2002).

The saccharide distribution in honeys of various floral origin (8 acacia, 8 mixed flower, 1 linden, 1 rape, 1 sunflower, and 1 pine sample) revealed that acacia honeys contain significantly higher fructose, sucrose, and minor oligosaccharides than samples from other floral sources (Horváth and Molmir-Perl 1997).

Heather honey was noted for the presence of erlose and nigerose, forest honey contained higher amounts of trehalose and melezitose, spike lavender honey was specified by isomaltose, while French lavender and thyme honeys were characterized by panose (Nozal et al. 2005), maltose being characteristic of both rhododendron and honeydew honeys (Senyuva et al. 2009).

Cotte et al. (2004b) reported that fir samples were characterized by a high content in trisaccharides, particularly raffinose melezitose and erlose in agreement with Sabatini et al. (1990); erlose was practically absent in rape and sunflower honey samples, enabling them to be distinguished from other botanical varieties; lavender and linden were characterized by intermediate erlose concentrations. Alcohol perseitol (D-glycero, D-galacto-heptitol) in spite of its low concentration could be used as a marker for avocado honey (Dvash et al. 2002).

Honey carbohydrate composition had also been used in characterizing its geographical origin. Maltose and raffinose were selected as two oligosaccharide parameters to be used in the classification of Turkish honey (Senyuva et al. 2009). Sanz et al. (2004) found the highest amounts of maltulose and turanose of honey, respectively, in 10 samples of honey from different regions of Spain.

Maltose was the major disaccharide present in 80 genuine Brazilian honey samples (mostly *Eucalyptus* spp., extrafloral, and multifloral honeys) (Da Costa Leite et al. 2000).

A study of 91 English honey samples for 40 oligosaccharides and a combination of hierarchical cluster analysis, PCA, and canonical discriminant analysis showed that it was possible, with some exceptions, to discriminate floral types, but unfortunately, these authors did not indicate which of the oligosaccharides contributed most to the discrimination (Goodall et al. 1995). Sugar profiles (determined by GC) were examined to test their suitability for characterizing Spanish unifloral honeys. Satisfactory results were obtained only for some honey types (Mateo and Bosch-Reig 1997). A study including 280 French honeys from seven monofloral varieties (acacia, chestnut, rape, lavender, fir, linden, and sunflower) was carried out with chromatographic analyses and multivariate statistical processing of the sugar contents. The PCA enabled zones of the different varieties to be differentiated (i.e., fir samples that were totally separated); even so, the complete distinction of each floral origin could not be obtained. The other origins were paired: acacia and chestnut honeys, lavender and linden, and rape and sunflower. This PCA thus indicates a trend. It is in fact possible to control an origin known in advance. On the contrary, it was impossible to use this PCA to determine the origin of an unknown sample. The authors concluded that, combined with other criteria, sugars are thus a relevant parameter for characterizing certain monofloral honeys (Cotte et al. 2004a).

In spite of contradictory results, it must be borne in mind that the storage duration may play an essential role.

Cavia et al. (2002) noted that increases in fructose and glucose contents were observed in most honey samples studied (30 samples) after 1 year. Their results were clearly different from those reported in previous papers by Donner (1977) where decreases in monosaccharides below their original values have been described.

It may be concluded that sugar composition is a more reliable indicator for honey classification and authentication only in the case of unifloral honeys with a very high amount of dominating plants; when the fraction of the dominating plant of nectar source reduces, the interpretation of the results of the measurement of saccharides becomes more difficult and almost unusable in determining the floral origin of such honeys (Kaškonien and Venskutonis 2010).

## OTHER COMPOUNDS AS POSSIBLE FLORAL ORIGIN MARKERS AND COMBINED DATA

Honey is a complex matrix, the minerals of which are minority components. The mineral content in floral honeys ranged from 0.020% to 1.028% with an average content of 0.169% (White 1978). Any marked deficiency of soil, rocks, or water in any particular element is reflected in the mineral composition of plants and hence that of nectar and pollen (Hernández et al. 2005). Therefore, the contents of metal ions in honey can help to identify its geographical origin, as this is consistent with the environmental conditions (Przybylowski and Wilczynska 2001). Honey has a rather low mineral content (typically 0.1%–0.2% in floral honey and 1% or higher in mellate honey) that varies widely depending on the particular botanical origin, pedoclimatic conditions, and extraction technique. The dominant element in honey is potassium followed by chlorine, sulfur, sodium, phosphorus, magnesium, silicon, iron, and copper (Ramos et al. 1999). In comparison with nectar honeys, honey dew honeys are higher in minerals, resulting in higher electrolytic conductivity. Potassium is the major metal followed by calcium, magnesium, sodium, sulfur, and phosphorus. Trace elements include iron, copper, zinc, and manganese (Lachman et al. 2007).

Nine metals were determined in 42 Spanish honey samples, which were divided into two categories: Galician and non-Galician honeys. Multivariate chemometric techniques such as cluster analysis, PCA, Bayesian methodology, partial least-squares (PLS) regression, and neural networks were applied to modeling classes on the basis of the chemical data. The results obtained indicated good performance in terms of classification and prediction for both the neural networks and PLS approaches. The metal profiles provided sufficient information to enable classification rules to be developed for identifying honeys according to their geographical origin (Latorre et al. 2000). Golob et al. (2005) detected up to 16 elements in Slovenian honey by total x-ray fluorescence spectroscopy and established statistically significant differences between different types of honey originating from acacia, flowers, lime, chestnut, spruce, fir, and forest honeydew. Different types of honey produced in the Canary Islands were characterized on the basis of their mineral contents. Overall, 10 metals were determined in 116 samples, 81 of which were from the Canaries and 35 from various other places. Iron, copper, zinc, magnesium, calcium, and strontium were determined by atomic absorption spectrophotometry and potassium, sodium, lithium, and rubidium by atomic emission spectrophotometry. A flow injection manifold was employed to analyze those samples requiring dilution. The chemometric processing of the spectroscopic results by various techniques (including PCA, cluster analysis, discriminant analysis, and logistic regression) allowed the accurate classification of the honey samples according to origin (Hernández et al. 2005). Like Slovak and Polish honeys, samples of Czech honeys had higher nickel levels than honeys originating from other parts of the world (Lachman et al. 2005). PCA and cluster analysis revealed that Canary honeys can be characterized in terms of their Na, K, Sr, Mg, Ca, and Cu contents (Hernández et al. 2005).

In fact, the mineral content of honey has been successfully correlated with its origin by using various statistical techniques, including PCA, PLS regression, neural networks, cluster analysis, linear discriminant analysis, the K nearest neighbor, and/or independent modeling of class analogy (Latorre et al. 1999, 2000; Rashed and Soltan 2004). Nevertheless, some risks should be carefully considered when using microelements for the authentication of honey. For instance, some minerals, such as Na and K, are of floral origin because they are accumulated in the plant cells and depend on the content of enzymes, while some metals, particularly Pb and Cd, may be present in honey due to the environmental contamination. Therefore, the concentration of such metals in honey may be dependent on contamination of the environment to a larger extent than to the floral and/or geographic origin of the honey (Kaškonien and Venskutonis 2010).

The electrical conductivity of honey is closely related to the concentration of mineral salts, organic acids, and proteins; it is a parameter that shows great variability according to the floral origin and is considered one of the best parameters for differentiating between honeys with different floral origins (Krauze and Zalewski 1991; Mateo and Bosch-Reig 1998; Terrab et al. 2002). However, it allows for the separation of only some varieties of nectar honeys among all the multifloral nectar and honeydew honeys (Kubisówa and Mastny 1976; Bańkowska-Penar and Pieczonka 1987; Popek 1998).

Přidal and Vorlová (2001) reported that classification of honey must be carried out not only by measuring the conductivity but also in relation to optical rotation and microscopic analysis, namely, in transition intervals of conductivity between the individual honey groups. By combination of element content and electrolytic conductivity, it was possible to divide the samples into two main groups according to their origin, namely, honeydew honeys and nectar honeys. Since the content of elements in honey not only depends on its nature but also reflects the environmental contamination of the locality as well, knowledge of element contents without conductivity does not allow separation of honey samples according to their origin (Lachman et al. 2005). According to Soria et al. (2004), color, electrical conductivity, acidity, ash content, and pH were the physicochemical parameters with higher discrimination power in the differentiation of nectar and honeydew honeys. Senyuva et al. (2009) showed that a combined data set of amino acids, volatiles, saccharides, and water activity measurements allows the floral origin of Turkish honey to be accurately predicted and thus provides a useful tool for authentication purposes.

Physicochemical parameters of 98 samples of Moroccan honeys were analyzed; nine parameters were measured, including water content, pH, acidity (free, lactonic, total, and lactonic acidity/free acidity ratio), hydroxymethylfurfural, diastase activity, and proline. In addition, characterization of the five unifloral honeys (*Eucalyptus* spp., *Citrus* sp., *Lythrum* sp., Apiaceae, and honeydew) by PCA and stepwise discriminant analysis was carried out. PCA showed that the cumulative variance was approximately 62%, and about 82% of the samples were correctly classified by using the stepwise discriminant analysis, with the best results being obtained or the eucalyptus and honeydew honeys (100% correct) (Terrab et al. 2002).

## CONCLUSION

The increasing interest in the therapeutic properties of certain honey varieties contributes to the demand of a reliable determination of their botanical origin. In addition to the different methods cited above, some others are suggested, involving measurements of dozen of parameters showing physical and chemical characteristics of honeys, with measured values statistically assessed with use of various analyses. As seen in this chapter, diverse contradictions exist between authors perhaps due to the differences in geographical origin or the experimental conditions. In other cases, the majority of methods suggested were only enabled to group investigated honeys into groups that, unfortunately, did not strictly correspond to the individual honey variety. As melissopalynology is the only method validated nowadays that suffer from many shortcomings, determining the honey botanical origin still needs a harmonized, implemented, and validated method.

## REFERENCES

Adams, R.J. and Smith, M.V. 1981. Seasonal pollen analysis of nectar from the hive and of extracted honey. *J Apic Res* 20(4):234–248.

Ahmed, A.K., Hoekstra, M.J., Hage, J.J., and Karim, R.B. 2003. Honey-medicated dressing: Transformation of an ancient remedy into modern therapy. *Ann Plas Surg* 50(2):143–147, discussion 147–148.

Al Somai, N., Coley, K.E., Molan, P.C., and Hancock, B.M. 1994. Susceptibility of *Helicobacter pylori* to the antibacterial activity of manuka honey. *J R Soc Med* 87(1):9–12.

Ali, A.T.M.M. 2003. Prevention of Ammonia-induced gastric lesions in rats by natural honey. *J Nutr Environ Med* 13(4):239–246.

Alissandrakis, E., Tarantilis, P.A., Harizanis, P.C., and Polissiou, M. 2005a. Evaluation of four isolation techniques for honey aroma compounds. *J Sci Food Agric* 85:91–97.

Alissandrakis, E., Kibaris, A.C., Tarantilis, P.A., Harizanis, P.C., and Polissiou, M. 2005b. Flavour compounds of Greek cotton honey. *J Sci Food Agric* 85:1444–1452.

Alissandrakis, E., Tarantilis, P.A., Harizanis, P.C., and Polissiou, M. 2007. Aroma investigation of unifloral Greek citrus honey using solid-phase microextraction coupled to gas chromatographic-mass spectrometric analysis. *Food Chem* 100:396–404.

Alissandrakis, E., Tarantilis, P.A., Pappas, C., Harizanis, P.C., and Polissiou, M. 2009. Ultrasound-assisted extraction gas chromatography-mass spectrometry analysis of volatile compounds in unifloral thyme honey from Greece. *Eur Food Res Technol* 229:365–373.

Alissandrakis, E., Tarantilis, P.A., Pappas, C., Harizanis, P.C., and Polissiou, M. 2011. Investigation of organic extractives from unifloral chestnut (*Castanea sativa* L.) and eucalyptus (*Eucalyptus globulus* Labill.) honeys and flowers to identification of botanical marker compounds. *LWT – Food Sci Technol* 44(4):1042–1051.

Al-Khalidi, A., Jawad, F.H., and Tawfiq, N.H. 1980. Effects of bees honey, Zahidi dates and its syrup on blood glucose and serum insulin of diabetics. *Nutr Rep Int* 21:631–643.

Al-Mamary, M., Al-Meeri, A., and Al-Habori, M. 2002. Antioxidant activities and total phenolic of different types of honey. *Nutr Res* 22:1041–1047.

Al-Waili, N.S. 2003. Intrapulmonary administration of natural honey solution, hyperosmolar dextrose or hypoosmolar distill water to normal individuals and to patients with type-2 diabetes mellitus or hypertension: Their effects on blood glucose level, plasma insulin and C-peptide, blood pressure and peaked expiratory flow rate. *Eur J Med Res* 8:295–303.

Al-Waili, N.S. 2004. Natural honey lowers plasma glucose, C-reactive protein, homocysteine, and blood lipids in healthy, diabetic, and hyperlipidemic subjects: Comparison with dextrose and sucrose. *J Med Food* 7:100–107.

Amiot, M.J., Aubert, S., Gonnet, M., and Tacchini, M. 1989. Les composés phénoliques des miels: *Étude préliminaire sur l'identification et la quantification par familles. Apidologie* 20:115–125.

Anapuma, D., Bhat, K.K., and Sapna, V.K. 2003. Sensory and physico-chemical properties of commercial samples of honey. *Food Chem* 83:183–191.

Andrade, P., Ferreres, F., and Amaral, M.T. 1997a. Analysis of honey phenolic acids by HPLC, its application to honey botanical characterization. *J Liq Chromatogr Relat Technol* 20:2281–2288.

Andrade, P., Ferreres, F., Gil, M.I., and Tomás-Barberán, F.A. 1997b. Determination of phenolic compounds in honeys with different floral origin by capillary zone electrophoresis. *Food Chem* 60:79–84.

Anklam, E. 1998. A review of the analytical methods to determine the geographical and the botanical origin of honey. *Food Chem* 63:549–562.

Anthony, S.M., Rieck, J.R., and Dawson, P.L. 2000. Effect of dry honey on oxidation in turkey breast meat. *Poult Sci* 79:1846–1850.

Azim, M.K. and Sajid, M. 2009. Evaluation of nematocidal activity in natural honey. *Pak J Bot* 41(6):3261–3264.

Baltuskevicius, A., Laiskonis, A., Vysniauskiene, D., Ceksteryte, V., and Racys, J. 2001. Use of different kinds of honey for hepatitis A treatment and for reduction of increased acidity of gastric juice. *Zemdirbyste, Mokslo Darbai* 76:173–180.

Bańkowska-Penar, H. and Pieczonka, W. 1987. Przewodnoć elektryczna miodów pszczelich i jej zmiany podczas skladowania. *Przemyst Spozywczy* 3:87–89.

Barral, M.P.G., Ponce-Díaz, M.C., and Venegas-Gallegos, C. 2010. Volatile compounds in honey produced in the central valley of ñuble province, Chile. *Chilean J Agric Res* 70(1):75–84.

Baroni, M.V., Chiabrando, G.A., Costa, C., and Wunderlin, D.A. 2002. Assessment of the floral origin of honey by SDS-page immunoblot techniques. *J Agric Food Chem* 50:1362–1367.

Baroni, M.V., Nores, M.L., Díaz, M. et al. 2006. Determination of volatile organic compound patterns characteristic of five unifloral honey by solidphase microextraction-gas chromatography-mass spectrometry coupled to chemometrics. *J Agric Food Chem* 54:7235–7241.

Bastos, C. and Alves, R. 2003. Compostos voláteis emméis florais. *Quim Nova* 26:90–96.

Batsoulis, A.N., Siatis, N.G., Kimbaris, A.C. et al. 2005. FT-Raman spectroscopic simultaneous determination of fructose and glucose in honey. *J Agric Food Chem* 53:207–210.

Beretta, G., Caneva, E., and Facino, R.M. 2007. Kynurenic acid in honey from arboreal plants: MS and NMR evidence. *Planta Med* 73:1592–1595.

Bertelli, D., Plessi, M., Sabatini, A.G., Lolli, M., and Grillenzoni, F. 2007. Classification of Italian honeys by mid-infrared diffuse reflectance spectroscopy (DRIFTS). *Food Chem* 101:1565–1570.

Bianchi, F., Careri, M., and Musci, M. 2005. Volatile norisoprenoids as markers of botanical origin of Sardinian strawberry-tree (*Arbutus unedo* L.) honey: Characterization of aroma compounds by dynamicheadspace extraction and gas chromatography–mass spectrometry. *Food Chem* 89:527–535.

Bicchi, C., Belliardo, F., and Fratinni, C. 1983. Identification of the volatile components of some piedmontese honeys. *J Apic Res* 22(2):130–136.

Bobis, O., Marghitas, L.A., Bonta, V., Dezmirean, D., and Maghear, O. 2007. Free phenolic acids, flavonoids and abscisic acid related to HPLC sugar profile in acacia honey. *Bull USAMV-CN* 64:179–185.

Bogdanov, S. 1984. Characterisation of antibacterial substances in honey. *Lebensmitt Wiss Technol* 17:74–76.

Bogdanov, S. 2009. Book of Honey, Chapter 5: Honeys Types. *Bee Product Science.* www.bee-hexagon.net.

Bogdanov, S., Ruoff, K., and Oddo, L.P. 2004. Physico-chemical methods of the characterization of unifloral honeys: A review. *Apidologie* 35:S4–S17.

Bogdanov, S. and Baumann, E. 1988. Determination of sugar composition of honeys by HPLC. *Mitt Geb Lebensmittelunters Hyg* 79:198–206.

Bogdanov, S., Jurendic, T., Sieber, R., and Gallmann, P. 2008. Honey for nutrition and health: A review. *J Am Coll Nutr* 27:677–689.

Bogdanov, S. and Martin, P. 2002. Honey authenticity. *Mitt Geb Lebensmittelunters Hyg* 93:232–254.

Bonaga, G., Giumanini, A.G., and Gliozzi, G. 1986. Chemical composition of chestnut honey: Analysis of the hydrocarbon fraction. *J Agric Food Chem* 34:319–326.

Bonvehi, J.S. and Coll, F.V. 2003. Flavour index and aroma profiles of fresh and processed honeys. *J Sci Food Agric* 83:275–282.

Bonvehi, J.S. 1988. Determinación de antranilato de metilo en la miel de citricos (*Citrus* sp.) del Levante Español, y su influencia en la actividad diastasica de la mile. *Alimentaria* 197:37–40.

Bonvehi, J.S., Pajuelo, A.G., and Galindo, J.G. 1987. Composition, phisico-chemical properties and pollen spectrum of some unifloral honeys from Spain. *Alimentaria* 24(185):61–84.

Bosi, G. and Battaglini, M. 1978. Gas chromatographic analysis of free and protein amino acids in some unifloral honeys. *J Apicult Res* 17:152–166.

Bouseta, A. Collin, S., and Doufour, J. 1992. Characteristic aroma profiles of unifloral honeys obtained with a dynamic headspace GC–MS system. *J Apic Res* 32:96–109.

Bouseta, A. and Collin, S. 1995. Optimized Likens–Nickerson methodology for quantifying honey flavors. *J Agric Food Chem* 43:1890–1897.

Bouseta, A., Scheirman, V., and Collin, S. 1996. Flavor and free amino acid composition of lavender and eucalyptus honeys. *J Food Sci* 61(4):683–694.

Bravo, L. 1998. Polyphenols: Chemistry, dietary sources, metabolism and nutritional significance. *Nutr Rev* 56:317–333.

Cabras, P., Angioni, A., Tuberoso, C. et al. 1999. Homogentisic acid: A phenolic acid as a marker of strawberry-tree (*Arbutus unedo*) honey. *J Agric Food Chem* 47:4064–4067.

Cajka, T., Hajslova, J., Pudil, F., and Riddellova, K. 2009. Traceability of honey origin based on volatiles pattern processing by artificial neural networks. *J Chromatogr A* 1216:1458–1462.

Carratù, B., Ciarrocchi, M., Mosca, M., and Sanzini, E. 2011. Free amino acids, oxalate and sulphate for honey characterization. *J ApiProduct ApiMedical Sci* 3(2):81–88.

Castro-Vázquez, L., Díaz-Maroto, M.C., González-Viñas, M.A., De La Fuente, E., and Pérez-Coello, M.S. 2008. Influence of storage conditions on chemical composition and sensory properties of citrus honey. *J Agric Food Chem* 56:1999–2006.

Castro-Vázquez, L., Diaz-Maroto, M.C., González-Vinas, M.A., and Pérez-Coello, M.S. 2009. Differentiation of monofloral citrus, rosemary, eucalyptus, lavender, thyme and heather honeys based on volatile composition and sensory descriptive analysis. *Food Chem* 112:1022–1030.

Castro-Vázquez, L., Diaz-Maroto, M.C., and Pérez-Coello, M.S. 2006. Volatile composition and contribution to the aroma of Spanish honeydew honeys. Identification of a new chemical marker. *J Agric Food Chem* 54:4809–4813.

Castro-Vázquez, L., Diaz-Maroto, M.C., and Pérez-Coello, M.S. 2007. Aroma composition and new chemical markers of Spanish citrus honeys. *Food Chem* 103:601–606.

Catapano, A.L. 1997. Antioxidant effect of flavonoids. *Angiology* 48(1):39–44.

Cavia, M.M., Fernández-Muinöa, M.A., Gömez-Alonsoa, E., Montes-Péreza, M.J., Huidobrob, J.F., and Sanchoa, M.T. 2002. Evolution of fructose and glucose in honey over one year: Influence of induced granulation. *Food Chem* 78:157–161.

Chepulis, L. and Starkey, N. 2008. The long-term effects of feeding honey compared with sucrose and a sugar-free diet on weight gain, lipid profiles, and DEXA measurements in rats. *J Food Sci* 73:H1–H7.

Cherchi, A., Spanedda, L., Tuberoso, C., and Cabras, P. 1994. Solid-phase extraction and high-performance liquid chromatographic determination of organic acids in honey. *J Chromatogr A* 669:59–64.

Chick, H., Shin, S., and Ustunol, Z. 2001. Growth and acid production by lactic acid bacteria and bifidobacteria grown in skim milk containing honey. *J Food Sci* 66:478–481.

*Codex Alimentarius*. 2001. Codex standard 12, Revised Codex Standard for Honey, Standards and Standard Methods 11:1–7.

Committee for the Promotion of Honey and Health 2008. Report to the Officers and Board of Directors. http://www.delta-business.com/.

Conway, P.L., Stern, R., and Tran, L. 2010. The Value-adding Potential of Prebiotic Components of Australian Honey Rural Industries Research and Development Corporation. RIRDC Publication No 09/179. RIRDC Project No. PRJ-000041.

Cotte, J.F., Casabianca, H., Chardon, S., Lheritier, J., and Grenier-Loustalot, M.F. 2004a. Chromatographic analysis of sugars applied to the characterisation of monofloral honey. *Anal Bioanal Chem* 380:698–705.

Cotte, J.F., Casabianca, H., Giroud, B., Albert, M., Lheritier, J., and Grenier-Loustalot, M.F. 2004b. Characterization of honey amino acid profiles using high-pressure liquid chromatography to control authenticity. *Anal Bioanal Chem* 378(5):1342–1350.

Cozzolino, D. and Corbella, E. 2005. The use of visible and near infrared spectroscopy to classify the floral origin of honey samples produced in Uruguay. *J Near Infrared Spectros* 13:63–68.

Cuevas-Glory, L.F., Pino, J.A., Santiago, L.S., and Sauri-Duch, E. 2007. A review of volatile analytical methods for determining the botanical origin of honey. *Food Chem* 103:1032–1043.

D'Arcy, B.R., Rintoul, G.B., Rowland, C.Y., and Blackman, A.J. 1997. Composition of Australian honey extractives. 1. Norisoprenoids, monoterpenes, and other natural volatiles from blue gum (*Eucalyptus leucoxylon*) and yellow box (*Eucalyptus melliodora*) honeys. *J Agric Food Chem* 45:1834–1843.

Da Costa Leite, J.M., Trugo, L.C., Costa, L.S.M., Quinteiro, L.M.C., Barth, O.M., Dutra, V.M.L., and De Maria, C.A.B. 2000. Determination of oligosaccharides in Brazilian honeys of different botanical origin. *Food Chem* 70:93–98.

Davidson, P.M., Sofos, J.N., and Brenem, A.L. 2005. *Antimicrobials in Foods*. 3rd ed. Marcel Dekker Inc., New York, pp. 291–306.

Davies, A.M.C. 1975. Amino acid analysis of honeys from eleven countries. *J Apic Res* 4(1):29–39.

Davies, A.M.C. 1976. The application of amino acid analysis to the determination of the geographical origin of honey. *J Food Technol* 11:515–523.

Davies, A.M.C. and Harris, R.G. 1982. Free amino acid analysis of honeys from England and Wales: Application to the determination of the geographical origin of honeys. *J Apic Res* 21:168–173.

De la Fuente, E., Martinez-Castro, I., and Sanz, J. 2005. Characterization of Spanish unifloral honeys by solid phase microextraction and gas chromatography–mass spectrometry. *J Sep Sci* 28:1093–1100.

De la Fuente, E., Valencia-Barrera, R.M., Martinez-Castro, I., and Sanz, J. 2007a. Occurrence of 2-hydroxy-5-methyl-3-hexanone and 3-hydroxy-5-methyl-2-hexanone as indicators of botanic origin in eucalyptus honeys. *Food Chem* 103:1176–1180.

De la Fuente, E., Sanz, M.L., Martinez-Castro, I., Sanz, J., and Ruiz-Matute, A.I. 2007b. Volatile and carbohydrate composition of rare unifloral honeys from Spain. *Food Chem* 105:84–93.

Deifel, A. 1985. Determination of sugars in honeys by gas chromatography. *Dtsch Lebensm Rundsch* 81:209–212.

Dilber, E., Kalyoncu, M., Yarış, N., and Okten, A. 2002. A Case of mad honey poisoning presenting with convulsion: Intoxication instead of alternative therapy. *Turk J Med Sci* 32:361–362.

Dimitrova, B., Gevrenova, R., and Anklam, E. 2007. Analysis of phenolic acids in honeys of different floral origin by solid-phase extraction and high-performance liquid chromatography. *Phytochem Anal* 18:24–32.

Doğan, A. and Kolankaya, D. 2005. Protective effect of Anzer honey against ethanol-induced increased vascular permeability in the rat stomach. *Exp Toxicol Pathol* 57(2):173–178.

Donarski, J.A., Jones, S.A., and Charlton, A.J. 2008. Application of cryoprobe H nuclear magnetic resonance spectroscopy and multivariate analysis for the verification of Corsican honey. *J Agric Food Chem* 56:5451–5456.

Donarski, J.A., Jones, S.A., Harrison, M., Driffield, M., and Charlton, A.J. 2010. Identification of botanical biomarkers found in Corsican honey. *Food Chem* 118:987–994.

Donner, L. 1977. The sugars of honey—A review. *J Sci Food Agric* 28:443–456.

Dvash, L., Afik, O., Shafir, S. et al. 2002. Determination by near-infrared spectroscopy of perseitol used as a marker for the botanical origin of avocado (*Persea americana* Mill.) honey. *J Agric Food Chem* 50:5283–5287.

English, H.K., Pack, A.R., and Molan, P.C. 2004. The effects of manuka honey on plaque and gingivitis: A pilot study. *J Int Acad Periodontol* 6:63–67.

Eraslan, E., Kanbur, M., Silici, S., and Karabacak, M. 2010. Beneficial effect of pine honey on trichlorfon induced some biochemical alterations in mice. *Ecotoxicol Environ Saf* 73(5):1084–1091.

Erejuwa, O.O., Sulaiman, S.A., Ab Wahab, M.S., Sirajudeen, K.N.S., Salleh, M.S.M., and Gurtu, S. 2001. Glibenclamide or metformin combined with honey improves glycemic control in streptozotocin-induced diabetic rats. *Int J Biol Sc* 7:244–252.

Escriche, I., Visquert, M., Juan-Borras, M., and Fito, P. 2009. Influence of simulated industrial thermal treatments on the volatile fractions of different varieties of honey. *Food Chem* 112:329–338.

Etzold, E. and Lichtenberg-Kraag, B. 2008. Determination of the botanical origin of honey by Fourier-transformed infrared spectroscopy: An approach for routine analysis. *Eur Food Res Technol* 227:579–586.

EU. Council Directive 2001/110/CE relating to honey. *Off J Eur Commun* 2002(L10):47–52.

Ezz El-Arab, A.M., Girgis, S.M., Hegazy, E.M., and Abd El-Khalek, A.B. 2006. Effect of dietary honey on intestinal microflora and toxicity of mycotoxins in mice. *BMC Complement Altern Med* 6:6.

Fahey, J.W. and Stephenson, K.K. 2002. Pinostrobin from honey and Thai ginger (*Boesenbergia pandurata*): A potent flavonoid inducer of mammalian phase 2 chemoprotective and anti-oxidant enzymes. *J Agric Food Chem* 50:7472–7476.

Ferreres, F., Tómas-Barberán, F.A., Gil, M.I., and Tomás-Lorente, F. 1991. An HPLC technique for flavonoid analysis in honey. *J Sci Food Agric* 56:49–56.

Ferreres, F., Ortiz, A., Silva, C., Garciá-Viguera, C., Tomás-Barberán, F.A., and Tomás-Lorente, F. 1992. Flavonoids of "La Alcarria" honey—A study of their botanical origin. *Z Lebensm Unters—For A* 194:139–143.

Ferreres, F., Garciá-Viguera, C., Tomás-Lorente, F., and Tomás-Barberán, F.A. 1993. Hesperetin: A marker of the floral origin of citrus honey. *J Sci Food Agric* 61:121–123.

Ferreres, F., Tomás-Barberáan, F.A., Soler, C., Garcia-Viguera, C., Ortiz, A., and Tomás-Lorente, F. 1994a. A simple extractive technique for honey flavonoid HPLC analysis. *Apidologie* 25:21–30.

Ferreres, F., Giner, J.M., and Tomás-Barberán, F.A. 1994b. A comparative study of hesperetin and methyl anthranilate as markers of the floral origin of citrus honey. *J Sci Food Agric* 65:371–372.

Ferreres, F., Andrade, P., and Tomás-Barberán, F.A. 1994c. Flavonoids from Portuguese heather honey. *Z Lebensm Unters—For A* 199:32–37.

Ferreres, F., Andrade, P., and Tomás-Barberán, F.A. 1996. Natural occurrence of abscisic acid in heather honey and floral nectar. *J Agric Food Chem* 44:2053–2056.

Forcone, A., Bravo, O., and Ayestarán, M.G. 2003. Intraannual variations in the pollinic spectrum of honey from the lower valley of the River Chubut (Patagonia, Argentina). *Span J Agric Res* 1(2):29–36.

Fossel, A. 1968. Pollenersatzmittel im mikroskopischen Befund von Frohtrachthonigen. *Z Bienenfor* 9(5):206–211.

Fukuda, M., Kobayashi, K., Hirono, Y. et al. 2009. Jungle honey enhances immune function and antitumor. *Evid Based Complement Alternat Med* 2011:1–7.

Ghashm, A.A., Othman, N.H., Khattak, M.N., Ismail, N.M., and Saini, R. 2010. Antiproliferative effect of Tualang honey on oral squamous cell carcinoma and osteosarcoma cell lines. *BMC Complement Altern Med* 10:49.

Gheldof, N., Wang, X.H., and Engeseth, N.J. 2002. Identification and quantification of antioxidant components of honeys from various floral sources. *J Agric Food Chem* 50:5870–5877.

Gheldof, N., Wang, X.H., and Engeseth, N.J. 2003. Buckwheat honey increases serum antioxidant capacity in humans. *J Agric Food Chem* 51:1500–1505.

Gil, M.I., Ferreres, F., Ortiz, A., Subra, E., and Tomás-Barberán, F.A. 1995. Plant phenolic metabolites and floral origin of rosemary honey. *J Agric Food Chem* 43:2833–2838.

Gilbert, J., Shepherd, M.J., Wallwork, M.A., and Harris, R.G. 1981. Determination of the geographical origin of honeys by multivariate analysis of gas chromatographic data on their free amino acid content. *J Apicult Res* 20:125–135.

Golob, T., Doberšek, U., Kump, P., and Nečěmer, M. 2005. Determination of trace and minor elements in Slovenian Honey by total reflection X-ray fluorescence spectroscopy. *Food Chem* 91:593–600.

Gómez-Caravaca, A.M., Gómez-Romero, M., Arráez-Román, D., Segura-Carretero, A., and Fernández-Gutiérrez, A. 2006. Advances in the analysis of phenolic compounds in products derived from bees. *J Pharm Biomed Anal* 41:1220–1234.

Goodall, I., Dennis, M.J., Parker, I., and Sharman, M. 1995. Contribution of high-performance liquid chromatographic analysis of carbohydrates to authenticity testing of honey. *J Chromatogr A* 706:353–359.

Grosch, W. 2001. Evaluation of the key odorants of foods by dilution experiments, aroma models and omission. *Chem Senses* 26:533–545.

Guidotti, M. and Vitali, M. 1998. Identification of volatile organic compounds present in different honeys through SPME and GC/MS. *Ind Aliment* 37:351–356.

Guyot, C., Bouseta, A., Scheiman, V., and Collins, S. 1998. Floral origin markers of chestnut and lime honeys. *J Agric Food Chem.* 46:625–633.

Guyot, C., Scheirman, V., and Collin, S. 1999. Floral origin markers of heather honeys: *Calluna vulgaris* and *Erica arborea*. *Food Chem* 64:3–11.

Guyot-Declerck, C., Renson, S., Bouseta, A., and Collin, S. 2002. Floral quality and discrimination of *Lavandula stoechas, Lavandula angustifolia*, and *Lavandula angustifolia* × *latifolia* honeys. *Food Chem* 79:453–459.

Hadjmohammadi, M.R., Nazari, S., and Kamel, K. 2009. Determination of flavonoid markers in honey with SPE and LC using experimental design. *Chromatographia* 69:1291–1297.

Häusler, M. and Montag, A. 1989. Isolation, identification and quantitative determination of the norisoprenoid (S)-(+)-dehydrovomifoliol in honey. *Z Lebensm Unters Forsch* 189:113–114.

Häusler, M. and Montag, A. 1991. Minorbestandteile des Honigs mit Aroma-Relevanz. IV. Vorkommen und trachtspezifische Verteilung des Aromastoffprekursors (*S*)-dehydrovomifoliol. (Aroma-relevant minor components of honey. IV. Occurrence and flower-specific distribution of the aroma precursor (*S*)-dehydrovomifoliol). *Deut Lebensm-Rundsch* 87:35–36.

Hermosin, I., Chicon, R.M., and Dolores Cabezudo, M. 2003. Free amino acid composition and botanical origin of honey. *Food Chem* 83:263–268.

Hernández, O.M., Fraga, J.M.G., Jiménez, A.I., Jiménez, F., and Arias, J.J. 2005. Characterization of honey from the Canary Islands: Determination of the mineral content by atomic absorption spectrophotometry. *Food Chem* 93:449–458.

Horváth, K. and Molmir-Perl, I. 1997. Simultaneous quantitation of mono-, di- and trisaccharides by GC-MS of their TMS ether oxime derivatives: II. In honey. *Chromatographia* 45:328–335.

Hussein, A.A., Rakha, M.K., and Nabil, Z.I. 2003. Anti-arrhythmic effect of wild honey against catecholamines cardiotoxicity. *J Med Sci* 3(2):127–136.

Iglesias, M.T., de Lorenzo, C., Polo, M.D., Martin-Álvarez, P.J., and Pueyo, E. 2004. Usefulness of amino acid composition to discriminate between honeydew and floral honeys. Application to honeys from a small geographic area. *J Agric Food Chem* 52:84–89.

Jawad, F.H., Al-Khalidi, A., and Tawfiq, N.H. 1981. Effects of bees honey, zahdi date and its syrup on blood glucose and serum insulin of normal subjects. *J Fac Med Baghdad* 23:169–180.

Jan Mei, S., Mohd Nordin, M., and Norrakiah, A.S. 2010. Fructooligosaccharides in honey and effects of honey on growth of *Bifidobacterium longum* BB 536. *Int Food Res J* 17:55–561.

Jerković, J., Mastelić, J., and Marijanović, Z. 2006. A variety of volatile compounds as markers in unifloral honey from Dalmatian sage (*Salvia officinalis* L.). *Chem Biodivers* 3:1307–1316.

Jerković, I., Mastelić, J., Marijanović, Z., Klein, Ž., and Jelić, M. 2007. Comparison of hydrodistillation and ultrasonic solvent extraction for the isolation of volatile compounds from two unifloral honeys of *Robinia pseudoacacia* L. and *Castanea sativa* L. *Ultrason Sonochem* 14:750–756.

Jerković, I., Tuberoso, C.I.G., Marijanović, Z., Jelić, M., and Kasum, A. 2009a. Headspace, volatile and semi-volatile patterns of *Paliurus spina-christi* unifloral honey as markers of botanical origin. *Food Chem* 112:239–245.

Jerković, I., Marijanović, Z., Kezić, J., and Gigić, M. 2009b. Headspace, volatile and semi-volatile organic compounds diversity and radical scavenging activity ultrasonic solvent extracts from *Amorpha fruticosa* honey samples. *Molecules* 14:2717–2728.

Jerković, I., Marijanović, Z., Malenica-Staver, M., and Lušić, D. 2010a. Volatiles from a rare Acer spp. honey sample from Croatia. *Molecules* 15(7):4572–4582.

Jerković, L., Hegić, G., Marijanović, Z., and Bubalo, D. 2010b. Organic extractives from *Mentha* spp. honey and the bee-stomach: Methyl syringate, vomifoliol, terpenediol I, hotrienol and other compounds. *Molecules* 15(4):2911–2924.

Kajiwara, S., Ganghi, H., and Ustulon, Z. 2002. Effect of honey on the growth of and acid production by human intestinal Bifidobacterium spp.: An *in vitro* comparison with commercial oligosaccharides and inulin. *J Food Prot* 65(1):214–218.

Kanematsu, H., Aoyama, M., Maruyama, T., and Niiya, L. 1982. Aminoacids analysis of honey with different geographical and botanical origin. *Jpn Soc Food Nutrition* 35:297–303.

Karabourniot, S. 2000. Detection of flora spectrum through honey microscopic analysis. *Apiacta* 2.

Karl-Gustav Bergner, K.G. and Hahn, H. 1972. Zum vorkommen und zur herkunft der freien aminosäuren in honig. *Apidologie* 3(1):5–34.

Kassim, M., Achoui, M., Mustafa, M.R., Mohd, M.A., and Yusoff, K.M. 2010. Ellagic acid, phenolic acids, and flavonoids in Malaysian honey extracts demonstrate *in vitro* anti-inflammatory activity. *Nut Res* 30(9):650–659

Kaškonien, V. and Venskutonis, P.R. 2010. Floral markers in honey of various botanical origins: A review. *Compr Rev Food Sci F* 9:620–634.

Kenjerić, D., Manić, M.L., Primorac, L., and Cacic, F. 2008. Flavonoid pattern of sage (*Salvia officinalis* L.) unifloral honey. *Food Chem* 110:187–192.

Kenjerić, D., Manić, M., Primorac, L., Bubalo, D., and Perl, A. 2007. Flavonoid profile of Robinia honeys produced in Croatia. *Food Chem* 102:683–690.

Khoo, Y.-T., Halim, A.S., Singh, K.-K.B., and Noor-Ayunie Mohamad, N.-A. 2010. Wound contraction effects and antibacterial properties of Tualang honey on full-thickness burn wounds in rats in comparison to hydrofibre. *J Complement Altern Med* 10:48.

Kilicoglu, B., Kismet, K., Kilicoglu, S.S., Erel, E., Gencay, O., Sorkun, K., Erdemli, E., Akhan, O., Akkus, M.A., and Sayek, I. 2008. Effects of honey as a scolicidal agent on the hepatobiliary system. *World J Gastroenterol* 14(13):2085–2088.

Kleerebezem, M. and Vaughan, E.E. 2009. Probiotic and gut lactobacilli and Bifidobacteria: Molecular approaches to study diversity and activity. *Annu Rev Microbiol* 63:269–290.

Korkmaz, A. and Kolankaya, D. 2009. Anzer honey prevents N-ethylmaleimide-induced liver damage in rats. *Exp Toxicol Pathol* 61:333–337.

Krauze, A. and Zalewski, R.L. 1991. Classification of honeys by principal component analysis on the basis of chemical and physical parameters. *Z Lebensm Unters Forsch* 192:19–23.

Kubisówa, S. and Mastny, V. 1976. Srounâni dvou metod diferencjujicich nektarove a medovicove medy. *Vedecke Prace Vyzkumneho Ustavu Vcelarskeho v Dole u Lbcic* 7:87–92.

Lachman, J., Kolihová, D., Miholová, D., Košăta, J., Titěra, D., and Kult, K. 2007. Analysis of minority honey components: Possible use for the evaluation of honey quality. *Food Chem* 101:973–979.

Ladas, S.D., Haritos, D.N., and Raptis, S.A. 1995. Honey may have a laxative effect on normal subjects because of incomplete fructose. *Am J Clin Nutr* 62:1212–1215.

Latorre, M.J., Pená, R., Pita, C., Botana, A., Garciá, S., and Herrero, C. 1999. Chemometric classification of honeys according to their type. II. Metal content data. *Food Chem* 66:263–268.

Latorre, M.J., Pená, R., García, S., and Herrero, C. 2000. Authentication of Galician (N.W. Spain) honeys by multivariate techniques based on metal content data. *Analyst* 125:307–312.

Liang, Y., Cao, W., Chen, W.-J., Xiao, X.-H., and Zheng, J.-B. 2009. Simultaneous determination of four phenolic components in citrus honey by high performance liquid chromatography using electrochemical detection. *Food Chem* 114:1537–1541.

Liu, S.M., Manson, J.E., Stampfer, M.J. et al. 2001. Dietary glycemic load assessed by food-frequency questionnaire in relation to plasma high-density-lipoprotein cholesterol and fasting plasma triacylglycerols in postmenopausal women. *Am J Clin Nutr* 73:560–566.

Louveaux, J., Maurizio, A., and Vorwohl, G. 1970. Methods of melissopalynology. *Bee World* 51:125–131.

Louveaux, J., Maurizio, A., and Vorwohl, G. 1978. Methods of melissopalynology. *Bee World* 59(4):139–157.

Lule, S.U. and Xia, W. 2005. Food phenolics, pros and cons: A review. *Food Rev Int* 21:367–388.

Lušić, D., Koprivnjak, O., Ćurić, D., Sabatini, A.G., and Conte, L.S. 2007. Volatile profile of Croatian lime tree (*Tilia* sp.), fir honeydew (*Abies alba*) and sage (*Salvia officinalis*) honey. *Food Technol Biotechnol* 45:156–165.

Mademtzoglou, D., Haza, A.I., Coto, A.L., and Morales, P. 2010. Rosemary, heather and heterofloral honeys protect towards cytotoxicity of acrylamide in human hepatoma cells. *Rev Complut Cienc Vet* 4(2):12–32.

Mahaneen, M., Sulaiman, S.A., Jaafar, H., Sirajudeen, K.N.S., Ismail, Z.I.M., and Islam, M.N. 2010. Effect of honey on testicular functions in rats exposed to cigarette smoke. *J ApiProduct ApiMedical Sci* 3(1):12–17.

Mannaş, D. and Altuğ, T. 2007. SPME/GC/MS and sensory flavour profile analysis for estimation of authenticity of thyme honey. *Int J Food Sci Technol* 42:133–138.

Marghitas, L.A., Dezmirean, D.S., Poco, C.B., Ilea, M., Bobis, O., and Gergen, I. 2010. The development of a biochemical profile of acacia honey by identifying biochemical determinants of its quality. *No Bo Hor Agrobo Cluj* 3(2):S84–S90.

Martos, I., Ferreres, F., and Tomás-Barberán, F.A. 2000a. Identification of flavonoid markers for the botanical origin of Eucalyptus honey. *J Agric Food Chem* 48:1498–1502.

Martos, I., Ferreres, F., Yao, L.H., D'Arcy, B.R., Caffin, N., and Tomas-Barberan, F.A. 2000b. Flavonoids in monispecificseucalyptus honeys from Australia. *Agri Food Chem* 48(10):474–4748.

Mateo, R. and Bosch-Reig, F. 1998. Classification of Spanish unifloral honeys by discriminant analysis of electrical conductivity, color, water content, sugar and pH. *J Agric Food Chem* 46:39–400.

Mateo, R. and Bosch-Reig, F. 1997. Sugar profiles of Spanish unifloral honeys. *Food Chem* 60:3–41.

Maurizio, A. 1975. Microscopy of honey. 1st ed. Crane, E. Ed. *Honey a Comprehensive Survey*. Heinemann, London, 25.

McGovern, D.P.B., Abbas, S.Z., Vivian, G., and Dalton, H.R. 1999. Manuka honey against *Helicobacter pylori*. *J R Soc Med* 83:127.

Medhi, B., Prakash, A., Avti, P.K., Saikia, U.N., Pandhi, P., and Khanduja, K.L. 2008. Effect of Manuka honey and sulfasalazine in combination to promote antioxidant defense system in experimentally induced ulcerative colitis model in rats. *Indian J Exp Biol* 48:58–590.

Medical journals buzzing over honey for wound care. Media release November 5, 2007. http://www medihoney.com/. Retrieved September 5, 2011.

Merken, H.M. and Beecher, G.R. 2000. Measurement of food flavonoids by high formance liquid chromatography: A review. *J Agr Food Chem* 4(3):57–599.

Michalkiewicz, A., Biesaga, M., and Pyrzyńska, K. 2008. Solid-phase extraction procedure for determination of phenolic acids and some flavonoids in honey. *J Chromatogr A* 1187:18–24.

Miyagawa, M., Fukuda, M., Hirono, Y. et al. 2010. Effect of Jungle honey on the chemotactic activity. *J ApiProduct ApiMedical Sci* (4):14–154.

Molan, P. 1998. The limitations of the methods of identifying the floral source of honeys. *Bee World* 79(2):59–68.

Montenegro, G., Gómez, M., Casaubon, G., Belancic, A., Mujica, A.M., and Peña, R.C. 2009. Analysis of volatile compounds in three unifloral native Chilean honeys. *Phyton* 78(1):61–65.

Moreira, R.F.A. and De Maria, C.A.B. 2005. Investigation of the aroma compounds from headspace and aqueous solution from the cambará (*Gochnatia velutina*) honey. *Flavour Fragr J* 20:13–17.

Moreira, R.F.A., Trugo, L.C., Pietroluongo, M., and De Maria, C.A.B. 2002. Flavor composition of cashew (*Anacardium occidentale*) and marmeleiro (*Croton* species) honeys. *J Agric Food Chem* 50:7616–7621.

Moore, O.A., Smith, L.A., Campbell, F., Seers, K., McQuay, H.J., and Moore, R.A. 2001. Systematic review of the use of honey as a wound dressing. *Compl Alternat Med* 1:2.

Motallebnejad, M., Akram, S., Moghadamnia, A., Moulana, Z., and Omidi, S. 2008. The effect of topical application of pure honey on radiation-induced mucositis: A randomized clinical trial. *J Contemp Dent Pract* 9:040–047.

Münstedt, K., Sheybani, B., Hauenschild, A., Brüggmann, D., Bretzel, R.G., and Winter, D. 2008. Effects of basswood honey, honey-comparable glucose-fructose solution, and oral glucose tolerance test solution on serum insulin, glucose, and C-peptide concentrations in healthy subjects. *J Med Food* 11(3):424–428.

Munuera, M. and Carrión, G.J.S. 1994. Pollen analysis of Citrus honeys of the Segura basin (Alicante and Murcia). *Alimentaria* 31:37–42

Mwipatayi, B.P., Angel, D., Norrish, J., Hamilton, M.J., Scott, A., and Sieunarine, K. 2004. The use of honey in chronic leg ulcers: A literature review. *Prim Int* 12(3):107–108, 110–112.

Nasuti, C., Gabbianelli, R., Falcioni, G., and Cantalamessa, F. 2006. Antioxidative and gastro-protective activities of anti-inflammatory formulations derived from chestnut honey in rats. *Nutr Res* 26:13–137.

National Honey Board (NHB). 1996. *Honey Information Kit of the Food and Beverage Industries*. Available from NHB, Longmont, CO.

Nisbet, C., Guler, A., Ciftci, G., and Yarim, G.F. 2009. The investigation of protein prophile of different botanical origin honey and density saccharose-adultered honey by SDS-Page Method. *Kafkas Univ Vet Fak Derg* 15(3):443–446.

Nisbet, H.O., Nisbet, C., Yarim, M., Guler, A., and Ozak, A. 2010. Effects of three types of honey on cutaneous wound healing. *Wounds* 22(11):275–283.

Nozal, M.J., Bernal, J.L., Toribio, L., Jimenez, J.J., and Martin, M.T. 2001. High-performance liquid chromatographic determination of methyl anthranilate, hydroxymethylfurfural and related compounds in honey. *J Chromatogr A* 917:95–103.

Nozal, M.J., Bernal, J.L., Toribio, L., Alamo, M., Diego, J.C., and Tapia, J. 2005. The use of carbohydrate profiles and chemometrics in the characterization of natural honeys of identical geographical origin. *J Agric Food Chem* 53:3095–3100.

Odeh, I., Abu-Lafi, S., Dewik, H., Al-Najjar, I., Imam, A., Dembitsky, V.M., and Hanuš, L.O. 2007. A variety of volatile compounds as markers in Palestinian honey from *Thymus capitatus, Thymelaea hirsuta,* and *Tolpis virgata. Food Chem* 101:1393–1397.

Öztaşan, N., Altinkaynak, K., Akçay, F., Göçer, F., and Dane, Ş. 2005. Effects of mad honey on blood glucose and lipid levels in rats with streptozocin-induced diabetes. *Turk J Vet Anim Sci* 29:109–1096.

Oduwole, O., Meremikwu, M.M., Oyo-Ita, A., and Udoh, E.E. 2010. Honey for acute cough in children. *Cochrane Database Syst Rev* 1: CD007094, doi: 10.1002/14651858. CD007094.pub3.

Onyesom, I. 2004. Effect of Nigerian citrus (*Citrus sinensis* Osbeck) honey on ethanol metabolism. *S Afr Med J* 94:98–986.

Orhan, F., Sekerel, B.E., Kocabas, C.N., Sackesen, C., Adalioglu, G., and Tuncer, A. 2003. Complementary and alternative medicine in children with asthma. *Ann Allerg Asthma Immunol* 9(6):611–615.

Osato, M.S., Reddy, S.G., and Graham, D.Y. 1999. Osmotic effect of honey on growth and viability of *Helicobacter pyloris. Dig Dis Sci* 44(3):462–464.

Overton, S.V. and Manura, J.J. 1999. Flavor and aroma in natural bee honey. *Scientific Instrument Services, Inc.* Application Note 25.

Ozlugedik, S., Genc, S., Unal, A., Elhan, A.H., Tezer, M., and Titiz, A. 2006. Can postoperative pains following tonsillectomy be relieved by honey? *Int J Pediatr Otorhinolaryngol* 70(11):1929–1934.

Palmer, J.K. and Brandes, W.B. 1974. Determination of sucrose, glucose, and fructose by liquid chromatography. *J Agric Food Chem* 22(4):709–712.

Paradkar, M.M. and Irudayaraj, J. 2002. Discrimination and classification of beet and cane inverts in honey by FT-Raman spectroscopy. *Food Chem* 76:231–239.

Paul, I.M., Beiler, J., Mc Monagle, A., Shaffer, M.L., Duda, L., and Berlin, C.M. Jr. 2007. Effect of honey, dextromethorphan, and no treatment on nocturnal cough and sleep quality for coughing children and their parents. *Arch Pediatr Adolesc Med* 161(12):114–1114.

Peretti, A., Carbini, L., Dazzi, E., Pittau, L., Spanu, P., and Manai, M. 1994. Uso razionale del mieleenell'alimentazione dei diabetici. *Clin Dietolog* 21:1–21.

Pérez, E., Rodríguez-Malaver, A.J., and Vit, P. 2006. Antioxidant capacity of Venezuelan honey in wistar rat homogenates. *J Med Food* 9(4):510–516.

Pérez, R.A., Sànchez-Brunete, C., Calvo, R.M., and Tadeo, J.L. 2002. Analysis of volatiles from spanish honeys by solid-phase microextraction and gas chromatography-mass spectrometry. *J Agric Food Chem* 50(9):2633–2637.

Persano Oddo, L. and Piro, R. 2004. Main European unifloral honeys: Descriptive sheet. *Apidologie* 35:38–51.

Persano Oddo, L., Piazza, M.G., Sabatini, A.G., and Accorti, M. 1995b. Characterization of unifloral honey. *Apidologie* 26:453–465.

Piasenzotto, L., Gracco, L., and Conte, L. 2003. Solid-phase microextraction (SPME) applied to honey quality control. *J Sci Food Agric* 83:1037–1044.

Pirini, A., Conte, L.S., Francioso, O., and Lercker, G. 1992. Capillary gas chromatographic determination of free amino acids in honey as a means of determination between different botanical sources. *J High Resolut Chromatogr* 15:165–170.

Pi-Sunyer, F.X. 2002. Glycemic index and disease. *Am J Clin Nutr* 76:290S–298S.

Plutowska, B., Chmiel, T., Dymerski, T., and Wardencki, W. 2011. A headspace solid-phase microextraction method development and its application in the determination of volatiles in honeys by gas chromatography. *Food Chem* 126:1288–1298.

Popek, S. 1998. Electrical conductivity as an indicator of the quality of nectar honeys. *Forum War* 1–4:75–79.

Přidal, A. and Vorlová, L. 2001. Honey and its physical parameters. *Czech Ani Sci* 4(10): 439–444.

Przybylowski, P. and Wilczynska, A. 2001. Honey as an environmental marker. *Food Chem* 74:289–291.

Pulcini, P., Allegrini, F., and Festuccia, N. 2006. Fast SPE extraction and LC-ESI-MS-MS analysis of flavonoids and phenolic acids in honey. *Apiacta* 41:21–27.

Pyrzynska, K. and Biesaga, M. 2009. Analysis of phenolic acids and flavonoids in honey. *TrAC, Trend Anal Chem* 28:893–902.

Radovic, B.S., Careri, M., Mangia, A., Musci, M., Gerboles, M., and Anklam, E. 2001. Contribution of dynamic headspace GC–MS analysis of aroma compounds to authenticity testing of honey. *Food Chem* 72:511–520.

Rady, H.M. and Yahya, S.M.M. 2011. Enhancement of the antitumor effect of honey and some of its extracts using adiponectin hormone. *Aust J Basic Appl Sci* 5(6):100–108.

Rakha, M.K., Nabil, Z.I., and Hussein, A.A. 2008. Cardioactive and vasoactive effects of natural wild honey against cardiac malperformance induced by hyperadrenergic activity. *J Med Food* 11(1):9–98.

Ramos, L.S., Pérez, M., and Ferreras, G. 1999. Aplicación de nuevas tecnologi ás en mieles canarias para sutipificación y control de calidad. Servicio de Publicaciones de la Caja General de Ahorros de Canarias, Tenerife.

Rashed, M.N. and Soltan, M.E. 2004. Major and trace elements in different types of Egyptian mono-floral and non-floral bee honeys. *J Food Comp Anal* 17:725–735.

Rezić, I., Horvat, A.J.M., Babić, S., and Kastelan-Macan, M. 2005. Determination of pesticides in honey by ultrasonic solvent extraction and thin-layer chromatography. *Ultrason Sonochem* 12:477–481.

Rice-Evans, C.A., Miller, N.J., and Paganga, G. 1996. Structure-antioxidant activity relationships of flavonoids and phenolic acids. *Free Radic Biol Med* 20(7):933–956.

Rothe, M., and Thomas, B. 1963. Aroma of bread. Evaluation of chemical taste analyses with the aid of threshold value. *Z Lebensm Unters Forsch* 119:302–310.

Rowland, Y.C., Blackman, A.J., D'Arcy, B.R., and Rintoul, G.B. 1995. Comparison of organic extractives found in leatherwood (*Eucryphia lucida*) honey and leatherwood flowers and leaves. *J Agric Food Chem* 43:753–763.

Rozaini, M.Z., Zuki, A.B.Z., Noordin, M., Norimah, Y., and Hakim, N. 2004. The effects of different types of honey on tensile strength evaluation of burn wound tissue healing. *Intern J Appl Res Vet Med* 2(4):290–296.

Ruiz-Matute, A.I., Brokl, M., Soria, A.C., Sanz, M.L., and Martinez-Castro, I. 2010. Gas chromatographic–mass spectrometric characterisation of tri- and tetrasaccharides in honey. *Food Chem* 120:637–642.

Ruoff, K. 2006. Authentication of the botanical origin of honey (Thesis). Diss. ETH No. 16857.

Sabatier, S., Amiot, M.J., Tacchini, M., and Aubert, S. 1992. Identification of flavonoids in sunflower honey. *J Food Sci* 57:773–774.

Sabatini, A.G. and Grillenzoni, F.V. 2002. Contenuto in prolina in differenti mieli uniflorali. Atti del convegno finale del Progetto Finalizzato AMA "Il ruolo della ricerca in apicoltura", pp. 260–261.

Sabatini, A.G., Persano Oddo, L., Piazza, M., Accorti, M., and Marcazzan, G. 1990. Glucide spectrum in the main Italian unifloral honeys. II. Di- and Trisaccharides. *Apicoltura* 6:63–70.

Sanchez, M.D., Huidobro, J.F., Mato, I., Muniategui, S., and Sancho, M.T. 2001. Correlation between proline content of honeys and botanical origin. *Dtsch Lebensm-Rundsch* 97(5):171–175.

Sanz, M.L., Sanz, J., and Martinez-Castro, I. 2004. Gas chromatographic–mass spectrometric method for the qualitative and quantitative determination of disaccharides and trisaccharides in honey. *J Chromatogr A* 1059:143–148.

Sanz, M.L., Polemis, N., Morales, V. et al. 2005. *In vitro* investigation into the potential prebiotic activity of honey oligosaccharides. *J Agric Food Chem* 53:2914–2921.

Schramm, D.D., Karim, M., Schrader, H.R., Holt, R.R., Cardetti, M., and Keen, C.L. 2003. Honey with high levels of antioxidants can provide protection to healthy human subjects. *J Agric Food Chem* 51:1732–1735.

Seif Eldin, A.R.M. and Elfadil, E.B. 2010. Identification of the floral origin of honey by amino acids composition. *Aust J Basic Appl Sci* 4(4):552–556.

Senanayake, M.J. 2006. *A Chemical Investigation of New Zealand Unifloral Honeys*. Thesis. Chapter 3: Extractable Organic Substances from NZ Kamahi Honey. University of Waikato, New Zealand, 90.

Senyuva, H.Z., Gilbert, J., Silici, S. et al. 2009. Profiling Turkish honeys to determine authenticity using physical and chemical characteristics. *J Agric Food Chem* 57:3911–3919.

Serra, J. and Ventura, F. 2003. Flavour index and aroma profiles of fresh and processed honeys. *J Sci Food Agric* 83:275–282.

Serrano, S., Villarejo, M., Espejo, R., and Jodral, M. 2004. Chemical and physical parameters of Andalusian honey: Classification of Citrus and Eucalyptus honeys by discriminant analysis. *Food Chem* 87:619–625.

Sesta, G., Piana, M.L., Persano Oddo, L., Lusco, L., and Paola Belligoli, P. 2008. Methyl anthranilate in *Citrus* honey. Analytical method and suitability as a chemical marker. *Apidologie* 39:334–342.

Shi, H., Noguchi, N., and Niki, E. 2001. Antioxidants in food. Pokorny, J., Yanishlieva, N., and Gordon, M. (Eds.), *Practical Applications*. Woodhead Publishing Ltd, Cambridge, U.K., p. 149.

Shi, H., Noguchi, N., and Niki, E. 2001. Introducing natural antioxidants, Part 3 Natural antioxidants in antioxidants in food. Pokorny, J., Yanishlieva, N., and Gordon, M. (Eds.), *Practical Applications*. CRC Press LLC, Boca Raton, FL.

Shimoda, M., Wu, Y., and Osajima, Y. 1996. Aroma compounds from aqueous solution of haze (*Rhus succedanea*) honey determined by adsorptive column chromatography. *J Agric Food Chem* 44:3913–3918.

Shin, H.S. and Ustunol, Z. 2005. Carbohydrate composition of honey from different floral sources and their influence on growth of selected intestinal bacteria: An *in vitro* comparison. *Food Res Int* 38:721–728.

Sulaiman, S.A., Hasan, H., Deris, Z.Z. et al. 2011. The benefit of Tualang honey in reducing acute respiratory symptoms among Malaysian Hajj pilgrims: A preliminary study. *J ApiProduct ApiMedical Sci* 3(1):38–44.

Soler, C., Gil, M.I., Garcia-Viguera, C., and Tomás-Barberán, F.A. 1995. Flavonoid patterns of French honeys with different floral origin. *Apidologie* 26:53–60.

Soria, A.C., González, M., de Lorenzo, C., Martinez-Castro, I., and Sanz, J. 2004. Characterization of artisanal honeys from Madrid (Central Spain) on the basis of their melissopalynological, physicochemical and volatile composition data. *Food Chem* 85:121–130.

Soria, A.C., Martínez-Castro, I., and Sanz, J. 2008. Some aspects of dynamic headspace analysis of volatile components in honey. *Food Res Int* 41:838–848.

Soria, A.C., Sanz, J., and Martínez-Castro, I. 2009. SPME followed by GC–MS: A powerful technique for qualitative analysis of honey volatiles. *Eur Food Res Technol* 228:579–590.

Soria, A.C., Martínez-Castro, I., and Sanz, J. 2003. Analysis of volatile composition of honey by solid phase microextraction and gas chromatography-mass spectrometry. *J Sep Sci* 26:793–801.

Stephens, J.M., Schlothauer, R.C., Morris, B.D. et al. 2010. Phenolic compounds and methylglyoxal in some New Zealand manuka and kanuka honeys. *Food Chem* 120:78–86.

Swallow, K.W. and Low, N.H. 1990. Analysis and quantitation of the carbohydrates in honey using high performance liquid chromatography. *J Agric Food Chem* 38:1828–1832.

Taguri, T., Tanaka, T., and Kouno, I. 2006. Antibacterial spectrum of plant polyphenols and extracts depending upon hydroxyphenyl structure. *Bio Phar Bull* 29:222–2235.

Tan, S.T., Wilkins, L.A., Holland, P.T., and McGhie, T.K. 1989. Extractives from New Zealand unifloral honeys. 2. Degraded carotenoids and other substances from heather honey. *Agri Food Chem* 3(5):1217–1221.

Tananaki, C.H., Thrasyvoulou, A., Giraudel, J.L., and Montury, M. 2007. Determination of volatile characteristics of Greek and Turkish pine honey samples and their classification by using Kohonen self organising maps. *Food Chem* 101:1687–1693.

Terrab, A., Recamales, A.F., Gonzalez-Miret, M.L., and Heredia, F.J. 2005. Contribution to the study of avocado honeys by their mineral contents using inductively coupled plasma optical emission spectrometry. *Food Chem* 92:305–309.

Terrab, A., Diez, M.J., and Heredia, F.J. 2002. Characterisation of Moroccan unifloral honeys by their physicochemical characteristics. *Food Chem* 79:373–379.

Thean, J.E. and Funderburk, W.C., Jr. 1997. High-pressure liquid chromatographic determination of sucrose in honey. *J Assoc Anal Off Chem* 60:83–841.

Tomás-Barberán, F.A., Martos, I., Ferreras, F., Radovic, B.S., and Anklam, E. 2001. HPLC flavonoid profiles as markers for the botanical origin of European unifloral honeys. *J Sci Food Agric* 81:485–496.

Tomás-Barberán, F.A., Ferreres, F., García-Viguera, C., and Tomás-Lorente, F. 1993a. Flavonoids in honey of different geographical origin. *Lebensm Unters Forsch* 196:3–44.

Truchado, P., Ferreres, F., Bortolotti, L., Sabatini, A.G., and Tomás-Barberán, F.A. 2008. Nectar flavonol rhamnosides are floral markers of acacia (*Robinia pseudacacia*) honey. *J Agric Food Chem* 56:8815–8824.

Tsao, R. and Deng, Z. 2004. Separation procedures for naturally occurring antioxidant hytochemicals. *J Chromator B Analt Technol Biomed Life Sci* 81:5–99.

Tsiapara, A.V., Jaakkola, M., Chinou, I., Graikou, K., Tolonen, T., Virtanen, V., and Moutsatsou, P. 2009. Bioactivity of Greek honey extracts on breast cancer (MCF-7), prostate cancer (PC-3) and endometrial cancer (Ishikawa) cells profile analysis of extracts. *Food Chem* 116(3):702–708.

Tuberoso, C.I.G., Bifulco, E., Jerković, I., Caboni, P., Cabras, P., and Floris, P. 2009. Methyelsyringate: A chemical marker of asphodel (*Asphodelus microcarpus* Salzm. et Viv.) monofloral honey. *J Agric Food Chem* 7(9):3895–3900.

Tuberoso, C.I.G., Bifulco, E., Caboni, P., Cottiglia, F., Cabras, P., and Floris, I. 2010. Floral markers of strawberry-tree (*Arbutus unedo* L.) honey. *J Agric Food Chem* 58:384–389.

Ustunol, Z. and Gandhi, H. 2001. Growth and viability of commercial *Bifidobacteria* spp. in honey-sweetened skim milk. *J Food Pron* 64(11):1775–1779.

Verzera, A., Campisi, S., Zappala, M., and Bonaccorsi, I. 2001. SPME-GC/MS analysis of honey volatile components for the characterization of different floral origin. *Am Lab* 33:18–21.

Vinson, J.A., Hao, Y., Su, X., and Zubik, L. 1998. Phenol antioxidant quantity and quality in foods: Vegetables. *J Agric Food Chem* 4:3630–3634.

Visser, R., Allen, M., and Shaw, J.G. 1988. The effect of heat on the volatile flavour fraction from a unifloral honey. *J Apic Res* 27:175–181.

Von der Ohe, W., Persano Oddo, L., Piana, L.P., Morlot, M., and Martin, P. 2004. Harmonized methods of melissopalynology. *Apidologie* 35:S18–S25.

Wang, H., Andrade, L., and Engeseth, J. 2002. Antimutagenic effect of various honeys and sugars against Trp-p-1. *J Agric Food Chem* 50(23):6923–6928.

Wang, W.J. 2011. Classification of monofloral honeys by voltammetric electronic tongue with chemometrics method. *Electrochem Acta* 56(13):497–4915.

We, J.W. Jr. 1978. *Honey (Advances in Food Research)*, vol. 24, Academic Press, New York.

Wei, Z., Wang, J., and Wang, Y. 2010. Classification of monofloral honeys from different floral origins and geographical origins based on rheometer. *J Food Eng* 96(3):469–479.

Weston, R.J. and Brocklebank, L.K. 1999. The oligosaccharide composition of some New Zealand honeys. *Food Chem* 64:33–37.

Weston, R.J., Brocklebank, L.K., and Lu, Y. 2000. Identification and quantitative levels of antibacterial components of some New Zealand honeys. *Food Chem* 70(4):427–435.

White, J.W. and Brya, V.M. Jr. 1996. Assessing citrus honey quality: Pollen and methyl anthranilate content. *J Agric Food Chem* 44:323–3425.

Wilkins, A.L., Lu, Y., and Tan, S.T. 1993. Extractives from New Zealand honeys. 4. Linalool derivatives and others components from nodding thistle (*Cardans nutans*) honey. *J Agric Food Chem* 41(6):873–878.

Wollgast, J. and Anklam, E. 2000. Review on polyphenols in *Theobroma cacao*: Changes in composition during the manufacture of chocolate and methodology for identification and quantification. *Food Res Int* 33:423–447.

Wolski, T., Tambor, K., Rybak-Chmielewska, H., and Kędzia, B. 2006. Identification of honey volatile components by solid phase microextraction (SPME) and gas chromatography/ mass spectrometry (GC/Mence). *J Apic Sci* 50(2):115–125.

Yao, L., Datta, N., Tomás-Barberán, F.A., Ferreres, F., Martos, I., and Singanusong, R. 2003. Flavonoids, phenolic acids and abscisic acid in Australian and New Zealand *Leptospermum* honeys. *Food Chem* 81:159–168.

Yao, L., Jiang, Y., Singanusong, R., Datta, N., and Raymont, K. 2004a. Phenolic acids and abscisic acid in Australian Eucalyptus honeys and their potential for floral authentication. *Food Chem* 86:169–177.

Yao, L., Jiang, Y., Singanusong, R. et al. 2004b. Flavonoid in Australian *Malaleuca, Guioa, Lophostemon, Banksia* and *Helianthus* honeys and their potential for floral authentication. *Food Res Int* 37:166–174.

Yao, L., Jiang, Y., Singanusong, R., Datta, N., and Raymont, K. 2005. Phenolic acids in Australian *Melaleuca, Guioa, Laphostemon, Banksia* and *Helianthus* honeys and their potential for floral authentication. *Food Res Int* 38:651–658.

Young, I.S. and Woodside, J.V. 2001. Antioxidants in health and disease. *J Clin Pathol* 54:176–186.

Yuzbasioglu, M.F., Kurutas, E.B., Bulbuloglu, E. et al. 2009. Administration of honey to prevent peritoneal adhesions in a rat peritonitis model. *Int J Surg* 7(1):54–57.

Zaid, S.S.M., Sulaiman, S.A., Sirajudee, K.N.M., and Othman, N.H. 2010. The effects of tualang honey on female reproductive organs, tibia bone and hormonal profile in ovariectomised rats—Animal model for menopause. *BMC Complement Altern Med* 10:82.

Zeina, B., Othman, O., and al-Assad, S. 1996. Effect of honey versus thyme on Rubella virus survival *in vitro*. *J Altern Complement Med* 2(3):345–348.

Zhou, Q., Wintersteen, C.L., and Cadwallader, K.R. 2002. Identification and quantification of aroma-active components that contribute to the distinct malty flavor of buckwheat honey. *J Agric Food Chem* 50:2016–2021.

Zürcher, K., Hadorn, H., and Strac, C.H. 1975. Determination of sugars by gas chromatography. *Mitt Geb Lebensmittelunters Hyg* 66:92–116.

# 6 Leptospermum (Manuka) Honey
## Accepted Natural Medicine

*Yasmina Sultanbawa*

## CONTENTS

## INTRODUCTION

The use of honey in healing dates back to ancient times. There has been growing interest in honey as a natural treatment, which has led to scientific investigations addressing its therapeutic properties. Research has revealed that particular honeys have effective healing properties. One such group of honeys is the *Leptospermum* honeys, which have been described as "the best natural antibiotic in the world" (Davis 2002). The *Leptospermum* honeys, known for their therapeutic properties, can be grouped as New Zealand manuka honey from *Leptospermum scoparium* and the Australian jelly bush honey from *Leptospermum polygalifolium*. This honey is produced by honeybees (*Apis mellifera*) feeding on *Leptospermum* flora (Chepulis and Francis 2013).

The therapeutic uses of honey are wide ranging and include management of wounds, ulcers, burns, oral mucositis, chemoprevention, dermatitis, and gangrene (Efem 1988; Al-Waili 2001; Jaganathan and Mandal 2009; Bardy et al. 2012).

However, honey is increasingly being used as a topical antibacterial agent for the treatment of surface infections, which include ulcers and bedsores, and those caused by burns, injuries, and surgical wounds (Allen et al. 1991; Carr 1998). The main body of clinical evidence generated for honey relates to ulcers, wound care, and burns. Honey-based wound care products have been registered in Australia,

New Zealand, European Union, Hong Kong, and the United States, with the relevant regulatory authorities. In most cases, these products use manuka honey from New Zealand or the equivalent honey produced from the *Leptospermum* species in Australia (Irish et al. 2011).

The antimicrobial component of *Leptospermum* honeys has been attributed to acidity, high osmolarity, and hydrogen peroxide. More recently, these honeys have been identified for their nonperoxide antimicrobial activity due to the presence of plant-derived compounds (Allen et al. 1991; Carr 1998). Methylglyoxal (MGO) has been identified as one of the key compounds that contribute to the nonperoxide antimicrobial activity in the New Zealand manuka (*L. scoparium*) honey (Adams et al. 2008, 2009).

## CHEMICAL COMPOSITION AND CHARACTERIZATION OF *LEPTOSPERMUM* HONEY

Honey is a natural nutritious food that is produced from the nectar and pollen of plants. However, the composition of honey varies widely and depends on the botanical origin of the nectar and also on the geographical location. Honey composition includes the sugars glucose, fructose, maltose, and sucrose; water; and other minor components such as proteins, organic acids, amino acids, flavonoids, vitamins, and minerals (Wang and Li 2011). The chemical composition of 18 *Leptospermum* honeys, including manuka and jelly bush, is given in Table 6.1 based on a study done by Mossel (2003).

Studies evaluating the antioxidant properties of honey are relatively recent, with the first one published in the 1990s. Among the phenolic compounds, phenolic acids (caffeic acid, coumaric acid, ferulic acid, ellagic acid, chlorogenic acid, gallic acid, and syringic acid) and flavonoids (chrysin, pinocembrin, pinobanksin, quercetin, kaempferol, luteolin, galangin, apigenin, hesperetin, and myricetin) have been

---

**TABLE 6.1**
**Chemical Composition of *Leptospermum* Honeys**

| Chemical Parameters | Units |
|---|---|
| Moisture | 20.21% |
| Ash | 0.23% |
| pH | 3.95 |
| Total acidity | 34.08 meq/kg |
| Fructose | 39.22% |
| Glucose | 26.63% |
| Sucrose | 0.605% |
| Turanose | 0.577% |
| Maltose | 0.587% |

*Source:* Mossel, B.L., Antimicrobial and quality parameters of Australian unifloral honey. Ph.D. thesis, University of Queensland, Australia, 2003.

---

identified as the most important in *A. mellifera* honeys (Bastos and Sampaio 2013). A study of the phenolic compounds in *Leptospermum* honeys revealed 15 flavonoids in the Australian jelly bush honey with an average content of 2.22 mg/100 g honey. The main flavonoids were myricetin, luteolin, and tricetin. The average of the phenolic acids in jelly bush honey was 5.15 mg/100 g with gallic and coumaric acids. Abscisic acid was quantified as twice the amount at 11.6 mg/100 g honey of the phenolic acids. In the New Zealand manuka honey, the total flavonoids amounted to 3.06 mg/100 g and are composed of quercetin, isorhamnetin, chrysin, luteolin, and an unknown flavanone. The phenolic acids were up to 14 mg/100 g honey, with gallic acid as the main component. A significant amount of abscisic acid (32.8 mg/100 g) was present in manuka honey (Yao et al. 2003). The *Leptospermum* honeys are within the concentration range for phenolic acids (16–113 mg/100 g) and flavonoids (1–16 mg/100 g) reported for other *Apis* honeys. Information from the current literature indicates that honey samples from different botanical origins are good radical scavengers and the antioxidant activity is positively associated with total polyphenol content (Bastos and Sampaio 2013). *Leptospermum* honeys and its effects in ameliorating a 2,4,6-trinitrobenzene sulfonic acid–induced colitis in rats were evaluated. The findings showed reduced colonic inflammation and the biochemical parameters (myeloperoxidase, lipid peroxidation, glutathione, and antioxidant enzymes) were reduced. These honeys restored lipid peroxidation and improved antioxidant parameters (Prakash et al. 2008). Another study reported on the ability of manuka honey to quench free radicals formed within 5 min of spiking the sample of honey in comparison with pasture honey, which showed quenching after 1 h, indicating the potential of using *Leptospermum* honeys to decrease inflammation that is present in chronic wounds (Henriques et al. 2006).

## BIOACTIVITY OF *LEPTOSPERMUM* HONEYS

The antimicrobial activity of honey has been extensively studied; acidity, osmolarity, and hydrogen peroxide have been identified as three components contributing to antimicrobial activity. Hydrogen peroxide is the most significant contributor and is produced by the enzyme glucose oxidase, which is added to the nectar during concentration by the honeybee (*A. mellifera*) from its hypopharyngeal gland (White et al. 1963). The other two antibacterial factors are osmolarity and acidity.

Honey is a saturated sugar solution and this high concentration of sugars leaves very little available water for the growth of microorganisms. Honey also contains many organic acids, mainly gluconic acid, produced from glucose by glucose oxidase and is characteristically acidic with a pH in the range of 3.2 to 4.5 (Davis 2002). The relative distribution of antimicrobial activity against *Staphylococcus aureus* and *Microcossus luteus* was studied by fractionating 10 different honeys into four basic groups: volatile, nonvolatile and nonpolar, acidic, and basic substances. The acidic fraction had the greatest inhibitory activity, whereas the volatiles were the weakest inhibitors. However, in manuka honey, 90% of the activity was found in the acidic fraction in comparison with the other honeys (sunflower, rape, lavender, mountain, blossom, and honeydew), which had 44% activity in the acidic fraction. In this study, the antibacterial activity correlated significantly with honey acidity but did not correlate with honey pH (Bogdanov 1997). Nonperoxide activity in honeys is observed when the hydrogen peroxide is removed by the addition of catalase, but this activity is not found in all honeys (Allen et al. 1991), suggesting that the variation in activity of New Zealand honeys might be attributable to the floral source. Honey from *L. scoparium* (manuka) demonstrated high antibacterial activity and this was shown to be due to a nonperoxide component.

Numerous studies have analyzed the antibacterial activities of Australian and New Zealand honeys and have revealed that several honey types contained significant levels of hydrogen peroxide antibacterial activity; however, only manuka and jelly bush honeys from the *Leptospermum* spp. gave consistent high levels of nonperoxide activity (Allen et al. 1991; Carter et al. 2010).

## NONPEROXIDE ACTIVITY OF *LEPTOSPERMUM* HONEYS

Many antibacterial phenolic acids, such as caffeic, ferulic, and syringic, and flavonoids, such as quercetin, isorhamnetin, and luteolin, have been identified in honey (Russell et al. 1990; Weston et al. 2000; Yao et al. 2003). The concentrations of these phenolic compounds in honey were far too low to have an antimicrobial effect. Mavric et al. (2008) were the first group from Dresden, Germany, to report that MGO was directly responsible for antibacterial activity. Three 1,2-dicarbonyl compounds (3-deoxyglucosulose, glyoxal, and MGO) were measured using reverse-phase high-performance liquid chromatography (HPLC) and ultraviolet detection in 50 commercial honey samples of various origin and 6 manuka honeys. The manuka honeys showed very high amounts of MGO ranging from 38 to 761 mg/kg, which was up to 100-fold higher than what had been reported for conventional honeys. Antibacterial activity of honey indicated a minimum inhibitory concentration (MIC) of 1.1 mM for MGO against both types of bacteria (*S. aureus* and *Escherichia coli*) using an agar well diffusion assay; MIC for glyoxal was 6.9 mM (*E. coli*) and 4.3 mM (*S. aureus*), respectively. 3-Deoxyglucosulose showed no inhibition in concentrations up to 60 mM. Most of the honey samples investigated in this study showed no inhibition in dilutions of 80% or below, whereas the samples of manuka honey exhibited antibacterial activity when diluted from 15% to 30%, which corresponded to MGO concentrations of 1.1 to 1.8 mM. This clearly demonstrates that the high antibacterial efficacy is due to the MGO present in manuka honey. A similar finding of MGO levels of 38 to 828 mg/kg was reported for 49 manuka honey samples (Adams et al. 2008). The question of whether MGO is formed by chemical or enzymatic means was addressed by Adams et al. (2009). During the isolation of MGO in manuka honeys, a second peak was observed in the HPLC trace, and this peak was only seen in manuka honey and not from other floral sources. The second peak was identified as dihydroxyacetone (DHA) and showed a linear relationship to nonperoxide antibacterial activity, but the correlation was less for MGO (Adams et al. 2008). Addition of DHA to clover honey in the presence of arginine or lysine amino acids resulted in the rapid formation of MGO, indicating that the conversion of DHA to MGO was nonenzymatic (Adams et al. 2009). Testing of freshly produced manuka honeys taken from the comb after depositions by the bees revealed low levels of MGO and high levels of DHA. Storage of these honeys at 37°C resulted in an increase in MGO and a decrease in DHA, and loss of both DHA and MGO was observed when manuka honeys were heated in excess of 50°C with a 10-fold increase in the hydroxymethylfurfural content (Adams et al. 2009). The relationship between DHA and MGO in manuka honeys was established by Atrott et al. (2012). DHA was determined in 6 fresh manuka honeys taken directly from the beehive and 18 commercial manuka honey samples, ranging from 600 to 2700 mg/kg and from 130 to 1600 mg/kg, respectively. The corresponding MGO contents varied from 50 to 250 mg/kg in fresh manuka honey and from 70 to 700 mg/kg in commercial manuka honey. This study reported a good linear correlation between DHA and MGO in commercial manuka honeys, which gave a mean ratio of DHA to MGO of 2:1. In comparison, the fresh manuka honeys gave a much higher ratio of DHA to MGO but stabilized to a ratio of 2:1 during storage. Heating of the manuka honey did not elevate the levels of MGO. Atrott and Henle (2009) reported a very good correlation ($r^2 = 0.91$) for MGO in 61 samples of manuka honey ranging from 189 to 835 mg/kg and the corresponding antibacterial activities of the samples, which were between 12.4% and 30.9% equivalent phenol concentration. This indicates the potential of using MGO as a tool for labeling the unique bioactivity of manuka honey. In addition to MGO, there are other markers that have been evaluated in *Leptospermum* honeys that could be used for identification and quality assessment of these honeys. Kato et al. (2012) isolated and characterized a novel glycoside of methyl syringate and named it "leptosin" after the genus *Leptospermum*. The concentration of leptosin was positively correlated to the unique manuka factor (UMF). This glycoside was only found in manuka and jelly bush honeys and could be used as a chemical marker for *Leptospermum* honeys.

## ANTIMICROBIAL ACTIVITY OF *LEPTOSPERMUM* HONEYS

*Leptospermum* honeys have been identified for its potent antimicrobial activity and are being increasingly used in wound care products globally. Several studies have reported on the in vitro antimicrobial activity of these honeys (Allen et al. 1991; Davis 2002; Irish et al. 2011) and a summary of their effects on pathogenic bacteria of clinical importance is given in Table 6.2.

The results for antimicrobial activity reported by several authors cannot be directly compared because different methods have been used. However, the broad-spectrum antibacterial activity of *Leptospermum* honeys is quite clear from the literature presented in Table 6.2. The resurgence in the interest of honey as a treatment for infections is due to the emergence of resistant strains of bacteria to the currently available antibiotics. Honey has been used in infection management for centuries, where these resistant strains have not been demonstrated, and this is very attractive as an antimicrobial treatment. In the study by George and Cutting (2007), a broad range of antibiotic-resistant bacteria were evaluated, and these included multiresistant *Acinetobacter baumannii*, *Enterobacter cloacae*, vancomycin-susceptible *Enterococcus faecalis*, extended-spectrum β-lactamase-producing isolates of *E. coli* and *Klebsiella pneumoniae*, *Pseudomonas aeruginosa*, methicillin-resistant *S. aureus* (MRSA), and vancomycin-resistant *E. faecalis* (VRE). All these clinical isolates of antibiotic-resistant bacteria were completely inhibited by *Leptospermum* honeys; *P. aeruginosa* required the highest concentration at 12% to 14% (v/v) and

**TABLE 6.2**
**In Vitro Testing of *Leptospermum* Honeys against Bacteria**

| *Leptospermum* Honeys | Tested Bacteria | References |
|---|---|---|
| Manuka | *Streptococcus mutans* | Badet and Quero 2011 |
| | *Streptococcus sobrinus* | |
| | *Lactobacillus rhamnosus* | |
| | *Actinomyces viscosus* | |
| | *Porphyromonas gingivalis* | |
| | *Fusobacterium nucleatum* | |
| Manuka and jelly bush | MRSA | Blair et al. 2009 |
| | *Staphylococcus aureus* | |
| | *Acinetobacter calcoaceticus* | |
| | *Citrobacter freundii* | |
| | *Enterobacter cloacae* | |
| | *Enterobacter aerogenes* | |
| | *Enterobacter agglomerans* | |
| | *Escherichia coli* | |
| | *Klebsiella pneumoniae* | |
| | *Morganella morganii* | |
| | *Serratia marcescens* | |
| Manuka | MRSA | Cooper et al. 2010, 2011 |
| | VRE | Jenkins et al. 2011 |
| | *E. coli* | |
| | *Pseudomonas aeruginosa* | |
| | *Staphylococcus epidermidis* | |
| Jelly bush | *S. aureus* | Irish et al. 2011 |
| Manuka | *S. aureus* | Jervis-Bardy et al. 2011 |
| Manuka | MRSA | Kilty et al. 2011 |
| | *P. aeruginosa* | |
| | *S. aureus* | |
| Manuka | *Bacillus subtilis* | Kwakman et al. 2011 |
| | *E. coli* | |
| | *Pseudomonas aeruginosa* | |
| | MRSA | |
| Manuka | *Campylobacter jejuni* | Lin et al. 2009 |
| | *Campylobacter coli* | |
| Manuka | *E. coli* | Lin et al. 2010, 2011 |
| | *Salmonella typhimurium* | |
| | *P. aeruginosa* | |
| | *E. aerogenes* | |
| | *E. cloacae* | |
| | *Shigella flexneri* | |
| | *Shigella sonnei* | |
| | *Yersinia enterocolitica* | |
| Manuka | *Alcaligenes faecalis* | Lusby et al. 2005 |

*(continued)*

**TABLE 6.2 (Continued)**
**In Vitro Testing of *Leptospermum* Honeys against Bacteria**

| *Leptospermum* Honeys | Tested Bacteria | References |
|---|---|---|
| | *C. freundii* | |
| | *E. coli* | |
| | *E. aerogenes* | |
| | *K. pneumoniae* | |
| | *Mycobacterium phlei* | |
| | *Salmonella california* | |
| | *Salmonella enteritidis* | |
| | *S. typhimurium* | |
| | *S. sonnei* | |
| | *S. aureus* | |
| | *S. epidermidis* | |
| Manuka | *Streptococcus pyogenes* | Maddocks et al. 2012 |
| Manuka | *Helicobacter pylori* | Somal et al. 1994; Ndip et al. 2007 |
| Manuka | MRSA | Sherlock et al. 2010 |
| | *P. aeruginosa* | |
| | *E. coli* | |
| *Leptospermum* spp. | *Acinetobacter baumannii* | George and Cutting 2007 |
| | *E. cloacae* | |
| | *E. faecalis* | |
| | *E. coli* | |
| | *K. pneumoniae* | |
| | *P. aeruginosa* | |
| | *S. aureus* | |
| | MRSA | |
| | VRE | |

MRSA was completely inhibited at 4% (v/v) concentration of honey. Jenkins and Cooper (2012a) have done studies to demonstrate the effectiveness of using combined therapies such as manuka honey and antibiotics. It is suggested that two or more antimicrobials with differing modes of action will decrease the growth or survival of these organisms. Another advantage of combined therapy is that less of each antimicrobial agent has to be given, thus reducing costs and the possibility of side effects. They have found synergistic relationships between oxacillin and manuka honey against MRSA and imipenem and manuka honey against MRSA but not toward *P. aeruginosa* (Jenkins and Cooper 2012b), and no synergistic interactions were found between vancomycin and manuka honey against vancomycin-sensitive *S. aureus* (Jenkins et al. 2012). Another important factor worth considering in the use of *Leptospermum* honey in wound applications is that you require a lower concentration to completely inhibit the bacteria and this would reduce the cost of the wound care product.

## Applications in Wound Care and Other Healing Benefits

Manuka honey is well known for its positive effects on health and wellness and there is literature to support the antimicrobial, anti-inflammatory, and immunostimulatory activities and wound-healing properties (Chepulis and Francis 2013). A 5.8-kDa component isolated from manuka honey stimulated the production of inflammatory cytokines, and these findings could lead to the development of novel therapeutics to improve wound healing for patients with acute and chronic wounds (Tonks et al. 2007). Systematic reviews of 43 studies were done to synthesize the evidence regarding honey's role in health care, and some of these studies also included *Leptospermum* honeys. The conclusion of this review stated that honey was a suitable alternative for wound healing, burns, and various skin conditions and also could potentially have a role within cancer (Bardy et al. 2008). This was considered to be a positive outcome to the honey industry. Manuka honey has also been found to be effective against *Helicobacter pylori*. This bacterium is a human gastric pathogen known to cause gastric or peptic ulcers and has been implicated in gastric cancer (Somal et al. 1994). Another study reported on manuka honey dressing as an effective treatment following surgical intervention for chronic or recurrent pilonidal sinus disease (Thomas et al. 2011). This study was done with *Leptospermum* honey dressings to reduce the incidence of wound infection after microvascular free tissue reconstruction for cancer of the head and neck. There was a reduction in duration of the hospital stay in the honey group (12 days) in comparison with the control (18 days), and the costs of standard and honey dressings were similar (Robson et al. 2012). The shorter stay in the hospital would significantly reduce the costs of the hospital and would be a comfort to the patient in terms of recovery. There are numerous studies that have been done on honey, and its healing and health benefits all indicate a positive effect; the trend toward acceptance within the medical fraternity is also increasing.

## Labeling of Nonperoxide *Leptospermum* Honeys

At present, there are two ways of labeling *Leptospermum* honeys to indicate the quality and bioactivity of these honeys to the consumer.

The most widely used one is the UMF, which was named by Professor P. Molan to indicate the antibacterial activity of manuka honey. The UMF units of antibacterial activity used commercially are for the nonperoxide antibacterial equivalent to that of w/v % phenol on a standardized antibacterial agar diffusion assay using *S. aureus* (NCTC 6571) as the control test organism. Therefore, manuka honey with a higher UMF has a greater antibacterial activity and these honeys would be more desirable for the development of wound care products or other therapeutic applications (Stephens 2006).

The other unit for labeling is the chemical marker MGO. As mentioned before, the MGO has a positive correlation to the antimicrobial activity and could be used as a tool for labeling of the bioactivity of *Leptospermum* honeys. The MGO values are used for labeling some of the *L. polygalifolium* honeys in Australia.

## Safety of *Leptospermum* Honeys

The safety of using *Leptospermum* honeys either as a topical wound application or ingesting it as food has been questioned due to the high levels of MGO present in these honeys. MGO could be a potential toxic metabolite that accumulates in various cell types. It can react with protein during processing, giving rise to unstable intermediate products that can degrade to advanced glycation end-products. They are implicated in a number of diseases such as renal disease, diabetes, and neurodegenerative and heart diseases (Wallace et al. 2010). A study was done to investigate the safety of consuming manuka honey (UMF 20+ about 800 mg/kg of MGO) on healthy individuals by knowing whether the honey caused allergic response, changed the microbiota in the gut, and affected the levels of N1-(carboxymethyl)-lysine (CML), a common advanced glycation end point. The results revealed that the honey did not change the CML levels or the gut microbiota and did not cause any allergic response (Wallace et al. 2010). Another study reported on the influence of in vitro simulated gastric and gastroduodenal digestion on the MGO content of manuka honey. MGO levels were determined before and after digestion and the effect of MGO on the proteins and enzymes was investigated. The results revealed that the MGO levels decreased as it reacts with digestive enzymes but did not influence the physiologic activity and therefore did not interfere with the digestion process (Daglia et al. 2013). These studies indicate the safety of using *Leptospermum* honeys and confirm the very long history that honey has as a safe food and traditional medicine.

## CONCLUSION

Honey from *L. scoparium* (manuka) and *L. polygalifolium* (jelly bush) has a broad range of antibacterial activities and this demonstrates its potential use in wound care and other novel therapeutic applications. The nonperoxide activity is a unique factor in these honeys and contributes to the potent antimicrobial activity. The key contributor to the nonperoxide component of *Leptospermum* honey is MGO, but this is not the only factor—other phytochemicals, glycosides, polysaccharides, peptides, and proteins have been also evaluated and shown to have efficacy. The possibility of using these honeys in combination with other antimicrobials from natural sources and current antibiotics opens an entire new field of study and gives opportunities to develop more potent and cost-effective antimicrobial agents for in vivo therapeutic applications.

## REFERENCES

Adams, C. J., Boult, C. H., Deadman, B. J., Farr, J. M., Grainger, M. N. C., Manley-Harris, M., and Snow, M. J. (2008). Isolation by HPLC and characterisation of the bioactive fraction of New Zealand manuka (*Leptospermum scoparium*) honey. *Carbohydrate Research*, 343, 651–659.

Adams, C. J., Manley-Harris, M., and Molan, P. C. (2009). The origin of methylglyoxal in New Zealand manuka (*Leptospermum scoparium*) honey. *Carbohydrate Research*, 344, 1050–1053.

Al-Waili, N. S. (2001). Therapeutic and prophylactic effects of crude honey on chronic sebor-rheic dermatitis and dandruff. *European Journal of Medical Research*, 6, 306–308.

Allen, K. L., Molan, P. C., and Reid, G. M. (1991). A survey of the antibacterial activity of some New Zealand honeys. *Journal of Pharmacy and Pharmacology*, 43, 817–822.

Atrott, J., Haberlau, S., and Henle, T. (2012). Studies on the formation of methylglyoxal from dihydroxyacetone in Manuka (*Leptospermum scoparium*) honey. *Carbohydrate Research*, 361, 7–11.

Atrott, J. and Henle, T. (2009). Methylglyoxal in manuka honey—Correlation with antibacte-rial properties. *Czech Journal of Food Sciences*, 27, S163–S165.

Badet, C. and Quero, F. (2011). The *in vitro* effect of manuka honeys on growth and adherence of oral bacteria. *Anaerobe*, 17, 19–22.

Bardy, J., Molassiotis, A., Ryder, W. D., Mais, K., Sykes, A., Yap, B., Lee, L., Kaczmarski, E., and Slevin, N. (2012). A double-blind, placebo-controlled, randomised trial of active manuka honey and standard oral care for radiation-induced oral mucositis. *British Journal of Oral and Maxillofacial Surgery*, 50, 221–226.

Bardy, J., Slevin, N. J., Mais, K. L., and Molassiotis, A. (2008). A systematic review of honey uses and its potential value within oncology care. *Journal of Clinical Nursing*, 17, 2604–2623.

Bastos, D. H. M. and Sampaio, G. R. (2013). Chapter 47—Antioxidant capacity of honey: Potential health benefit. In *Bioactive Food as Dietary Interventions for Diabetes*, 90–137. Academic Press, San Diego, CA.

Blair, S. E., Cokcetin, N. N., Harry, E. J., and Carter, D. A. (2009). The unusual antibacte-rial activity of medical-grade *Leptospermum* honey: Antibacterial spectrum, resistance and transcriptome analysis. *European Journal of Clinical Microbiology and Infectious Diseases*, 28, 1199–1208.

Bogdanov, S. (1997). Nature and origin of the antibacterial substances in honey. *LWT—Food Science and Technology*, 30, 748–753.

Carr, A. C. (1998). Therapeutic properties of New Zealand and Australian tea trees (*Leptospermum* and *Melaleuca*). *New Zealand Pharmacy*, 18, 1–5.

Carter, D. A., Blair, S. E., and Irish, J. (2010). An investigation into the therapeutic properties of honey. Rural Industries and Development Corporation, Union Offset Printing, Canberra, Australia.

Chepulis, L. and Francis, E. (2013). The glycaemic index of Manuka honey. *e-SPEN Journal*, 8, e21–e24.

Cooper, R., Jenkins, L., and Rowlands, R. (2011). Inhibition of biofilms through the use of manuka honey. *Wounds UK*, 7, 24–32.

Cooper, R. A., Jenkins, L., Henriques, A. F. M., Duggan, R. S., and Burton, N. F. (2010). Absence of bacterial resistance to medical-grade manuka honey. *European Journal of Clinical Microbiology and Infectious Diseases*, 29, 1237–1241.

Daglia, M., Ferrari, D., Collina, S., and Curti, V. (2013). Influence of *in vitro* simulated gas-troduodenal digestion on methylglyoxal concentration of manuka (*Lectospermum sco-parium*) honey. *Journal of Agricultural and Food Chemistry*, 61, 2140–2145. Dx.doi.org/10.1021/jf304299d.

Davis, C. (2002). The use of Australian honey in moist wound management. *A Report for the Rural Industries Research and Development Corporation*, Union Offset Printing, Canberra, Australia.

Efem, S. E. (1988). Clinical observations on the wound healing properties of honey. *The British Journal of Surgery*, 75, 679–681.

George, N. M. and Cutting, K. F. (2007). Antibacterial honey (Medihoney™): *In vitro* activ-ity against clinical isolates of MRSA, VRE, and other multiresistant gram-negative organisms including *Pseudomonas aeruginosa*. *Wounds—A Compendium of Clinical Research and Practice*, 19, 231–236.

Henriques, A., Jackson, S., Cooper, R., and Burton, N. (2006). Free radical production and quenching in honeys with wound healing potential. *Journal of Antimicrobial Chemotherapy*, 58, 773–777.

Irish, J., Blair, S., and Carter, D. A. (2011). The antibacterial activity of honey derived from Australian flora. *PLoS ONE*, 6, e18229.

Jaganathan, S. K. and Mandal, M. (2009). Antiproliferative effects of honey and of its polyphenols: A review. *Journal of Biomedicine and Biotechnology*, 1, 1–13.

Jenkins, R., Burton, N., and Cooper, R. (2011). Manuka honey inhibits cell division in methicillin-resistant *Staphylococcus aureus*. *Journal of Antimicrobial Chemotherapy*, 66, 2536–2542.

Jenkins, R. and Cooper, R. (2012a). Improving antibiotic activity against wound pathogens with manuka honey *in vitro*. *PLoS ONE*, 7, 1–9.

Jenkins, R. E. and Cooper, R. (2012b). Synergy between oxacillin and manuka honey sensitizes methicillin-resistant *Staphylococcus aureus* to oxacillin. *Journal of Antimicrobial Chemotherapy*, 67, 1405–1407.

Jenkins, R., Wootton, M., Howe, R., and Cooper, R. (2012). Susceptibility to manuka honey of *Staphylococcus aureus* with varying sensitivities to vancomycin. *International Journal of Antimicrobial Agents*, 40, 88–89.

Jervis-Bardy, J., Foreman, A., Bray, S., Tan, L., and Wormald, P.-J. (2011). Methylglyoxal-infused honey mimics the anti-*Staphylococcus aureus* biofilm activity of manuka honey: Potential implication in chronic rhinosinusitis. *Laryngoscope*, 121, 1104–1107.

Kato, Y., Umeda, N., Maeda, A., Matsumoto, D., Kitamoto, N., and Kikuzaki, H. (2012). Identification of a novel glycoside, leptosin, as a chemical marker of manuka honey. *Journal of Agricultural and Food Chemistry*, 60, 3418–3423.

Kilty, S. J., Duval, M., Chan, F. T., Ferris, W., and Slinger, R. (2011). Methylglyoxal: (Active agent of manuka honey) *in vitro* activity against bacterial biofilms. *International Forum of Allergy and Rhinology*, 1, 348–350.

Kwakman, P. H. S., Velde, A. A. T., De Boer, L., Vandenbroucke-Grauls, C. M. J. E., and Zaat, S. A. J. (2011). Two major medicinal honeys have different mechanisms of bactericidal activity. *PLoS ONE*, 6, 1–7.

Lin, S. M., Molan, P. C., and Cursons, R. T. (2009). The *in vitro* susceptibility of *Campylobacter* spp. to the antibacterial effect of manuka honey. *European Journal of Clinical Microbiology and Infectious Diseases*, 28, 339–344.

Lin, S. M., Molan, P. C., and Cursons, R. T. (2010). The post-antibiotic effect of manuka honey on gastrointestinal pathogens. *International Journal of Antimicrobial Agents*, 36, 467–468.

Lin, S. M., Molan, P. C., and Cursons, R. T. (2011). The controlled *in vitro* susceptibility of gastrointestinal pathogens to the antibacterial effect of manuka honey. *European Journal of Clinical Microbiology and Infectious Diseases*, 30, 569–574.

Lusby, P. E., Coombes, A. L., and Wilkinson, J. M. (2005). Bactericidal activity of different honeys against pathogenic bacteria. *Archives of Medical Research*, 36, 464–467.

Maddocks, S. E., Lopez, M. S., Rowlands, R. S., and Cooper, R. A. (2012). Manuka honey inhibits the development of *Streptococcus pyogenes* biofilms and causes reduced expression of two fibronectin binding proteins. *Microbiology*, 158, 781–790.

Mavric, E., Wittmann, S., Barth, G., and Henle, T. (2008). Identification and quantification of methylglyoxal as the dominant antibacterial constituent of Manuka (*Leptospermum scoparium*) honeys from New Zealand. *Molecular Nutrition and Food Research*, 52, 483–489.

Mossel, B. L. (2003). Antimicrobial and quality parameters of Australian unifloral honey. Ph.D. Thesis, University of Queensland, Australia.

Ndip, R. N., Malange Takang, A. E., Echakachi, C. M., Malongue, A., Akoachere, J.-F. T. K., Ndip, L. M., and Luma, H. N. (2007). *In vitro* antimicrobial activity of selected honeys on clinical isolates of *Helicobacter pylori*. *African Health Sciences*, 7, 228–232.

Prakash, A., Medhi, B., Avti, P. K., Saikia, U. N., Pandhi, P., and Khanduja, K. L. (2008). Effect of different doses of manuka honey in experimentally induced inflammatory bowel disease in rats. *Phytotherapy Research*, 22, 1511–1519.

Robson, V., Yorke, J., Sen, R. A., Lowe, D., and Rogers, S. N. (2012). Randomised controlled feasibility trial on the use of medical grade honey following microvascular free tissue transfer to reduce the incidence of wound infection. *British Journal of Oral and Maxillofacial Surgery*, 50, 321–327.

Russell, K. M., Molan, P. C., Wilkins, A. L., and Holland, P. T. (1990). Identification of some antibacterial constituents of New Zealand manuka honey. *Journal of Agricultural and Food Chemistry*, 38, 10–13.

Sherlock, O., Dolan, A., Athman, R., Power, A., Gethin, G., Cowman, S., and Humphreys, H. (2010). Comparison of the antimicrobial activity of Ulmo honey from Chile and Manuka honey against methicillin-resistant *Staphylococcus aureus*, *Escherichia coli* and *Pseudomonas aeruginosa*. *BMC Complementary and Alternative Medicine*, 10, 1–5.

Somal, N. A., Coley, K. E., Molan, P., and Hancock, B. M. (1994). Susceptibility of *Helicobacter pylori* to the antibacterial activity of manuka honey. *Journal of Royal Society of Medicine*, 87, 9–12.

Stephens, J. M. D. C. (2006). The factors responsible for the varying levels of UMF® in manuka (*Leptospermum scoparium*) honey. Ph.D Thesis, University of Waikato, New Zealand.

Thomas, M., Hamdan, M., Hailes, S., and Walker, M. (2011). Manuka honey as an effective treatment for chronic pilonidal sinus wounds. *Journal of Wound Care*, 20, 528.

Tonks, A. J., Dudley, E., Porter, N. G., Parton, J., Brazier, J., Smith, E. L., and Tonks, A. (2007). A 5.8-kDa component of manuka honey stimulates immune cells via TLR4. *Journal of Leukocyte Biology*, 82, 1147–1155.

Wallace, A., Eady, S., Miles, M., Martin, H., Mclachlan, A., Rodier, M., Willis, J., Scott, R., and Sutherland, J. (2010). Demonstrating the safety of manuka honey UMF (R) 20+ in a human clinical trial with healthy individuals. *British Journal of Nutrition*, 103, 1023–1028.

Wang, J. and Li, Q. X. (2011). Chapter 3—Chemical composition, characterization, and differentiation of honey botanical and geographical origins. In *Advances in Food and Nutrition Research*, ed L. T. Steve, pp. 609–619. Academic Press, London, UK.

Weston, R. J., Brocklebank, L. K., and Lu, Y. (2000). Identification and quantitative levels of antibacterial components of some New Zealand honeys. *Food Chemistry*, 70, 427–435.

White, J. W., Jr., Subers, M. H., and Schepartz, A. I. (1963). The identification of inhibine, the antibacterial factor in honey, as hydrogen peroxide and its origin in a honey glucose-oxidase system. *Biochimica et Biophysica Acta*, 73, 57–70.

Yao, L., Datta, N., Tomás-Barberán, F. A., Ferreres, F., Martos, I., and Singanusong, R. (2003). Flavonoids, phenolic acids and abscisic acid in Australian and New Zealand *Leptospermum* honeys. *Food Chemistry*, 81, 159–168.

# 7 Honey in Burn and Wound Management

*Laïd Boukraâ*

## CONTENTS

## INTRODUCTION

All civilizations on the globe have relied on natural therapeutic agents to meet their primary health care needs at some point in time. Honey is one of the oldest topical wound-healing herbals and has been used for thousands of years. The use of honey as a wound dressing extends some 4500 years into the past. The ancient Egyptians were among the earliest recorded beekeepers and regularly used honey as a primary wound treatment. Honey has even been found in Egyptian tombs to help preserve body parts. Ancient papyrus documents have recorded that honey was used as an integral part of the "Three Healing Gestures" that included cleaning the wound, applying a salve made from honey, lint (vegetable fiber), and grease (animal fat), and bandaging the wound. These three steps of ancient wound care are very similar to how wounds are still treated today. Honey and honey dressings applied directly to the skin were commonly used to relieve pain, promote wound healing, and treat sores, boils, cuts, abrasions, insect bites, burns, and skin disorders. Despite the long history of honey being used for medical conditions, it largely fell out of favor in conventional medical practice during the era of modern antibiotics in the 1970s. However, due to the development of antibiotic-resistant wound infections, the use of honey for wound care has undergone a renaissance in the last few years. Now, the use of honey in wound care is regaining popularity again, as researchers are determining exactly how honey can help fight serious skin infections. According to their findings, certain types of honey might be "more effective" than antibiotics. Until the first part of the 20th century, honey dressings were part of everyday wound care practice. The misuse of antibiotics, the emergence of resistant bacteria, and an increasing interest in therapeutic honey have provided an opportunity for honey to be reestablished as a broad-spectrum, antibacterial agent that is nontoxic to human tissue. The present chapter discusses the usefulness of honey in wound management and skin care.

## MOLECULAR MECHANISM UNDERLYING THE ROLE OF HONEY IN WOUND-HEALING PROCESS

One way in which honey may work is through its stimulation of an inflammatory response in leukocytes (Abuharfeil et al. 1999; Tonks et al. 2001, 2003). There are also suggestions that a small amount of bacteria may be beneficial to the wound-healing process (Edwards and Harding 2004), and many reports suggest that honey could contain osmophilic bacteria that stimulate an inflammatory response (Molan 2002a,b), as inflammation triggers the cascade of cellular events that give rise to the production of growth factors that control angiogenesis and proliferation of fibroblasts and epithelial cells (Denisov and Afanas'ev 2005). Honey may influence the activation of various cellular and extracellular matrix components and cells. Fukuda et al. (2009) described the properties of a protein (MRJP1) from jungle honey as an effective component that could have potential therapeutic effect for the treatment of wounds. It was reported that incubation of human fibroblasts from normal skin or chronic wounds in the presence of the immunostimulatory component of honey enhances cellular proliferation (Tonks and Tonks 2008). Other authors have found that concentration of honey as low as 0.1% stimulates the proliferation of lymphocytes in cell culture and activates phagocytes from blood (Abuharfeil et al. 1999; Molan 2002b). Immunomodulatory effects were demonstrated in vitro by cytokine release from the monocytic cell line Mono Mac 6 (MM6) and human peripheral monocytes after incubation with honey (Tonks et al. 2003). Honey could also induce interleukin (IL)-6, IL-1β, and tumor necrosis factor (TNF)-α release (Tonks et al. 2001). Tonks et al. (2007) discovered a 5.8-kDa component of manuka honey that stimulates the production of TNF-α in macrophages via Toll-like receptor 4. These molecules also induced the production of IL-1β and IL-6 (Tonks and Tonks 2008). Honey can also stimulate angiogenesis, granulation, and epithelialization in animal models (Bergman et al. 1983; Gupta et al. 1992). In wound and burn management, hydration is an important factor for optimal wound healing (Atiyeh et al. 2005). Moisture retaining dressings provide a protective barrier, prevent eschar formation, reduce dermal necrosis seen in wounds that have been allowed to dry, and significantly accelerate wound reepithelialization (Jurjus et al. 2007). Healing under both wet and moist environments is significantly faster than under dry conditions (Eaglstein et al. 1987). Thus, it is suggested that honey dressings do not stick to the surface of wounds as they sit on a layer of diluted honey without any growth of new tissue into the dressing (Tonks et al. 2007). The physical properties of honey explain its effectiveness as a wound dressing (Caskey 2002; Bray et al. 2009), and the osmolarity of honey draws fluid out from tissues to create a moist healing environment (Molan 2006). Several aspects of healing such as cell proliferation and migration are supported by redox signaling where low-level reactive oxygen species (ROS) produced by nonphagocytic oxidases serve as messenger molecules (Sen and Roy 2008) and induce Nrf2, a transcription factor implicated in the transactivation of genes encoding antioxidant enzymes. However, an intermediate amount of ROS triggers an inflammatory response through the activation of nuclear factor (NF)-κB and AP-1. The transcription factor NF-κB is crucial in a series of cellular processes, such as inflammation, immunity, cell proliferation, and apoptosis (Gloire et al. 2006). Wound fluid from healing tissues contains the highest concentration of hydrogen peroxide ($H_2O_2$) compared with all other body fluids (Roy et al. 2006; Ojha et al. 2008), and this serves to stimulate the growth of fibroblasts and epithelial cells to repair the damage (Ojha et al. 2008). The hydrogen peroxide produced in honey would also be a factor responsible for the rapid rate of healing observed when wounds are dressed with honey. It has been shown that the optimal dilution at which the honey will produce maximal amounts of hydrogen peroxide is between 40% and 60% (Henriques et al. 2006). Honey provides such a controlled delivery of hydrogen peroxide, with the enzymatic production giving a slow release achieving equilibrium concentrations of 20 to 95 μmol/L (Tonks and Tonks 2008). Prolonged exposure to elevated levels of ROS causes cell damage and may eventually inhibit healing of both acute and chronic wounds. Typically, burn injuries show excessive activity of free radicals (Wan and Evans 1999; Subrahmanyam et al. 2003). It is likely that reducing levels of free radicals and other oxidants in the wound bed will aid wound management. Natural honey has been shown to reduce the production of ROS in endotoxin primed MM6 cells (Tonks et al. 2001) and serves as a source of natural antioxidants, which are effective in reducing the risk of different inflammatory processes (National Honey Board 2003). Honey has a wide range of phytochemicals including polyphenols (Saravana and Mahitosh 2009). Important groups of polyphenols found in honey are flavonoids (chrysin, pinocembrin, pinobanksin, quercetin, kaempferol, luteolin, galangin, apigenin, hesperetin, and myricetin), phenolic acids (caffeic, coumaric, ferrulic, ellagic, and chlorogenic), organic acids, enzymes, and vitamins (Gheldof et al. 2002).

Honey has been used since ancient times for wound repair, but the subjacent mechanisms are almost unknown. Very recently, Ranzato et al. (2012) tried to elucidate the modulatory role of honey in an in vitro model of HaCaT keratinocyte reepithelialization by using acacia, buckwheat, and manuka honeys. Scratch wound and migration assays showed similar increases in reepithelialization rates and chemoattractant effects in the presence of different types of honey (0.1%, v/v). However, the use of kinase and calcium inhibitors suggested the occurrence of different mechanisms. All honeys activated cyclin-dependent kinase 2, focal adhesion kinase, and rasGAP SH3 binding protein 1. However, vasodilator-stimulated phosphoprotein, integrin-$\beta_3$, cdc25C, and p42/44 mitogen-activated protein kinase showed a variable activation pattern. Reepithelialization recapitulates traits of epithelial-to-mesenchymal transition (EMT) and the induction of this process was evaluated by a polymerase chain reaction array, revealing marked differences among honeys. Manuka honey induced a few significant changes in the expression of EMT-regulatory genes, while the other two honeys acted on a wider number of genes and partially showed a common profile of upregulation and downregulation. In conclusion, these findings have shown that honey-driven wound repair goes through the activation of keratinocyte reepithelialization, but the ability of inducing EMT varies sensibly among honeys, according to their botanical origin (Ranzato et al. 2012). Table 7.1 summarizes the biological activities and their rationale.

### TABLE 7.1
### Summary of Bioactivity of Honey and Its Rationale

| Bioactivity of Honey | Suggested Rationale |
|---|---|
| Prevention of cross-contamination | Viscosity of honey provides a protective barrier |
| Provides a moist wound-healing environment | Osmolarity draws fluid from underlying tissues |
| Dressings do not adhere to wound surface Tissue does not grow into dressings | Viscous nature of honey provides an interface between wound bed and dressing |
| Promotes drainage from wound | Osmotic outflow sluices the wound bed |
| Removes malodour | Bacterial preference for sugar instead of protein (amino acids) means that lactic acid is produced in place of malodorous compounds |
| Promotes autolytic debridement | Bacterial preference for sugar instead of protein (amino acids) means that lactic acid is produced in place of malodorous compounds |
| Stimulates healing | Bioactive effect of honey |
| Anti-inflammatory | Number of inflammatory cells reduced in honey-treated wounds |
| Managing infection | Antiseptic properties found to be effective against a range of microbes including multiresistant strains |

## SCIENTIFIC EVIDENCE FOR THE ROLE OF
## HONEY IN WOUND MANAGEMENT

Empirical evidence established honey as a treatment for wounds and sores in ancient times. Today, an extensive body of scientific literature on the wound-healing capabilities of honey confirms its value as both an antimicrobial agent and a promoter of healing (Molan 2006). A multitude of wound types have successfully been treated with honey dressings. There have only been a few cases reported where improvement did not occur: a Buruli ulcer, a small group in which only a small amount of honey was applied, two cases with immunodepression, one who stopped treatment because of a painful reaction to honey, one burn that had only a good initial response, and an ulcer complicated by the presence of varicose veins (Molan 2001). Clearly, the antimicrobial activity in honey that prevents and treats infections is fundamental to its wound-healing properties. However, scientific evidence points to a more diverse role for honey in the process (Efem et al. 1992; Phuapradit and Saropala 1992; Subrahmanyam 1993; Cooper 2001). Observed therapeutic effects attributed to using honey as a wound dressing include rapid healing, stimulation of the healing process, clearance of infection, cleansing action on wounds, stimulation of tissue regeneration, reduction of inflammation, and the comfort of the dressings due to lack of adhesion to the tissues (Molan 2001). Healing is a complex, dynamic process that involves many systems and cell types. Molecular and cellular components are responsible for the degradation and repair of tissues that occur during healing (Jones 2001). Although the exact mechanisms for all the observed effects of honey when applied to wounds, burns, and skin ulcers are yet to be defined, recent research clarifies and elucidates some possible explanations.

The clinical evidence in support of the effectiveness of honey in wound care has been comprehensively reviewed (Vermeulen et al. 2005). This review summarizes the findings of 17 randomized controlled trials involving a total of 1965 participants and 5 clinical trials of other forms involving 97 participants treated with honey. All of these found that honey was more effective than the conventional wound care practices used as controls, other than one trial on burns in which, only in respect of control of infection, early surgical tangential incision was found to be more effective than dressing the wounds with honey. The review also summarizes the findings of 16 trials on a total of 533 wounds on experimental animals, where again honey was found to be more effective than the controls in assisting wound healing. There is also a summarized large amount of evidence in the form of case studies that have been reported. Ten publications have reported on multiple cases, totaling 276 cases. There are also 35 reports of single cases (Vermeulen et al. 2005). This evidence is far greater than that for modern wound dressings. Perhaps the most heavily advertised wound dressings are the nanocrystalline silver dressings, but if the PubMed database is searched for evidence to support their use, it can be seen that there is in fact very little good clinical evidence that has been published other than two recent trials. A conclusion reached in a recent systematic review of publications on the use of advanced dressings in the treatment of pressure ulcers has found that their generalized use in the treatment of pressure ulcers is not supported by good research evidence (Zumla and Lulat 1989). There is now a general movement toward evidence-based medicine. In this, decisions should be made on the basis of the available evidence. If randomized controlled trials of the highest quality have not been conducted, then it is necessary to consider evidence of a lower quality. Advertising, other than that which presents good clinical evidence, should not be allowed to influence decisions (Molan 2006).

## EVIDENCE IN ANIMAL STUDIES

In one experimental study (Postmes et al. 1997), comparisons were made between honey and silver sulfadiazine, and between honey and sugar, on standard deep dermal burns, $7 \times 7$ cm, made on Yorkshire pigs. Epithelialization was achieved within 21 days with honey and sugar, whereas it took 28 to 35 days with silver sulfadiazine. Granulation was clearly seen to be suppressed initially by treatment with silver sulfadiazine. In all honey-treated wounds, the histological appearance of biopsy samples showed less inflammation than in those treated with sugar and silver sulfadiazine, and a weak or diminished actin staining in myofibroblasts suggesting a more advanced stage of healing. In another study on experimental burns (Burlando 1978), superficial burns created with a red-hot pin (15 mm²) on the skin of rats were treated with honey or with a sugar solution with a composition similar to honey. Healing was seen histologically to be more active and advanced with honey than with no treatment or the sugar solution. The time taken for complete repair of the wound was significantly less ($P < 0.01$) with honey than with no treatment or with the sugar solution, and necrosis was never so serious. Treatment with honey gave a clearly seen attenuation of inflammation and exudation and a rapid regeneration of outer epithelial tissue and rapid cicatrization.

In another experimental study on buffalo calves (Gupta et al. 1992), full-thickness skin wounds, $2 \times 4$ cm, were made after infecting the area of each wound by subcutaneous injection of *Staphylococcus aureus* 2 days prior to wounding. Topical application of honey, ampicillin ointment, and saline as a control were compared as treatment for the wounds. Clinical examination of the wounds and histomorphological examination of biopsy samples showed that honey gave the fastest rate of healing compared with the other treatments, the least inflammatory reaction, the most rapid fibroblastic and angioblastic activity in the wounds, the fastest laying down of fibrous connective tissue, and the fastest epithelialization.

An experimental study carried out using mice (Bose 1982) also compared honey with saline dressings on wounds made by excising skin ($10 \times 10$ mm) down to muscle. Histological examination showed that the thickness of granulation tissue and the distance of epithelialization from the edge of the wound were significantly greater, and the area of the wound significantly smaller, in those treated with honey ($P < 0.001$). None showed gross clinical infection (honey or control). In another study on rats (Kandil et al. 1987; El-Banby et al. 1989), a 10-mm-long incision was made in the skin of each rat and the wounds were treated topically or orally with floral honey, honey from bees fed on sugar, or saline. A statistically significant increase in the rate of healing was seen with the treatment with floral honey compared with the saline control, this being greater with oral than with topical administration. The treatment with honey from bees that fed on sugar, while initially giving a greater rate of healing, after 9 days gave results no better than those obtained with the saline control. The granulation, epithelialization, and fibrous tissue seen histologically reflected the rate of healing, which is measured as a decrease in wound length. The infiltration of granulation tissue with chronic inflammatory cells was greatest in wounds treated with honey from bees that fed on sugar, less in those treated topically with floral honey, and least in those treated orally with floral honey.

Oral and topical application of honey were compared in another study on rats (Suguna et al. 1992), in which full-thickness $2 \times 2$ cm skin wounds were made on the backs of the rats by cutting away the skin. The rats were treated with topical application of honey to the wound, oral administration of honey, or intraperitoneal administration of honey, or untreated as a control. After 7 days of treatment, tritiated proline was injected subcutaneously to serve as an indicator of collagen synthesis in the subsequent 24-h period. Both the quantity of collagen synthesized and the degree of cross-linking of the collagen in the granulation tissue were found to have increased significantly compared with the untreated control as a result of treatment with honey ($P < 0.001$). Systemic treatment gave greater increases than topical treatment, with the intraperitoneal route giving a better result than the oral route. In a similarly conducted study following this (Suguna et al. 1993), the rats were treated in the same way, but different parameters were studied to assess healing. The granulation tissue that had formed was excised from the wounds for biochemical and biophysical measurement of wound healing. The content of DNA, protein, collagen, hexosamine, and uronic acid and the tensile strength, stress–strain behavior, rate of contraction, and the rate of epithelialization were found to have increased significantly as a result of treatment with honey ($P < 0.05$ to $P < 0.001$). Systemic treatment gave greater increases than topical treatment, with the intraperitoneal route giving the best results.

## Evidence in Clinical Studies

What was effectively a form of crossover trial was conducted in a study (Efem 1988) of 59 patients with recalcitrant wounds and ulcers, 47 of which had been treated for what clinicians deemed a "sufficiently long time" (1 month to 2 years) with conventional treatment (such as Eusol toilet and dressings of Acriflavine, Sofra-Tulle, or Cicatrin or systemic and topical antibiotics) with no signs of healing, or the wounds were increasing in size. The wounds were of varied etiology, such as Fournier's gangrene, burns, cancrum oris and diabetic ulcers, traumatic ulcers, decubitus ulcers, sickle cell ulcers, and tropical ulcers.

Microbiological examination of swabs from the wounds showed that 51 wounds with bacteria present became sterile within 1 week and the others remained sterile. In one of the cases, a Buruli ulcer, treatment with honey was discontinued after 2 weeks because the ulcer was rapidly increasing in size. The outcomes of the 58 other cases were reported as follows: "showed remarkable improvement following topical application of honey." Some general observations reported for the outcomes from honey treatment of these recalcitrant wounds were that sloughs and necrotic and gangrenous tissues separated so that they could be lifted off painlessly within 2 to 4 days in Fournier's gangrene, cancrum oris, and decubitus ulcers (but it took much longer in other types). Sloughs and necrotic tissue were rapidly replaced with granulation tissue and advancing epithelialization. Surrounding edema subsided, weeping ulcers dehydrated, and foul smelling wounds were rendered odorless within 1 week. Burn wounds treated early healed quickly, not becoming colonized by bacteria.

A similar study, but with less detail given, was carried out on 40 patients, half of which had been treated with another antiseptic that had failed (Ndayisaba et al. 1993). The wounds were of mixed etiology: surgical, accidental, infective, trophic, and burns; the average size of the wounds was 57 cm². One third of the wounds were purulent; the rest were red with a whitish coat. The number of microorganism isolates from the wounds dropped from 48 to 14 after 2 weeks of treatment. Seven of the patients had necrotic tissue excised after treatment with honey, and three of these had skin grafts. It was noted that the honey delimited the boundaries of the wounds and cleansed the wounds rapidly to allow this. Of the 33 patients treated only with honey dressings, 29 healed successfully, with good quality healing, on an average time of 5 to 6 weeks. Of the four cases where successful healing was not achieved, two were attributed to the poor general quality of the patients who were suffering from immunodepression: one was withdrawn from treatment with honey because of a painful reaction to the honey and one burn remained stationary after a good initial response.

## Evidence in Clinical Trials

Twenty consecutive cases of Fournier's gangrene managed conservatively with systemic antibiotics (oral amoxicillin/clavulanic acid and metronidazole) in addition to daily topical application of honey were compared retrospectively with 21 similar cases of Fournier's gangrene managed by the orthodox method (wound debridement, wound excision, secondary suturing, and in some cases scrotal plastic reconstruction in addition to receiving a mixture of systemic antibiotics dictated by sensitivity results from cultures) (Efem 1993). The microorganisms cultured in both treatment groups were similar. Although the average duration of hospitalization was slightly longer, topical application of honey showed distinct advantages over the orthodox method. Three deaths occurred in the group treated by the orthodox method, whereas no deaths occurred in the group treated with honey. The need for anesthesia and expensive surgical operation was obviated with the use of honey. Response to treatment and alleviation of morbidity were faster in the group treated with honey. Although some of the bacteria isolated from honey-treated patients were not sensitive to the antibiotics used, the wounds became sterile within 1 week.

The usefulness of honey dressings as an alternative method of managing abdominal wound disruption was assessed in a prospective trial over 2 years compared retrospectively with patients of a similar age over the preceding 2 years (Phuapradit and Saropala 1992). Fifteen patients whose wound disrupted after caesarean section were treated with honey application and wound approximation by micropore tape instead of the conventional method of wound dressing with subsequent resuturing. (The comparative group, 19 patients, had had their dehisced wounds cleaned with hydrogen peroxide and Dakin solution and packed with saline-soaked gauze prior to resuturing under general anesthesia.)

It was noted that with honey dressings, slough and necrotic were replaced by granulation and advancing epithelialization within 2 days, and foul-smelling wounds were made odorless within 1 week. Excellent results were achieved in all the cases treated with honey, thus avoiding the need to resuture, which would have required general anesthesia. Eleven of the cases completely healed within 7 days, and all 15 within 2 weeks.

The required period of hospitalization was 2 to 7 days (mean 4.5) compared with 9 to 18 days (mean 11.5) for the comparative group. Two from the comparative group had their wounds become reinfected, and one developed hepatocellular jaundice from the anesthetic.

A retrospective study of 156 burn patients treated in a hospital over a period of 5 years found that the 13 cases treated with honey had a similar outcome to those treated with silver sulfadiazine (Adesunkanmi and Oyelami 1994).

A prospective randomized controlled trial was carried out to compare honey-impregnated gauze with OpSite as a cover for fresh partial thickness burns in two groups of 46 patients. Wounds dressed with honey-impregnated gauze showed significantly faster healing compared with those dressed with OpSite (mean 10.8 vs. 15.3 days; $P < 0.001$). Less than half as many of the cases became infected in the wounds dressed with honey-impregnated gauze compared with those dressed with OpSite ($P < 0.001$).

Another prospective randomized clinical study was carried out to compare honey impregnated gauze with amniotic membrane dressing for partial thickness burns (Subrahmanyam 1994).

Forty patients were treated with honey-impregnated gauze and 24 were treated with amniotic membrane. The burns treated with honey healed earlier compared with those treated with amniotic membrane (mean 9.4 vs. 17.5 days; $P < 0.001$). Residual scars were noted in 8% of patients treated with honey-impregnated gauze and in 16.6% of cases treated with amniotic membrane ($P < 0.001$).

Honey was compared with silver sulfadiazine–impregnated gauze for efficacy as a dressing for superficial burn injury in a prospective randomized controlled trial that was carried out with a total of 104 patients (Subrahmanyam 1991). In the 52 patients treated with honey, 91% of the wounds were rendered sterile within 7 days. In the 52 patients treated with silver sulfadiazine, 7% showed control of infection within 7 days. Healthy granulation tissue was observed earlier in patients treated with honey (mean, 7.4 vs. 13.4 days). The time taken for healing was significantly shorter with the honey-treated group ($P < 0.001$): of the wounds treated with honey, 87% healed within 15 days compared with 10% of those treated with silver sulfadiazine. Better relief of pain, less exudation, less irritation of the wound, and a lower incidence of hypertrophic scar and postburn contracture were noted with the honey treatment. The honey treatment also gave acceleration of epithelialization at 6 to 9 days, a chemical debridement effect, and removal of offensive smell.

In another prospective randomized controlled trial comparing honey with silver sulfadiazine–impregnated gauze on comparable fresh partial thickness burns (Subrahmanyam 1998), histological examination of biopsy samples from the wound margin as well as clinical observations of wound healing were made to assess relative effects on wound healing in two groups of 25 patients. The time taken for healing was significantly shorter with the honey-treated group ($P < 0.001$). Of the wounds treated with honey, 84% showed satisfactory epithelialization by the 7th day and

100% by the 21st day. In wounds treated with silver sulfadiazine, epithelialization occurred by the 7th day in 72% of the patients and in 84% of patients by 21 days. Histological evidence of reparative activity was seen in 80% of wounds treated with honey dressing by the 7th day, with minimal inflammation. Of the wounds treated with silver sulfadiazine, 52% showed reparative activity, with inflammatory changes, by the 7th day. Reparative activity reached 100% by 21 days with the honey dressing and 84% with silver sulfadiazine. In honey-dressed wounds, early subsidence of acute inflammatory changes, better control of infection, and quicker wound healing were observed, while in the wounds treated with silver sulfadiazine, sustained inflammatory reaction was noted even on epithelialization. No skin grafting was required for the wounds treated with honey, but four of the wounds treated with silver sulfadiazine converted to deep wounds and required skin grafts.

Honey was also compared with boiled potato peel as a cover for fresh partial-thickness burns in another prospective randomized controlled trial (Subrahmanyam 1996a). Of the 40 patients treated with honey who had had positive swab cultures at the time of admission, 90% had their wounds rendered sterile within 7 days. All of the 42 patients treated with boiled potato peel dressings who had had positive swab cultures at the time of admission had persistent infection after 7 days. Of the wounds treated with honey, 100% healed within 15 days compared with 50% of the wounds treated with boiled potato peel dressings. The mean times to heal, 10.4 days with honey versus 16.2 days with boiled potato peel, were significantly different ($P < 0.001$).

Clinical observations from human trials reported that honey-debrided wounds (Efem 1988; Molan 2002a) facilitated formation of granulating tissue (Subrahmanyam 1993), improved epithelialization (Molan 2002b), and reduced inflammation (Subrahmanyam 1998). This was consistent with what was reported in the animal study data previously discussed; however, clinical observation without the support of histological evidence is limited. Although animal studies report accelerated healing time with the use of medical-grade honey (Saber 2010), results in humans have been varied. Recently, three small ($n = 40$, each study) randomized, single-blind (examiner) controlled trial (Motallebnejad et al. 2008; Khanal et al. 2010) and one small ($n = 40$) randomized, nonblinded controlled trial (Rashad et al. 2009) demonstrated that honey may have some protective effects against radiation-induced mucositis in head-and-neck cancer patients undergoing therapy.

In a randomized, double-blind, controlled trial, honey dressing showed no difference in healing time compared with hydrogel dressings in patients who sustained abrasions or minor lacerations (Ingle et al. 2006). In a randomized, double-blind controlled trial (McIntosh and Thomson 2006) and a randomized single-blind controlled trial (Marshall et al. 2005), patients who sustained toenail avulsions showed no differences in mean healing times when honey was compared with paraffin gauze and iodoform gauze, respectively. A meta-analysis of these three studies confirmed no statistical difference in mean time to healing between honey and conventional dressing in these minor acute wounds (Jull et al. 2008).

In several randomized controlled trials, using honey on minor burns (superficial to partial-thickness burns) shows accelerated healing time compared with conventional dressings, such as silver sulfadiazine dressing (Bangroo et al. 2005; Mashhood et al. 2006) and transparent polyurethane film dressing. In addition, honey was found to be superior to nonconventional dressings, such as potato peels (Subrahmanyam 1996a) and amniotic membrane (Subrahmanyam 1994). However, the strength of these studies has been questioned because of the absence of the description of how randomization was achieved (Wijesinghe et al. 2009). Meta-analyses of these trials showed the use of honey to accelerate healing of minor burns compared with the previously mentioned comparators (Moore et al. 2001).

Other histological and clinical studies of wound healing were performed in comparable cases of fresh partial-thickness burns treated with honey dressing or mafenide acetate in two groups of 50 randomly allocated patients. Of the patients with honey-treated wounds, 84% showed satisfactory epithelialization by day 7 and 100% by day 21. In wounds treated with mafenide acetate, epithelialization occurred by day 7 in 72% of cases and in 84% by day 21. Histological evidence of reparative activity was observed in 80% of wounds treated with honey dressing by day 7 with minimal inflammation. Fifty-two percent of the mafenide acetate–treated wounds showed reparative activity with inflammatory changes by day 7. Reparative activity reached 100% by day 21 with the honey dressing and 84% with mafenide acetate. Thus, in honey-dressed wounds, early subsidence of acute inflammatory changes, better control of infection, and quicker wound healing were observed, while in mafenide acetate–treated wounds, a sustained inflammatory reaction was noted even on epithelialization. Further in vitro studies and animal research need to be done to identify other components of honey involved in antibacterial, debriding, and anti-inflammatory properties. This may shed more light for the reason behind the variation between the different types of honey and may lead to the standardization of the type of honey used in therapy.

## ROLE OF MANUKA HONEY IN WOUND HEALING

With high-activity manuka honey being available commercially, especially that which has been sterilized by γ-irradiation (a process that does not reduce the activity; Molan and Allen 1996), there have been several clinical cases published where the results have been remarkable. Three, using manuka honey with a "Unique Manuka Factor" (UMF) rating of 12, have reported healing wounds infected with methicillin-resistant *S. aureus* (MRSA) (Dunford et al. 2000a; Betts and Molan 2001). Another case, also using manuka honey with a UMF rating of 12, reported rapidly healing widespread serious skin ulcers resulting from meningococcal septicemia that were heavily infected with *Pseudomonas*, *S. aureus*, and *Enterococcus* and had not responded to all modern conventional treatments over a period of 9 months in intensive care (Dunford et al. 2000b). Also, Cooper et al. (2001) have reported a case of hidradenitis suppurativa that had been giving recurrent abscesses for 22 years and had given a nonhealing wound for the past 3 years that had had three attempts at surgical removal of infected tissue and a wide range of antibiotics, antiseptics, and wound dressings. This healed (with no recurrence of infection in the 2 years since) within 1 month by dressing with manuka honey with a UMF rating of 13. Another case reported was of a large wound from surgical removal of an area of necrotizing fasciitis, which was heavily infected with *Pseudomonas* after surgery and so a skin graft could not be applied: this was rapidly cleared of infection by application of a dressing of manuka honey with a UMF rating of 12 and then successfully skin grafted (Robson et al. 2000). Betts and Molan (2001) have reported a trial using manuka honey with a UMF rating of 12 on a wider range of types of infected wounds (venous leg ulcers, leg ulcers of mixed etiology, diabetic foot ulcers, pressure ulcers, unhealed graft donor sites, abscesses, boils, pilonidal sinuses, and infected wounds from lower limb surgery). Infection was rapidly cleared and all wounds were healed successfully other than the ones where there was an underlying failure in arterial blood supply creating nonviable tissue.

A prospective observational multicenter study was conducted using Medihoney dressings in 10 hospitals: 9 in Germany and 1 in Austria. Wound-associated parameters were monitored systematically at least three times in all patients. Data derived from the treatment of 121 wounds of various etiologies over a period of 2 years were analyzed. Almost half of the patients were younger than 18 years old, and 32% of the study population were oncology patients. Overall, wound size decreased significantly during the study period and many wounds healed after relatively short time periods. Similarly, perceived pain levels decreased significantly, and the wounds showed noticeably less slough/necrosis. In conclusion, the findings show honey to be an effective and feasible treatment option for professional wound care. In addition, the study showed a relationship between pain and slough/necrosis at the time of recruitment and during wound healing (Biglari et al. 2012). Table 7.2 gives some examples on the successful results achieved by using honey as a dressing.

## TABLE 7.2
## Examples of Honey Dressings That Have Achieved Successful Results after Other Treatments Failed

| Wound | Outcome | Reference |
|---|---|---|
| Porous nonadherent dressing placed between graft and honey dressing | *Pseudomonas* eliminated; donor sites healed faster; better cosmetic results | Robson et al. 2000 |
| 36-month-old surgical wound in the axilla | Healed in 1 month | Cooper et al. 2001 |
| Skin lesions infected with *Pseudomonas*, *S. aureus*, and *Enterococcus* | Healed within 10 weeks | Dunford et al. 2000a |
| Skin-graft failure of lower leg cavity wound; infected with *Pseudomonas*, *S. aureus*, and MRSA: donor-site infection | Healed in 8 weeks: elimination of wound odor | Dunford et al. 2000b |
| Hydroxyurea-induced leg ulcer infected with MRSA treated with UMF 12 manuka honey under DuoDERM (ConvaTec) | Infection cleared in 14 days; healed in 21 days | Natarajan et al. 2001 |
| 4 × 4 cm nonhealing traumatic wound | Granulation and epithelialization visible within 1 week; complete healing in 6 weeks | Dunford et al. 2000a,b |

*(continued)*

## TABLE 7.2 (Continued)
## Examples of Honey Dressings That Have Achieved Successful Results after Other Treatments Failed

| Wound | Outcome | Reference |
|---|---|---|
| Ulcer from back of knee to ankle, infected with *S. pyogenes* and *S. aureus* treated with Medihoney | Elimination of odor; reduction in pain; reduced bleeding at dressing change; 80% reduction in size in 15 weeks | Stewart 2002 |
| Bilateral leg ulcers, extending 18 cm up from ankle on the inner and outer surfaces of an 88-year-old patient, treated with Medihoney | Healed in 6 weeks | Richards 2002 |
| Chronic leg ulcers (20 years), Medihoney compared with Aquacel | Cleaner wound bed; infection and exudate cleared in 10 days | Alcaraz and Kelly 2002 |
| 25-year history of venous ulceration and recurrent infection treated with UMF 10+ manuka honey | Rapid removal of odor; eczema cleared after 10 days | Kingsley 2001 |
| 14 cases of gangrene in the genitals and perineum; honey was applied directly to the wounds, which were covered with a honey-soaked compress | Average debridement time: 5.2 days; granulation: 9.4 days; healing: 28.7 days | Anoukoum et al. 1998 |
| Fournier's gangrene, postsurgical debridement | Healing in six patients | Gürdal et al. 2003 |
| Retrospective review of 50 cases of lactational breast abscesses incised, drained, and packed daily with honey | Good response | Efem 1995 |
| Skin excoriation due to ostomy bag leakage | Rapid epithelialization | Aminu et al. 2000 |
| Soaked ribbon gauze; 16 acute traumatic, 23 complicated surgical, and 21 chronic nonresponding wounds treated with time HoneySoft (Mediprof) | Two wounds did not heal but did not worsen; others healed in a mean of 3 weeks (range, 1–28 weeks) | Ahmed et al. 2003 |

## TYPES OF WOUNDS AND BURNS CURED BY HONEY

### INFECTED WOUNDS AND BURNS

Of course, infected wounds and burns are more difficult to manage clinically. Honey has been evaluated recently for its usefulness in dealing with these conditions. Research in the 1990s found honey to be effective in healing infected nonhealing skin wounds (McInerney 1990; Somerfield 1991). Studies on Fournier's gangrene treated with topical unprocessed honey showed rapid improvement with decreased edema and discharge, rapid regeneration with little or no scarring, wound debridement, and reduced mortality (Efem 1991; Hajase et al. 1996). Animal studies with buffalo calves compared honey to ampicillin and nitrofurazone in treating infection and found that honey decreased infection and healing time and was generally more effective (Gupta et al. 1992; Kumar et al. 1992).

## WOUND INFECTION AND BIOFILMS

Microbial resistance to honey has never been reported, which makes it a very promising topical antimicrobial agent. Indeed, the in vitro activity of honey against antibiotic-resistant bacteria and the reported successful application of honey in the treatment of chronic wound infections that were not responding to antibiotic therapy have attracted considerable attention.

Wounds that are infected with *Streptococcus pyogenes* often fail to respond to treatment. This is largely due to the development of biofilms that may be difficult for antibiotics to penetrate, in addition to problems of antibiotic resistance. The results of the team working in Cardiff University (United Kingdom) showed that very small concentrations of honey prevented the start of biofilm development and that treating established biofilms grown in Petri dishes with honey for 2 h killed up to 85% of bacteria within them. The latest study reveals that honey can disrupt the interaction between *S. pyogenes* and the human protein fibronectin, which is displayed on the surface of damaged cells. "Molecules on the surface of the bacteria latch onto human fibronectin, anchoring the bacteria to the cell. This allows infection to proceed and biofilms to develop," explained Dr. Sarah Maddocks who led the study. "We found that honey reduced the expression of these bacterial surface proteins, inhibiting binding to human fibronectin, therefore making biofilm formation less likely. This is a feasible mechanism by which manuka honey minimizes the initiation of acute wound infections and also the establishment of chronic infections."

Ongoing work in Dr. Maddocks' laboratory is investigating other wound-associated bacteria including *Pseudomonas aeruginosa* (PA) and MRSA. Manuka honey has also been shown to be effective at killing these bacteria. "There is an urgent need to find innovative and effective ways of controlling wound infections that are unlikely to contribute to increased antimicrobial resistance. No instances of honey-resistant bacteria have been reported to date, or seem likely," said Dr. Maddocks. "Applying antibacterial agents directly to the skin to clear bacteria from wounds is cheaper than systemic antibiotics and may well complement antibiotic therapy in the future. This is significant as chronic wounds account for up to 4% of health care expenses in the developed world" (Maddocks et al. 2012).

Alandejani et al. (2009) used a biofilm model to assess antibacterial activity of honey against 11 methicillin-susceptible *S. aureus* (MSSA), 11 MRSA, and 11 PA isolates. Honeys were tested against both planktonic and biofilm-grown bacteria. They found that honey was effective in killing 100% of the isolates in the planktonic form. The bactericidal rates for the Sidr and manuka honeys against MSSA, MRSA, and PA biofilms were from 63% to 82%, from 73% to 63%, and from 91% to 91%, respectively. These rates were significantly higher ($P = 0.001$) than those seen with single antibiotics commonly used against *S. aureus*.

## SURGICAL WOUNDS

Bulman (1955) used honey as surgical dressing for vulvectomies because of its bactericidal capabilities. He also noted success in treating ulcerations following radical surgery for carcinoma of the breast and varicose veins with honey. In 1970, other researchers reported using undiluted honey following radical operations for carcinoma of the vulva, resulting in no infections, minimal debridement, and reduced hospital stays (Cavanagh et al. 1970). In the 1980s, a number of studies used mice to investigate honey in surgical wound healing and found that there were more granulation, smaller wounds, and more rapid healing (Kandil et al. 1987). Other research evaluated the use of honey and microtape for wound closure in women with wound disruption following caesarean sections and showed healing within 2 weeks (Phuapradit and Saropala 1992).

## PRESSURE SORES AND SKIN ULCERS

Numerous studies have demonstrated that the use of unprocessed/undiluted honey resulted in rapid debridement and epithelialization, quick recovery, wound cleanliness, improved taking of grafts, and ease of dressing changes. Table 7.3 summarizes some of the key studies on the use of honey to treat pressure sores and skin ulcers.

## BURNS

Management of burn victims requires reestablishment of a barrier that will protect the internal environment from external contaminants but also help hold in and regulate fluids and tissues under repair. Honey may be able to heal burns as well as or better than conventional dressings (Postmes et al. 1997). A series of studies by Subrahmanyam in India has shown that dressings with pure, unprocessed, undiluted

---

**TABLE 7.3**

**Some Clinical Findings Related to the Use of Honey for Pressure Sores and Ulcers**

| Findings | References |
|---|---|
| Rapid debridement, rapid epithelialization; quick recovery | Efem 1988 |
| Increased granulation, wound cleanliness, and prompt taking of grafts | Farouk et al. 1988 |
| Manuka honey helped heal leg ulcers also treated with antibiotics | Wood et al. 1997 |
| Successful treatment of wounds with honey under dry dressing in accident and emergency room departments | Blomfield 1993 |
| Treatment of pressure ulcer in orthopedic patients; accelerated healing time | Weheida et al. 1991 |

honey obtained from hives had advantages over standard medical treatments such as OpSite (Subrahmanyam 1993), silver sulfadiazine, and traditional, low-cost treatments such as boiled potato peels, but not over early tangential excision and skin grafting of moderate burns (Subrahmanyam 1999). Results from the studies comparing different dressings demonstrated that honey is an effective dressing that speeds healing, sterilizes wounds, reduces pain with enhanced formation of granulation tissue, and lessens inflammation and scarring. Its viscous quality protects the surface from infection and scraping. Other benefits are the ease of dressing changes and its lower cost.

### FOURNIER'S GANGRENE

According to a study conducted by Subrahmanyam (2004), 30 patients admitted with the diagnosis of Fournier's gangrene were randomly allocated to two groups, one treated by honey dressing and the other by Eusol dressing. All patients were treated with broad-spectrum antibiotics and underwent debridement and delayed closure as required. In 14 patients treated with honey dressing, healthy granulation appeared in 4 patients in 1 week and in all patients within 3 weeks. In 16 patients treated with Eusol dressing, healthy granulation appeared in 1 week and by 4 weeks in remaining patients. Secondary suturing and skin grafting was required in 9 patients in each group. Mean hospital stay was 28 days (range, 9–40 days) in the honey-treated group and 32 days (range, 12–52 days) in the Eusol-treated group.

## TYPES OF DRESSINGS

Medical honey for wound healing comes in a variety of forms: pure-form honey, tube honey, honey ointment, a variety of honey-impregnated fiber dressing, and honey-impregnated calcium alginate dressings. Honey has many beneficial properties including analgesia, anti-inflammatory, antibacterial, antiviral, antifungal, antioxidant, immunostimulant, debridement, and deodorizing actions. Honey also has the ability to nourish and moisten the skin and decrease scarring. For these reasons, honey should not be looked at as a generic substance. From a medical perspective, choosing the right type of honey that has been appropriately produced, tested, processed, and packaged is critical for optimal treatment outcome. Topical honey has been shown to be safe and effective and can be combined with conventional medications and therapies in the clinic and at home to decrease pain and promote overall quality of life.

For wound and burn management, many kinds of temporary dressings have been designed in a membranous shape to provide a physical barrier and contribute to an adequate environment for epithelial regeneration. Furthermore, on the basis of strategies of tissue engineering and wound repair, several approaches involving the use of growth factors, matrix materials, epidermal and dermal cell inoculations, and complex skin substitutes have been explored (Sha Huang and Xiaobing 2010). Topical application of growth factors to wound has shown good capability to speed up tissue repair in animal models (Mustoe et al. 1991; Pierce et al. 1992).

Moreover, human recombinant platelet-derived growth factor that directly interferes and favors the repair process has given good results in the healing of diabetic patient ulcers (Steed 1998). Modern hydrocolloid wound dressings are presently favored as moist dressing, although such wound dressings are expensive. Foams, gels, and alginates are also available for treating chronic wounds. Although moist wound care enhances the healing process through tissue regrowth, such moist conditions favor the growth of infecting bacteria (Caskey 2004). Recently, researchers have focused on the incorporation of antibiotics into the membranes to prevent infections, because sustaining a sufficient drug concentration at the site of infection is important for the treatment of an infected wound. Therefore, different types of medicated collagen dressings with antibiotics have been developed (Lee et al. 2002). Health care professionals are aware that wound dressings should be judged on effectiveness, safety, and cost. As such, honey that is to be used for medicinal purposes has to be free of residual herbicides, pesticides, heavy metals, and radioactivity. Also, it has to be sterilized by γ-radiation to prevent wound infection (wound botulism). Furthermore, glucose oxidase in honey has to be controlled during processing to maintain the potency for infection prevention without doing harm to the wound tissues. Besides these primary conditions, the application of honey should be easy (Emsen 2007). Understanding the scientific basis of the anti-inflammatory properties of honey could potentially lead to the development of novel therapeutic agents with a view to rationalizing and optimizing its use for wound therapy. It seems that the optimum environment would be an intermediate gelatinous environment between moist and dry such as seen under highly vapor-permeable dressings (Bernabei et al. 1999; Braddock et al. 1999). Formulation of honey-based gel compositions containing protein growth factors and/or debriding enzymes is provided to achieve rapid and optimal wound healing (Gunwar 2005). Moreover, in many patented products, honey has been combined with fatty ester, wax, and wax-like compound to form an ointment that could be applied on wounds (Van Den Berg and Hoekstra 2009; Vandeputte 2009; Caskey 2010; Cotton 2010).

## WHAT MAKES HONEY IDEAL FOR WOUND CARE?

### WOUND BED PREPARATION

Wound bed preparation may be viewed as management of the wound in order to accelerate endogenous healing. The bioactivity of honey aligns closely with the concept of wound bed preparation. The physiology of healing in acute wounds is a carefully controlled series of events that ensures that healing progresses in a timely fashion. However, in chronic wounds, this orderly sequence is disrupted and the repair process is delayed. If wound bed preparation is to be successful, the impediments to healing must be recognized and addressed, implying appropriate management of exudate, devitalized tissue, and associated bioburden. The appropriate application of honey dressings offers a way forward in managing potential wound-related barriers to healing (Bogdanov 2012). Table 7.4 shows the comparison in terms of wound healing between honey and conventional treatment.

**TABLE 7.4**

**Mean Healing Time Comparison between Honey and Conventional Products for Wounds**

|                | Healing Time with Other Products (days) | Healing Time with Honey (days) | % Difference + *P* Value (Statistically Significant) |
|----------------|:---:|:---:|:---:|
| Skin tears     | 24 | 17 | 24% (*P* = 0.490) |
| Burns          | 22 | 16 | 27% (*P* = 0.045) |
| Venous ulcers  | 62 | 38 | 39% (*P* = 0.001) |
| Pressure ulcers| 93 | 78 | 17% (*P* = 0.027) |
| Diabetes       | 48 | 37 | 23% (*P* = 0.260) |

*Source:* Nestjones, D. and Vandeputte, J., *Wounds UK, 8*(2), 106–112, 2012.

## DEALING WITH EXUDATE

Betts and Molan's (2001) in vivo pilot study reported that honey helps reduce the amount of wound exudate. This is most likely a consequence of honey's anti-inflammatory properties. Inflammation increased vessel permeability, which increases fluid movement into soft tissue, subsequently increasing surface exudate. A decrease in inflammatory cells has been found (histologically) in animal models following application of honey in full-thickness burns. It follows that reducing inflammation lessens exudate production and dressing change frequency, which may conserve resources in terms of dressings used, staff time, and unnecessary disturbance of the patient and the wound bed.

## DEVITALIZED TISSUE

The osmotic pull of honey draws lymph from the deeper tissues and constantly bathes the wound bed. Lymph fluid contains proteases that contribute to the debriding activity of honey. In addition, the constant sluicing of the wound bed is believed to help remove foreign body contamination (Molan 2002a). Molan (2005) has suggested that the most likely explanation for honey's debriding activity involves the conversion of inactive plasminogen to plasmin, an enzyme that breaks down the fibrin that tethers slough and eschar to the wound bed. Stephen-Haynes (2004), who presented the results of three patient case studies and an additional five patients who benefited from management of wound malodor, attested to the clinical impact of honey in debridement. Malodor is known to occur in a variety of wounds in conjunction with slough and necrotic tissue; it is a particular concern when managing fungating lesions. Malodorous substances such as ammonia and sulfur compounds are produced when bacteria metabolize protein. Because honey provides bacteria an alternative source of energy (glucose), these noxious compounds are no longer produced and wound malodor is avoided.

## REDUCING THE RISK OF MACERATION

Macerated periwound skin can be a problem in some wounds and is often related to the dressing used (Cutting and White 2002). The osmotic action of honey, previously mentioned, has been shown in previous reviews of the literature to reduce the risk of maceration; honey draws moisture rather than donates it (Molan 2002a). Thus, periwound skin is protected from overhydration.

## PREVENTING INFECTION

An ample variety of microorganisms may colonize the burn wound, proliferate on and within the eschar, progress in depth, and initiate a systemic infection that remains a major cause of death among people with burns (Bauer and Yurt 2005; Soares de Macedo and Santos 2005) and any means of preventing will lead to higher survival rates in burned patients. Prevention and treatment of burn wound infection include proper wound dressing (Edwards et al. 2003), surgical debridement, and systemic and topical antimicrobial therapy (Monafo and Bessey 2002). Third-degree burn wound eschar is avascular and frequently several millimeters distant from the patient's microvasculature. Therefore, systemically administered antimicrobial agents may not achieve therapeutic levels by diffusion to the wound, where microbial numbers are usually very high. In addition, systemic antibiotics can lead to the development of drug-resistant respiratory and urinary tract infections. Many studies (Natarajan et al. 2001; Cooper et al. 2002; Lusby et al. 2002; Henriques et al. 2006; Simon et al. 2009) have shown that application of honey to severely infected cutaneous wounds is capable of clearing infection from the wound and improving healing. Honey contains an enzyme that produces hydrogen peroxide when diluted (White et al. 1963). It has been reported that hydrogen peroxide is more effective when supplied by continuous generation with glucose oxidase than when added in isolation (Pruitt and Reiter 1985). Additional nonperoxide antibacterial factors have been identified (Allen et al. 1991). Methylglyoxal has been recently isolated and identified as the dominant antibacterial fraction of manuka honey (Adams et al. 2008, 2009; Mavric et al. 2008). The most frequently isolated bacteria from burns and wounds, namely, *S. aureus* and *P. aeruginosa*, have been found to be sensitive to honey action (Boukraa and Amara 2008; Boukraa et al. 2008). Table 7.5 lists the common infectious agents isolated from wounds that have been sensitive to honey.

## ANTI-INFLAMMATORY EFFECT

Normal burn healing is a complex process in which damaged tissue is removed and gradually replaced by restorative tissue during an overlapping series of events that include inflammation, cell proliferation, and tissue remodeling (Figure 7.1). Topical application of honey to burn wounds and other wounds has been found to be effective in controlling infection and producing a clean granulating bed (Bulman 1995; Subrahmanyam 1996b). Tonks et al. (2003) suggested that the wound-healing

## TABLE 7.5
## Most Common Wound-Infecting Bacteria and Fungi Found to Be Sensitive to Honey

| Bacterial Strains | References |
| --- | --- |
| *S. aureus* | Attipou et al. 1998; Molan 2002a,b; Moore et al. 2001; Cooper et al. 2002; Davis 2002; Fox 2002; Lusby et al. 2005; Kwakman et al. 2008; Morris 2008; Alandejani et al. 2009; Simon et al. 2009; Alvarez-Suarez et al. 2010; Mandal and Mandal 2011; Stephen Haynes and Callaghan 2011; Subrahmanyam et al. 2001; Subrahmanyam 2004; Visavadia et al. 2008 |
| MRSA | Allen et al. 2000;; Natarajan et al. 2001; Cooper et al. 2002; Davis 2002; Blaser et al. 2007; George and Cutting 2007; Kwakman et al. 2008; Maeda et al. 2008; Visavadia et al. 2008; Yasunori et al. 2008; Alandejani et al. 2009; Simon et al. 2009; Boukraa and Sulaiman 2010; Mandal and Mandal 2011; Stephen Haynes and Callaghan 2011 |
| Coagulase-negative staphylococci | French et al. 2005; Kwakman et al. 2008; Simon et al. 2009; Mandal and Mandal 2011 |
| *P. aeruginosa* | Attipou et al. 1998; Subrahmanyam et al. 2001, 2003; Cooper et al. 2002; Davis 2002; Wilkinson and Cavanagh 2005; George and Cutting 2007; Mullai and Menon 2007; Kwakman et al. 2008; Alandejani et al. 2009; Boukraa and Sulaiman 2010; Sherlock et al. 2010; Mandal and Mandal 2011 |
| *S. pyogenes* | Molan 1992a,b; Willix et al. 1992; Attipou et al. 1998; Bogdanov 2011; Mandal and Mandal 2011 |
| *Burkholderia cepacia* | Cooper et al. 2000 |
| *Escherichia coli* | Willix et al. 1992; Attipou et al. 1998; Subrahmanyam et al. 2001; Cooper et al. 2002; Davis 2002; Badawy et al. 2004; Lusby et al. 2005; Wilkinson and Cavanagh 2005; Adeleke et al. 2006; Kwakman et al. 2008; Mandal et al. 2010; Sherlock et al. 2010; Stephen Haynes and Callaghan 2011 |
| *Klebsiella oxytoca* | Kwakman et al. 2008; Mandal and Mandal 2011 |
| *Alcaligenes faecalis* | Molan 1992a,b; Lusby et al. 2005 |
| *Serratia marcescens* | Molan 1992a,b; Willix et al. 1992; Lusby et al. 2005 |
| *Proteus mirabilis* | Willix et al. 1992; Attipou et al. 1998; Cooper 2001; Subrahmanyam et al. 2001; Subrahmanyam and Ugane 2004; Lusby et al. 2005; Mandal and Mandal 2011 |
| *Candida albicans* | Obaseiki-Ebor and Afonya 1984; Toth et al. 1987; Molan 1992a,b; Thenussen et al. 2001; Koç et al. 2011; Stephen Haynes and Callaghan 2011 |
| Dermatophytes | Sheikh et al. 1995; Brady et al. 1997 |

effect of honey may in part be related to the release of inflammatory cytokines from surrounding tissue cells, mainly monocytes and macrophages. The findings show that natural honeys can induce IL-6, IL-1β, and TNF-α release. Artificial honey only induces release of these cytokines to a negligible extent (Tonks et al. 2001). Honey has been shown to have mitogenic activity on human B and T lymphocytes (Abuharfeil et al. 1999) and a human myeloid cell line (Watanabe et al. 1998). Proteins present in honey will be highly glycosylated because of high sugar content.

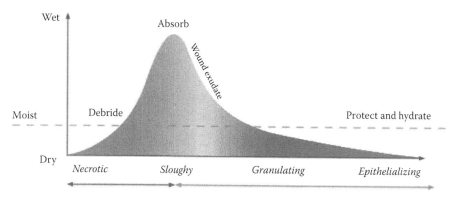

The wound-healing progression model

**FIGURE 7.1** (**See color insert.**) Wound-healing process.

Glycosylated proteins have been shown to activate a number of cell types including monocytic cells (Brownlee 1995). It is possible that these may affect MM6 activation. It has been reported, in a recent study, that honey significantly stimulated the rate of burn wound healing as the rate of wound contraction at day 7 was greater compared with untreated control groups (Norimah et al. 2007). Similarly, Suguna et al. (1993) and Aljadi et al. (2000) reported that honey hastens wound healing by accelerating wound contractions.

## REGULATING OXIDATIVE STRESS

Levels of several antioxidants are measurably lower in burn victims, including β-carotene and vitamins A, C, and E. For this reason, antioxidant therapy using vitamins C and E and carotenoids is often part of the treatment of burns (Nguyen et al. 1993). Another important point is that the regions of interest (ROIs) formed during wound healing cause more inflammation and tissue destruction (Cochrane 1991). The enzyme catalase present in honey has an antioxidant property; thus, honey may have a role as an antioxidant in thermal injury (Subrahmanyam 1996b). The nutrient contents of honey, such as glucose and fructose, improve local substrate supply and may help promote epithelialization. Furthermore, the increase in vitamin C and other antioxidants caused by honey in the blood concentration is particularly important for granulation tissue development and wound healing (Al-Waili 2003; Schramm et al. 2003). The reduction in ROIs seen in the presence of honey may serve to limit tissue damage by activated macrophages during the healing process. Natural honey has also been shown to reduce the production of ROS in endotoxin primed MM6 cells (Senyuva et al. 1997). The harmful effects of hydrogen peroxide are further reduced because honey sequesters and inactivates the free iron, which catalyzes the formation of oxygen free radicals produced by hydrogen peroxide (Bunting 2001), and its antioxidant components help to mop up oxygen free radicals (Frankel et al. 1998).

A study was conducted in the Netherlands in order to select honey for use in wound-healing products. Different samples were compared for their capacity to reduce levels of ROS in vitro. Although most honey samples were shown to be active, significant differences were observed, with the highly active honey exceeding the activities of samples with minor effects by factors of 4 to 30. Most pronounced activities were found for American buckwheat honey from the state of New York. Phenolic constituents of buckwheat honey were shown to have antioxidant activity (Van Den Berg et al. 2008).

## PROVIDING MOISTURE

It has been shown that wounds epithelialize more rapidly in a moist environment (Davies 1984), which depends on the qualities of dressing such as adherence, occlusiveness, and limitations of water vapor transport. Honey, because of its high viscosity, forms a physical barrier, creating a moist environment that appears to be helpful and accelerate wound healing. Honey-impregnated gauze acts as a viscous barrier to wound invasion and fluid loss, although this is not well understood (Quinn et al. 1985). Furthermore, honey as a topical agent does not adhere to the surface. Since there was no difficulty in removing the honey gauze and not much pain during dressing, these factors, associated with easy availability, make honey suitable as cover for burns in their management.

## DEODORIZING AND DEBRIDING

Malodor is a common feature of chronic wounds and is attributed to the presence of anaerobic bacterial species that produce malodorous compounds from decomposed serum and tissue proteins (Bowler et al. 2001). It is probably more than just antimicrobial action that is responsible for the rapid deodorizing of wounds that is observed when honey dressings are used. Honey provides a rich source of glucose that bacteria metabolize in preference to amino acids, resulting in the production of a nonodorous metabolite, lactic acid (White and Molan 2005). Honey facilitates the debridement of wounds by the autolytic action of tissue proteases. It creates a moist wound environment by drawing out lymph fluid from the wound tissues through its strong osmotic action. This provides a constant supply of proteases at the interface of the wound bed and the overlying necrotic tissue, which may help to explain the rapid debridement brought about by honey. This action also washes the surface of the wound bed from beneath, explaining the frequent observations that honey dressings remove debris such as foreign bodies with the dressing (Molan 2002a) and the painless lifting off of slough and necrotic tissue (Subrahmanyam 1998). The activation of proteases by hydrogen peroxide liberated by honey may also offer an explanation for the observed rapid debridement (White and Molan 2005). Honey is known to deodorize wounds rapidly (White and Molan 2005). Van der Weyden (2003) reported that the use of honey alginate on patients with pressure ulcers led to quick and complete healing as well as having a deodorizing and anti-inflammatory effect. Similarly, Stephen-Haynes (2004) reported upon clinical cases where the use of honey resulted in the debridement of wounds in three patients and the management of odor in five patients. There is clearly a need for the control of malodorous wounds within primary care, particularly with pressure ulcers, and honey has proven effective at achieving this goal (Scanlon and Stubbs 2002; Booth 2003). In addition, Hampton (2004) identified the importance of the control of malodor with patients with fungating wounds and recognized the role that honey could play in treating these patients. This debriding action of a honey dressing may also contribute to the lowering of a wound's bacterial load by removal of dead tissue. Dead tissue is well known to provide an excellent medium for bacterial growth and increase the risk of infections if left in the wound (Leaper 1994).

## Pain Management and Reduction in Scarring

While several clinical outcomes that have been reported in relation to the use of honey are attributed to its therapeutic mechanisms, the management of pain (Molan 2002a,b; White and Molan 2005) (although some patients complain of stinging or drawing sensation) (Pieper 2009) and minimizing scarring are also reported (Dunford et al. 2000a; Molan 2001; Stephen Haynes 2005). A study carried out on patients with burns has shown that application of antioxidants to mop up free radicals reduces inflammation (Tanaka et al. 1995). Honey has a significant content of antioxidants that perform this function (Frankel et al. 1998), and this may account for the fact that honey dressings prevent partial-thickness burns from converting to full-thickness burns requiring plastic surgery (Subrahmanyam 1998).

## GETTING HONEY ACCEPTED AS WOUND HEALER

Honey does have significant potential to assist with wound healing and this has been demonstrated repeatedly (Molan 1998; Dunford et al. 2000b; Natarajan et al. 2001). As the potential for honey as a topical wound dressing is further recognized by the health care community, there continues to be a search for other honeys that can be used in this way. These newly identified honeys may be advantageous due to enhanced antibacterial activity, local production (and therefore ready availability), greater selectivity, or broad spectrum of activity against medically important organisms.

The establishment by research that there are bioactive components in honey, and the wide dissemination of this knowledge, has led to a general acceptance that honey is a respectable therapeutic agent and to a rapidly increasing uptake of its usage by clinicians as well as by the general public. The finding that there are multiple bioactive components involved in the therapeutic action makes it a much more attractive option to use the natural product rather than to attempt to identify individual active components and use synthesized copies of those.

Despite lack of promotional support from large corporations, interest in the use of honey in wound management has increased in recent years. However, a clinical profile in wound care commensurate with other modalities has not been achieved despite offering similar indications of use and an increase in research activity and clinical reports. It is observed that "the therapeutic potential of uncontaminated, pure honey is grossly underutilized" (Zumla and Lulat 1989). Clinicians need reassurance that any health-related agent is safe and meets its stated therapeutic purpose. Therefore, it is important to emphasize that, although natural in origin, the honey used in wound care should be of medical-grade standard and not sourced from honey destined for the supermarket shelf. Medical-grade honey is filtered, γ-irradiated, and produced under carefully controlled standards of hygiene to ensure that a standardized honey is produced (White et al. 2005).

## BENEFITS OF WOUND DRESSING WITH HONEY

Some of the features of honey-based wound dressings include the following:

- Its antibacterial properties clear infections rapidly and prevent new microbial growth.
- Unlike many other topical antiseptics, honey causes no tissue damage.
- Reduces inflammation, which leads to less swelling and pain relief.
- Increases blood circulation to the wound.
- Reduces wound odor by preventing ammonia production.
- Honey provides a moist healing environment; a moist healing environment aids healing in most wound types.
- Stimulates tissue healing and thus wound closure by hastening granulation and epithelialization. The acidity of honey might be the cause of the stimulation of the healing process.
- Promotes rapid healing with minimal scarring.
- Honey makes direct contact with the wound's surface, but the dressing does not stick, so there is no pain or tissue damage when dressings are removed.
- Its antioxidant activity decreases the amount of destructive free radicals.
- Its osmotic effect draws dead tissue, exudate, and dirt from the wound.
- Its debriding action allows easy detachment of scabs and dead tissue.
- Honey is more effective than silver sulfadiazine and a polyurethane film dressing for the treatment of burns.

## PRACTICAL CONSIDERATIONS FOR HONEY APPLICATION ON WOUNDS

According to Professor Peter Molan from Waikato University (New Zealand), the following points should be observed when applying honey on wounds and burns:

(a) Start using honey on a wound from the onset.
(b) Use only honey that has been selected for use in wound care.
(c) Use dressings that will hold sufficient honey in place on a wound to get a good therapeutic effect.
(d) Ensure that honey is in full contact with the wound bed.
(e) If a nonadhering dressing is used between the honey dressing and the wound bed, it must be sufficiently porous to allow the active components of the honey to diffuse through it.
(f) Ensure that honey dressings extend to cover any area of inflammation surrounding wounds.

(g) Use a suitable secondary dressing to prevent leakage of honey.
(h) Change the dressings frequently enough to prevent the honey from being washed away or excessively diluted by wound exudate.
(i) When using honey to debride hard eschar (slough), scoring and softening the eschar by soaking with saline will allow better penetration of the honey.

The successful use of honey for wound healing requires that an adequate amount be used and it must be kept in close contact with the wound bed. The severity, location, and level of wound exudate are the primary determining factors for choosing the appropriate type of honey product or dressing to use for a particular wound or condition.

Most commercial honey dressings contain 20 mL (25–30 g) of honey per slow release 10 × 10 cm dressing. If too little honey is used, it will be quickly diluted by exudate to the point where it becomes ineffective. The frequency of dressing changes also depends on the amount of exudation present. The higher the level of exudation, the greater the dilution of honey and the more frequently the dressing changes. Wounds with low or mild exudation may only require dressing changes every 3 to 5 days, whereas moderate exudation typically requires once-daily dressing changes. Wounds with heavy exudation may initially require twice-daily dressing changes and should be closely monitored.

## ADVERSE EFFECTS AND LIMITATIONS

As a medical treatment, honey is rather innocuous. Other than occasional stinging when applied to wounds and redness in the eye, no adverse effects have been reported. In theory, wound botulism from naturally occurring *Clostridium botulinum* spores is possible, but in practice this has never been reported. Since high heat is known to inactivate the antimicrobial factors in honey, pasteurization or other heat treatments are not sterilization options. However, treatment with γ-irradiation will kill the spores while leaving the components responsible for antimicrobial activity intact (Molan 2001; Molan and Allen 1996).

Allergic reactions to honey are rare (Kiistala et al. 1995) and have been attributed in some cases to a reaction to specific pollen in the honey (Helbling et al. 1992; Bauer et al. 1996). Honey processed for use in wound care is passed through fine filters that remove most of the pollen. In more than 500 published reports on the clinical usage of honey in open wounds (Al-Waili and Saloom 1999; Subrahmanyam 1999; Robson et al. 2000; Dunford et al. 2000a,b; Betts and Molan 2001; Natarajan et al. 2001), there have been no adverse reactions noted other than a localized stinging sensation described by some patients. This may be due to the acidity of honey as it has not been reported when the acidity is neutralized (Betts and Molan 2001). A transient stinging sensation was also observed in 102 cases in a trial of honey for ophthalmological use (Emarah 1996), although this was never severe enough to stop treatment. In papers describing the application of honey to open wounds, it has been reported to be soothing and to relieve pain (Subrahmanyam 1993), be nonirritating (Bulman 1955; Cavanagh et al. 1970; Subrahmanyam 1996a,b), be pain free on application (McInerney 1990), and have no adverse effects (Ndayisaba et al. 1993). A number of histological studies examining wound tissues also support the safe use of honey (Bergman et al. 1983; El-Banby et al. 1989; Gupta et al. 1992; Postmes et al. 1997).

## CONCLUSION

Wound healing is a complex process; the ultimate aim of this process is to repair the barrier properties of the skin as quickly as possible. Despite the advances in the development of wound dressings, no single dressing is suitable for all types of wounds, and often, different types are needed during the healing of a single wound. Additionally, the healing process may be delayed as a result of disease conditions, such as in the case of diabetic foot ulcers. Therefore, a continual search for new drugs and dressings capable of interacting with the damaged tissue to speed up the healing process is carried out by physicians. In this regard, one question remains unanswered: which "molecule(s)" from honey is best suited for wound and burn management? Thus, understanding the mechanism of anti-inflammatory action seems crucial since the composition of this natural product varies widely from one type of honey to another. Whichever honey is used for medicinal purposes, consideration needs to be given to its quality. Natural products may have a large variance in therapeutic components depending on their origin. Thus, the floral source of honey plays an important role on its biological properties. In consequence, it would not be surprising that the provenance of honey could determine its healing properties. It is also possible that the mixing of bee products affects their biological activity since those with lower activities may mask the higher activity of others. Moreover, bee products that are to be used for medicinal purposes have to meet certain criteria. As such, they have to be free of residual herbicides, pesticides, heavy metals, and radioactivity. They have also to be sterilized to prevent secondary infections. The problem of antibiotic residues should also be highlighted. As long as beekeepers continue to use antibacterial drugs to control bee diseases, the risk of antibiotic residues in some bee products will remain. Although authorities have allowed a maximum residue limit for each molecule, consumers want honey free of residues because it is perceived as a pure, natural product.

## REFERENCES

Abuharfeil N, Al-Oran R, and Abo-Shehada M. 1999. The effect of bee honey on proliferative activity of human B- and T-lymphocytes and the activity of phagocytes. *Food Agric Immunol*, 11:169–177.

Adams CJ, Boult CH, Deadman BJ et al. 2008. Isolation by HPLC and characterization of the bioactive fraction of New Zealand manuka (*Leptospermum scoparium*) honey. *Carbohydr Res*, 343:651–659.

Adams CJ, Manley-Harris M, and Molan PC. 2009. The origin of methylglyoxal in New Zealand manuka (*Leptospermum scoparium*) honey. *Carbohydr Res*, 344:1050–1053.

Adeleke OE, Olaitan JO, and Okepekpe EI. 2006. Comparative antibacterial activity of honey and gentamicin against *Escherichia coli* and *Pseudomonas aeruginosa. Annals Burn Fire Disasters*, v.19(4).

Adesunkanmi K and Oyelami OA. 1994. The pattern and outcome of burn injuries at Wesley Guild Hospital, Ilesha, Nigeria: A review of 156 cases. *J Trop Med Hyg*, 97(2):108–112.

Ahmed AK, Hoekstra MJ, Hage JJ, and Karim RB. 2003. Honey-medicated dressing: Transformation of an ancient remedy into modern therapy. *Ann Plast Surg*, 50(2):143–147, discussion 147–148.

Alandejani T, Marsan J, Ferris W, Slinger R, and Chan F. 2009. Effectiveness of honey on *Staphylococcus aureus* and *Pseudomonas aeruginosa* biofilms. *Otolaryngol Head Neck Surg*, 141:114–118.

Alcaraz A and Kelly J. 2002. Treatment of an infected venous leg ulcer with honey dressings. *Br J Nurs*, 11(13):859–860, 862, 864–866.

Aljadi AM, Kamaruddin MY, Jamal AM, and Mohd Yassim MY. 2000. Biochemical study on the efficacy of Malaysian honey on inflicted wounds: An animal model. *Med J Islamic Acad Sci*, 13(3):125–132.

Allen KL, Hutchinson G, and Molan PC. 2000. The potential for using honey to treat wounds infected with MRSA and VRE. First World Wound Healing Congress, Melbourne, Australia.

Allen KL, Molan PC, and Reid GM. 1991. A survey of the antibacterial activity of some New Zealand honeys. *J Pharm Pharmacol*, 43(12):817–822.

Alvarez-Suarez JM, Gonzalez-Parma AM, Santos-Buelga C, and Battino M. 2010. Antioxidant characterization of native monofloral Cuban honeys. *J Agric Food Chem*, 58: 9817–9824.

Al-Waili NS. 2003. Effects of daily consumption of honey solution on haematological indices and blood levels of minerals and enzymes in normal individuals. *J Med Food*, 6:135–140.

Al-Waili NS and Saloom KY. 1999. Effects of topical honey on post-operative wound infections due to gram positive and gram negative bacteria following caesarean sections and hysterectomies. *Eur J Med Res*, 4(3):126–130.

Aminu SR, Hassan AW, and Babayo UD. 2000. Another use of honey. *Trop Doct*, 30:250–251.

Anoukoum T, Attipou KK, Aylte A et al. 1998. Le traitement des gangrènes périnéales et de la sphere genitale par du miel. *Tunis Med*, 76(5):132–135.

Atiyeh BS, Gunn SW, and Hayek SN. 2005. New technologies for burn wound closure and healing-review of the literature. *Burns*, 31:944–956.

Attipou K, Anoukoum T, Ayite A, Missoho K, and James K. 1998. Traitement des plaies au miel; Expérience du CHU de lomé. *Med Afr Noire*, 45(11):658–660.

Badawy OF, Shafii SS, Tharwat EE, and Kamal AM. 2004. Antibacterial activity of bee honey and its therapeutic usefulness against *Escherichia coli* O157:H7 and *Salmonella typhimurium* infection. *Rev Sci Tech*, 23:1011–1022.

Bangroo AK, Katri R, and Chauhan S. 2005. Honey dressing in pediatric burns. *J Indian Assoc Pediatr Surg*, 10:172–175.

Bauer GJ and Yurt RW. 2005. Burns. In Mandel GL, Bennett JE, and Dolin R (eds). *Principles and Practice of Infectious Diseases*, 6th ed. Churchill-Livingstone, Philadelphia, PA.

Bauer L, Kohlich A, Hirschwehr R et al. 1996. Food allergy to honey: Pollen or bee products? Characterization of allergenic proteins in honey by means of immunoblotting. *J Allergy Clin Immunol*, 97(1 Pt 1):65–73.

Bergman A, Yanai J, Weiss J, Bell D, and David M. 1983. Acceleration of wound healing by topical application of honey. An animal model. *Am J Surg*, 145:374–376.

Bernabei R, Landi F, Bonini S et al. 1999. Effect of topical application of nerve-growth factor on pressure ulcers. *Lancet*, 354(9175):307.

Betts JA and Molan PC. 2001. A pilot trial of honey as a wound dressing has shown the importance of the way that honey is applied to wounds. Paper of the European Wound Management Association Conference, Dublin, Eire.

Biglari B, vd Linden PH, Simon A, Aytac S, Gerner HJ, and Moghaddam A. 2012. Use of Medihoney as a non-surgical therapy for chronic pressure ulcers in patients with spinal cord injury. *Spinal Cord*, 50(2):165–169. doi:10.1038/sc.2011.8.

Blaser G, Santos K, Bode U, Vetter H, and Simon A. 2007. Effect of medical honey on wounds colonised or infected with MRSA. *J Wound Care*, 16:325–328.

Blomfield R. 1993. Honey for decubitus ulcers. *JAMA*, 224(5):905.

Bogdanov S. 2011. *Functional and Biological Properties of the Bee Products: A Review*. Bee Product Science, available at www.bee-hexagon.net.

Bogdanov S. 2012. *Honey in Medicine*. Bee Product Science, available at www.bee-hexagon. net.

Booth S. 2003. Are honey and sugar paste alternatives to topical antiseptics? *J Wound Care*, 13(1):31–33.

Bose B. 1982. Honey or sugar in treatment of infected wounds? *Lancet*, 1:963.

Boukraa L and Amara K. 2008. Synergistic action of starch on the antibacterial activity of honey. *J Med Food*, 11(1):195–198.

Boukraa L, Benbarek H, and Aissat S. 2008. Synergistic action of starch and honey against *Pseudomonas aerugenosa* in correlation with diastase number. *J Altern Complement Med*, 14(2):181–184.

Boukraa L and Sulaiman SA. 2010. Honey use in burn management: Potentials and limitations. *Forsch Komplementmed*, 17:74–80.

Bowler P, Duerden B, and Armstrong D. 2001. Wound microbiology and associated approaches in wound management. *Clin Micro Rev*, 14(2):244–269.

Braddock M, Campbell CJ, and Zuder D. 1999. Current therapies for wound healing: Electrical stimulation, biological therapeutics, and the potential for gene therapy. *Int J Dermatol*, 38:808–817.

Brady NF, Molan PC, and Harfoot CG. 1997. The sensitivity of dermatophytes to the antimicrobial activity of manuka honey and other honey. *Pharmac Sci*, 2:1–3.

Bray R, Patel C, and Caskey PR. 2009. Wound dressings. US20090012440.

Brownlee M. 1995. Advanced protein glycosolation diabetes and aging. *Annu Rev Med*, 46:223–234.

Bulman MW. 1955. Honey as a surgical dressing. *Midd Hosp J*, 55(6):188–189.

Bunting CM. 2001. The production of hydrogen peroxide by honey and its relevance to wound healing. MSc thesis, University of Waikato, New Zealand.

Burlando F. 1978. Sull'azione terapeutica del miele nelle ustioni. *Minerva Dermatol*, 113:699–706.

Caskey PR. 2002. The use of honey in wound dressings. WO02000269.

Caskey PR. 2004. Honey in wound dressings. US20040127826.

Caskey PR. 2010. Use of honey in dressings. US7714183.

Cavanagh D, Beazley J, and Ostapowicz R. 1970. Radical operation for carcinoma of the vulva: A new approach to wound healing. *J Obstet Gynaecol Br Commonw*, 77:1037–1040.

Cochrane CG. 1991. Cellular injury by oxidants. *Am J Med*, 91(3C):23S–30S.

Cooper R, Wigley P, and Burton NF. 2000. Susceptibility of multi-resistant strains of *Burkholderia cepacia* to honey. *Lett Appl Microbiol*, 31(1):20–24.

Cooper RA. 2001. How does honey heal wounds? In Munn P and Jones R (eds). *Honey and Healing*. International Bee Research Association, Cardiff, U.K.

Cooper RA, Molan RC, Krishnamoorthy L, and Harding KG. 2001. Manuka honey used to heal a recalcitrant surgical wound. *Euro J Clin Microbiol Infect Dis*, 20:758–759.

Cooper RA, Molan PC, and Harding K. 2002. The sensitivity to honey of Gram-positive cocci of clinical significance isolated from wounds. *J Appl Microbiol*, 93:857–863.

Cotton S. 2010. Honey wound dressing. WO2010010399.

Cutting K and White R. 2002. Maceration of the skin: 1. The nature and causes of skin maceration. *J Wound Care*, 11:275–278.

Davies JWL. 1984. Synthetic materials for covering burn wounds. Progress towards perfection. I. Short term materials. *Burns*, 10:94–103.

Davis C. 2002. *The Use of Australian Honey in Moist Wound Management*. A report for the Rural Industries Research and Development Corporation. RIRDC Web Publication No W05/159.

Denisov ET and Afanas'ev IB. 2005. *Oxidation and Antioxidants in Organic Chemistry and Biology*. Taylor & Francis, London, 932.

Dunford C, Cooper RA, and Molan PC. 2000a. Using honey as a dressing for infected skin lesions. *Nurs Times*, 96(14 Suppl):7–9.

Dunford C, Cooper RA, White RJ, and Molan PC. 2000b. The use of honey in wound management. *Nurs Stand*, 15(11):63–68.

Eaglstein WH, Mertx PM, and Falanga V. 1987. Occlusive dressings. *Am Fam Physician*, 35:211–216.

Edwards JV, Bopp AF, Batistie SL, and Goynes WR. 2003. Human neutrophil elastase inhibition with a novel cotton-alginate wound dressing formulation. *J Biomed Mater Res*, 66:433–440.

Edwards R and Harding KG. 2004. Bacteria and wound healing. *Curr Opin Infect Dis*, 17:91–96.

Efem SE. 1995. Breast abscesses in Nigeria: Lactational versus non-lactational. *J R Coll Surg Edinb*, 40(1):25–27.

Efem SEE. 1988. Clinical observations on the wound healing properties of honey. *Br J Surg*, 75:679–681.

Efem SEE. 1991. Recent advances in the management of Fournier's gangrene: Preliminary observations. *Surgery*, 113(2):200–204.

Efem SEE, Udoh KT, and Iwara CI. 1992. The antimicrobial spectrum of honey and its clinical significance. *Infection*, 20(4):228–229.

El-Banby M, Kandil A, Abou-Sehly G, El-Sherif ME, and Abdel-Wahed K. 1989. Healing effect of floral honey and honey from sugar-fed bees on surgical wounds (animal model). Fourth International Conference on Apiculture in Tropical Climates. Cairo: International Bee Research Association, London, pp. 46–49.

Emarah MH. 1996. A clinical study of the topical use of bee honey in the treatment of some occular diseases. *Bull Islamic Med*, 2(5):422–425.

Emsen IM. 2007. A different and safe method of split thickness skin graft fixation: Medical honey application. *Burns*, 33:782–787.

Fox C. 2002. Honey as a dressing for chronic wounds in adults. *Br J Community Nurs*, 7(10):530–534.

Farouk A, Hassan T, Kashif H, Khalid SA, Mutawali I, and Wadi M. 1988. Studies on Sudanese bee honey: Laboratory and clinical evaluation. *Int J Crude Drug Res*, 26(3):161–168.

Frankel S, Robinson GE, and Berenbaum MR. 1998. Antioxidant capacity and correlated characteristics of 14 unifloral honeys. *J Apic Res*, 37(1):27–31.

French VM, Cooper RA, and Molan PC. 2005. The antibacterial activity of honey against coagulase-negative *staphylococci*. *J Antimicrob Chemother*, 56:228–231.

Fukuda M, Kobayashi K, Hirono Y, Miyagawa M, Ishida T, Ejiogu EC et al. 2009. Jungle honey enhances immune function and antitumor activity. *Evid Based Complement Alternat Med*, 2011, 908743, 8 pp. doi:10.1093/ecam/nen086.

George NM and Cutting KF. 2007. Antibacterial honey (Medihoney™): *In vitro* activity against clinical isolates of MRSA, VRE, and other multiresistant gram-negative organisms including *Pseudomonas aeruginosa*. *Wounds*, 19(10):A10.

Gheldof N, Wang XH, and Engeseth NJ. 2002. Identification of antioxidant components of honeys from various floral sources. *J Agric Food Chem*, 50:5870–5877.

Gürdal M, Yiicebas E, Tekin A et al. 2003. Predisposing factors and treatment outcome in Fournier's gangrene: Analysis of 28 cases. *Urol Int*, 70(4):286–290.

Gloire G, Legrand-Poels S, and Piette J. 2006. NF-kB activation by reactive oxygen species: Fifteen years later. *Biochem Pharmacol*, 72:1493–1505.

Gunwar S. 2005. Honey based gel formulations. WO2005077402.

Gupta SK, Singh H, Varshney AC, and Prakash P. 1992. Therapeutic efficacy of honey in infected wounds in buffaloes. *Indian J Anim Sci*, 62(6):521–523.

Hajase MJ, Simonin JE, Bihrle R, and Coogan CL. 1996. Genital Fournier's gangrene: Experience with 38 patients. *Adult Urology*, 47(5):734–739.

Hampton S. 2004. Managing symptoms of fungating wounds. *J Comm Nurs*, 18(10):22–28.

Helbling A, Peter C, Berchtold E, Bogdanov S, and Muller U. 1992. Allergy to honey: Relation to pollen and honey bee allergy. *Allergy*, 47(1):41–49.

Henriques A, Jackson S, Cooper R, and Burton N. 2006. Free radical production and quenching in honeys with wound healing potential. *J Antimicrob Chemother*, 58:773–777.

Ingle R, Levin J, and Polinder K. 2006. Wound healing with honey-a randomised controlled trial. *S Afr Med J*, 96:831–835.

Jones KP. 2001. The role of honey in wound healing and repair. In Munn P and Jones R (eds). *Honey and Healing*. International Bee Research Association, Cardiff, U.K.

Jull AB, Rodgers A, and Walker N. 2008. Honey as a topical treatment for wounds. *Cochrane Database Syst Rev*, (4):CD005083.

Jurjus A, Atiyeh BS, Abdallah IM, Jurjus RA, Hayek SN, Abou Jaoude M et al. 2007. Pharmacological modulation of wound healing in experimental burns. *Burns*, 33:892–907.

Kandil A, Elbanby M, Abd-Elwahed K, Sehly GA, and Ezzet N. 1987. Healing effect of true floral and false nonfloral honey on medical wounds. *J Drug Res Egypt*, 17(1–2):71–76.

Khanal B, Baliga M, and Uppal N. 2010. Effect of topical honey on limitation of radiation-induced oral mucositis: An intervention study. *Int J Oral Maxillofac Surg*, 39(12):1181–5.

Kiistala R, Hannuksela M, Makinen-Kiljunen S, Niinimuki A, and Haahtela T. 1995. Honey allergy is rare in patients sensitive to pollens. *Allergy*, 50(10):844–847.

Kingsley A. 2001. The use of honey in the treatment of infected wounds: Case studies. *Br J Nurs*, 10(22 Suppl):S13-6, S18, S20.

Koç AN, Silici S, Kasap F, Hörmet-Öz HT, Mavus-Buldu H, and Ercal BD. 2011. Antifungal activity of the honeybee products against *Candida* spp. and *Trichosporon* spp. *J Med Food*, 14(1/2):128–134.

Kumar A, Sharma V, Gupta K, Singh H, Prakash P, and Singh SP. 1992. Efficacy of some indigenous drugs in tissue repair in buffaloes. *Indian Vet J*, 70(1):42–44.

Kwakman PHS, Johannes PC, Van den Akker et al. 2008. Medical-grade honey kills antibiotic-resistant bacteria *in vitro* and eradicates skin colonization. *Clin Infect Dis*. 1;46(11):1677–82.

Leaper D. 1994. Prophylactic and therapeutic role of antibiotics in wound care. *Am J Surg*, 167(1A):15S–20S.

Lee JE, Park JC, Lee HK, Oh SH, and Suh H. 2002. Laminin modified infection-preventing collagen membrane containing silver sulfadiazine-hyaluronan microparticles. *Artif Organs*, 26:521–528.

Lusby PE, Coombes A, and Wilkinson JM. 2002. Honey: A potent agent for wound healing? *J Wound Ostomy Continence Nurs*, 29:295–300.

Lusby PE, Coombes AL, and Wilkinson JM. 2005. Bactericidal activity of different honeys against pathogenic bacteria. *Arch Med Res*, 36:464–467.

Maddocks SE, Lopez MS, Rowlands RS, and Cooper RA. 2012. Manuka honey inhibits the development of *Streptococcus pyogenes* biofilms and causes reduced expression of two fibronectin binding proteins. *Microbiology*, 158(Pt 3):781–790.

Maeda Y, Loughrey A, Earle JA, Millar, BC, Rao JR, Kearns A et al. 2008. Antibacterial activity of honey against community associated methicillin-resistant *Staphylococcus aureus* (CA-MRSA). *Complement Ther Clin Pract*, 14(2):77–82.

Mandal DM and Mandal S. 2011. Honey: Its medicinal property and antibacterial activity. *Asian Pac J Trop Biomed*, 1691(11)60016-6:154–160.

Mandal S, Mandal DM, Pal NK, and Saha K. 2010. Antibacterial activity of honey against clinical isolates of *Escherichia coli*, *Pseudomonas aeruginosa* and *Salmonella enterica* serovar Typhi. *Asian Pacific J Trop Med*, 3(12):961–964.

Marshall C, Queen J, and Manjooran J. 2005. Honey vs povidine iodine following toenail surgery. *Wounds*, 1(1):10–18.

Mashhood AA, Khan TA, and Sami AN. 2006. Honey compared with 1% silver sulfadiazine cream in the treatment of superficial and partial thickness burns. *J Pak Assoc Dermatol*, 16:14–19.

Mavric E, Wittmann S, Barth G, and Henle T. 2008. Identification and quantification of methylglyoxal as the dominant antibacterial constituent of Manuka (*Leptospermum scoparium*) honeys from New Zealand. *Mol Nutr Food Res*, 52:483–489.

McInerney RJ. 1990. Honey—A remedy rediscovered. *J R Soc Med*, 83(2):127.

McIntosh CD and Thomson CE. 2006. Honey dressing versus paraffin tulle gras following toenail surgery. *J Wound Care*, 15:133–136.

Molan PC. 1992a. The antibacterial activity of honey. 1. The nature of the antibacterial activity. *Bee World*, 73(1):5–28.

Molan PC. 1992b. The antibacterial activity of honey. 2. Variation in the potency of the antibacterial activity. *Bee World*, 73(2):59–76.

Molan PC, Allen KL. 1996. The effect of gamma-irradiation on the antibacterial activity of honey. *J Pharm Pharmacol*, 48 1206–1209.

Molan PC. 2002. *Manuka Honey as a Medicine*. Honey Research Unit, University of Waikato, Hamilton, New Zealand.

Molan PC. 2002. Re-introducing honey in the management of wounds and ulcers—Theory and practice. *Ostomy Wound Manage*, 48:28–40.

Molan PC. 1998. The evidence for honey promoting wound healing. *Aust J Wound Manag*, 6:148–158.

Molan PC. 2006. The evidence supporting the use of honey as a wound dressing. *Int J Low Extrem Wounds*, 5(1):40–54.

Molan PC. 2005. Mode of action. In White R, Cooper R, and Molan P (eds). *Honey: A Modern Wound Management Product*. Wounds, Aberdeen, U.K., pp. 1–23.

Molan PC. 2001. Why honey is effective as a medicine. 1. Its use in modern medicine. In Munn P and Jones R (eds). *Honey and Healing*. Inter Bee Res Assoc, Cardiff, U.K.

Monafo WW and Bessey PQ. 2002. Wound care. In Herendon DN (ed). *Topical Burn Care*. Saunders, pp. 109–119.

Moore OA, Smith LA, Campbell F, Seers K, McQuay HJ, and Moore RA. 2001. Systematic review of the use of honey as a wound dressing. *BMC Complement Altern Med*, 1:2.

Morris C. 2008. The use of honey in wound care and the Mesitran product range. *Wounds UK*, 4(3):84–87.

Motallebnejad M, Akram S, Moghadamnia A, Moulana Z, and Omidi S. 2008. The effect of topical application of pure honey on radiation-induced mucositis: A randomized clinical trial. *J Contemp Dent Pract*, 9(3):40–47.

Mullai V and Menon T. 2007. Bactericidal activity of different types of honey against clinical and environmental isolates of *Pseudomonas aeruginosa*. *J Altern Complement Med*. 13(4):439–441.

Mustoe TA, Pierce GF, Morishima C, and Deuel TF. 1991. Growth factor induced acceleration of tissue repair through direct and inductive activities in a rabbit dermal ulcer model. *J Clin Invest*, 87:694–703.

Natarajan S, Williamson D, Grey J, Harding KG, and Cooper RA. 2001. Healing of an MRSA-colonized, hydroxyurea- induced leg ulcer with honey. *J Dermatolog Treat*, 2:33–36.

National Honey Board. 2003. Honey health and therapeutic qualities, available at www.honey.com.

Ndayisaba G, Bazira L, Habonimana E, and Muteganya D. 1993. Clinical and bacteriological results in wounds treated with honey. *J Orthop Surg*, 7(2):202–204.

Nestjones D and Vandeputte J. 2012. Clinical evaluation of Melladerm® Plus: A honey-based wound gel. *Wounds UK*, 8(2):106–112.

Nguyen T, Cox C, Traber D et al. 1993. Free radical activity and loss of plasma antioxidants, vitamin E, and sulfhydryl groups in patients with burns: The 1993 Moyer Award. *J Burn Care Rehabil*, 14:602–609.

Norimah Y, Ainul Hafiza AH, Rozaini MZ, and Md Zuki AB. 2007. Development of honey hydrogel dressing for enhanced wound healing. *Radiat Phys Chem*, 76:1767–1770.

Obaseiki-Ebor EE and Afonya TCA. 1984. *In vitro* evaluation of the anticandidiasis activity of honey distillate (HY-1) compared with that of some antimycotic agents. *J Pharm Pharmacol*, 36(8):283–284.

Ojha N, Roy S, He G, Biswas S, Velayutham M, Khanna S et al. 2007. Assessment of wound-site redox environment and the significance of Rac2 in cutaneous healing. *Free Radic Biol Med*, 138–149.

Phuapradit W and Saropala N. 1992. Topical application of honey in treatment of abdominal wound disruption. *Aust N Z J Obstet Gynaecol*, 32(4):381–384.

Pieper B. 2009. Honey-based dressings and wound care: An option for care in the United States. *J Wound Ostomy Continence Nurs*, 36(1):60–66.

Pierce GF, Tarpley JE, Yanagihara D, Mustoe TA, Fox GM, and Thomason A. 1992. Platelet-derived growth factor (BB homodimer), transforming growth factor-beta 1, and basic fibroblast growth factor in dermal wound healing. Neovessel and matrix formation and cessation of repair. *Am J Pathol*, 140:1375–1388.

Postmes TJ, Bosch MMC, Dutrieux R, van Baare J, and Hoekstra MJ. 1997. Speeding up the healing of burns with honey. An experimental study with histological assessment of wound biopsies. In Mizrahi A and Lensky Y (eds). *Bee Products: Properties, Applications and Apitherapy*. Plenum Press, New York, pp. 27–37.

Pruitt KM and Reiter B. 1985. Biochemistry of peroxidase system: Antimicrobial effects. In Pruitt KM, Tenovuo JO (eds). *The Lactoperoxidase System: Chemistry and Biological Significance*. Marcel Dekker, New York, pp. 144–178.

Quinn KJ, Evans JH, Courtney JM et al. 1985. Nonpressure treatment of hypertrophic scars. *Burns*, 12:102.

Ranzato E, Martinotti S, and Burlando B. 2012. Epithelial mesenchymal transition traits in honey-driven keratinocyte wound healing: Comparison among different honeys. *Wound Repair Regen*, 20:778–785.

Rashad UM, Al-Gezawy SM, El-Gezawy E, and Azzaz AN. 2009. Honey as topical prophylaxis against radiochemotherapy-induced mucositis in head and neck cancer. *J Laryngol Otol*, 123(2):223–228.

Richards L. 2002. Healing infected recalcitrant ulcers with antibacterial honey. Paper presented the 4th Australian Wound Management Association Conference, Adelaide, Australia.

Robson V, Ward RG, and Molan PC. 2000. The use of honey in split skin grafting. 10th Conference of the European Wound Management Association, Harrogate, U.K.

Roy S, Khanna S, Nallu K, Hunt TK, and Sen CK. 2006. Dermal wound healing is subject to redox control. *Molec Ther*, 13:211–220.

Saber A. 2010. Effect of honey versus intergel in intraperitoneal adhesion prevention and colonic anastomotic healing: A randomized controlled study in rats. *Int J Surg*, 8:121–127.

Saravana KJ and Mahitosh M. 2009. Antiproliferative effects of honey and of its polyphenols: A review. *J Biomed Biotech*, 2009, 830616.

Scanlon E and Stubbs N. 2002. To use or not to use? The debate on the use of antiseptics in wound care. *Br J Comm Nurs*, 7(9)(suppl):8–20.

Schramm DD, Karim M, Schrader HR, Holt RR, Cardetti M, and Keen CL. 2003. Honey with high levels of antioxidants can provide protection to healthy human subjects. *J Agric Food Chem*, 51:1732–1735.

Sen CK and Roy S. 2008. Redox signals in wound healing. *Biochim Biophys Acta*, 1780:1348–1361.

Senyuva C, Yucel A, Erdamar S, Seradjmir M, and Ozdemir C. 1997. Fate of alloplastic materials placed under a burn scar: An experimental study. *Burns*, 23:484.

Sha Huang A and Xiaobing F. 2010. Naturally derived materials-based cell and drug delivery systems in skin regeneration. *J Contr Release*, 142:149–159.

Sheikh D, Zaman S, Naqvi SB, Sheikh MR, and Ali G. 1995. Studies on the antimicrobial activity of honey. *Pakistan J Pharm Sci*, 8(1):51–62.

Sherlock O, Dolan A, Athman R et al. 2010. Comparison of the antimicrobial activity of Ulmo honey from Chile and Manuka honey against methicillin-resistant *Staphylococcus aureus*, *Escherichia coli* and *Pseudomonas aeruginosa*. *BMC Complement Altern Med*, 10:47.

Simon A, Traynor K, Santos K, Blaser G, Bode U, and Molan P. 2009. Medical honey for wound care-still the "latest resort"? *Evid Based Complement Alternat Med*, 6:165–173.

Soares de Macedo JL and Santos JB. 2005. Bacterial and fungal colonization of burn wounds. *Mem Inst Oswaldo Cruz*, 100:535–539.

Somerfield SD. 1991. Honey and healing. *J R Soc Med*, 84(3):179.

Steed DL. 1998. Modifying the wound healing response with exogenous growth factors. *Clin Plast Surg*, 25:397–405.

Stephen-Haynes J. 2004. Evaluation of honey impregnated tulle dressing in primary care. *Brit J Community Nurs*, (*Wound Care* Supplement):S21–S27.

Stephen Haynes J. 2005. Implications of honey dressings within primary care. In Molan P, Cooper R, White R (eds). *Honey: A Modern Wound Management Product*. Wounds, Aberdeen, U.K., Chapter 3.

Stephen Haynes J and Callaghan R. 2011. Properties of honey: Its mode of action and clinical outcomes. *Wounds UK*, 7(1):50–57.

Stewart D. 2002. Therapeutic honey used to reduce pain and bleeding associated with dressing changes. Paper presented at the 4th Australian Wound Management Association Conference, Adelaide, Australia.

Subrahmanyam M. 1991. Topical application of honey in treatment of burns. *Br J Surg*, 78(4):497–498.

Subrahmanyam M. 1993. Honey impregnated gauze versus polyurethane film (OpSite) in the treatment of burns-a prospective randomised study. *Br J Plast Surg*, 46:322–323.

Subrahmanyam M. 1994. Honey-impregnated gauze versus amniotic membrane in the treatment of burns. *Burns*, 20(4):331–333.

Subrahmanyam M. 1996a. Honey dressing versus boiled potato peel in the treatment of burns: A prospective randomized study. *Burns*, 22(6):491–493.

Subrahmanyam M. 1996b. Addition of antioxidants and polyethylene glycol 4000 enhances the healing property of honey in burns. *Ann Burns Fire Disast*, 9:93–95.

Subrahmanyam M. 1998. A prospective randomised clinical and histological study of superficial burn wound healing with honey and silver sulfadiazine. *Burns*, 24(2):157–161.

Subrahmanyam M. 1999. Early tangential excision and skin grafting of moderate burns is superior to honey dressing: A prospective randomised trial. *Burns*, 25:729–731.

Subrahmanyam M and Ugane SP. 2004. Honey dressing beneficial in treatment of Fournier's gangrene. *Indian J Surg*, 66(2):75–77.

Subrahmanyam M, Sahapure AG, Nagane NS, Bhagwat VR, and Ganu JV. 2001. Effects of topical application of honey on burn wound healing. *Ann Burns Fire Disasters*, 20(3):137–139.

Subrahmanyam M, Shahapure AG, Naganer NS, Bhagwat VR, and Ganu JV. 2003. Free radical control-the main mechanism of the action of honey in burns. *Ann Burns Fire Disasters*, 16:135–137.

Suguna L, Chandrakasan G, Ramamoorthy U, and Thomas JK. 1993. Influence of honey on biochemical and biophysical parameters of wounds in rats. *J Clin Biochem Nutr*, 14:91–99.

Suguna L, Chandrakasan G, and Thomas JK. 1992. Influence of honey on collagen metabolism during wound healing in rats. *J Clin Biochem Nutr*, 13:7–12.

Tanaka H, Hanumadass M, Matsuda H, Shimazaki S, Walter RJ, and Matsuda T. 1995. Haemodynamic effects of delayed initiation of antioxidant therapy (beginning two hours after burn) in extensive third-degree burns. *J Burn Care Rehab*, 16(6):610–615.

Thenussen F, Grobler S, and Gedalia I. 2001. The antifungal action of three South African honeys on Candida albicans. *Apidologie*, 32:371–379.

Tonks AJ and Tonks A. 2008. TLR4 ligand isolated from honey. WO2008117019.

Tonks A, Cooper RA, Price AJ, Molan PC, and Jones KP. 2001. Stimulation of TNF-alpha release in monocytes by honey. *Cytokine*, 14:240–242.

Tonks AJ, Cooper RA, Jones KP, Blair S, Parton J, and Tonks A. 2003. Honey stimulates inflammatory cytokine production from monocytes. *Cytokine*, 21:242–247.

Tonks AJ, Dudley E, Porter NG, Parton J, Brazier J, Smith EL et al. 2007. A 5.8-kDa component of manuka honey stimulates immune cells via TLR4. *J Leukoc Biol*, 82:1147–1155.

Toth G, Lemberkovics E and Kutasi-Szabo J. 1987. The volatile components of some Hungarian honeys ant their antimicrobial effects. *Am Bee J*, 127:496–497.

Van den Berg AJ, van den Worm E, van Ufford HC, Halkes SB, Hoekstra MJ, and Beukelman CJ. 2008. An *in vitro* examination of the antioxidant and anti-inflamatory properties of buckwheat honey. *J Wound Care*, 17(4):172–4, 176–8.

Van Den BA JJ and Hoekstra MJ. 2009. Wound dressings incorporating honey. US20090304780.

Van der Weyden. 2003. The use of honey for the treatment of two patients with pressure ulcers. *Br J Comm Nurs*, 14–20.

Vandeputte J. 2009. Wound care treatment product. US20090291122.

Vermeulen H, Ubbink DT, Goossens A et al. 2005. Systematic review of dressings and topical agents for surgical wounds healing by secondary intention. *Br J Surg*, 92(6):665–672.

Visavadia BG, Honetsett J, and Danford MH. 2008. Manuka honey dressing: An effective treatment for chronic wound infectons. *Br J Oral Maxillofacial Surg*, 46(1):55–56.

Wan KC and Evans JH. 1999. Free radical involvement in hypertrophic scar formation. *Free Rad Biol Med*, 26:603–608.

Watanabe K, Shinmoto H, Kobori M et al. 1998. Stimulation of cell growth in the U-937 human myeloid cell line by honey royal jelly protein. *Cytotechnology*, 26:23–27.

Weheida SM, Nagubib HH, El-Banna M, and Marzouk S. 1991. Comparing the effects of two dressing techniques on healing of low grade pressure ulcers. *J Med Res Instit*, 12(2):259–278.

White JW, Subers MH, and Schepartz AI. 1963. The identification of inhibine, the antibacterial factor in honey, as hydrogen peroxide and its origin in a honey glucose-oxidase system. *Biochim Biophys Acta*, 73:57–70.

White R and Molan P. 2005. A summary of published clinical research on honey in wound management. In: White R, Cooper R, Molan P. *Honey: A Modern Wound Management Product*. Wounds, Aberdeen, U.K., 160 pp.

Wijesinghe M, Weatherall M, Perrin K, and Beasley R. 2009. Honey in the treatment of burns: A systematic review and meta-analysis of its efficacy. *N Z Med J*, 122(1295):47–60.

Wilkinson JM and Cavanagh HM. 2005. Antibacterial activity of 13 honeys against *Escherichia coli* and *Pseudomonas aeruginosa*. *J Med Food*, 8:100–103.

Willix DJ, Molan PC, and Harfoot CJ. 1992. A comparison of the sensitivity of wound-infecting species of bacteria to the antibacterial activity of manuka honey and other honey. *J Appl Bacteriol*, 73(3):388–394.

Wood B, Rademacher M and Molan PC. 1997. Manuka honey: A low cost leg ulcer dressing. *N Z Med J*, 110(1040):107.

Yasunori M, Loughreya A, Earle JAP, Millar BC, Raod JR, and Kearnse A. 2008. Antibacterial activity of honey against community-associated methicillin resistant *Staphylococcus aureus* (CA-MRSA). *Complement Ther Clin Pract*, 14:77–82.

Zumla A and Lulat A. 1989. Honey—A remedy rediscovered. *J R Soc Med*, 82(7):384–385.

# 8 Honey for Gastrointestinal Disorders

*Fatiha Abdellah and Leila Ait Abderrahim*

## CONTENTS

## INTRODUCTION

Since ancient times, people have speculated about honey's curative properties. The ancient Egyptians, Assyrians, Chinese, Greeks, and Romans all used honey, in combination with other herbs and on its own, to treat the diseases of the gut (Molan 1999). Until recently, there was little scientific evidence to support therapeutic uses of honey. Lately, however, many studies have shown that honey has valid medical use because of its antibacterial, anti-inflammatory, and antioxidant activities. It can inhibit the growth of a wide range of bacteria, fungi, protozoa, and viruses (Molan 1992; Blair and Carter 2005). In more recent times, its role in the treatment of burns, gastrointestinal (GI) disorders, asthma, infected and chronic wounds, skin ulcers, cataracts, and other eye ailments has been reported (Molan 1992; Marcucci 1995; Castaldo and Capasso 2002; Orhan et al. 2003). Pure honey is bactericidal for many pathogenic organisms, including enteropathogens, such as *Salmonella* spp., *Shigella* spp., enteropathogenic *Escherichia coli*, and other gram-negative organisms (Cavanah et al. 1968; Jeddar et al. 1985), and is a readily available source of glucose and fructose. It has been reported to contain about 200 substances: a complex mixture of sugars but also small amounts of other constituents such as minerals, proteins, vitamins, organic acids, flavonoids, phenolic acids, enzymes, and other phytochemicals (White 1979).

Infections of the intestinal tract are common throughout the world, affecting people of all ages. Diarrhea and gastroenteritis are major causes of death and health problems in many developing countries. Loss of water and electrolytes from the body can lead to severe dehydration, which can be fatal in young children, especially those already in poor health and malnourished. The infectious diarrhea exacerbates nutritional deficiencies in various ways, but as in any infection the calorific demand is increased (Jeddar et al. 1985; Farthing et al. 2008; Borhany 2011; Diskin 2012). The increasing interest in the use of alternative therapies is the result of the development of antibiotic resistance in bacteria becoming a major problem and because people are experiencing the sometimes severe side effects of many pharmaceuticals, which in the currently prevailing ambience of chemophobia may be sufficient to give rise to an aversion to all synthetic drugs (Molan 1999). Honey has been credited for solving problems ranging from sore throats to GI distress. Some honey benefit claims are more substantially supported with research than others.

## HUMAN GI SYSTEM

The digestive tract (Figure 8.1) is a long muscular tube that begins with the mouth leading to the pharynx, esophagus, stomach, and the small and large intestines and terminating at the anus. It is associated with several accessory digestive organs, such as the pancreas, liver, and gallbladder (Ahmed et al. 2007). As ingested, food is slowly propelled through this tract; the gut assimilates calories and nutrients that are essential for the establishment and maintenance of normal bodily functions (Siegel et al. 2008).

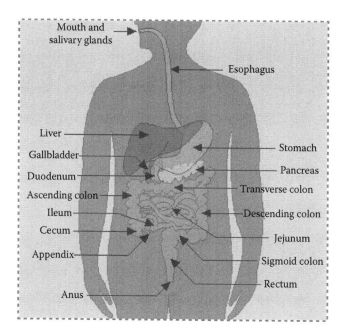

**FIGURE 8.1**   Human digestive system.

## DIGESTIVE PROCESS

*The start of the process—the mouth*: The digestive process begins in the mouth. Food is partly broken down by the process of chewing and by the chemical action of salivary enzymes (these enzymes are produced by the salivary glands and break down starches into smaller molecules).

*On the way to the stomach—the esophagus*: After being chewed and swallowed, the food enters the esophagus. The esophagus is a long tube that runs from the mouth to the stomach. It uses rhythmic, wave-like muscle movements (called peristalsis) to force food from the throat into the stomach. This muscle movement gives us the ability to eat or drink even when we are upside-down.

*In the stomach*—The stomach is a large, sack-like organ that churns the food and bathes it in a very strong acid (gastric acid). Food in the stomach that is partly digested and mixed with stomach acids is called chime (Maton 1997).

*In the small intestine*—After being in the stomach, food enters the duodenum, the first part of the small intestine. It then enters the jejunum and then the ileum (the final part of the small intestine). Digestive enzymes are produced by the inner wall of the small intestine to help in the breakdown of food. Also, the pancreas secretes hormones into the small intestine to assist in digestion and into the blood to regulate blood glucose levels. Enzymes in pancreatic juice include the carbohydrate-digesting enzyme: pancreatic amylase; several protein-digesting enzymes: trypsin, chymotrypsin, and carboxypeptidase; and lipid-digesting enzymes: lipase.

The liver has several important functions. It acts as a mechanical filter by filtering blood that travels from the intestinal system. It detoxifies several metabolites. In addition, the liver has synthetic functions, such as plasma protein production. It plays an important role in the digestion and processing of food. Liver cells produce bile, a greenish-yellow fluid that aids the digestion of fats and the absorption of fat-soluble nutrients. Bile is delivered to the small intestine (at the duodenum) through the bile duct; when there is no food to digest, extra bile is stored in the gallbladder located beneath the liver (Highleyman and Franciscus 2012).

*In the large intestine*—After passing through the small intestine, food passes into the large intestine. In the large intestine, some of the water and electrolytes are removed from the food. Many microbes (bacteria such as *Bacteroides*, *Lactobacillus acidophilus*, *E. coli*, and *Klebsiella*) in the large intestine help in the digestion process. These bacteria are beneficial because they can help crowd out potentially pathogenic bacteria from the digestive tract; they also play an important role in regulating the immune system. The first part of the large intestine is called the cecum. Food then travels upward in the ascending colon. The food travels across the abdomen in the transverse colon and goes back down the other side of the body in the descending colon and then through the sigmoid colon. Solid waste is then stored in the rectum until it is excreted via the anus (Maton 1997).

## HONEY AND GI DISORDERS

GI disorders afflict a lot of people in the world. Whereas some GI disorders may be controlled by diet and pharmaceutical medications, others are poorly moderated by conventional treatments (National Organization for the Reform of Marijuana Laws 2012). Functional GI disorders are the most commonly presented illnesses seen in primary care or gastroenterology. They are those in which the bowel looks normal but does not work properly. They are the most common problems affecting the colon and rectum and include constipation and irritable bowel syndrome (IBS). Structural disorders are those in which the bowel looks abnormal and does not work properly. Sometimes, the structural abnormality needs to be removed surgically. The most common structural disorders are those affecting the anus as well as diverticular disease and cancer. Upper GI complaints include chest pain, chronic and recurrent abdominal pain, dyspepsia, lump in the throat, halitosis, nausea and vomiting, and rumination. Some upper GI complaints represent functional illness. Lower GI complaints include constipation, diarrhea, gas and bloating, abdominal pain, and rectal pain or bleeding (Greenberger 2012).

According to the Muslim Holy Hadith, dating back to the 8th century AD, the prophet Mohamed recommended honey against diarrhea. Also, the Roman physician Celsus used honey as a cure for diarrhea. Honey has continued as a medicine into present-day folk medicine. Preparations of honey have been proposed as potentially useful for many conditions of the GI tract (GIT), including periodontal and other oral diseases, dyspepsia, and gastritis, and as part of ORT (Molan 1999; Ghosh and Playford 2003).

## HONEY FOR ORAL HEALTH

### Periodontal Diseases

Any inherited or acquired disorder of the tissues surrounding and supporting the teeth (periodontium) can be defined as a periodontal disease. These diseases may be of developmental, inflammatory, traumatic, neoplastic, genetic, or metabolic origin. However, the term periodontal disease usually refers to the common inflammatory disorders of gingivitis and periodontitis that are caused by pathogenic microflora in the biofilm or dental plaque that forms adjacent to the teeth on a daily basis (Pihlstrom et al. 2005).

The antimicrobial characteristics of honey may make it a helpful treatment for periodontal disease, gingivitis, and mouth ulcers in some cases. Honey having an anti-inflammatory activity raises the possibility of it being useful as a therapeutic agent for periodontitis. Its antibacterial activity would provide "anti-infective" therapy of periodontal disease, by removing the etiologic factor, and the anti-inflammatory activity would block the direct cause of the erosion of the connective tissues and bone (Molan 2001a). Another beneficial feature of using honey to treat periodontal disease would be its well-established stimulation of the growth of granulation tissue and epithelial cells, which would aid in repair of the damage done by infecting bacteria and by the free radicals from the inflammatory response to them (Altman 2010).

### Stomatitis

Stomatitis is an inflammation of the mucous lining of any of the structures in the mouth, which may involve the cheeks, gums, tongue, lips, and the roof or floor of the mouth. The inflammation can be caused by conditions in the mouth itself, such as poor oral hygiene, poorly fitted dentures, mouth burns, cheek biting, and jagged teeth, or by conditions that affect the entire body, such as medications, allergic reactions, or infections such as herpetic infections, gonorrhea, measles, leukemia, AIDS, and lack of vitamin C. Other systemic diseases associated with stomatitis include inflammatory bowel disease (IBD) and Behçet's syndrome. Bad breath (halitosis) may also accompany the condition (Swartout-Corbeil and Knight 2006).

Aphthous stomatitis, also known as recurrent aphthous ulcers or canker sores, is a specific type of stomatitis that presents with shallow, painful ulcers that are usually located on the lips, cheeks, gums, or the roof or floor of the mouth. It has the most incidences of all oral mucosal diseases. Exposure to heavy metals can cause stomatitis. Thrush, a fungal infection, is a type of stomatitis (El-Haddad 2009).

Ulcerous stomatitis, common to childhood, is indicated by ulceration commencing on or near the gums, more frequently in the lower than in the upper jaw, and usually on one side only, and spreading over the entire mouth. Ulceration speedily occurs, exposes the necks of the teeth, and extends to the mucous membrane of the mouth generally. Ulcers of the mouth are often due to syphilis and dyspepsia and are aggravated by the irritation resulting from the irregular edges of fractured and decayed teeth. When neglected, such ulcers may assume the appearance of epithelioma, especially when induration of the glands beneath the jaw is present (Gorgas 1901).

A study done by El-Haddad (2009) demonstrated that the topical application of commercial honey seemed to be very effective in treatment of minor recurrent aphthous stomatitis, while current therapy does not provide a satisfactory means for cure.

Honey can be used to treat stomatitis, ulcers or "boils" inside the mouth or on the tongue, or inflamed gums. To treat these conditions, smear the individual ulcers, boils, or aphthae with honey—or simply trickle a good spoonful of honey in the mouth and then swirl it around the mouth. Honey penetrates the tissues very quickly, and it seems that it is precisely when diluted that its curative power is activated (Hermyan 2010).

There is much debate whether honey is harmful to teeth. Some reports show a cariogenic effect of honey (Shannon et al. 1979; Bowen and Lawrence 2005), while others claim that the effect of honey is less cariogenic than sucrose (Decaix 1979; Emarah 1982). Due to its antibacterial activity, honey ingestion inhibits the growth of bacteria that cause caries (Steinberg et al. 1996; Molan 2001b) and might have a carioprotective effect (Edgar and Jenkins 1974; Sela et al. 1998). It was also shown that manuka honey has a positive effect against dental plaque development and gingivitis and thus can be used in place of refined sugar in the manufacture of candy (Molan 2001b). According to electron microscopic studies, ingestion of honey does not cause erosion of tooth enamel as observed after drinking fruit juice (pH 3.5). Ten minutes after consumption of fruit juice, tooth erosion was seen, while 30 min after honey ingestion, the erosion was only very weak. This effect can be explained only partially by the calcium, phosphorous, and fluoride levels of honey; other colloidal honey components have to also be responsible (Grobler et al. 1994).

### Halitosis (Oral Malodor)

Halitosis is a frequent or persistent unpleasant breath malodor. The majority (some 85%–90%) of the cases of breath odors come from the oral cavity itself. Other sources may include nasopharyngeal or upper respiratory conditions, metabolic disorders, systemic diseases, and other external sources. In the oral cavity, most of the odor is due to putrefactive microbial activity (Sterer and Rosenberg 2011). It most often results from fermentation of food particles by anaerobic gram-negative bacteria in the mouth such as *Porphyromonas gingivalis* or *Fusobacterium nucleatum*, producing volatile sulfur compounds (VSCs) such as hydrogen sulfide and methyl mercaptan. Causative bacteria may be present in areas of periodontal disease. The causative organisms reside deep in periodontal pockets around teeth. In patients with healthy periodontal tissue, these bacteria may proliferate on the dorsal posterior tongue. GI disorders rarely cause halitosis because the esophagus is normally collapsed (Murchison 2012).

Manuka honey is well known for a pronounced antibacterial activity, which is higher than in any other honey. It contains a typical antibacterial substance "methylglyoxal (MGO)" and shows the strong antibacterial activity against oral bacteria, by which it can be a promising functional food for oral care. A study done by Shiga et al. (2010) reported the decrease of halitosis by intake of manuka honey or acacia honey in 10 healthy subjects (average age: 36 years old, 5 men and 5 women). The concentrations of VSCs and anaerobic bacteria in the mouth decreased after intake of both honeys, but the decreasing effect of manuka honey was markedly stronger than that of acacia honey. Manuka honey showed a 127 times higher amount of MGO than acacia honey by reverse-phase high-performance liquid chromatography analysis. Because there is a strong relationship between the MGO level in honey and the antibacterial activity, the results indicate that the decreasing effect of halitosis by manuka honey would originate from the strong antibacterial activity of MGO.

## Honey for Pharyngitis (Sore Throat)

Pharyngitis is an inflammation of the pharynx that can lead to a sore throat. The most common bacterial cause of pharyngitis, GABHS, is also known as *Streptococcus pyogenes*. Viruses (such as the flu or mononucleosis), other nonstreptococcal bacteria (*Mycoplasma* or *Haemophilus*), fungal infections, or irritants such as pollutants or chemical substances can also cause pharyngitis (Wilson 2008).

Manuka honey with its anti-inflammatory, antiviral, antibacterial, and antifungal powers is used more and more for a sore throat. The best honey rating for a sore throat is a unique manuka factor rating of 10 to 18, as these ratings are recommended for most therapeutic uses. To treat a sore throat, gargling honey is a very effective preventive and curative measure. Honey coats the lining of the throat, destroying harmful germs and soothing it at the same time (Hermyan 2010). Also a way to remedy throat problems involves taking 1 glass of skim milk in which has been poured 1 teaspoon of butter, 1 teaspoon of honey, and 1 teaspoon of turmeric powder. It should be drunk hot. A cup of tea with honey is also very soothing.

## Honey for Cough

In a study involving 105 children aged 2 to 18 years with upper respiratory tract infections of 7 days or less and nighttime coughing, a single nighttime dose of buckwheat honey was an effective alternative treatment for symptomatic relief of nocturnal cough and sleep difficulty compared with a single dose of dextromethorphan (DM). The dose of honey used was 1/2 teaspoon for 2- to 5-year-olds, 1 teaspoon for 6- to 11-year-olds, and 2 teaspoons for 12- to 18-year-olds. Buckwheat honey was chosen in this study because of its high antioxidant properties. The same study shows that honey is more effective than a chemical anti-cough syrup. These anti-cough properties of honey are related to its capacity to dilute bronchial secretions and to improve the function of the bronchial epithelium (Paul et al. 2007).

Researchers from the Pennsylvania State College of Medicine (United States) asked parents to give either honey, honey-flavored DM, or no treatment to the children. The first night, the children did not receive any treatment. The following night, they received a single dose of buckwheat honey, honey-flavored DM, or no treatment 30 min before bedtime. The trial was partially blind, as parents could not distinguish between the honey and the medication. Significant symptom improvements were seen in the honey-supplemented children compared with the no treatment group and DM-treated group, with honey consistently scoring the best and no-treatment scoring the worst.

## Honey for Hiccups

Hiccups are caused by swallowing air, either as a result of eating too quickly, drinking carbonated beverages, or smoking. Drinking alcohol, excitement, and switching between hot and cold drinks have also been said to cause hiccups. Hiccups can be annoying and even painful at times. There are numerous home remedies that can get rid of this condition, several of which feature honey (Kurz 2012).

Combine 1 teaspoon of honey, 1 teaspoon of ginger juice, 1 teaspoon of lime juice, and a pinch of pepper. Stir the mixture together in a cup or small bowl until it is uniform in color and texture. Dip a spoon into the mixture and lick it every few minutes until the hiccups have disappeared.

## Honey for Gastroesophageal Reflux Disease

Acid reflux or gastroesophageal reflux disease (GERD) is one of the most commonly occurring diseases affecting the upper GIT. During gastroesophageal reflux, gastric contents (chyme) passively move up from the stomach into the esophagus. Among the contents is acid that causes irritation, and possible inflammation, which might cause permanent damage to the esophagus. This usually feels like heartburn or leaves a taste of bile or acid in the mouth. It can be caused by the relaxation of the muscle around the top of the stomach, which may cause a "hiatus hernia" or because the stomach is being pressed on due to pregnancy or obesity. This chronic injury is called esophagitis (Siegel et al. 2008).

Although there is no clinical evidence for the medical use of manuka honey on acid reflux, abundant anecdotal reports show that it does at least seem to offer heartburn relief. Common medicines are known to offer relief, but these have side effects, whereas manuka honey does not.

To treat acid reflux, the recommended method of use is to eat 1 tablespoon of raw organic honey before each meal. The honey can also be taken with a small piece of fruit or bread. No liquids should be taken for at least 20 min afterward, as this can dilute the honey and reduce its efficacy. The thickness of the honey coats the esophagus and the opening of the stomach and keeps more acid from coming up. The use of honey stimulates the tissue on the sphincter and helps it to regrow and hence reduces the chances of acid reflux (Sarfaraz 2012).

## Honey for Malabsorption

Malabsorption is inadequate assimilation of dietary substances due to defects in digestion, absorption, or transport. It can affect macronutrients (e.g., proteins, carbohydrates, and fats), micronutrients (e.g., vitamins and minerals), or both, causing excessive fecal excretion, nutritional deficiencies, and GI symptoms. Malabsorption may be global, with impaired absorption of almost all nutrients, or partial (isolated), with malabsorption of only specific nutrients (Ruiz 2012).

The effects of honey and its carbohydrate constituents (glucose, fructose, and raffinose) on calcium absorption in rats were investigated in acute and chronic feeding studies done by Ariefdjohan et al. (2008). It was observed that rats given honey, in the acute study, showed an increase in calcium absorption. Although a positive dose–response effect of honey and its carbohydrate constituents on calcium absorption was observed, this effect disappeared upon long-term feeding in rats, implying that adaptation had occurred.

In certain cases, consumption of relatively large amounts of honey (50–100 g) can lead to a mild laxative effect in individuals with insufficient absorption of honey fructose. Supplementation of honey in concentrations of 2, 4, 6, and 8 g/100 g to protein fed to rats improved the protein and lipid digestibility (Sirnik et al. 1978).

## HONEY FOR DYSPEPSIA, GASTRITIS, AND GASTRIC ULCER

Dyspepsia, also known as upset stomach or indigestion, refers to a condition of impaired digestion. It is a medical condition characterized by chronic or recurrent pain in the upper abdomen, upper abdominal fullness, and feeling full earlier than expected when eating. It can be accompanied by bloating, belching, nausea, or heartburn. Other symptoms include irregular bowel movements and constipation. Dyspepsia is a common problem and is frequently associated with GERD or gastritis. In a small minority, it may be the first symptom of peptic ulcer disease and occasionally cancer (National Institute for Health and Clinical Excellence 2004; Talley and Vakil 2005).

Gastritis is a stomach inflammation without an ulcer or sore. It is usually caused by medications, including corticosteroids, nonsteroidal anti-inflammatory drugs (NSAIDs), cancer drugs, or antibiotics, or by drinking alcohol or experiencing stress or some trauma. In the elderly, poor HCl secretion allows for bacterial growth such as *Helicobacter pylori* and may be another cause of gastritis. A long-term effect of gastritis is vitamin B12 deficiency, which mimics that of senility (Connor and Connor 2010).

A stomach ulcer or gastric ulcer, also called a peptic ulcer, is small erosion (hole) in the GIT. The most common type, duodenal, occurs in the first 12 in. of small intestine beyond the stomach. Duodenal ulcers are almost always benign, while stomach ulcers may become malignant. Within the past decade, it has become accepted that gastric ulcers primarily result from altered mucosal defenses, whereas duodenal ulcers are associated with increased acid production. It has become clear that *H. pylori* plays a vital role in peptic ulcer development in both sites. Ninety-five percent of duodenal ulcers and 80% of stomach ulcers are caused by *H. pylori*, which may be transmitted from person to person through contaminated food and water. The remaining 20% of gastric ulcers and 5% of duodenal ulcers are caused by NSAIDs. A complex relationship exists between host defense mechanisms, the presence of elevated acid, pepsin levels, and *H. pylori* (Siegel et al. 2008).

In the traditional medicine of some parts of the world, honey has been used to treat dyspepsia and stomach ulcers. There are numerous reports of this treatment being used successfully in clinics in Russia in modern times and a recent report of a clinical trial in Egypt, which established that this traditional remedy is in fact effective. However, there has been no explanation of how honey works in this treatment, which has prevented the treatment from being considered seriously by many in the medical profession (Sharma 2006).

Honey is a potent inhibitor of the causing agent of peptic ulcers and gastritis, *H. pylori*. In vitro studies of *H. pylori* isolates have been shown to be inhibited by a 20% solution of honey. Even isolates that exhibited a resistance to other antimicrobial agents were susceptible (Ali et al. 1991; Osato et al. 1999). In a clinical study, the administration of a bland diet and 30 mL of honey three times a day was found to be an effective remedy in 66% of patients and offered relief to a further 17%, while anemia was corrected in more than 50% of the patients (Salem 1981). Clinical and animal studies have shown that honey reduces the secretion of gastric acid. The possibility that the healing effect of honey on the stomach may be through its acting on *H. pylori* was suggested by Niaz Al Somai at the University of Waikato. A study in 1994 showed that the manuka antibacterial factor in raw active manuka honey completely halted the growth of the bacteria at concentrations as low as 5%, but the hydrogen peroxide components in other honeys did not, even at concentrations as high as 50%. Additionally, gastric ulcers have been successfully treated by the use of honey as a dietary supplement. An 80% recovery rate of 600 gastric ulcer patients treated with oral administration of honey has been reported. Radiologic examination showed that ulcers disappeared in 59% of patients receiving honey (Kandil et al. 1987). In rats, honey acted against experimentally induced gastric ulcers. Honey intake in rats prevented indomethacin-induced gastric lesions in rats by reducing the ulcer index, microvascular permeability, and myeloperoxidase activity of the stomach. In addition, honey has been found to maintain the level of nonprotein sulfhydryl compounds (e.g., glutathione) in gastric tissue subjected to factors inducing ulceration (Nasuti et al. 2006). The effect of honey under clinical conditions on more than 40 gastric ulcer patients was studied in a Russian hospital. It was found that ingestion of 120 mL of 33% honey solution by gastric ulcer patients improves the microcapillary blood circulation, which can beneficially influence the gastric ulcers. Ingestion of 120 mL of 33% warm honey solution decreases the acidity of the gastric juice, while the ingestion of the same amount and concentration of a cold honey solution increased the acidity of the gastric juice. The sleep of the gastric ulcer patients was also improved by ingestion of 50 g of honey before sleep. In order to decrease gastric juice acidity, the author recommends the intake of warm honey solution 40 to 60 min before eating. The function of the gallbladder is improved by the ingestion of cool solution of 100 mL of 50% honey (13°C–15°C) (Dubtsova 2009). There are reports on healing of patients suffering from gastritis, duodenitis, and duodenal ulcers by intake of 30 mL of honey (Salem 1981). Another author concludes that ingestion of warm honey in combination with propolis is a good way to treat gastric ulcers (Dubtsova 2009). In a previous study, it was found that honey administered subcutaneously or orally before oral administration of ethanol affords protection against gastric damage and reverses changes in pH induced by ethanol (Ali 1991).

The anti-ulcerous capacity of honey has been attributed to the presence of phenolic compounds, particularly the flavonoids (Gracioso et al. 2002; Batista et al. 2004; Hiruma-Lima et al. 2006). The action mechanism of these compounds varies. According to Speroni and Ferri (1993), flavonoids increase the mucosal content of prostaglandins, which enhances the protective effect on the gastric mucosa, thus preventing ulceration. Honey is not involved on prostaglandin production but has a stimulatory effect on the sensory nerves in the stomach that respond to capsaicin (Ali

1995; Al-Swayeh and Ali 1998). Vilegas et al. (1999) also mention how the flavonoids increase the mucosal content of prostaglandins and have an important inhibitory effect on acid secretions, preventing the formation of peptic ulcers. Other researchers argue that ulcers are related with reactive oxygen species, and flavonoids inhibit lipid peroxidation, which considerably increases glutathione peroxidase activity (Martin et al. 1998; Young et al. 1999; Duarte et al. 2001). Another mechanism of the gastro-protective effect of honey has been suggested by Beretta et al. (2010). It involves the salivary reduction of nitrate $\left(NO_3^-\right)$ to nitrite $\left(NO_2^-\right)$ and the intragastric formation of nitric oxide (NO), the latter involved in the preservation of the gastric mucosa capillaries and in boosting mucous production. The antigastric ulcer and antigastritis effect of honey can be explained by its antibacterial, anti-inflammatory, and anti-oxidant action as well as with its inhibitory effect on the acidity of the gastric juice. The positive effect of honey on nutrition function is also due to its prebiotic effect.

## HONEY FOR GASTROENTERITIS

Gastroenteritis (often called "gastro" or "stomach flu," gastric flu, tummy bug) is an inflammation of the digestive system involving both the stomach and the small intestine. Most cases are infectious, although gastroenteritis may occur after inges-tion of drugs and chemical toxins. Acquisition may be foodborne, waterborne, or via person-to-person spread. It may be caused by bacteria or parasites from spoiled food or unclean water or eating food that irritates the stomach lining and causes emotional upsets such as anger, fear, or stress (Borhany 2011). Gastroenteritis is typically charac-terized by vomiting, watery diarrhea, abdominal cramps, and nausea. Headache and low-grade fever may also be reported. Gastroenteritis can cause dehydration, which sometimes results in a loss of sugar and salts that the body needs to work normally. It is more common in winter and affects people of all ages. Viruses, particularly rotavi-rus, and the bacteria *E. coli* and *Campylobacter* are the primary causes of gastroen-teritis. There are, however, many other infectious agents that can cause this syndrome (norovirus, adenovirus, and astrovirus, *Salmonella, Shigella, Clostridium difficile, Yersinia enterocolitica, Giardia, Cryptosporidium*, and *Cyclospora*) (Boyce 2012).

The primary treatment of gastroenteritis in both children and adults is rehydration. This is preferably achieved by oral rehydration therapy (ORT), although intravenous delivery may be required if there is a decreased level of consciousness or dehydration is severe. Complex carbohydrate–based ORT, such as those made from wheat or rice, may be superior to simple sugar-based ORT (Ingle and Hinge 2012).

A clinical study of honey treatment in infantile gastroenteritis was reported by Haffejee and Moosa (1985). They found that by replacing the glucose in the standard electrolyte-containing oral rehydration solution (ORS) recommended by the World Health Organisation/UNICEF as well as the solution of electrolyte with honey, the mean recovery times of patients (aged 8 days to 11 years) were significantly reduced. Honey was found to shorten the duration of diarrhea in patients with bacterial gastroenteritis caused by organisms such as *Salmonella, Shigella*, and *E. coli*. The researchers concluded that the high sugar content in honey promoted electrolyte and water reabsorption in the gut. At the same time, honey does not result in osmotic diarrhea due to its ratio of fructose to glucose being greater than 1:1.

Another study done by Abdulrhman et al. (2010) was carried out to verify whether the addition of honey in ORS could affect the duration of symptoms of acute gastroenteritis in infants and children. In the honey-treated group, the frequencies of vomiting and diarrhea were significantly reduced compared with the control group. Also, the recovery time, defined as the number of hours from initiation of treatment to when normal soft stools are passed, with the patient showing normal hydration and satisfactory weight gain, was significantly shortened after honey ingestion. It was concluded that honey added to ORS-promoted rehydration of the body and sped the recovery from vomiting and diarrhea.

Honey was tested for its relative antibacterial potency against all the bacterial species that commonly cause gastroenteritis, comparing manuka honey and a honey with the usual hydrogen peroxide activity, and also with an artificial honey, to assess how much of the antibacterial activity was due simply to the acidity and the osmotic effect of the sugar in honey. With some of the species of bacteria, the assessment was repeated with additional strains obtained from clinical isolates supplied by medical and animal health laboratories to see if there was any variation in sensitivity between different strains of a species. The results (Table 8.1) showed that honey

## TABLE 8.1

## Minimum Inhibitory Concentration of Honeys in Nutrient Agar Plates (%, v/v) Giving Partial Inhibition, Bacteriostatic Activity, and Bactericidal Activity against Various Strains of Bacteria That Cause Gastroenteritis

| Bacterial Strain | Manuka Honey With Catalase | | | Pasture Honey | | |
|---|---|---|---|---|---|---|
| | PI | BS | BC | PI | BS | BC |
| *Escherichia coli* 916 | 6% | 7% | 10% | 5% | 6% | 6% |
| *Escherichia coli* ex AHL | 6% | 7% | 10% | — | 6% | 6% |
| *Escherichia coli* K88+ | 6% | 6% | — | 10% | 7% | 6% |
| *S. enteritis* 3484 | 7% | 8% | 10% | 4% | — | 6% |
| *Salmonella hadar* 326 | 6% | 7% | 10% | — | 6% | 6% |
| *Salmonella infantis* 93 | 7% | 8% | 10% | 6% | 7% | 10% |
| *Salmonella typhimurium* 298 | 6% | 7% | 8% | — | 6% | 8% |
| *Salmonella typhimurium* 1739 | 6% | 7% | 9% | — | 6% | 7% |
| *Salmonella typhimurium* ex WH | — | 5% | 10% | — | 5% | 10% |
| *Shigella boydii* 2616 | 6% | 7% | 10% | — | 5% | 6% |
| *Shigella flexneri* 983 | 6% | 7% | 10% | — | 6% | 6% |
| *Shigella sonnei* 86 | 6% | 7% | 10% | — | 5% | 5% |
| *Shigella sonnei* ex WH | 5% | 6% | 10% | — | 6% | 10% |
| *Vibrio cholerae* | 5% | 7% | 10% | 6% | 7% | 10% |
| *Vibrio parahaemolyticus* | 5% | 6% | 10% | — | 4% | 6% |
| *Yersinia enterocolitica* | 10% | 11% | 15% | 7% | 8% | 9% |

*Source:*  Ko, Y.-J. 2005. Investigating the sensitivity of enteropathogenic bacteria to the antibacterial activity of honey. Thesis (M.Sc. Biological science). University of Waikato, New Zealand.

*Note:*  BC, bactericidal activity; BS, bacteriostatic activity; PI, partial inhibition.

## TABLE 8.2
## Minimum Inhibitory Concentration (% v/v) of Tualang Honey by Visual Inspection and Spectrophotometric Measurement against Selected Strains of Wound and Enteric Microorganism

| | Tualang Honey | |
| --- | --- | --- |
| Microorganism | MICv | MICs |
| *Streptococcus pyogenes* | 12.5 | 12.5 |
| Coagulase-negative *staphylococci* | 12.5 | 12.5 |
| Methicillin-resistant *Staphylococcus aureus* | 12.5 | 12.5 |
| *Streptococcus agalactiae* | 25 | 25 |
| *Staphylococcus aureus* | 25 | 25 |
| *Stenotrophomonas maltophilia* | 12.5 | 12.5 |
| *Acinetobacter baumanii* | 12.5 | 12.5 |
| *Salmonella enterica* serovar *typhi* | 25 | 25 |
| *Pseudomonas aeruginosa* | 25 | 25 |
| *Proteus mirabilis* | 25 | 25 |
| *Shigella flexneri* | 25 | 25 |
| *Escherichia coli* | 25 | 25 |
| *Enterobacter cloacae* | 25 | 25 |

*Note:* MICs, MIC by spectrophotometric measurement; MICv, MIC by visual inspection.

with an average level of hydrogen peroxide activity is bacteriostatic at 4% to 8% (v/v) and bactericidal at 5% to 10% (v/v). The nonperoxide activity of an average manuka honey is bacteriostatic at 5% to 11% (v/v) and bactericidal at 8% to 15% (v/v). Activity (just bacteriostatic) was not seen with artificial honey unless it was at 20% to 30% (v/v), clearly showing the importance of factors other than sugar and acidity (Ko 2005).

Another study done in Nigeria to evaluate the antimicrobial effect of natural honey on diarrhea-causing bacteria (*E. coli*, *Campylobacter jejuni*, *Salmonella enterocolitis*, and *Shigella dysenteriae*) revealed that samples of natural honey used were effective in inhibiting the growth of all the organisms tested.

A study done by Kaur-Kirnpal et al. (2011) evaluated the antibacterial properties of Tualang (*Koompassia excelsa*) honey against wound and enteric bacteria; the results of this study are presented in Table 8.2.

## HONEY AND IBS

IBS, also called spastic colon, irritable colon, or nervous stomach, is a functional bowel disorder in which abdominal pain or discomfort is associated with defecation or a change in bowel habits. Bloating and distension are commonly associated features. Often, people with IBS have alternating constipation and diarrhea. It has a prevalence of 12% worldwide (Mertz 2003; Quigley et al. 2009).

Active manuka honey is used for the entire GIT from acid reflux, heartburn, upset stomach, stomach ulcer, duodenum ulcer, IBS, diverticulitis, and mild forms of diverticulitis. The osmotic effect of this particular type of honey helps stool to move by affecting the way fluids move into the colon (Buonanotte 2009). People have reported being able to conquer IBS completely with the combination of a good diet and taking 1 teaspoon of raw manuka honey dissolved in a cup of warm water three times daily on an empty stomach (Isaacs 2010).

Researchers at the Chandigarh Postgraduate Institute of Medical Education and Research (India) have discovered that eating regular doses of manuka honey can help in cases of IBS and ulcerative colitis. They induced the digestive ailments in experimental rats and then fed the animals with manuka honey. At examination, the rats that received the honey showed greatly reduced levels of inflammation in the bowel and improved values for cell changes and antioxidant levels. The dose used in the studies was 5 g/kg of body weight daily (Williams 2012). A drop of dill oil to 1 teaspoon of honey may also be useful as a home remedy. Take after every meal.

## HONEY FOR CONSTIPATION

Constipation is the passage of small amounts of hard, dry stool fewer than three times per week or a significant change in one's usual routine, accompanied by straining, and feelings of being bloated, or having abdominal fullness. Persistence of these symptoms for 3 months or longer is defined as chronic constipation. It sometimes causes problems such as fissures, hemorrhoids, arthritis, high blood pressure, cataract, appendicitis, and more. Constipation can be attributed to at least one of several factors: poor eating habits, lack of exercise, not eating enough fiber or drinking enough fluids, side effects of medicines, or even poisoning by heavy metals (Folden et al. 2002).

Constipation can be treated by increasing the intake of water. Apart from these, constipation can also be treated with honey. Using honey for constipation is one of the many home remedies one can use. Many modern and ancient writings have mentioned that honey has a variety of healing properties, which include being a mild laxative.

The mild laxative properties of honey are used for the treatment of constipation in Eastern Europe, China, and the Near East (Cutting 2007).

It has been scientifically proven recently that raw honey absorbs water and can also hold a lot of water. This combination helps honey to keep the fecal matter soft and wet when it passes through the digestive system. Hence, it acts like a lubricant stimulating the bowels for the passage of fecal matter. Sugar molecules in honey can change into other sugars, for example, fructose can change to glucose. Therefore, in spite of its large content of acid, honey can be easily digested especially by people who have sensitive stomachs. Honey helps the intestines and kidneys perform better, thus helping combat constipation. It can be used in several ways for constipation (www.ApitherapyHealth.com).

Take 1 tablespoon of honey three times a day. Consuming 10 g of honey in the morning on an empty stomach and not eating anything for at least an hour gives relief from constipation. Also, mixing 2 tablespoons of lemon juice and 1 teaspoon of honey into 1 glass of warm water and drinking 1 glass in the morning and one in the evening brings back regularity. Add 1 teaspoon of honey to 1 glass of warm milk and consume it two times a day. This aids in curing severe constipation.

## HONEY FOR ANAL DISORDERS

Hemorrhoidal disease and anal fissure are common disorders of the anorectal area. In addition to surgery, many procedures and preparations have been used for the treatment of hemorrhoids (Al-Waili et al. 2006).

### Hemorrhoids

Hemorrhoids, or piles, are swollen and inflamed veins around the anus or in the lower rectum. They can cause mild itching or discomfort and often bleeding. Less often, the blood inside a hemorrhoid forms a clot, or thrombus, causing a great deal of pain. They are caused by chronic excess pressure from straining during a bowel movement, persistent diarrhea, or pregnancy. Hemorrhoids may be located just inside the anal canal (internal hemorrhoid) or surrounding the anal opening (external hemorrhoids) (University Health Services 2009; National Institutes of Health 2010).

### Anal Fissures

Anal fissures are splits or cracks in the skin at the opening of the anus, leaving exposed some of the muscle fibers of the anal canal. Pain results from recurrent opening of the wound when the bowels are open and this is often accompanied by bleeding. In addition, the inner circle of muscle in the anal canal (called the internal sphincter) goes into spasm: this makes the pain worse and can prevent healing. The most common cause of an anal fissure is the passage of very hard or watery stools (Colorectal Surgery 2010). Initial treatment for anal fissures includes pain medicine, dietary fiber to reduce the occurrence of large, bulky stools, and sitz baths (sitting in a few inches of warm water).

A study done by Al-Waili et al. (2006) demonstrated that the mixture of natural honey, olive oil, and beeswax was effective for the management of anal fissure and hemorrhoids. The improvement was probably due to the antimicrobial, anti-inflammatory, antioxidant, and healing properties of the mixture according to the properties of the ingredients (Table 8.3). Honey reduces prostaglandin concentration and enhances NO production. NO, which has antimicrobial activity, can accelerate wound healing, improves microcirculation of the flap, and increases its survival rates. Exogenous NO has been shown useful in decreasing the internal anal sphincter tone. NO donors have been used for treatment of chronic anal fissure and acute strangulation of prolapsed internal hemorrhoids.

A mixture of onion and honey can be surprisingly effective in treating hemorrhoids. Honey is an antibacterial and antifungal, while onion can have a soothing property. Combining these two ingredients helps the hemorrhoids shrink and reduces pain. Just mash an onion and mix together with 1 tablespoon of honey. After cleaning the area, dab the mixture onto hemorrhoids as an at-home cream (Table 8.2).

**TABLE 8.3**

**Some Physical and Biological Properties of the Mixture Ingredients**

| Honey | Beeswax | Olive Oil |
|---|---|---|
| Increases NO | Decreases prostaglandins | Increases NO |
| Decreases prostaglandins | Antioxidant effect | Decreases prostaglandins |
| Liberates $H_2O_2$ | Anti-inflammatory effect | Antioxidant effect |
| Acidity effect | Healing property | Anti-inflammatory effect |
| Osmolality effect | Decreases leukotriene B4 | Healing property |
| Antioxidant effect | | Decreases leukotriene B4 |
| Anti-inflammatory effect | | Modulate cytokine production |
| Healing property | | Antimicrobial effect |
| Antimicrobial effect | | |
| Modulate cytokine production | | |

## Anal Abscesses and Fistula

Anal abscesses and fistulas are the result of an anal gland infection. Anorectal abscesses vary depending on the site of the abscess and can be classified as perianal, ischiorectal, intersphincteric, or supralevator (Rickard 2005). Treatment includes draining the abscess, usually under local anesthesia. Often, a fistula follows an abscess draining. Symptoms of anorectal abscesses and fistulas include pain, swelling, drainage, bleeding, constipation, general feeling of being ill, and rarely urinary difficulties. People with IBD may also develop an anal abscess and fistula (Sseitzinger 2008).

A case study done by Vlcekova et al. (2011) followed a 55-year-old male who had been suffering from ongoing perianal fistulas for 10 years. The patient tested positive for *Staphylococcus aureus*, *Staphylococcus haemolyticus*, and *E. coli*. After 10 days of treatment with local γ-irradiated honeydew honey, either diluted with sterile physiologic solution or undiluted, purulent discharge was reduced, and after 24 days, epithelization had begun. After 5 months, fistulas in the left region of the buttocks had disappeared.

## HONEY FOR IBD

IBD is a general classification of inflammatory processes that affect the large and small intestines. Ulcerative colitis and Crohn's disease together make up IBD. Ulcerative colitis involves the mucosa and submucosa of the colon. Crohn's disease or regional enteritis involves all layers of the gut. When the inflammation occurs in the rectum and lower part of the colon, it is called ulcerative proctitis. If the entire colon is affected, it is called pancolitis. If only the left side of the colon is affected, it is called limited or distal colitis (Siegel et al. 2008; National Institutes of Health 2010). Ulcerative colitis and Crohn's diseases are chronic conditions that can last years to decades. Men and women are affected equally. They most commonly begin during adolescence and early adulthood, but they also can begin during childhood and later in life. It can be difficult to diagnose an ulcerative colitis because its symptoms are similar to other intestinal disorders and to Crohn's disease. Ulcerative colitis may also cause problems such as arthritis, inflammation of the eye, liver disease, and osteoporosis. It is not known why these problems occur outside the colon. Scientists think these complications may be the result of inflammation triggered by the immune system. About 5% of people with ulcerative colitis develop colon cancer. The risk of cancer increases with the duration of the disease and how much the colon has been damaged (Schoenfeld and Wu 2011).

A study carried out by Prakash et al. (2008) to evaluate the effect of different doses of manuka honey in experimentally induced IBD in rats showed a significant protection with manuka honey. All the treated groups showed reduced colonic inflammation and all the biochemical parameters, and also morphologic and histologic scores, were significantly reduced in the treated group. Manuka honey at different doses restored lipid peroxidation as well as improved the antioxidant parameters. In the inflammatory model of colitis, oral administration of manuka honey 5 g/kg and manuka honey 10 g/kg body weight significantly reduced the colonic inflammation.

As a home remedy, to properly treat colitis, cider vinegar and honey can be used. Both have been used effectively in the treatment of colitis. Take 2 teaspoons of cider vinegar and honey and mix in water. Be sure to drink this three times a day.

## Honey for Liver Care

A healthy liver is essential for proper functioning. Damage in the liver can lead to disturbance in body functions. It plays a crucial role in removing toxicants or unwanted substances introduced in the body by long chemical-based medicine. Liver functioning gets disturbed because of harmful infectious particles, accidents, severe illness, and inadequate rest.

Chronic liver diseases are the biggest problem. An estimated 30% of the world's total population have been infected with the hepatitis B virus and more than 350 million people are inflicted by chronic hepatitis B virus. About 170 million people have hepatitis C. Chronic liver diseases from viral hepatitis might lead one fifth of the patients to liver cirrhosis, which might end in the need of a liver transplantation. Again, 20% of the cirrhotic patients are at risk of developing hepatocellular carcinoma, which calls for liver resection. A much smaller number of liver patients, about 12,000 per year, get fulminant hepatic failure or acute liver diseases (Stolzke 2008).

Honey works to harmonize the liver system, neutralize toxins, and relieve pain (Brody 2011). Many complications have been attributed to oxidative damage. Obstructive jaundice from a blocked common bile duct results in severe oxidative damage and inflammation of liver tissue. Results of a study done by Erguder et al. (2008) showed that honey, with its antioxidant activity, has a potential protective effect on liver damage induced by obstruction of the common bile duct.

A mixture of honey and bhringaraj (*Eclipta alba*) is effective in treating liver cirrhosis and provides good liver care. Make natural solution using juice of the medicinal herb bhringaraj (2 tablespoons) and honey (2 tablespoons). This is a dose that should be taken two times in a day (half in the morning and half in the evening). This is one of the important home remedies of liver care.

The average 1:1 ratio of fructose to glucose found in commonly available honey is the key to one of its most significant health benefits. Fructose derivatives in the liver activate the release of glucokinase from the liver mitochondria. Glucokinase is necessary for the conversion of glucose to glycogen in the liver. In other words, the presence of fructose facilitates or accelerates the conversion of glucose to glycogen. Insulin becomes in effect a fat-storing hormone that rapidly ushers glucose out of the bloodstream and into the cells. Depending on the immediate energy needs of muscle cells, some of the glucose is converted to glycogen to be used for fuel. Most of the glucose from a typical high-carbohydrate meal (excluding one with honey) is stored in the cells as fat. With fructose present, as in honey, glucose is stored as glycogen in the liver, where it is available for use by the brain, the heart, the kidneys, and the red blood cells during periods of rest and recovery.

A positive effect of honey on hepatitis A patients was found after ingestion of clover and rape honey, causing a decrease of alanine aminotransferase activity (by 9–13 times) and of bilirubin production by 2.1 to 2.6 times (Baltuskevicius et al. 2001).

### HONEY FOR PANCREAS DISEASES

The pancreas is an organ that has two important functions in digestion: the production of enzymes to digest food and the production of hormones to control blood sugar.

There are a variety of disorders of the pancreas including acute pancreatitis, chronic pancreatitis, hereditary pancreatitis, and pancreatic cancer. The evaluation of pancreatic diseases can be difficult due to the inaccessibility of the pancreas (National Pancreas Foundation).

In a study done by Erejuwa et al. (2010), on diabetic rats, treatment with Tualang honey had significantly reduced blood glucose levels and restored superoxide dismutase and catalase activities. These results suggest that the hypoglycemic effect of Tualang honey might be attributed to its antioxidative effect on the pancreas.

## HONEY AS A PREBIOTIC

Prebiotics are food ingredients that are supposed to stimulate the activity and potentially alter the composition of the gut flora by providing energy to selected species of the microbial community. For a food ingredient to be classified as a prebiotic, it must fulfill the following criteria: (1) it should either be hydrolyzed or absorbed in the upper part of the GIT; (2) it should be selectively fermented by one or a limited number of potentially beneficial bacteria commensal to the colon (e.g., bifidobacteria and lactobacilli), which are stimulated to grow and/or become metabolically activated; and (3) prebiotics must be able to alter the colonic microflora toward a healthier composition (Gibson 1998). Attempts have been made to increase the number of *Bifidobacterium* and *Lactobacillus* bacterial strains, which are considered to have health-promoting properties. Studies have demonstrated positive effects on calcium and other mineral absorption (Scholz-Ahrens and Schrezenmeir 2007), immune system effectiveness, bowel pH, reduction of colorectal cancer risk, inflammatory bowel disorders, hypertension, and intestinal regularity (Geier et al. 2006; Hedin et al. 2007; Lomax and Calder 2009). Recent human trials have reinforced the role of prebiotics in preventing and possibly stopping early-stage colon cancer (Munjal et al. 2009). It has been argued that many of these health effects emanate from increased production of short-chain fatty acids (SCFAs) by the stimulated beneficial bacteria. Thus, food supplements specifically enhancing the growth of SCFA-producing intestinal bacteria (such as *Clostridia* and *Bacteroides* spp.) are widely recognized to be beneficial. It has been argued that prebiotics are beneficial to Crohn's disease through production of SCFAs to nourish the colon walls and beneficial to ulcerative colitis through reduction of hydrogen sulfide gas due to reduction of sulfate-producing bacteria, which do not thrive in the slightly acidic environment that SCFAs create.

The immediate addition of substantial quantities of prebiotics to the diet may result in a temporary increase in gas, bloating, or bowel movement. Human colonic bacteria substrates are relatively stable. Productions of SCFA and fermentation quality are reduced during long-term diets of low fiber intake (El Oufir et al. 1996). Until bacterial flora are gradually established to habilitate or restore intestinal tone, nutrient absorption will be impaired and colonic transit time temporarily increased with an immediate addition of higher prebiotic intake (Givson et al. 1996).

Important effects of honey on human digestion have been linked to oligosaccharides. These honey constituents have prebiotic effects, similar to that of fructo-oligosaccharide (FOS) (Sanz et al. 2005). According to an in vitro study on five bifidobacteria strains, honey has a growth-promoting effect similar to that of fructose and glucose oligosaccharides. It has been found that the activity of certain species of *Bifidobacterium* in the colon can be stimulated by FOSs (Kajiwara et al. 2002). The effect of honey was similar to that of commercial FOS, gluco-oligosaccharide, and inulin. It has been established that different honeys contain specific oligosaccharides. The oligosaccharide panose was the most active one, and isomaltose and melezitose would also prove stimulatory to *Bifidobacterium* spp. Equally relevant is the fact that the lactobacilli in the small intestine could well be affected advantageously by the sugars and/or other components in honey (Haddadin et al. 2007). Researchers from Michigan State University have shown that adding honey to dairy products such as yogurt can enhance the growth, activity, and viability of these beneficial bacteria (Altman 2010).

Although the stimulatory role of oligosaccharides on the gut flora(s) has received the most attention, it has been speculated that the same components in honey could inhibit the development of pathogens such as *H. pylori* or *S. aureus* in the body. More specifically, it has been proposed that the oligosaccharides could become attached to the cell walls of the bacteria and prevent adhesion to human tissues.

The benefits of FOS are numerous. They help carry moisture through the digestive system and therefore promote healthy bowel movements. Growth of the beneficial bacteria encourages the production of SCFAs, lowering pH, inhibiting unhealthy bacteria, and detoxifying carcinogens in the diet. FOS also helps to lower blood sugar and cholesterol levels by reducing absorption of carbohydrates and fats into the bloodstream and nourishes the cells in the gut, which is important for digestive well-being and the prevention of constipation. Improving digestion is the key to establishing good health. When in good health, we get rid of wastes and toxins through regular bowel movements and eliminate the buildup of unhealthy microorganisms and internal toxins, providing a strong and intact intestinal barrier to prevent the leaking of undigested food fragments into the bloodstream (http://www.benefits-of-honey.com/milk-and-honey.html).

## SOME HONEY-BASED MIXTURES FOR GI DISORDERS

The use of honey in medicinal mixtures has a long tradition. Research showed that the employment of honey in mixtures prevails over the use of pure honey (Tables 8.4 and 8.5).

### HONEY AND LEMON

Honey and lemon have been combined by many cultures for countless generations. Combining pure, raw honey with fresh lemon juice is a popular weight-loss remedy, although there are many other benefits to be had, including enhancing the digestion. Honey and lemon share some of the same properties that benefit the GI system, but they contrast in other ways, especially their sweet and sour tastes. The acidity of lemon juice can aid the ability of stomach acid to chemically digest food, which not only provides more nutrients to absorb in the intestines but also reduces symptoms of indigestion, bloating, and heartburn. Lemon juice also displays antimicrobial properties and is a rich source of vitamin C, which combine to reduce the incidence of infections throughout the GIT. It also contains calcium, magnesium, and potassium. Honey and lemon both possess antimicrobial and antioxidant properties, and they are good sources of nutrients. A better approach is to use slightly warm water to dissolve the honey before adding lemon juice. The ratios sometimes used are 1 tablespoon of raw honey to 2 tablespoons of lemon juice in an 8-ounce glass of warm water. Honey and lemon can also be added to tea, but the tannins and caffeine in black tea leaves may lead to GI upset (Bond 2011).

### HONEY AND CINNAMON

The combination of honey and cinnamon has been used in both oriental and Ayurvedic medicine for centuries. Cinnamon is one of the oldest spices known to mankind and honey's popularity has continued throughout history. The two ingredients have a long history as a home remedy. On January 17, 1995, *Weekly World News, Canada* published a list of ailments/diseases that can be treated with a honey/cinnamon mixture. Cinnamon's essential oils and honey's enzyme that produces hydrogen peroxide qualify as the two "antimicrobial" foods with the ability to help stop the growth of bacteria as well as fungi. Both are used not just as a beverage flavoring and medicine but also as an embalming agent and are used as alternatives to traditional food preservatives due to their effective antimicrobial properties. People

**TABLE 8.4**

**Some Traditional Mixtures for the Treatment of Digestive System Disorders**

| Substances | Indication | References |
|---|---|---|
| Honey and lemon | Throat problems, GI upset | Bond 2011 |
| Honey and cinnamon | Gastric ulcer, mouth ulcer, indigestion, and upset stomach | http://www.leaflady.org/honey.htm |
| Honey, olive oil, and beeswax | Hemorrhoids and anal fissure | Al-Waili et al. 2006 |
| Honey and onion | Hemorrhoids | http://www.ehow.com/way_5365888_homemade-remedies-hemorrhoids.html |
| Honey and propolis | Gastric ulcer | Dubtsova 2009 |
| Ginger juice and honey | Digestive disorders causing diarrhea | http://www.speedyremedies.com/home-remedies-for-common-digestive-disorders.html |
| Pomegranate juice and honey | Dyspepsia | http://www.speedyremedies.com/home-remedies-for-common-digestive-disorders.html |
| Honey apple cider vinegar | Colitis, cleanse the colon and stimulate the immune system | http://www.ehow.co.uk/way_5392778_apple-remedy-cure-baldness.html |
| Honey and tea | Sore throat | http://www.ehow.co.uk/way_5392778_apple-remedy-cure-baldness.html |
| Fresh leaf juice of drumsticks, and honey and 1 glass of tender coconut water | Colitis | http://www.vegrecipes4u.com/health-benefits-of-drumstick.html#.U11N48UsCvQ |
| Honey and milk | Constipation | http://www.benefits-of-honey.com/milk-and-honey.html |
| Honey with milk, butter, and turmeric | Pharyngitis | www.botanical-online.com |
| Honey, ginger juice, limejuice, and a pinch of pepper | Hiccups | http://www.ehow.com/how_5607281_cure-hiccups-honey.html |
| Dill oil and honey | IBS | http://voices.yahoo.com/irritable-bowel-syndrome-it-3521975.html?cat=68 |
| Honey and juice of bhringaraj | Liver cirrhosis, provides good liver care | http://www.mickeymehtahbf.com/blog/2012/01/10/cirrhosis-liver-care-part/ |

**TABLE 8.5**

**Explaining the Use of Honey in Medicine**

| Therapeutic and Health-Enhancing Use | Biological Rationale |
|---|---|
| Therapy of digestive diseases such as peptic ulcers and gastritis | Antibacterial and anti-inflammatory effects |
| Against children's diarrhea | Antibacterial and anti-inflammatory effects |
| Improvement of gut microbial health and of digestion | Prebiotic effect |
| Improvement of immune reaction of the body | Immunoactivating effect |
| Long-term ingestion of honey can reduce the risk of human cancer | Anticancerogenic effect |
| Positive glycemic nutritional effect; can be used as a sweetener for people with diabetes type 2 and also probably type 1 | Some honeys have a low glycemic index (e.g., acacia honey); other fructose-rich honeys such as thyme, chestnut, heather, and tupelo are good alternatives |
| Use for the treatment of radiation-induced mucositis | Antibacterial and anti-inflammatory effects |
| Positive effect of honey ingestion on hepatitis A patients | Anti-inflammatory effect |
| Improvement of cough in children | Contact soothing effect, sweet substances, as a sweetener honey causes reflex salivation and increases airway secretions, which may lubricate the airway and remove the trigger that causes a dry, nonproductive cough |

*Source:* Cutting, K., *Ostomy Wound Management, 53*(11), 49–54, 2007.

have claimed that the mixture is a natural cure for many diseases and a formula for many health benefits (http://www.epatienthealthcare.com/tag/colds/).

*Gas*: According to the studies done in India and Japan, it is revealed that if honey is taken with cinnamon powder, the stomach is relieved of gas.

*Immune system*: Daily use of honey and cinnamon powder strengthens the immune system and protects the body from bacterial and viral attacks. Constant use of honey strengthens the white blood corpuscles to fight bacterial and viral diseases.

*Indigestion*: Cinnamon powder sprinkled on 2 tablespoons of honey taken before food relieves acidity and digests the heaviest of meals.

*Upset stomach*: A mixture of honey and cinnamon is also beneficial in treating various stomach problems. A mixture prepared by equal quantity of honey and cinnamon powder can help in relieving pain caused by gas or bladder infection. If taken regularly, the mixture is also helpful in facilitating proper digestion of food.

*Bad breath*: In order to maintain fresh breath throughout the day, the people of South America gargle with 1 teaspoon of honey and cinnamon powder mixed in hot water first thing upon awakening.

*Gastric disturbances*: Honey taken with cinnamon powder cures stomach aches and can help heal stomach ulcers. According to studies done in India and Japan, honey mixed with cinnamon powder relieves gas by removing acidity and helps with digestion and absorption of nutrients. Honey mixed with cinnamon can also help cure bladder infections by destroying germs in the bladder (Patel 2012).

## HONEY AND APPLE CIDER

Apple cider vinegar and honey have been proven over time to help heal the body in several ways. Going back several hundred years, apples have continuously been used for medicinal purposes as well as for cleaning. Apple cider vinegar is usually fermented and contains many vitamins, minerals, and other elements that help promote a healthy body. Some of those are magnesium, potassium, iron, and calcium. All these substances clean out the body by helping multiple areas rid themselves of toxins, impurities, and inflammation. Honey is a carbohydrate compound that helps create and move energy through the body. It also promotes cell oxidation and the breaking down of harmful bacteria to keep the body clean and balanced.

Mix 1 tablespoon of honey and 1 tablespoon of apple cider vinegar, add the mixture to a cup of boiling water, stir, and drink. This will help cleanse the colon. Drinking a cup of this mixture daily should help regulate the digestive tract.

With the combination of apples and honey in the system, there are several internal ailments that can slowly be cured. It all begins when these two ingredients are taken and they stimulate the immune system. Possibly the most important thing is that they are both used as disinfectants and antiseptics to kill harmful bacteria in the digestive tract that are responsible for causing food poisoning. Also, for body chemical and energy balance, apples and honey help restore pH levels. That means having greater resistance to the common cold and flu, ulcers, headaches, and even yeast infections. Apple cider vinegar and honey also make a good fiber additive that helps lower cholesterol and controls blood sugar. The combination also helps detoxify the liver to promote overall health and aid in weight loss. Inflammation in joints that could cause arthritis and osteoporosis is lowered, so bone mass can stay strong (Bailey 2012).

## REFERENCES

Abdulrhman, M.A., Mekawy, M.A., Awadalla, M.M. and Mohamed, A.H. 2010. Bee honey added to the oral rehydration solution in treatment of gastroenteritis in infants and children. *Journal of Medicinal Food* 13(3):605–609.

Ahmed, N., Dawson, M., Smith, C. and Wood, E. 2007. *Biology of Disease*. Taylor & Francis, Boca Raton, FL.

Ali, A.T., Chowdhury, M.N. and Al-Humayyd, M.S. 1991. Inhibitory effect of natural honey on *Helicobacter pylori*. *Tropical Gastroenterology* 12:139–143.

Ali, A.T.M. 1995. Natural honey accelerates healing of indomethacin-induced antral ulcers in rats. *Saudi Medical Journal* 16(2):161–166.

Ali, A.T.M.M. 1991. Prevention of ethanol-induced gastric lesions in rats by natural honey, and its possible mechanism of action. *Scandinavian Journal of Gastroenterology* 26(3):281–288.

Al-Swayeh, O.A. and Ali, A.T. 1998. Effect of ablation of capsaicin-sensitive neurons on gastric protection by honey and sucralfate. *Hepato Gastroenterology* 45(19):297–302.

Altman, N. 2010. *The Honey Prescription: The Amazing Power of Honey as Medicine.* Healing Arts Press. Inner Traditions, Bear & Company, Rochester, VT.

Al-Waili, N.S., Saloom, K.S., Al-Waili, T.N. and Al-Waili, A.N. 2006. The safety and efficacy of a mixture of honey, olive oil, and beeswax for the management of hemorrhoids and anal fissure. *Scientific World Journal* 6:1998–2005.

Ariefdjohan, M.W., Martin, B.R., Lachcik, P.J. and Weaver, C.M. 2008. Acute and chronic effects of honey and its carbohydrate constituents on calcium absorption in rats. *Journal of Agricultural and Food Chemistry* 56(8):2649–2654.

Baltuskevicius, A., Laiskonis, A., Vysniauskiene, D., Ceksteryte, V. and Racys, J. 2001. Use of different kinds of honey for hepatitis A treatment and for reduction of increased acidity of gastric juice. *Zemdirbyste, Mokslo Darbai* 76:173–180.

Batista, L.M., de Almeida, A.B., de Pietro Magri, L., Toma, W., Calvo, T.R., Vilegas, W., and Souza-Brito, A.R. 2004. Gastric antiulcer activity of *Syngonanthus arthrotrichus* SILVEIRA. *Biological and Pharmaceutical Bulletin* 27(3):328–332.

Beretta, G., Gelmini, F., Lodi, V., Piazzalunga, A. and Facino, R.M. 2010. Profile of nitric oxide (NO) metabolites (nitrate, nitrite and N-nitroso groups) in honeys of different botanical origins: Nitrate accumulation as index of origin, quality and of therapeutic opportunities. *Journal of Pharmaceutical and Biomedical Analysis* 53(3):343–349.

Blair, S.E. and Carter, D.A. 2005. The potential for honey in the management of wounds and infections. *Australian Infection Control* 10(1):24–31.

Bond, O. 2011. Honey and Lemon for Digestion. Available at http://www.livestrong.com/article/467515-honey-and-lemon-for-digestion/.

Borhany, S.A. 2011. Honey: Perpetual medicine of the Quran, for all diseases. The News International, Pakistan.

Boyce, T.G. 2012. *Overview of Gastroenteritis.* The Merck Manual, Whitehouse Station, NJ.

Bowen, W.H. and Lawrence, R.A. 2005. Comparison of the cariogenicity of cola, honey, cow milk, human milk, and sucrose. *Pediatrics* 116(4):921–926.

Brody, L. 2011. *Bees, Honey and the Liver.* Harmony Healing Center, Sebastopol, USA. Available at http://harmonyhealingcentersebastopol.com/201103/spring-newsletter-bees-honey-liver/.

Buonanotte, F. 2009. *Manuka Honey for the Treatment of Irritable Bowel Syndrome.* Available at http://www.articlesbase.com/.

Castaldo, S. and Capasso, F. 2002. Propolis, an old remedy used in modern medicine. *Fitoterapia* 73(1):S1–S6.

Cavanah, D., Beazley, J. and Ostapowicz, F. 1968. Radical operation for carcinoma of the vulva; a new approach to wound healing. *Journal of Obstetrics and Gynaecology of the British Commonwealth* 77:1037–1040.

Colorectal Surgery. 2010. *Anal Fissures.* Cambridge University Hospitals NHS Foundation Trust. Cambridge, U.K. Available at www.cuh.org.uk.

Connor, J.G. and Connor, B. 2010. Gastrointestinal disorders. Compassionate Acupuncture and Healing Arts, U.K.

Cutting, K. 2007. Honey and contemporary wound care: An overview. *Ostomy Wound Manage* 53(11):49–54.

Decaix, C. 1976. Comparative study of sucrose and honey. *Le Chirurgien-dentiste de France* 46(285–286):59–60.

Diskin, A. 2012. *Gastroenteritis in Emergency Medicine.* American Academy of Emergency Medicine. Milwaukee, WI.

Duarte, J., Galisteo, M., Angeles-Ocete, M., Perez-Vizcaino, F., Zarzuelo, A. and Tamargo, J. 2001. Effects of chronic quercetin treatment on hepatic oxidative status of spontaneously hypertensive rats. *Molecular and Cellular Biochemistry* 221(1/2):155–160.

Dubtsova, E. 2009. *Clinical Studies with Bee Products for Therapy of Some Nutritional Diseases* (in Russian). Central Moscow Institute of Gastroenterology, Moscow, Russia, pp. 1–38.

Edgar, W.M. and Jenkins, G.N. 1974. Solubility-reducing agents in honey and partly-refined crystalline sugar. *British Dental Journal* 136:7–14.

El-Haddad, S. 2009. *Honey: A New Treatment for Recurrent Minor Aphthous Stomatitis.* 6th International Arab Apicultural Conference. Available at http://www.saudibi.com/files/file/c.pdf.

El Oufir, L., Flourié, B., Bruley desVarannes, S. et al. 1996. Relations between transit time, fermentation products, and hydrogen consuming flora in healthy humans. *Gut* 38(6):870–877.

Emarah, M.H. 1982. A clinical study of the topical use of bee honey in the treatment of some ocular diseases. *Bulletin of Islamic Medicine* 2(5):422–425.

Erejuwa, O.O., Sulaiman, S.A., Wahab, M.S., Sirajudeen, K.N., Salleh, M.S. and Gurtu, S. 2010. Antioxidant protection of Malaysian tualang honey in pancreas of normal and streptozotocin-induced diabetic rats. *Annales d'Endocrinologie (Paris)* 71(4):291–296.

Erguder, B.I., Kilicoglu, S.S., Namuslu, M., Kilicoglu, B., Devrim, E., Kismet, K., Durak, I. 2008. Honey prevents hepatic damage induced by obstruction of the common bile duct. *World Journal of Gastroenterology* 14(23):3729–3732.

Farthing, M., Lindberg, G., Dite, P. et al. 2008. *World Gastroenterology Organisation Practice Guideline: Acute Diarrhea.* World Gastroenterology Organisation, Munich, Germany.

Folden, S.L., Backer, J.H., Gilbride, J.A. et al. 2002. Guidelines for the management of constipation in adults. Rehabilitation Nursing Foundation. Available at www.rehabnurse.org.

Geier, M.S., Butler, R.N. and Howarth, G.S. 2006. Probiotics, prebiotics and synbiotics: A role in chemoprevention for colorectal cancer? *Cancer Biology and Therapy* 5(10):1265–1269.

Ghosh, S. and Playford, R.J. 2003. Bioactive natural compounds for the treatment of gastrointestinal disorders. *Clinical Science* 104:547–556.

Gibson, G.R. 1998. Dietary modulation of the human gut microflora using prebiotics. *British Journal of Nutrition* 80:S209–S212.

Givson, G.R., Willems, A., Reading, S. and Collins, M.D. 1996. Fermentation of non-digestible oligosaccharides by human colonic bacteria. Symposium 2. *Proceedings of the Nutrition Society* 55(3):899–912.

Gorgas, F.J.S. 1901. *Dental Medicine. A Manual of Dental Materia Medica and Therapeutics.* P. Blakiston's Son & Co. Philadelphia, PA.

Gracioso, J.D.S., Vilegas, W., Hiruma-Lima, C.A. and Souza Brito, A.R. 2002. Effects of tea from *Turnera ulmifolia L.* on mouse gastric mucosa support the Turneraceae as a new source of antiulcerogenic drugs. *Biological and Pharmaceutical Bulletin* 25(4):487–491.

Greenberger, N.J. 2012. *Overview of Gastrointestinal Symptoms.* The Merck Manual. Merck Sharp & Dohme Corp., Whitehouse Station, NJ.

Grobler, S.R., Du Toit, I.J. and Basson, N.J. 1994. The effect of honey on human tooth enamel in vitro observed by electron microscopy and microhardness measurements. *Archives of Oral Biology* 39:147–153.

Haddadin, M.S.Y., Nazer, I., Abu Raddad, Sara' Jamal, I. and Robinson, R.K. 2007. Effect of honey on the growth and metabolism of two bacterial species of intestinal origin. *Pakistan Journal of Nutrition* 6(6):693–697.

Haffejee, I.E. and Moosa, A. 1985. Honey in the treatment of infantile gastroenteritis. *British Medical Journal* 290:1866–1867.

Hedin, C., Whelan, K. and Lindsay, J.O. 2007. Evidence for the use of probiotics and prebiotics in inflammatory bowel disease: A review of clinical trials. *Proceedings of the Nutrition Society* 66(3):307–315.

Hermyan, H. 2010. Honey—A Remedy for Stomatitis. Available at http://suite101.com/article/honey-a-remedy-for-stomatitis-a197800.

Highleyman, L. and Franciscus, A. 2012. An introduction to the liver. HCSP fact sheet. A publication of the Hepatitis C Support Project.

Hiruma-Lima, C.A., Calvo, T.R., Rodrigues, C.M., Andrade, F.D., Vilegas, W. and Brito, A.R. 2006. Antiulcerogenic activity of *Alchornea castaneaefolia*: Effects on somatostatin, gastrin and prostaglandin. *Journal of Ethnopharmacology* 104(1–2):215–224.

Ingle, S.B. and Hinge, C.R. 2012. Gastroenteritis overview. *International Journal of Pharma and Bio Sciences* 3(2):607–613.

Isaacs, T. 2010. Researchers Discover the Cause: IBS is not in the Mind. *Natural News Network.* Available at http://www.naturalnews.com/029980_IBS_causes.html

Jeddar, A., Kharsany, A., Ramsaroop, U.G., Bhamiee, A., Haffejee, I.E. and Moosa, A. 1985. The anti-bacterial action of honey: An in vitro study. *South African Medical Journal* 67:257–258.

Kajiwara, S., Gandhi, H. and Ustunol, Z. 2002. Effect of honey on then growth of and acid production by human intestinal *Bifidobacterium* spp: An in vitro comparison with commercial oligosaccharides and inulin. *Journal of Food Protection* 65:214–218.

Kandil, A., El-Banby, M., Abdel-Wahed, G.K., Abdel-Gawwad, M. and Fayez, M. 1987. Curative properties of true floral and false non-floral honeys on induced gastric ulcers. *Journal of Drug Research* 17:103–106.

Kaur-Kirnpal, B.S., Tan, H.-T., Boukraa, L. and Gan, S.-H. 2011. Different solid phase extraction fractions of Tualang (*Koompassia excelsa*) honey demonstrated diverse antibacterial properties against wound and enteric bacteria. *Journal of Api Product and Api Medical Science* 3(1):59–65.

Ko, Y.-J. 2005. Investigating the sensitivity of enteropathogenic bacteria to the antibacterial activity of honey. Thesis (M.Sc. Biological science). University of Waikato, New Zealand.

Lomax, A.R. and Calder, P.C. 2009. Prebiotics, immune function, infection and inflammation: A review of the evidence. *British Journal of Nutrition* 101(5):633–658.

Marcucci, M.C. 1995. Propolis: Chemical composition, biological properties and therapeutic activity. *Apidologie* 26:83–99.

Martin, M.J., Casa, C., Alarcon-de-la-Lastra, C., Cabeza, J., Villegas, I. and Motilva, V. 1998. Antioxidant mechanisms involved in gastroprotective effects of quercetin. *Z. Naturforsch. C* 53(1/2):82–88.

Maton, A. 1997. *Human Biology and Health*, 3rd ed: Prentice Hall Science. Englewood Cliffs, NJ.

Mertz, H.R. 2003. Irritable bowel syndrome. *New England Journal of Medicine* 349:2136–2146.

Molan, P.C. 1992. The antibacterial nature of honey. The nature of the antibacterial activity. *Bee World* 73(1):5–28.

Molan, P.C. 1999. Why honey is effective as a medicine. Its use in modern medicine. *Bee World* 80(2):80–92.

Molan, P.C. 2001. Honey for oral health. *Journal of Dental Research* 80(special issue):1–130.

Molan, P.C. 2001a. The potential of honey to promote oral wellness. *General Dentistry* 49(6):584–589.

Munjal, U., Glei, M., Pool-Zobel, B.L. and Scharlau, D. 2009. Fermentation products of inulin-type fructans reduce proliferation and induce apoptosis in human colon tumour cells of different stages of carcinogenesis. *British Journal of Nutrition* 102(5):663–671.

Murchison, D.F. 2012. *Halitosis (Fetor Oris; Oral Malodor).* The Merck Manual, Merck Sharp & Dohme Corp., Whitehouse Station, New Jersey.

Nasuti, C., Gabbianelli, R., Falcioni, G. and Cantalamessa, F. 2006. Antioxidative and gastroprotective activities of anti-inflammatory formulations derived from chestnut honey in rats. *Nutrition Research* 26(3):130–137.

National Institutes of Health. 2010. Hemorrhoids. The National Digestive Diseases Information Clearinghouse. U.S. Department of Health and Human Services. NIH Publication 11:3021.

NICE. 2004. *Dyspepsia: Management of Dyspepsia in Adults in Primary Care.* National Institute for Health and Clinical Excellence, London. Available at http://www.nice.org.uk/nicemedia/pdf/CG017NICEguideline.pdf.

NORML. 2012. Gastrointestinal disorders. The National Organization for the Reform of Marijuana Laws. Available at www.norml.org.

NPF Brochure (The National Pancreas Foundation). 2010. *Common Disorders of the Pancreas.* Boston, USA. Available at http://www.pancreasfoundation.org

Orhan, F., Sekerel, B.E., Kocabas, C.N., Sackesen, C., Adalioğlu, G. and Tuncer, A. 2003. Complementary and alternative medicine in children with asthma. *Annals of Allergy, Asthma, and Immunology* 90:611–615.

Osato, M.S., Reddy, S.G. and Graham, D.Y. 1999. Osmotic effect of honey on growth and viability of *Helicobacter pylori. Digestive Diseases and Sciences* 44:462–464.

Paul, I.M., Beiler, J., Mcmonagle, A., Shaffer, M.L., Duda, L. and Berlin, C.M. 2007. Effect of honey, dextromethorphan, and no treatment on nocturnal cough and sleep quality for coughing children and their parents. *Archives of Pediatrics and Adolescent Medicine* 161(12):1140–1146.

Pihlstrom, B.L., Michalowicz, B.S. and Johnson, N.W. 2005. Periodontal diseases. *Lancet* 366:1809–1820.

Prakash, A., Medhi, B., Avti, P.K., Saikia, U.N., Pandhi, P. and Khanduja, K.L. 2008. Effect of different doses of Manuka honey in experimentally induced inflammatory bowel disease in rats. *Phytotherapy Research* 22(11):1511–1519.

Quigley, E., Fried, M., Gwee, K.A. et al. 2009. Irritable bowel syndrome: A global perspective. World Gastroenterology Organisation Global Guideline.

Rickard, M.J.F.X. 2005. Review article: Anal abscesses and fistulas. *ANZ Journal of Surgery* 75:64–72.

Ruiz, A.R. 2012. Overview of Malabsorption. The Merck Manual. Merck Sharp & Dohme Corp., Whitehouse Station, NJ.

Salem, S.N. 1981. Treatment of gastroenteritis by the use of honey. *Islamic Medicine* 1:358–362.

Sanz, M.L., Polemis, N., Morales, V. et al. 2005. In vitro investigation into the potential prebiotic activity of honey oligosaccharides. *Journal of Agricultural and Food Chemistry* 53:2914–2921.

Scholz-Ahrens, K.E. and Schrezenmeir, J. 2007. Inulin and oligofructose and mineral metabolism: The evidence from animal trials. *Journal of Nutrition* 137:2513S–2523S.

Schoenfeld, A. and Wu, G.Y. 2011. Ulcerative colitis. MedicineNet.com. Available at http://www.medicinenet.com/ulcerative_colitis/page4.htm.

Sela, M.O., Shapira, L., Grizim, I., Lewinstein, I., Steinberg, D., Gedalia, I. and Grobler, S.R. 1998. Effects of honey consumption on enamel microhardness in normal versus xerostomic patients. *Journal of Oral Rehabilitation* 25(8):630–634.

Shannon, I.L., Edmonds, E.J. and Madsen, K.O. 1979. Honey: Sugar content and cariogenicity. *Journal of Dentistry for Children* 46(1):29–33.

Sharma, R. 2006. Improve Your Health With Honey. Diamond Pocket Books (P) Ltd., New Delhi, India.

Shiga, H., Jo, A., Terao, K., Nakano, M., Ohshima, T. and Maeda, N. 2010. Decrease of halitosis by intake of manuka honey. The Preliminary Program for IADR General Session.

Siegel, M.A., Jacobson, J.J. and Braun, R.J. 2008. Diseases of the gastrointestinal tract. In *Burket's Oral Medicine: Diagnosis and Treatment*, Burket, L.W., Greenberg, M.S., Glick, M. and Ship, J.A. eds. People's Medical Publishing House.

Sirnik, V., Koch, V. and Golob, T. 1978. The influence of honey on the digestibility of nutritive substances for albin rats (L'influence du miel sur la digestibilité des substances nutritives chez le rat albinos). *IIIe Symposium International d'Apitherapie,* 11–15 Septembre 1978, Portoroz, *Yougoslavie, Apimondia Bukarest*, pp. 286–290.

Speroni, E. and Ferri, S. 1993. Gastroprotective effects in the rat of a new flavonoid derivative. *Acta Horticulturae* 332:249–252.

Sseitzinger, S. 2008. *A Patient's Guide. Fistula Causes, Symptoms and Treatment Options.* Cook Medical, Bloomington, IN.

Steinberg, D., Kaine, G. and Gedalia, I. 1996. Antibacterial effect of propolis and honey on oral bacteria. *American Journal of Dentistry* 9(6):236–239.

Sterer, N. and Rosenberg, M. 2011. *Breath Odors: Origin, Diagnosis, and Management,* 1st ed. Springer.

Stolzke, Tilo. 2008. Indications in Hepatology and Liver Diseases. Sananet-GmbH. Available at www.sananet.com.

Swartout-Corbeil, D. and Knight, J. 2006. Stomatitis. Gale Encyclopedia of Children's Health: Infancy through Adolescence. Available at http://www.encyclopedia.com/doc/1G2-3447200543.html.

Talley, N.J. and Vakil, N. 2005. Guidelines for the management of dyspepsia. *American Journal of Gastroenterology* 100(10):2324–2337.

University Health Services. 2009. *Hemorrhoids and Anal Fissures.* University of California, Berkeley, CA.

Vilegas, W., Sanommiya, M., Rastrelli, L. and Pizza, C. 1999. Isolation and structure elucidation of two new flavonoid glycosides from the infusion of *Maytenus aquifolium* leaves. Evaluation of the antiulcer activity of the infusion. *Journal of Agricultural and Food Chemistry* 47(2):403–406.

Vlcekova, P., Krutakova, B., Takac, P., Kozanek, M., Salus, J. and Majtan, J. 2011. Alternative treatment of gluteofemoral fistulas using honey: A case report. *International Wound Journal* 9(1):100–103.

White, J.W. 1979. Composition of honey. In E. Crane (Ed.), *Honey: A Comprehensive Survey,* pp. 157–158. Heinemann, London.

Williams, D. 2011. My All-Natural Treatment Plan for Ulcerative Colitis. Available at http://www.drdavidwilliams.com/ulcerative-colitis-natural-treatments. Accessed on October 2012.

Wilson, A. 2008. Pharyngitis. In *Essential Infectious Disease Topics for Primary Care,* Skolnik, N.S. and Albert, R.H. eds. Humana Press, Totowa, NJ.

Young, J.F., Nielsen, S.E., Haraldsdottir, J. et al. 1999. Effect of fruit juice intake on urinary quercetin excretion and biomarkers of antioxidative status. *American Journal of Clinical Nutrition* 69(1):87–94.

## INTERNET SITES

Bailey, J. Apple Remedy and Cure for Baldness. http://www.ehow.co.uk/way_5392778_apple-remedy-cure-baldness.html. Accessed on October, 2012.

Kurz, A. How to Cure Hiccups with Honey. http://www.ehow.com/how_5607281_cure-hiccups-honey.html. Accessed on October, 2012.

Patel, R. Honey and Cinnamon as a Medication. http://www.leaflady.org/honey.htm. Accessed on October, 2012.

Sarfaraz, I. Honey Cure for Acid Reflux. http://www.ehow.com/way_5701379_honey-cure-acid-reflux.html. Accessed on October, 2012.

www.ApitherapyHealth.com

www.EnchantedLearning.com

http://www.epatienthealthcare.com/tag/colds/

http://www.benefits-of-honey.com/milk-and-honey.html

# 9 Honey for Cardiovascular Diseases

*Tahira Farooqui and Akhlaq A. Farooqui*

## CONTENTS

## INTRODUCTION

Honey, "a sweet and viscous fluid with a unique flavor" (Farooqui 2009), is a plant product that is produced by honeybees. Honeybees gather nectar and sap from various flowers and trees and mix them with enzymes (transglucosylase and glucose oxidase) that are secreted from their own glands (Molan 2006). Transglucosylase breaks the glycosidic bond of sucrose and therefore catalyzes the hydrolysis of sucrose into glucose and fructose. Glucose oxidase catalyzes the oxidation of some of the glucose, forming gluconic acid and hydrogen peroxide ($H_2O_2$). Honeybees evaporate most of the water from the nectar by fanning the honeycomb through their wings, resulting in the concentrated solution of honey. Antimicrobial properties of honey are due to its high osmolarity, low pH, and $H_2O_2$ contents (Molan 2006). Honey is commonly used by humans as a sweetener; however, it also offers many medicinal uses described in traditional medicine. The therapeutic properties of honey depend on phytochemicals derived from nectar and sap. Honey is efficacious in various medical and surgical procedures due to the presence of a wide range of phenolic constituents that exert many biological activities, including antioxidant, anti-inflammatory, and immunomodulatory activities. Thus, medicinal properties of honey depend on its chemical composition, which varies depending on the botanical source, geographical area, seasonal collection time, and production methods (Ball 2007).

Honey gets its sweetness from the monosaccharides, "simple" 6-carbon sugars such as fructose and glucose. Other sugars include maltose (a 12-carbon sugar composed of two glucose molecules) and sucrose (a 12-carbon sugar composed of a glucose and a fructose molecule). Unlike table sugar, honey contains acids, minerals, vitamins, and amino acids in varying quantities. Honey has a distinctive flavor and is 40% denser than water with a density of 1.4 kg/L. Most microorganisms do not grow in honey because of its low water content and high acidity (pH 3.9). Honey also contains a natural resinous substance (propolis), which is collected by honeybees mainly from poplar bud exudates. Propolis protects honeybees from bacterial/viral infections. It is utilized in cosmetic and nutraceutical formulations. The antimicrobial properties of propolis make it a valid agent for treating upper respiratory tract infections (De Vecchi and Drago 2007; Farooqui and Farooqui 2010). Propolis extracts have a wide multispectrum of activities, including anti-inflammatory, anesthetic, healing, vasoprotective, antioxidative, antitumoral, anti-ulcer, hepatoprotective, cardioprotective, and neuroprotective properties (De Vecchi and Drago 2007; Farooqui and Farooqui 2010, 2012a,b,c). Propolis is rich in flavonoids; therefore, it exerts strong antioxidant activities. In rat heart mitochondria, polyphenols extracted from propolis show free radical scavenging activity and protect against the peroxidative damage induced by doxorubicin toxicity (Alyane et al. 2008). This observation suggests that the cardioprotective effects of propolis are due to its polyphenols, which include flavonoids and phenolic acids. The consumption of natural honey reduces cardiovascular risk factors, such as total cholesterol, low-density lipoprotein (LDL), triacylglycerole, body weight, fasting blood glucose, and C-reactive protein, particularly in subjects with elevated cardiovascular risk factors (Yaghoobi et al. 2008), supporting the view that honey exerts therapeutic activity against cardiovascular diseases.

Cardiovascular diseases contribute to more than one third of deaths in the United States and high mortality rates around the world. It has been speculated that cardiovascular diseases will be the primary cause of death throughout the globe by the arrival of 2020. Accumulating evidence suggests that reactive oxygen species (ROS)-mediated oxidative stress, which alters many functions of endothelium, may be the major contributing factor to cardiovascular diseases and death (Steinberg 1997; Cai and Harrison 2000; Griendling et al. 2000; Heitzer et al. 2001; Tousoulis et al. 2011). Therefore, polyphenols of plant origin are exploited for the treatment of acute and chronic free radical-mediated diseases (Doner 1977; Halliwell 2007; Benguedouar et al. 2008; Rakha et al. 2008; Jaganathan et al. 2010; Omotayo et al. 2010). The purpose of this overview is to describe the chemical composition, risk factors of cardiovascular diseases, and cardioprotective effects of honey and to discuss the potential molecular mechanism(s) underlying the therapeutic effects of honey for the treatment of cardiovascular diseases.

## COMPOSITION OF HONEY

As stated above, the chemical composition of honey is based on the botanical and geographical origins of the nectar. Carbohydrates are the major contents of honey, comprising about 95% of its dry weight (Siddiqui 1970; Doner 1977; Bogdanov et al. 2008). Honey carbohydrates include mainly monosaccharides (38% fructose and 31% glucose), 5% disaccharides, and 3% oligosaccharides (sucrose, maltose, trehalose, and turanose) (Figure 9.1). Due to its high carbohydrate content, honey is considered as an excellent source of energy. Depending on the botanical origin, the glycemic index of honey varies from 32 to 91 (Bogdanov 2010).

Important minor contents of honey are proteins (mainly enzymes, free amino acids), minerals and other minute quantities of constituents such as vitamins and polyphenols (Table 9.1) that account for honey's health promoting effects (Bogdanov 2010; Vorlova and Přidal 2002; Sánchez et al. 2001). Honey contains several enzymes such as α-glucosidase (invertase), α- and β-amylase (diastase), glucose oxidase, catalase, and acid phosphatase, which support digestion and assimilation (Vorlova and Přidal 2002). The major enzymes in honey are α-glucosidase, glucose oxidase, and a mixture of α- and β-amylases (diastase), which have been added to honey by bees during honey production (Table 9.2). The excretion of α-glucosidase is needed during the process of honey ripening and depends on the age and physiological stage of honeybee, condition of colony, temperature, and intensity or type of honey flow (Sánchez et al. 2001). Amylase and glucose oxidase are expressed in the hypopharyngeal glands of forager bees. These carbohydrate-metabolizing enzymes are needed to process nectar into honey (Shepartz and Subers 1964; Ohashi et al. 1999; Babacan and Rand 2005, 2007). Catalase and acid phosphatase in honey are thought to be derived from the plant source: nectar and pollen. Catalase decomposes $H_2O_2$ into $H_2O$ and $O_2$. Therefore, the presence of catalase in honey may produce a decrease in the antibacterial activity produced by glucose oxidase activity (Huidobro et al. 2005). Furthermore, variation in $H_2O_2$ production may be related to the presence or absence of catalase in the nectar or in the pollen from a particular plant species. Acid phosphatase is an enzyme that removes phosphate groups from food molecules

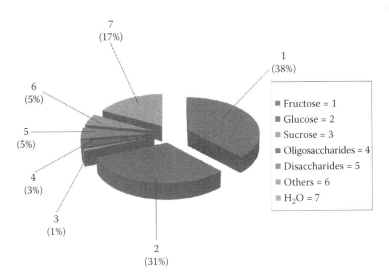

**FIGURE 9.1**   **(See color insert.)** General composition of honey.

**TABLE 9.1**

## Minor Contents of Honey

| Enzymes | Amylase | α-Glucosidase | Glucose oxidase | Catalase | Acid phosphatase | | | | | |
|---|---|---|---|---|---|---|---|---|---|---|
| Amino acids | Cysteine | Poline | Glutamine | Lysine | Tryptophan | | | | | |
| Minerals | Ca | Cu | Fe | Mg | Mn | K | Zn | P | Cr | Se |
| Vitamins | Ascorbic acid (C) | Thiamine (B1) | Riboflavin (B2) | Niacin (B3) | Pantothenic acid (B5) | Pyridoxine (B6) | Vitamin (K) | | | |
| Polyphenols | Apigenin | Chrysin | Galangin | Quercetin | Kaempferol | Pinocembrin | Acacetin | CA | CAPE | |

*Source:* Adapted from Bogdonov, S. et al., *J Am Coll Nutr* 27:677–689, 2008; Wang, J. and Li, Q.X., *Adv Food Nutr Res* 62:89–137, 2011.

**TABLE 9.2**
**Major and Minor Enzymes in Honey**

| Enzyme | Function | Source | References |
|--------|----------|--------|-----------|
| α-Glucosidase[a] (invertase) | Decomposes sucrose into fructose and glucose | Bees | Sánchez et al. 2001 |
| Amylase[a] (diastase) | Converts starch, glycogen, and other polysaccharides into smaller sugars | Bees | Shepartz and Subers 1964; Ohashi et al. 1999; Babacan and Rand 2005, 2007 |
| Glucose oxidase[a] | Converts glucose into gluconic acid and $H_2O_2$ | Bees | Shepartz and Subers 1964; Ohashi et al. 1999; Babacan and Rand 2005, 2007 |
| Catalase[b] | Breaks down $H_2O_2$ into $H_2O$ and $O_2$ | Nectar and pollen | Huidobro et al. 2005 |
| Acid phosphatase[b] | Catalyzes removal of aphosphate groups from other molecules during digestion | Nectar and pollen | Huidobro et al. 2005; Alonso-Torre et al. 2006 |

[a] Major enzymes introduced to honey by bees.
[b] Other enzymes found in honey are in lesser amounts.

during digestion. Values of acid phosphatase in honey have been related to honey's fermentation (Alonso-Torre et al. 2006).

As stated above, honey also contains phenolic acid, including caffeic acid, caffeic acid phenyl esters (CAPE), and a wide range of flavonoids (Table 9.3). Therefore, honey exerts antioxidant, anti-inflammatory, antiatherogenic, antithrombotic, immunomodulating, and analgesic activities (Gómez-Caravaca et al. 2006). Considerable differences in the composition and content of phenolic compounds have been found in different unifloral honeys (Amiot et al. 1989). Single and multiple components can well indicate the botanical and geographical origins of the honey (Wang and Li 2011). Flavonoids are incorporated into honey from propolis, nectar, or pollen. Flavonoids (such as quercetin, luteolin, kaempferol, apigenin, chrysin, galangin, acacetin, pinocembrin, pinobanksin, caffeic acid, and CAPE) have been reported as specific markers for the botanical origin of the honey (Tomás-Barberán et al. 1993; Blasa et al. 2007; Bogdanov et al. 2008; Viuda-Martos et al. 2008; Jaganathan and Mandal 2009). In a recent analysis of flavonoids in honey performed by high-performance liquid chromatography coupled with coulometric electrode array detection and electrospray ionization mass spectrometry, it has been shown that galangin, kaempferol, quercetin, isorhamnetin, and luteolin are detected in all honeys, whereas hesperetin is found only in lemon and orange honeys and naringenin in lemon, orange, rhododendron, rosemary, and cherry blossom honeys, suggesting that flavonoid contents and profiles of different types of honeys (Table 9.3) may vary greatly (Ferreres et al. 1993, 1994a,b, 1996, 1998; Cherchi et al. 1994; Tomás-Barberán et al. 2001; Hamdy et al. 2009; Petrus et al. 2011). Due to

**TABLE 9.3**

**Phenolic Compounds Are Markers of the Floral Origins in Particular Types of Honey**

| Flavonoids and Phenolic Acids | Honey Types | References |
| --- | --- | --- |
| Pinocembrin, pinobanksin, and chrysin | Most European honey samples | Tomás-Barberán et al. 2001 |
| Galangin, kaempferol, quercetin, isorhamnetin, and luteolin | All honeys | Petrus et al. 2011 |
| Hesperetin | Lemon and orange honeys | Ferreres et al. 1993, 1994a, 1994b; Hamdy et al. 2009; Petrus et al. 2011 |
| Naringenin | Lemon, orange, rhododendron, rosemary, and cherry blossom honeys | Petrus et al. 2011 |
| Kaempferol | Rosemary honey | Cherchi et al. 1994; Ferreres et al. 1998 |
| Quercetin | Sunflower honey | Tomás-Barberán et al. 2001 |
| Caffeic, p-coumaric, and ferulic acids | Chestnut honey | Cherchi et al. 1994 |
| Abscisic acid | Heather honey | Ferreres et al. 1996 |

differences in botanical and geographical origins, flavonoid content, and flavonoid concentration, important issues are raised regarding the quality and safety of honey when considering its therapeutic activity.

## CLASSIFICATION OF HONEY FLAVONOIDS

In general, flavonoids are composed of 15 carbons, two benzene rings (A and C) connected by a six-member ring (B). Flavonoids occur as aglycones, glycosides, and methylated derivatives. The main flavonoid subgroups are flavonol, flavone, flavan-3-ol, flavanone, isoflavone, and anthocyanidin (Jaganathi and Crozier 2010). Flavonoid classification is based on the modifications in the central C-ring (Figure 9.2). Flavones have a backbone of 2-phenyl-1-benzopyran-4-one. This class includes compounds such as chrysin, apigenin, and luteolin. Flavonols are structurally similar to flavones, except the presence of a hydroxyl group at the 3-position in the C-ring (3-hydroxyflavone backbone). Examples of flavonols are compounds such as galangin, quercetin, and kaempferol. Flavanones differ from flavonols by the absence of a hydroxyl group at the 3-position in the C-ring and a C2–C3 double bond. Natural honey contains all subclasses of flavonoids, including flavonol, flavone, flavan-3-ol, flavanone, isoflavone, and anthocyanidin.

Flavanones are generally glycosylated by a disaccharide at the seventh position of the A-ring. Flavanones include pinostrobin, pinocembrin, hesperetin, pinobanksin, and naringenin. Flavan-3-ols belong to a structurally most complex subclass

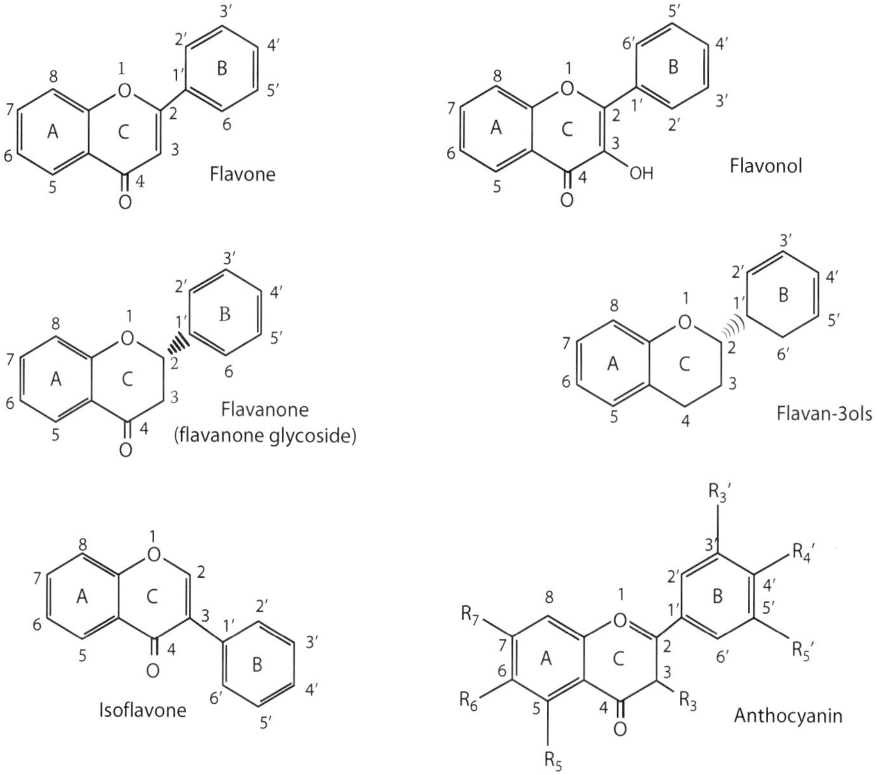

**FIGURE 9.2**  Backbone of flavonoids.

of flavonoids, which range from the simple monomers (+)-catechin and its isomer (−)-epicatechin to the oligomeric and polymeric proanthocyanidins (Jaganathi and Crozier 2010). In isoflavones, the B-ring is attached at the 3-position rather than the 2-position of the C-ring. The primary examples are genistein and daidzein. The anthocyanidins are aglycones of anthocyanin glycosides (sugar is attached at the 3-position on the C-ring or the fifth position of the A-ring) based on the flavylium ion or 2-phenylchromenylium, a type of oxonium ion. Due to this positive charge, anthocyanidins differ from other flavonoids. 3-Deoxyanthocyanidins lack hydroxyl group at the 3-position in the C ring. The stability of anthocyanidins is dependent on pH. They are colored at a low pH and colorless at a high pH (Woodward et al. 2009). Examples of anthocyanidins include cyanidin, pelargonidin, delphindin, luteolinidin, malvidin, and petunidin.

The most important flavonoids of honey, which are derived from propolis, include apigenin, galangin, chrysin, quercetin, CAPE, luteolin, pinocembrin, pinobanksin, acacetin, and kaempferol (Figure 9.3). Emerging evidence suggests that honey exhibits a broad spectrum of biological activities, implicating many beneficial effects; however, only a few of them have been supported by clinical and experimental evidence

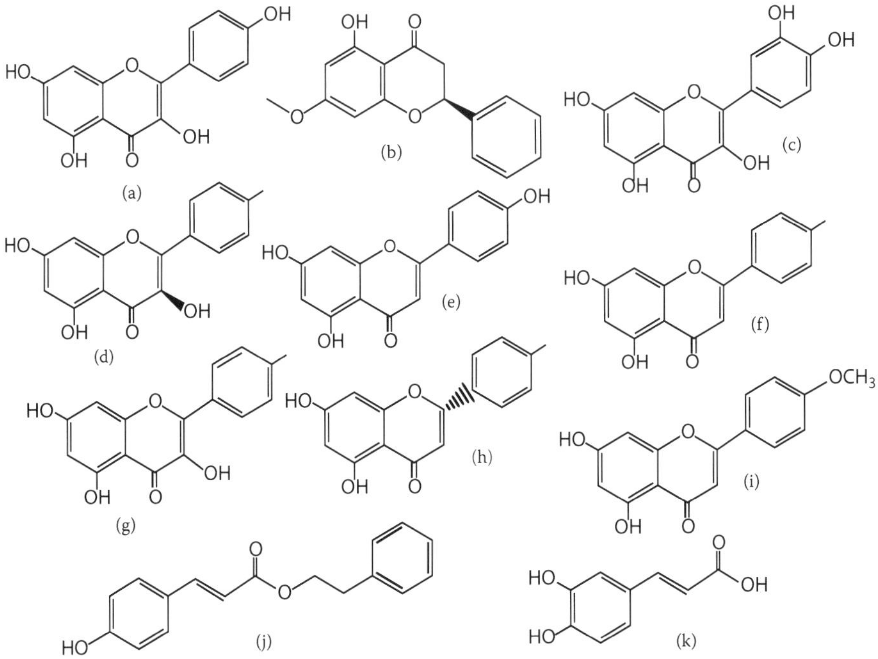

**FIGURE 9.3**  Major flavonoids of honey. Kaempferol (a), pinostrobin (b), quercetin (c), pinobanksin (d), apigenin (e), chrysin (f), galangin (g), pinocembrin (h), acacetin (i), CAPE (j), and caffeic acid (k).

(Estevinho et al. 2008; Rakha et al. 2008; Viuda-Martos et al. 2008; Alvarez-Suarez et al. 2010; Khalil et al. 2010; Omotayo et al. 2010; Khan et al. 2011).

## BIOLOGICAL ACTIVITIES OF HONEY

Biological activities of honey are mainly attributed to flavonoids and phenolic acids. These activities (Table 9.4) may markedly vary in different lots of honey depending on its flavonoid constituents. The mechanism of action of honey is not fully understood. However, honey has been widely used throughout the world as a "healing medicine" and its curative power is possibly based on the following biological activities.

### ANTIBACTERIAL ACTIVITY OF HONEY

Flavonoid and phenolic acids in honey exert antibacterial activity (Wahdan 1998; Cushnie and Lamb 2005). Honey has hygroscopic properties, meaning that it can shrink the bacteria due to its hyperosmolar properties (Molan 2001, 2006; Molan and Betts 2008). In the biological assays, *Staphylococcus aureus* is found to be one of the most sensitive microorganisms to the antimicrobial activity of honey extracts, whereas other microorganisms, such as *Bacillus subtilis*, *Staphylococcus lentus*,

## TABLE 9.4
## Cardioprotective Effects of Flavonoids from Honey

| Flavonoids | Group | Cardioprotective Effects | References |
|---|---|---|---|
| Naringenin | Flavanone | Antiatherogenic | Mulvihill and Huff 2010 |
| Pinocembrin and hesperetin | Flavanone | Vasorelaxing | Calderone et al. 2004 |
| Acacetin, apigenin, chrysin, and luteolin | Flavone | Vasorelaxing | Calderone et al. 2004 |
| Apigenin | Flavone | Cardiomyocyte cytoprotective effect | Psotová et al. 2004 |
| Quercetin and kaempferol | Flavonol | Cardiomyocyte cytoprotective effect | Psotová et al. 2004 |
| Quercetin | Flavonol | Reduces blood pressure in hypertensive subjects | Edwards et al. 2007; Egert et al. 2009 |
| Luteolin and chrysin | Flavone | Reduction of blood pressure in hypertensive subjects | Edwards et al. 2007 |
| Quercetin | Flavonol | Chelation of intracellular iron and suppression of OH• radical production, chain breaking antioxidant | Lakhanpal and Rai 2008; Vlachodimitropoulou et al. 2011 |
| Luteolin and chrysin | Flavone | Chelation of intracellular iron and suppression of OH• radical production | Vlachodimitropoulou et al. 2011 |
| Hesperetin | Flavanone | Antiplatelet activity | Jin et al. 2008 |
| Syringetin | O-methylated flavonol | Antiplatelet activity | Bojic et al. 2011 |
| Acacetin | O-methylated flavone | Inhibition of atrial fibrillation | Gui-Rong et al. 2008 |
| Kaempferol | Flavonol | Endothelium-independent and endothelium-dependent relaxation of coronary arteries | Xu et al. 2006 |
| Apigenin | Flavone | Protection of endothelium-dependent relaxation of aorta against oxidative stress | Jin et al. 2009 |
| CAPE | Caffeic acid phenethyl ester | Inhibition of lipid peroxidation and induction in the expression of antioxidant enzymes in diabetic heart | Okutan et al. 2005 |
| | | Antiarrhythmic activity | Huang et al. 2005 |

*Klebsiella pneumoniae*, and *Escherichia coli*, are moderately sensitive (Estevinho et al. 2008). Honey reduces infection and enhances wound healing in burns, ulcers, and other diabetic wounds with its dressings (Eddy et al. 2008; Makhdoom et al. 2009). The antimicrobial activity of some honey preparations also depends on the endogenous $H_2O_2$, which is produced by the bee-derived glucose oxidase (Brudzynski

2006). Therefore, endogenous $H_2O_2$ levels of honey can be used as a strong predictor of its antibacterial activity. Furthermore, the low pH of honey is another important factor that may be responsible for inhibiting and blocking the growth of many pathogenic bacteria. In the case of $H_2O_2$-dependent honey preparations, antibacterial effects depend on the stability of glucose oxidase activity, which is modulated by several factors, such as pH, light, and temperature; however, nonperoxide antimicrobial activity is more closely associated with the floral source (White and Subers 1964; Irish et al. 2011). Collective evidence suggests that the low water activity and acidic nature of honeys make them unsuitable media for bacterial growth; therefore, many (types of) honeys can be used as potential antibacterial agents to fight infections.

## ANTI-INFLAMMATORY ACTIVITY OF HONEY

The anti-inflammatory properties of honey are due to the presence of propolis. As mentioned earlier, honey is rich in flavonoids that exert its anti-inflammatory activity. However, the molecular mechanisms of how the flavonoids exhibit their anti-inflammatory activity have not yet been clarified. The anti-inflammatory action of honey has been shown to reduce edema as well as amount of exudates by downregulating the inflammatory process. Honey reduces both pain and pressure around the tissue by downregulating the inflammatory process (Lee et al. 2003). Flavonoids, such as chrysin and quercetin, display similar anti-inflammatory activity via different mechanisms because chrysin suppresses the lipopolysaccharide (LPS)-induced cyclooxygenase (COX)-2 protein and its mRNA expression (Woo et al. 2005), whereas quercetin inhibits inducible nitric oxide (NO) synthase (NOS) expression by decreasing NO production in endotoxin/cytokine-stimulated microglia (Kao et al. 2010). The anti-inflammatory effect of CAPE is through its inhibition of LPS-induced upregulation of the tumor necrosis factor (TNF)-$\alpha$ and interleukin (IL)-8 production, resulting in suppression of I$\kappa$B$\alpha$ degradation (Song et al. 2008). Low concentrations of CAPE suppress osteoclastogenesis and bone resorption through the inhibition of nuclear factor (NF)-$\kappa$B activation and nuclear factor of activated T-cell activity (Ang et al. 2009), implicating that honey can be used for the treatment of osteolytic bone diseases. Recently, it has been reported that structurally related flavonoids (such as apigenin and luteolin) significantly inhibit TNF-$\alpha$–induced NF-$\kappa$B transcriptional activation without altering the degradation of I$\kappa$B proteins and the nuclear translocation and DNA-binding activity of NF-$\kappa$B p65 (Funakoshi-Tago et al. 2011). The suppression of NF-$\kappa$B activation mediated by these flavonoids is due to the inhibition of the transcriptional activation of NF-$\kappa$B.

Apigenin and luteolin slightly inhibit TNF-$\alpha$–induced c-Jun N-terminal kinase (JNK) activation but show no effect on TNF-$\alpha$–induced activation of extracellular signal-regulated kinase (ERK) and p38. However, fisetin enhances and sustains activation of ERK and JNK but not p38 in response to TNF-$\alpha$. Furthermore, the administration of apigenin and luteolin markedly inhibits acute carrageenan-induced paw edema in mice. However, fisetin fails to have such effect, suggesting that a slight structural difference in flavonoids may cause differences in anti-inflammatory responses. The effects of honey and its extracts have been recently demonstrated on inflammation in rats. It is reported that honey and honey extract not only inhibit edema and pain in inflamed tissues but also produce potent inhibitory activities against NO and prostaglandin $E_2$ (Kassim et al. 2010a,b).

## IMMUNOMODULATORY ACTIVITY OF HONEY

Immunity is designed to work as the body defense system against invading germs and microorganisms. Therefore any dysfunction in the immune system may lead to infections and diseases. White blood cells (leukocytes) belong to the immune defense system. There are two types of leukocytes: (1) phagocytes that play a role in chewing up the invading organisms and (2) lymphocytes (B and T lymphocytes) that help the body in remembering, recognizing, and destroying the previous invaders.

Honey impacts on various aspects of immunity by exerting immunomodulatory (immunostimulatory and immunosuppressive) activities. Polyphenols (flavonoids and related compounds) are certainly involved with immunomodulatory activities, but other compounds in honey may also synergistically contribute to the overall immunoregulatory properties. Honey has been shown to stimulate the proliferation of both B and T lymphocytes, inducing antibody production during primary and secondary immune responses (Abuharfeil et al. 1999; Al-Waili and Haq 2004). Honey stimulates the release of signaling proteins in certain white blood cells and thereby upregulates the immune response. Honey also stimulates TNF-α secretion from murine macrophages, whereas the deproteinized honey does not exert any effect on the release of TNF-α, suggesting that its immunostimulatory effect depends on its major royal jelly and honey glycoprotein content—apalbumin1 (Majtán et al. 2006). Previously, it has been demonstrated that a variety of honey types can stimulate human monocytic cells to release inflammatory cytokines (e.g., TNF-α, IL-1, and IL-6) that activate the immune response to infection (Tonks et al. 2001). In addition to stimulation of these leukocytes, honey supplies glucose—a substrate for glycolysis, which is the major mechanism for energy production in the macrophages, thereby allowing them to function in damaged tissue and exudates where the oxygen supply is often poor (Molan 2001). Furthermore, acidity of honey may also assist in the bacteria-destroying action of macrophages, as an acid pH inside the phagocytotic vacuole is involved in killing ingested bacteria (Molan 2001).

Initially, it was believed that honey-induced cytokine release is due to the presence of microbes or some other contaminants present in the honey. However, now it is confirmed that a 5.8-kDa protein component of honey is responsible for the stimulation of innate immune cells via Toll-like receptor (TLR) 4 (Tonks et al. 2007). TLR4 can detect liposaccharides from gram-negative bacteria; thereby, this receptor plays a role in pathogen recognition and activation of innate immunity. Blocking of the TLR4 but not the TLR2 receptor inhibits honey-stimulated TNF-α production significantly in human monocytes, confirming involvement of the TLR4 receptor. The oral intake of honey augments antibody production in primary and secondary immune responses against thymus-dependent and thymus-independent antigens (Al-Waili and Haq 2004).

In the case of a cold, both the immune and inflammatory systems gear up to increase mucus secretions, resulting in the running nose and the stuffy feeling. The infection into the cell lining of lungs produces the cough. Due to antimicrobial, anti-inflammatory, and immunomodulatory activities, honey stimulates the immune system and helps the body in fighting the infection, eliminating the virus or bacteria from the system. Therefore, honey has been used to cure children's nightly cough induced by upper respiratory tract infection, sleep difficulty, and the common cold (Paul et al. 2007; Pourahmad and Sobhanian 2009; Shadkam et al. 2010). Furthermore, honey suppresses induction of ovalbumin-specific humoral antibody responses against different allergens (Duddukuri et al. 1997), implicating its protective use in various health conditions.

Collectively, the immunomodulatory effects of honey on phagocytosis and/or blast transformation of leukocytes depend on chemical composition rather than on the concentration. It may provide the basis to be used as a novel therapeutic agent for patients with immune diseases. Further research is needed to compare immunomodulatory effects in honeys of different origins.

## ANTIOXIDANT ACTIVITY OF HONEY

The human body produces ROS, which are capable of oxidizing lipid, proteins, and nucleic acids. The oxidized products (lipid peroxides, oxidized proteins, and nucleic acids) can damage cellular membranes. Many chronic diseases are associated with increased oxidative damage caused by an imbalance between free radical production and antioxidant level. Therefore, it is suggested that antioxidants may play an important role in the improvement of the human health and ROS-mediated chronic diseases can be treated with antioxidants. Due to the presence of its polyphenolic constituents (i.e., catechins, isoflavones, anthocyanidins, and phenolic acids), honey is gaining increased attention in promoting overall health. Based on the chemical nature of polyphenolic constituents and their plant source, the quality of honey can be judged (Herrero et al. 2005). Furthermore, the classification of honey antioxidants is also established based on their mechanism of actions: (1) some honey flavonoids terminate the oxidation chain reaction by donating hydrogen or electrons to free radicals, (2) others produce a synergistic effect due to oxygen scavenging and iron chelating activities, and (3) some honey flavonoids prevent the oxidation reaction via decomposition of lipid peroxides into stable end-products by secondary metabolites (Rajalakhmi and Narasimham 1996). The phenolic contents of the Buckwheat and Tualang honeys are relative compared with other kinds of honey. Therefore, both the above-mentioned honeys exert high antioxidant activities compared with other honey sources tested (van den Berg et al. 2008; Kishore et al. 2011).

The hypoglycemic effect of honey may be associated with the decrease in oxidative stress in kidneys of streptozotocin-induced diabetic rats (Griendling et al. 2000). Furthermore, the oral administration of carbon tetrachloride ($CCl_4$) induces severe hepatic and kidney injury due to oxidative stress (El Denshary et al. 2011). The combined treatment with $CCl_4$ plus honey and/or Korean ginseng extract significantly protects Sprague Dawley rats against the severe $CCl_4$-mediated hepatic and renal toxic effects, suggesting that the protective effect of honey and/or Korean ginseng extract may be related to their antioxidant activities (El Denshary et al. 2011). Oxidation of LDL plays a key role in vascular damage and in modulation of several endothelial properties, including NO production and expression of adhesion molecules in cardiovascular diseases (Ramasamy et al. 1998; Grassmann et al. 2011). The oxidation of LDL is induced by macrophages, and this process is promoted by metal ions such as copper and iron. Several studies have shown that certain flavonoids with potent antioxidant properties inhibit oxidation of human LDL. The antioxidant properties of flavonoids are significantly higher than α-tocopherol (Frankel et al. 1993). Flavonoids significantly inhibit copper-catalyzed oxidation of LDL (Meyer et al. 1998; Vinson et al. 2001).

Accumulating evidence suggests that daily consumption of honey may reduce the risk of cardiovascular diseases, in part because of their antioxidant properties. Although the exact molecular mechanisms of antioxidant activity of the phenolic compounds present in honey and related beehive products are not yet fully understood, it has been suggested that free radical scavenging activity, hydrogen donation, and the interference with propagation reactions or inhibition of enzymatic systems are involved in initiation reactions, such as metallic ion chelation (Middleton et al. 2000; Viuda-Martos et al. 2008).

## Cardiovascular Diseases

Cardiovascular diseases are a leading cause of death not only in the United States but also all over the world. The rise in the incidence of cardiovascular diseases can be attributed to an unhealthy lifestyle (such as lack of physical activity, fast food, and smoking) and genetic risk factors (including hypertension, hypercholesterolemia, hyperhomocysteinemia, and diabetes mellitus) (Black 1992; Mozaffarian et al. 2008). Cardiovascular risk factors that cannot be changed are age, gender, and heredity. However, there are cardiovascular risk factors that can be changed (Figure 9.4). There are several types of cardiac diseases, such as atherosclerosis, coronary artery

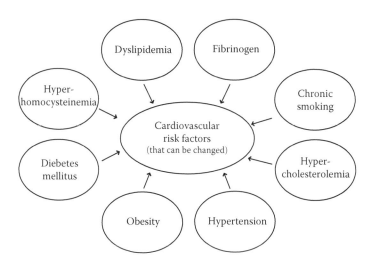

**FIGURE 9.4**   Risk factors of cardiovascular diseases.

disease, acute myocardial infarction, cardiac arrhythmia, congestive heart failure, ischemic heart disease, and cardiomegaly, which are associated with increased oxidative stress (Dhalla et al. 2000; Mozaffarian et al. 2008; Sugiura et al. 2011).

Atherosclerosis, a pathologic condition in which deposits of fatty substances, cholesterol, calcium, and other substances build up in the inner lining of an artery in the form of plaque, is an active process associated with vascular cell activation, inflammation, and thrombosis. Inflammatory processes in vascular wall destabilize atherosclerosis and produce lipid mediators and biomarkers that may provide insights into molecular mechanism of pathogenesis of atherosclerosis and its relationship with inflammatory reactions. Atherosclerosis, a chronic inflammatory disease, in particular, is largely based on the functional and structural changes in systemic vasculature (Perticone et al. 2001; Libby 2002; Vanhoutte et al. 2009). Atherosclerosis is characterized by the formation of arterial lesions or plaques as a result of an inflammatory response to endothelial injury (Li and Chen 2005). The plaque primarily is composed of macrophages, LDLs, and neutral lipids, with subsequent calcification and ulceration appearing around the outer part of mature plaques (Matsushita et al. 2000; Rosenfeld 2000). The development of a vulnerable plaque and the subsequent ischemic events result in the profound loss of vascular homeostasis. Atherosclerosis may eventually lead to artery enlargement (arterial stenosis), which causes an insufficient supply of blood to associated organs, and may ultimately result in an arterial rupture. In healthy vascular systems, vascular endothelium provides NO that is essential for the maintenance of homeostasis. However, increased vascular ROS production triggers atherosclerosis, causing oxidative stress, activation of proinflammatory signaling, and breakdown of vasoprotective NO in the vascular endothelium (Judkins et al. 2010), which results in endothelial dysfunction. Impaired NO bioavailability as a consequence of oxidative stress, and increased levels of angiotensin (an oligopeptide that stimulates the release of aldosterone from the adrenal cortex producing oxidative stress and endothelial dysfunction in blood vessels), cause blood vessels to constrict, resulting in atherosclerosis (Mason 2011). Increased oxidative stress promotes vascular inflammation, which leads to endothelial damage; therefore, derivatives of reactive oxygen metabolites are used as oxidative stress markers for detecting endothelial damage in patients with early stage atherosclerosis (Mozaffarian et al. 2008).

The vascular endothelium plays a key role in the regulation of vascular homeostasis. Therefore, alterations in endothelial functions contribute to the pathogenesis of cardiovascular diseases. The diverse risk factors of cardiovascular diseases associated with endothelial dysfunction include dyslipidemia (disorder of lipoprotein metabolism), hypertension, diabetes mellitus, smoking, hyperhomocysteinemia, hypercholesterolemia (Figure 9.4), and many other conditions such as systemic inflammation, infectious processes, postmenopausal state, physical inactivity, and aging (not described here). The genetic variation in the activity of antioxidant enzymes such as NO synthase may influence the endothelial function. In addition, the environmental factors (such as diet enriched in ω-6 fatty acids and lack of physical activity) may influence endothelial function and intake of polyphenols may help in reducing the risk for cardiovascular diseases (Vita 2005).

Hypertension is considered as one of the leading risk factors in the development of cardiovascular disease. Uncontrolled and prolonged elevation of blood pressure increases risk of serious health problems, including a variety of changes in the myocardial structure, coronary vasculature, and conduction system of the heart. These changes in turn can lead to the development of left ventricular hypertrophy (thickening of the myocardium of the left ventricle), coronary artery disease (arteries that supply blood to the heart muscle become hardened and narrowed), various conduction system diseases, systolic and diastolic dysfunctions of the myocardium that manifest clinically as angina or myocardial infarction (interruption of blood supply to part of the heart causing heart cells to die), cardiac arrhythmias (palpitations or irregular heartbeat; e.g., atrial fibrillation), and congestive heart failure (inadequate delivery of oxygen-rich blood to the body) (Standridge 2005; Law et al. 2009). Collective evidence suggests that ROS-mediated oxidative stress plays a key role in the pathogenesis of hypertension associated with cardiovascular diseases (Rodrigo et al. 2011; Schulz et al. 2011). The treatment plan for the hypertension such as modified diet and lifestyle modifications are currently believed to be more effective than pharmacologic therapies in the management of high blood pressure.

Obesity, the most common nutritional disorder, is usually associated with cardiovascular diseases. Adipocytokines, such as leptin, are produced in adipose tissue. Therefore, endocrine activity of adipose tissue is responsible for vascular impairment, prothrombotic tendency, and low-grade chronic inflammation associated with cardiovascular events (Anfossi et al. 2010). Furthermore, "metabolic syndrome" defined on the basis of combining abnormal conditions (such as obesity, arterial hypertension, chronic hyperglycemia [diabetes mellitus], and atherogenic dyslipidemia together with a prothrombotic and proinflammatory state), is associated with increased risk of cardiovascular complications (Vacca et al. 2011).

## CARDIOPROTECTIVE EFFECTS OF HONEY

To date, a substantial number of studies have reported the efficacy of flavonoids in heart diseases. Flavonoids reduce the risk of cardiovascular diseases in acute and short-term interventions in healthy volunteers and in risk population groups (Al-Waili 2004). Some studies are in favor of beneficial effects of increased intake of honey flavonoids on the cardiovascular system (Hertog et al. 1993, 1995, 1997a; Knekt et al. 1996, 2002; Yochum et al. 1999; Arts et al. 2001; Geleijnse et al. 2002; Al-Waili 2004; Beretta et al. 2007; Hooper et al. 2008; Benguedouar et al. 2008; Rakha et al. 2008; Ahmad et al. 2009; Massignani et al. 2009; Punithavathi and Stanely Mainzen Prince 2011), while data from other studies indicate that honey has no effect on heart disease. Results on the effect of honey in heart disease are conflicting (Rimm et al. 1996; Hertog 1997b; Sesso et al. 2003; Lin et al. 2007); therefore, more studies especially on a large human population are urgently required.

Native honey from multifloral origin shows a protective effect in vivo against acute and chronic free radical-mediated diseases (Beretta et al. 2007). Natural wild honey also exerts its cardioprotective and therapeutic effects against epinephrine-induced cardiac disorders and vasomotor dysfunction (Rakha et al. 2008). A good correlation has been observed between radical scavenging activity and total pheno-lic contents of honey (Estevinho et al. 2008; Rakha et al. 2008; Viuda-Martos et al. 2008; Alvarez-Suarez et al. 2010; Khan et al. 2011). Honey exhibits a wide range of cardioprotective effects, including vasodilatory, antithrombotic, maintaining the role of vascular homeostasis, improving lipid profile, and many others (Figure 9.5). Therefore, proper consumption of honey can play an important protective role in treating heart diseases.

Honey flavonoids decrease the risk of coronary heart disease by the follow-ing major actions: (1) improving coronary vasodilatation, (2) decreasing the abil-ity of platelets conversion to clot, (3) preventing oxidation of LDLs, (4) increasing high-density lipoprotein (HDL), and (5) improving endothelial function. In addi-tion, honey has high levels of antioxidant compounds (such as caffeic acid, CAPE, chrysin, galangin, quercetin, acacetin, kaempferol, pinocembrin, pinobanksin, and apigenin). These polyphenols may also contribute to stabilization of the atheroma plaque and therefore add to the cardioprotective effects of honey (Table 9.4), support-ing its cardioprotective properties (Khalil and Sulaiman 2010).

However, wild honey from the nectar of some species of rhododendron may be poisonous. The poisoning is due to the presence of a toxin (grayanotoxin), which is a naturally occurring sodium channel toxin and may cause life-threatening bradycar-dia, hypotension, and altered mental status (Dubey et al. 2009). Similar incidences have occurred with the ingestion of honey produced widely in northern parts of Turkey. This honey can be toxic because grayanotoxins and romedotoxins are pro-duced by plants of the Ericaceae family (Akinci et al. 2008; Okuyan et al. 2010). Therefore, patients admitted to emergency with bradycardia and hypotension should be checked carefully, because there may be a case of intoxication caused by consum-ing leaves and flowers of the Rhododendron or Ericaceae family.

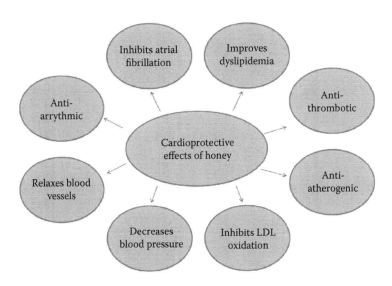

**FIGURE 9.5**  Cardioprotective effects of honey.

Blood vessels deliver oxygen and nutrients to every part of the body including blood vessel growth. Therefore, blood vessels also nourish diseases such as angiogenesis (Carmeliet and Jain 2011). Polyphenols have been shown to reduce atherosclerotic lesions through mechanisms including the downregulation of inflammatory and angiogenic factors (Daleprane et al. 2012).

Last, diabetes mellitus is often associated with cardiovascular diseases. Two most commonly prescribed antidiabetic drugs (glibenclamide and metformin) in combination with Tualang honey offer additional protection for the pancreas of streptozotocin-induced diabetic rats against oxidative stress and damage. These drugs produce no significant effects on lipid peroxidation and on antioxidant enzymes, except glutathione peroxidase in diabetic rats. In contrast, the combination of these drugs and honey significantly upregulates catalase activity and downregulates glutathione peroxidase activity, while significantly reduces levels of lipid peroxidation, suggesting that Tualang honey potentiates the effect of antidiabetic drugs to protect the pancreas against oxidative stress and damage (Erejuwa et al. 2010).

Collective evidence supports the view that honey can be used as a therapeutic agent in combating cardiovascular problems.

## Molecular Mechanism Underlying Cardioprotective Effects of Honey

Both human and experimental animal models of cardiovascular diseases exhibit high oxidative stress: (1) by promoting free radical generation, (2) by reducing endogenous levels of antioxidants, and (3) by depleting free radical scavenging enzymes. If free radicals are not destroyed, they can be toxic due to their propensity to react with biological molecules such as lipids, proteins, and DNA, resulting in ROS-mediated pathogenesis of a variety of disorders, including cardiovascular diseases.

Proatherogenic agents, such as oxidized LDL (OxLDL), proinflammatory cytokines (TNF-$\alpha$, interferon [IFN]-$\gamma$, also known as type II IFN or immune IFN, IL-1, and IL-6), and a peptide hormone angiotensin II, stimulate intracellular ROS generation producing proliferation and gene endothelin-1 expression in cardiac fibroblasts (Lo et al. 1996; Cheng et al. 2003; Watanabe et al. 2003). Therefore, cytokine receptor antagonists plus flavonoids may be beneficial in preventing the formation of excessive cardiac fibrosis. High levels of OxLDL induce ROS generation (Cominacini et al. 1998). The predominant source of vascular superoxide production is NADPH oxidases. These enzymes generate ROS in the artery wall in conditions, such as hypertension, hypercholesterolemia, diabetes, aging, and atherosclerosis. Increased superoxide production contributes to reduced NO bioactivity, resulting in the endothelial dysfunction. Isoforms of NADPH oxidases are constitutively expressed in each of the predominant cell types of the vascular wall (Csanyi et al. 2009). NADPH oxidases are important contributors to vascular oxidative stress, endothelial dysfunction, and vascular inflammation (Drummond et al. 2011). Reduced endothelial vasorelaxations and increased vascular NADPH oxidase activity are associated with increased risk of atherosclerosis (Cai and Harrison 2000; Griendling et al. 2000; Heitzer et al. 2001). Diabetes and hypercholesterolemia, risk factors of cardiovascular diseases, are also independently associated with increased NADPH-dependent superoxide production. LDL oxidation contributes to atherogenesis (Steinberg 1997).

ROS induces endothelial cell damage and vascular smooth muscle growth, which are responsible for hypertension, atherosclerosis, restenosis (reoccurrence of stenosis—a narrowing of a blood vessel that leads to restricted blood flow), and heart failure (Abe and Berk 1998; Yoshizumi et al. 1998; Griendling et al. 2000).

Honey flavonoids exert their therapeutic effects through the improvement of cardiovascular risk factors, such as endothelial function, inhibition of LDL oxidation, reduction in blood pressure, and improvement of conditions such as dyslipidemia and hyperinsulinemia (Figure 9.6). For example, treatment with naringenin, a flavonoid found in honey, corrects dyslipidemia, hyperinsulinemia, and obesity, resulting in

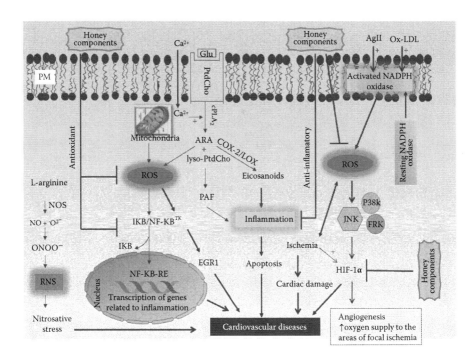

**FIGURE 9.6**  A hypothetical scheme showing antioxidant and anti-inflammatory activity of flavonoids found in honey. These effects are associated with cardioprotective effects of honey. The flavonoids found in honey reduce formation of ROS and eicosanoids in endothelial cells. Honey flavonoids also inhibit the oxidation of LDL cholesterol and thus prevent the formation of atherosclerotic plaque. Enhancement of HIF-1α expression causes angiogenesis within ischemic patients, promoting vessel proliferation needed for oxygenation. Honey flavonoids downregulate expression of HIF-1α and thus exert antiangiogenic effects by inhibiting proliferation and protect the heart against ischemia. ARA, arachidonic acid; COX-2, cyclooxygenase-2; cPLA2, cytosolic phospholipase A2; EGR1, early growth response protein 1; ERK, extracellular signal-regulated kinase; HIF-1, hypoxia-induced factor-1; JNK, c-Jun N-terminal kinase; lyso-PtdCho, lyso-phosphatidylcholine; NO, nitric oxide; NOS, nitric oxide synthase; NF-κB, nuclear factor-κB; Ox-LDL, oxidized low-density lipoprotein; NOO⁻, peroxynitrite; PAF, platelet activating factor; PtdCho, phosphatidylcholine; RNS, reactive nitrogen species; ROS, reactive oxygen species. (Modified from Farooqui, T. and Farooqui, A.A., *Curr Nutr Food Sci* 7(4):232–252, 2012a.)

attenuation of atherosclerosis in a mouse model of cardiovascular disease (Mulvihill and Huff 2010). Honey also contains flavones (acacetin apigenin, chrysin, luteolin, 5-hydroxyflavanone, 5-methoxyflavanone, and 7-hydroxyflavanone) and flavanones (hesperetin, pinocembrin, 4′-hydroxyflavanone, and 6-hydroxyflavanone), which exert vasorelaxing effects in blood vessels (Calderone et al. 2004). It is reported that the vasodilatory activity of hesperetin, luteolin, 5-hydroxyflavone, and 7-hydroxyflavanone can be antagonized by tetraethyl ammonium chloride, indicating the possible involvement of calcium-activated potassium channels. Moreover, iberiotoxin clearly antagonizes the effects of 5-hydroxyflavanone, indicating the probable importance of a structural requirement (the hydroxy group in position 5) for possible interactions with large-conductance, calcium-activated potassium channels (Calderone et al. 2004). Accumulating evidence suggests that honey flavonoids act as free radical scavenging and chelating agents and produce anti-inflammatory, anti-ischemic, vasodilating, and chemoprotective effects. Flavonoids provide cytoprotection in the cardiomyocyte model of cardiovascular diseases in a concentration-dependent manner. The extent of cardioprotection varies from one flavonoid to another. It is highest for baicalein and lowest for kaempferol (baicalein > luteolin congruent with apigenin > quercetin > kaempferol) (Psotová et al. 2004). Quercetin supplementation reduces blood pressure in hypertensive rodents (Edwards et al. 2007) and reduces systolic blood pressure and plasma OxLDL concentrations in overweight subjects with a high cardiovascular disease risk phenotype (Egert et al. 2009). A number of other flavonoids, such as luteolin, chrysin, and 3,6-dihydroxyflavone, chelate intracellular iron and suppress OH• production in Madin Darby canine kidney cells (Vlachodimitropoulou et al. 2011). It is shown that the most effective chelation is provided by the B-ring found in both flavones and flavonols. These flavonoids can be used as effective protective agents against oxidative stress mediated by metal ion overload. These studies suggest that quercetin, luteolin, chrysin, and 3,6-dihydroxyflavone supplementation may provide protection against cardiovascular diseases (Edwards et al. 2007; Lakhanpal and Rai 2008; Egert et al. 2009; Vlachodimitropoulou et al. 2011).

Platelet aggregation is a central mechanism in the pathogenesis of acute coronary syndromes, including myocardial infarction, and unstable angina. Flavonoids have been shown to exert beneficial effect in the prevention of cardiovascular diseases, which can be attributed, at least partially, to their antiplatelet effects. Hesperetin, a flavonoid found in honey, has antiplatelet aggregation activity. It inhibits phospholipase γ2 phosphorylation and interferes with COX-1 activity (Jin et al. 2008). These effects contribute to the beneficial effects of honey on cardiovascular disease. In another study, investigators analyzed approximately 30 flavonoid aglycones for their antiaggregatory activity (ranging from 0.119 to 122 μM). The most potent flavonoids were 3,6-dihydroxyflavone (0.119 μM) and syringetin (O-methylated flavonol) (Bojic et al. 2011).

Acacetin (flavone), an atrium-selective agent, selectively inhibits atrial ultrarapid delayed rectifier $K^+$ current and the transient outward $K^+$ current ($IC_{50}$ 3.2 and 9.2 μmol/L, respectively) and prolongs action potential duration in human atrial myocytes (Gui-Rong et al. 2008). Kaempferol at high concentrations (100 μM) produces significant relaxation in isolated porcine coronary artery rings, whereas at low concentration (10 μM) it is devoid of significant vascular relaxation effects but enhances endothelium-dependent and endothelium-independent relaxations that are unrelated to its antioxidant property (Xu et al. 2006). Apigenin protects endothelium-dependent relaxation of the aorta against oxidative stress in isolated rat aortic rings (Jin et al. 2009).

The risk for cardiovascular disease is significantly high in diabetes mellitus, because it increases ROS-mediated oxidative stress in cardiac tissue. Hyperglycemia induces oxidative stress by stimulating the mitochondrial pathway, NADPH oxidase, and xanthine oxidase. In addition, hyperglycemia up-regulates the expression of reactive nitrogen species (RNS), which may react with superoxide forming peroxynitrite that increases nitrosative stress, promoting cardiovascular diseases (Zhu et al. 2011).

CAPE, an active component of propolis, exerts an antiarrhythmic effect in rats subjected to myocardial ischemia and ischemia-reperfusion injury (Massignani et al. 2009). CAPE ameliorates the oxidative stress via its antioxidant and anti-inflammatory properties (Okutan et al. 2005), which are attributed to suppression of prostaglandin and leukotriene synthesis (Rossi et al. 2002). It is also demonstrated that CAPE completely and specifically blocks the activation of NF-κB (Natarajan et al. 1996), a transcription factor that plays a pivotal role in controlling chronic diseases associated with oxidative stress and inflammation, such as cardiovascular disease. NF-κB family proteins include several members (p50, p52, p65, RelB, and cRel) that are located in the cytoplasm of most resting cells in an inactive state bound to inhibitory molecules of κB (IκB). Upon stimulation by ROS, and proinflammatory cytokines (IL-1β and TNF-α), which are markedly increased in cardiovascular diseases, IκB of the NF-κB/IκB heterodimer is phosphorylated via activated kinases and subsequently degraded, while the activated "free" NF-κB migrate to the nucleus. NF-κB interacts with specific DNA sequences and induces the expression of genes for proinflammatory cytokines. CAPE and flavonoids in honey inhibit the activation of NF-κB leading to the beneficial effect of honey in a rat model of vascular injury and cardiovascular disease (Maffia et al. 2002). Detailed investigations have indicated that CAPE induces bradycardia and hypotension in rats (Iraz et al. 2005). CAPE-induced bradycardia can be completely abolished by bilateral vagotomy and attenuated by atropine, a muscarinic receptor antagonist, suggesting that CAPE's effect on the heart rate is due to both indirect cardiac muscarinic activation through the vagus nerve and direct activation of endothelial vascular muscarinic receptors (parasympathetic nervous system). However, CAPE-induced hypotensive effect can be altered neither by vagotomy nor atropine (Iraz et al. 2005), suggesting that blood pressure is controlled by different physiologic mechanisms. Further experiments are necessary to clarify the underlying mechanism(s) responsible for CAPE-mediated hypotensive response. In addition, pretreatment of CAPE (0.1 and 1 g/kg) not only reduces both the incidence and duration of ventricular tachycardia and ventricular fibrillation but also decreases the mortality during the myocardial ischemia and reperfusion in rats, suggesting that it exerts antiarrhythmic effects (Huang et al. 2005).

Administration of propolis (a component of honey) in diabetic rats decreases levels of blood glucose, malondialdehyde, NO, NOS, total cholesterol, triglyceride, LDL cholesterol, and very LDL cholesterol in serum of fasting rats, and increases serum levels of HDL cholesterol and superoxide dismutase, suggesting that honey components regulate the metabolism of blood glucose and blood lipid leading to a decrease in outputs of lipid peroxidation and scavenging the free radicals in rats with diabetes mellitus (Zhu et al. 2011).

Other redox-sensitive transcription factors that are associated with heart disease include the activator protein (AP)-1 and hypoxia-inducible factor (HIF)-1. AP-1, which modulates cell differentiation and growth, and consists of homodimers and heterodimers of the c-Jun and c-Fos protein families (Kunsch and Medford 1999). Oxidative stimuli such as OxLDL, $H_2O_2$, or lipid peroxidation products activate AP-1 in both vascular smooth muscle cells and endothelial cells. Although the mechanisms are largely unknown, the ROS-mediated AP-1 activation is possibly dependent on the nuclear protein Ref-1 and the JNK cascade (Kunsch and Medford 1999). CAPE has been reported to inhibit activities of AP-1 transcription factor (Abdel-Latif et al. 2005).

In vascular diseases, HIF-1 regulates vascular remodeling by interfering in cell proliferation and angiogenesis (Hanze et al. 2007). Vascular endothelial growth factor gene expression is mediated by HIF-1 in the vascular wall (Hanze et al. 2007). Under hypoxic conditions, HIF-1 induces a certain arterial remodeling characterized by thickening of the vessel wall and a reduction in the lumen area in the pulmonary circulation, while its exact role in atherogenesis is still not clearly understood (Hanze et al. 2007). Honey components have been reported to inhibit activities of the AP-1 transcription factor (Roos et al. 2011).

Despite above studies on biological effects of honey and the proposed molecular mechanism (Figure 9.6) associated with beneficial effects on cardiovascular diseases, further studies are needed, particularly with regard to understanding the specific mechanisms of action through which natural honey intake may not only be involved in suppression of oxidative stress and inflammation but also associated with modulation of glycemic responses.

## CONCLUSION

For many centuries, honey has been known for its good taste and its beneficial effects on health. Recent research has revealed that honey does indeed exert beneficial cardioprotective effects, which are mainly mediated by its flavonoids. The beneficial effects of honey are most likely due to increased bioavailability of NO. This may explain the improvement in endothelial function, reduction in platelet aggregation, and beneficial effects on blood pressure, insulin resistance, and blood lipid profiles. These processes lead to vasodilatation. Despite the negative results of some studies, epidemiologic studies suggest that dietary consumption of honey is associated with a reduced risk of cardiovascular diseases. In summary, increased production of ROS supports the oxidative modification of LDLs, promoting atherosclerosis by increasing their incorporation into the arterial intima, an essential step in atherogenesis. In addition, the presence of oxidative stress is reported in a wide variety of cardiovascular diseases. In addition to pharmacologic therapy, honey consumption will add therapeutic effects in plaque stability, vasomotor function, and the tendency to thrombosis, improving hypertension, atherosclerosis, atherogenesis, ischemic heart disease, cardiomyopathies, and congestive heart failure due to antioxidative effect of its flavonoids.

## REFERENCES

Abe J and Berk BC. 1998. Reactive oxygen species as mediators of signal transduction in cardiovascular disease. *Trends Cardiovasc Med* 8:59–64.

Abdel-Latif MM, Windle HJ, Homasany BS, Sabra K, and Kelleher D. 2005. Caffeic acid phenethyl ester modulates Helicobacter pylori-induced nuclear factor-kappa B and activator protein-1 expression in gastric epithelial cells. *Br J Pharmacol* 146:1139–1147.

Abuharfeil N, Al-Oran R, and Abo-Shehada M. 1999. The effect of bee honey on the proliferative activity of human B- and T-lymphocytes and the activity of phagocytes. *Food Agric Immun* 11:169–177.

Ahmad A, Khan RA, and Mesaik MA. 2009. Anti-inflammatory effect of natural honey on bovine thrombin-induced oxidative burst in phagocytes. *Phytother Res* 23:801–808.

Akinci S, Arslan U, Karakurt K, and Cengel A. 2008. An unusual presentation of mad honey poisoning: Acute myocardial infarction. *Int J Cardiol* 129:e56–e58.

Alonso-Torre SR, Cavia MM, Fernández-Muiño MA, Moreno G, Huidobro JF, and Sancho MT. 2006. Evolution of acid phosphatase activity of honeys from different climates. *Food Chem* 97:750–755.

Alvarez-Suarez JM, Tulipani S, Díaz D et al. 2010. Antioxidant and antimicrobial capacity of several monofloral Cuban honeys and their correlation with color, polyphenol content, and other chemical compounds. *Food Chem Toxicol* 48:2490–2499.

Al-Waili NS and Haq A. 2004. Effect of honey on antibody production against thymus-dependent and thymus-independent antigens in primary and secondary immune responses. *J Med Food* 7:491–494.

Al-Waili NS. 2004. Natural honey lowers plasma glucose, C-reactive protein, homocysteine, and blood lipids in healthy, diabetic, and hyperlipidemic subjects: Comparison with dextrose and sucrose. *J Med Food* 7:100–107.

Alyane M, Kebsa LB, Boussenane HN, Rouibah H, and Lahouel M. 2008. Cardioprotective effects and mechanism of action of polyphenols extracted from propolis against doxorubicin toxicity. *Pak J Pharm Sci* 21:201–209.

Amiot MJ, Aubert S, Gonnet M, and Tacchini M. 1989. Les composés phónoliques des miels: étude préliminaire sur l'identification et la quantifi cation par familles. *Apidologie* 20:115–125.

Ang ES, Pavlos NJ, Chai LY et al. 2009. Caffeic acid phenethyl ester, an active component of honeybee propolis attenuates osteoclastogenesis and bone resorption via the suppression of RANKL-induced NF-kappaB and NFAT activity. *J Cell Physiol* 221:642–649.

Anfossi G, Russo I, Doronzo G, Pomero A, and Trovati M. 2010. Adipocytokines in atherothrombosis: Focus on platelets and vascular smooth muscle cells. *Mediators Inflamm* 2010:174341.

Arts IC, Jacobs DR Jr, Harnack LJ, Gross M, and Folsom AR. 2001. Dietary catechins in relation to coronary heart disease death among postmenopausal women. *Epidemiology* 12:668–675.

Babacan S and Rand AG. 2007. Characterization of honey amylase. *J Food Sci* 72:C050–C055.

Babacan S and Rand AG. 2005. Purification of amylase from honey. *J Food Sci* 70:C413–C418.

Ball DW. 2007. The chemical composition of honey. *J Chem Educ* 84:1643–1646.

Benguedouar L, Boussenane HN, Wided K, Alyane M, Rouibah H, and Lahouel M. 2008. Efficiency of propolis extract against mitochondrial stress induced by antineoplasic agents (doxorubicin and vinblastin) in rats. *Indian J Exp Biol* 46:112–119.

Beretta G, Orioli M, and Facino RM. 2007. Antioxidant and radical scavenging activity of honey in endothelial cell cultures (EA.hy926). *Planta Med* 73:1182–1189.

Black HR. 1992. Cardiovascular risk factors. In: *Heart Book*, Zaret BL, Moser M, Cohen LS (Eds.), Yale University, School of Medicine, pp. 32–36.

Bogdanov S. 2010. Nutritional and functional properties of honey. *Vopr Pitan* 79:4–13.

Bogdanov S, Jurendic T, Sieber R, and Gallmann P. 2008. Honey for nutrition and health: A review. *J Am Coll Nutr* 27:677–689.

Bojic M, Debeljak Z, Tomicic M, Medic-Saric M, and Tomic S. 2011. Evaluation of antiaggregatory activity of flavonoid aglycone series. *Nutr J* 10:73 (in press).

Blasa M, Candiracci M, Accorsi A, Piacentini MP, and Piatti E. 2007. Honey flavonoids as protection agents against oxidative damage to human red blood cells. *Food Chem* 104:1635–1640.

Brudzynski K. 2006. Effect of hydrogen peroxide on antibacterial activities of Canadian honeys. *Can J Microbiol* 52:1228–1237.

Cai H and Harrison G. 2000. Endothelial dysfunction in cardiovascular diseases: The role of oxidant stress. *Circ Res* 87:840–844.

Calderone V, Chericoni S, Martinelli C et al. 2004. Vasorelaxing effects of flavonoids: Investigation on the possible involvement of potassium channels. *Naunyn Schmiedebergs Arch Pharmacol* 370:290–298.

Carmeliet P and Jain RK. 2011. Molecular mechanisms and clinical applications of angiogenesis. *Nature* 473:298–307.

Cheng TH, Cheng PY, Shih NL, Chen IB, Wang DL, and Chen JJ. 2003. Involvement of reactive oxygen species in angiotensin II-induced endothelin-1 gene expression in rat cardiac fibroblasts. *J Am Coll Cardiol* 42:1845–1854.

Cherchi, Spanedda L, Tuberoso C, and Cabras P. 1994. Solid-phase extraction and high performance liquid chromatographic determination of organic acids in honey. *J Chromatogr A* 669:59–64.

Cominacini L, Garbin U, Pasini AF et al. 1998. Oxidized low-density lipoprotein increases the production of intracellular reactive oxygen species in endothelial cells: Inhibitory effect of lacidipine. *J Hypertens* 16:1913–1919.

Csanyi G, Taylor WR, and Pagano PJ. 2009. NOX and inflammation in the vascular adventitia. *Free Radic Biol Med* 47:1254–1266.

Cushnie TP and Lamb AJ. 2005. Antimicrobial activity of flavonoids. *Int J Antimicrob Agents* 26:343–356.

Daleprane JB, da Silva Freitas V, Pacheco A et al. 2011. Anti-atherogenic and anti-angiogenic activities of polyphenols from propolis. *J Nutr Biochem* (in press).

De Vecchi E and Drago L. 2007. Propolis' antimicrobial activity: What's new? *Infez Med* 15:7–15.

Dhalla NS, Temsah RM, and Netticadan T. 2000. Role of oxidative stress in cardiovascular diseases. *J Hypertens* 18:655–673.

Doner I.W. 1977. The sugars of honey—A review. *J Sci Food Agric* 28:443–456.

Dubey L, Maskey A, and Regmi S. 2009. Bradycardia and severe hypotension caused by wild honey poisoning. *Hellenic J Cardio* 50:426–428.

Duddukuri GR, Kumar PS, Kumar VB, and Athota RR. 1997. Immunosuppressive effect of honey on the induction of allergen-specific humoral antibody response in mice. *Int Arch Allergy Immunol* 114:385–388.

Drummond GR, Selemidis S, Griendling KK, and Sobey CG. 2011. Combating oxidative stress in vascular disease: NADPH oxidases as therapeutic targets. *Nat Rev Drug Discov* 10:453–471.

Eddy JJ, Gideonsen MD, and Mack GP. 2008. Practical considerations of using topical honey for neuropathic diabetic foot ulcers: A review. *WMJ* 107:187–190.

Edwards RL, Lyon T, Litwin SE, Rabovsky A, Symons JD, and Jalili T. 2007. Quercetin reduces blood pressure in hypertensive subjects. *J Nutr* 137:2405–2411.

Egert S, Bosy-Westphal A, Seiberl J et al. 2009. Quercetin reduces systolic blood pressure and plasma oxidised low-density lipoprotein concentrations in overweight subjects with a high-cardiovascular disease risk phenotype: A double-blinded, placebo-controlled cross-over study. *Br J Nutr* 102:1065–1074.

El Denshary ES, Al-Gahazali MA, Mannaa FA, Salem HA, Hassan NS, and Abdel-Wahhab MA. 2011. Dietary honey and ginseng protect against carbon tetrachloride-induced hepatonephrotoxicity in rats. *Exp Toxicol Pathol* 64(7–8):753–760.

Erejuwa OO, Sulaiman SA, Wahab MS, Salam SK, Salleh MS, and Gurtu S. 2010. Antioxidant protective effect of glibenclamide and metformin in combination with honey in pancreas of streptozotocin-induced diabetic rats. *Int J Mo Sci* 11:2056–2066.

Estevinho L, Pereira AP, Moreira L, Dias LG, and Pereira E. 2008. Antioxidant and antimicrobial effects of phenolic compounds extracts of Northeast Portugal honey. *Food Chem Toxicol* 46:3774–3779.

Farooqui T. 2009. Honey: An Anti-Aging Remedy to Keep you Healthy in a Natural Way. Available at http://www.scienceboard.net/community/perspectives.228.html.

Farooqui T and Farooqui AA. 2010. Molecular mechanism underlying the therapeutic activities of propolis: A critical review. *Curr Nutr Food Sci* 6:186–199.

Farooqui T and Farooqui AA. 2012a. Health benefits of honey: Implications for treating cardiovascular diseases. *Curr Nutr Food Sci* 7(4):232–252.

Farooqui T and Farooqui AA. 2012b. Beneficial effects of propolis on human health and neurological diseases. *Front Biosci* (Elite Ed) E4:779–793.

Farooqui T and Farooqui AA. 2012c. Propolis: Implications for the treatment of neurodegenerative diseases. In: *Beneficial Effects of Propolis on Human Health and Chronic Diseases*, (Farooqui T & Farooqui AA, eds), Nova Science Publishers, v.2, pp. 13–35.

Ferreres F, Andrade P, and Tomás-Barberán FA. 1996. Natural occurrence of abscisic acid in heather honey and floral nectar. *J Agric Food Chem* 44:2053–2056.

Ferreres F, Blazquez MA, Gil MI, and Tomás-Barberán FA. 1994a. Separation of honey flavonoids by micellar electrokinetic capillary chromatography. *J Chromatogr A* 669:268–274.

Ferreres F, Garcia–Viguera C, Tomas-Lorente F, and Tomás-Barberán FA. 1993. Hesperetin: A marker of the floral origin of citrus honey. *J Sci Food Agric* 61:121–123.

Ferreres F, Giner JM, and Tomás-Barberán FA. 1994b. A comparative study of hesperetin and methyl anthranilate as markers of the floral origin of citrus honey. *J Sci Food Agric* 65:371–372.

Ferreres F, Juan T, Perez-Arquillue C, Herrera-Marteache A, Garcia-Viguera C, and Tomás-Barberán FA. 1998. Evaluation of pollen as a source of kaempferol in rosemary honey. *J Sci Food Agric* 77:506–510.

Frankel EN, Kanner J, German JB, Parks E, and Kinsella JE. 1993. Inhibition of oxidation of human low-density lipoprotein by phenolic substances in red wine. *Lancet* 341:454–457.

Funakoshi-Tago M, Nakamura K, Tago K, Mashino T, and Kasahara T. 2011. Anti-inflammatory activity of structurally related flavonoids, Apigenin, Luteolin and Fisetin. *Int Immunopharmacol* 11(9):1150–1159.

Gómez-Caravaca AM, Gómez-Romero M, Arráez-Román D, Segura-Carretero A, and Fernández-Gutiérrez A. 2006. Advances in the analysis of phenolic compounds in products derived from bees. *J Pharm Biomed Anal* 41:1220–1234.

Geleijnse JM, Launer LJ, Van der Kuip DA, Hofman A, and Witterman JC. 2002. Inverse association of tea and flavonoid intakes with incident myocardial infarction: The Rotterdam Study. *Am J Clin Nutr* 275:880–886.

Grassmann J, Schneider D, Weiser D, and Elstner EF. 2001. Antioxidative effects of lemon oil and its components on copper induced oxidation of low density lipoprotein. *Arzneimittelforschung* 51:799–805.

Griendling KK, Sorescu D, and Ushio-Fukai. 2000. NAD(P)H oxidase: Role in cardiovascular biology and disease. *Circ Res* 86:494–501.

Gui-Rong L, Wang HB, Qin GW et al. 2008. Acacetin, a natural flavone, selectively inhibits human atrial repolarization potassium currents and prevents atrial fibrillation in dogs. *Circulation* 117:2449–2457.

Halliwell B. 2007. Dietary polyphenols: Good, bad, or indifferent for your health? *Cardiovasc Res* 73:341–347.

Hamdy AA, Ismail HM, Al-Ahwal Ael-M, and Gomaa NF. 2009. Determination of flavonoid and phenolic Acid contents of clover, cotton and citrus floral honeys. *J Egypt Public Health Assoc* 84:245–259.

Hanze J, Weissmann N, Grimminger F, Seeger W, and Rose F. 2007. Cellular and molecular mechanisms of hypoxia-inducible factor driven vascular remodeling. *Thromb Haemost* 97:774–787.

Heitzer T, Schlinzig T, Krohn K, Meinertz T, and Munzel T. 2001. Endothelial dysfunction, oxidative stress, and risk of cardiovascular events in patients with coronary artery disease. *Circulation* 104:2673–2678.

Herrero M, Ibáñez E, and Cifuentes A. 2005. Analysis of natural antioxidants by capillary electromigration methods. *J Separation Sci* 28:883–897.

Hertog MG, Feskens EJ, Hollman PC, Katan MB, and Kromhout D. 1993. Dietary antioxidant flavonoids and risk of coronary heart disease: The Zutphen Elderly Study. *Lancet* 342:1007–1011.

Hertog MG, Feskens EJ, and Kromhout D. 1997a. Antioxidant flavonols and coronary heart disease risk. *Lancet* 349:699.

Hertog MG, Sweetnam PM, Fehily AM, Elwood C, and Kromhout D. 1997b. Antioxidant flavonols and ischemic heart disease in a Welsh population of men: The Caerphilly Study. *Am J Clin Nutr* 65:1489–1494.

Hertog MGL, Kromhout D, Aravanis C et al. 1995. Flavonoid intake and longterm risk of coronary heart disease and cancer in the seven countries study. *Arch Intern Med* 155:381–386.

Hooper L, Kroon PA, Rimm EB et al. 2008. Flavonoids, flavonoid-rich foods, and cardiovascular risk: A meta-analysis of randomized controlled trials. *Am J Clin Nutr* 88:38–50.

Huang SS, Liu SM, Lin SM et al. 2005. Antiarrhythmic effect of caffeic acid phenethyl ester (CAPE) on myocardial ischemia/reperfusion injury in rats. *Clin Biochem* 38:943–947.

Huidobro JF, Sánchez MP, Muniategui S, and Sancho MT. 2005. Catalase in honey. *J AOAC Int* 88:800–804.

Iraz M, Fadillioglu E, Tasdemir S, and Erdogan S. 2005. Role of vagal activity on bradicardic and hypotensive effects of caffeic acid phenethyl ester (CAPE). *Cardiovasc Toxicol* 5:391–396.

Irish J, Blair S, and Carter DA. 2011. The antibacterial activity of honey derived from Australian flora. *PLoS One* 6:e18229.

Jaganathan SK and Mandal M. 2009. Antiproliferative effects of honey and of its polyphenols: A review. *J Biomed Biotechnol* 2009:830616.

Jaganathan SK, Mandal SM, Jana SK, Das S, and Mandal M. 2010. Studies on the phenolic profiling, anti-oxidant and cytotoxic activity of Indian honey: *In vitro* evaluation. *Nat Prod Res* 24:1295–1306.

Jaganathi IB and Crozier A. 2010. Dietary flavonoids and phenolic compounds. In: *Plant Phenolics and Human Health: Biochemistry, Nutrition, and Pharmacology*, Fraga (Ed.), John Wiley & Sons, Inc, pp. 1–50.

Jin BH, Quin LB, Chen S et al. 2009. Apigenin protects endothelium-dependent relaxation of rat aorta against oxidative stress. *Eur J Pharmacol* 16:200–205.

Jin YR, Han XH, Zhang YH et al. 2008. Antiplatelet activity of hesperetin, a bioflavonoid, is mainly mediated by inhibition of PLC- γ2 phosphorylation and cyclooxygenase-1 activity. *J Cell Biochem* 104:1–14.

Judkins CP, Diep H, Broughton BR et al. 2010. Direct evidence of a role for Nox2 in superoxide production, reduced nitric oxide bioavailability, and early atherosclerotic plaque formation in ApoE-/- mice. *Am J Physiol Heart Circ Physiol* 298:H24–H32.

Kao TK, Ou YC, Raung SL, Lai CY, Liao SL, and Chen CJ. 2010. Inhibition of nitric oxide production by quercetin in endotoxin/cytokine-stimulated microglia. *Life Sci* 86:315–321.

Kassim M, Achoui M, Mansor M, and Yusoff KM. 2010a. The inhibitory effects of Gelam honey and its extracts on nitric oxide and prostaglandin $E_2$ in inflammatory tissues. *Fitoterapia* 81:1196–1201.

Kassim M, Achoui M, Mustafa MR, Mohd MA, and Yusoff KM. 2010b. Ellagic acid, phenolic acids, and flavonoids in Malaysian honey extracts demonstrate *in vitro* anti-inflammatory activity. *Nutr Res* 30:650–659.

Khalil MI, Sulaiman SA, and Boukraa L. 2010. Antioxidant properties of honey and its role in preventing health disorder. *The Open Nutraceuticals J* 3:6–16.

Khalil ML and Sulaiman SA. 2010. The potential role of honey and its polyphenols in preventing heart disease: A review. *Afr J Tradit Comlement Altern Med* 7:315–321.

Khan MS, Devaraj H, and Devaraj N. 2011. Chrysin abrogates early hepatocarcinogenesis and induces apoptosis in N-nitrosodiethylamine-induced preneoplastic nodules in rats. *Toxicol Appl Pharmacol* 251:85–94.

Kishore RK, Halim AS, Syazana MS, and Sirajuddin KN. 2011. Tualang honey has higher phenolic content and greater radical scavenging activity compared with other honey sources. *Nutr Res* 31:322–325.

Knekt P, Jarvinen R, Reunanen A, and Maatela J. 1996. Flavonoid intake and coronary mortality in Finland: A cohort study. *BMJ* 312:478–481.

Knekt P, Kumpulainen J, Jarvinen R et al. 2002. Flavonoid intake and risk of chronic diseases. *Am J Clin Nutr* 76:560–568.

Kunsch C and Medford RM. 1999. Oxidative stress as a regulator of gene expression in the vasculature. *Circ Res* 85:753–766.

Lakhanpal P and Rai DK. 2008. Role of quercetin in cardiovascular diseases. *Internet J Med Update* 3:31–49.

Law MR, Morris JK, and Wald NJ. 2009. Use of blood pressure lowering drugs in the prevention of cardiovascular disease: Meta-analysis of 147 randomised trials in the context of expectations from prospective epidemiological studies. *BMJ* 338:b1665.

Lee JY, Ebel H, Friese M, Schillinger G, Schröder R, and Klug N. 2003. Influence of TachoComb in comparison to local hemostyptic agents on epidural fibrosis in a rat laminectomy model. *Minim Invasive Neurosurg* 46:106–109.

Libby P. 2002. Inflammation in atherosclerosis. *Nature* 420:868–874.

Li JJ and Chen JL. 2005. Inflammation may be a bridge connecting hypertension and atherosclerosis. *Med Hypotheses* 64:925–929.

Lin J, Rexrode KM, Hu F et al. 2007. Dietary intakes of flavonols and flavones and coronary heart disease in US women. *Am J Epidemiol* 165:1305–1313.

Lo YY, Wong JM, and Cruz TF. 1996. Reactive oxygen species mediate cytokine activation of c-Jun NH2-terminal kinases. *J Biol Chem* 271:15703–15707.

Maffia P, Ianaro A, Pisano B et al. 2002. Beneficial effects of caffeic acid phenethyl ester in a rat model of vascular injury. *Br J Pharmacol* 136:353–360.

Majtán J, Kovácová E, Bíliková K, and Simúth J. 2006. Immunostimulatory effect of the recombinant apalbumin 1-major honeybee royal jelly protein-on TNFalpha release. *Int Immunopharmacol* 6:269–278.

Makhdoom A, Khan MS, Lagahari MA, Rahopoto MQ, Tahir SM, and Siddiqui KA. 2009. Management of diabetic foot by natural honey. *J Ayub Med Coll Abbottabad* 21: 103–105.

Mason RP. 2011. Optimal therapeutic strategy for treating patients with hypertension and atherosclerosis: Focus on olmesartan medoxomil. *Vasc Health Risk Manag* 7:405–416.

Massignani JJ, Lemos M, Maistro EL et al. 2009. Antiulcerogenic activity of the essential oil of Baccharis dracunculifolia on different experimental models in rats. *Phytother Res* 23:1355–1360.

Matsushita M, Nishikimi N, Sakurai T, and Nimura Y. 2000. Relationship between aortic calcification and atherosclerotic disease in patients with abdominal aortic aneurysm. *Int Angiol* 19:276–279.

Meyer AS, Heinonen M, and Frankel EN. 1998. Antioxidant interactions of catechin, cyanidin, caffeic acid, quercetin, and ellagic acid on human LDL oxidation. *Food Chem* 61:71–75.

Middleton E, Kandaswami C, and Theoharides TC. 2000. The effects of plants flavonoids on mammalian cells: Implications for inflammation, heart disease, and cancer. *Pharmacol Rev* 52:673–751.

Molan PC and Betts JA. 2008. Using honey to heal diabetic foot ulcers. *Adv Skin Wound Care* 21:313–316.

Molan PC. 2001. Potential of honey in the treatment of wounds and burns. *Am J Clin Dermatol* 2:13–19.

Molan C. 2006. The evidence supporting the use of honey as a wound dressing. *Int J Extrem Wounds* 5:40–54.

Mozaffarian D, Wilson PW, and Kannel WB. 2008. Beyond established and novel risk factors: Lifestyle risk factors for cardiovascular disease. *Circulation* 117:3031–3038.

Mulvihill EE and Huff MW. 2010. Antiatherogenic properties of flavonoids: Implications for cardiovascular health. *Can J Cardiol* 26(Suppl A):17A–21A.

Natarajan K, Singh S, Burke TR Jr, Grunberger D, and Aggarwal BB. 1996. Caffeic acid phenethyl ester is a potent and specific inhibitor of activation of nuclear transcription factor NF-kappa B. *Proc Natl Acad Sci U S A* 93:9090–9095.

Ohashi K, Natori S, and Kubo T. 1999. Expression of amylase and glucose oxidase in the hypopharyngeal gland with an age-dependent role change of the worker honeybee (Apis mellifera L.). *Eur J Biochem* 265:127–133.

Okutan H, Ozcelik N, Yilmaz HR, and Uz E. 2005. Effects of caffeic acid phenethyl ester on lipid peroxidation and antioxidant enzymes in diabetic rat heart. *Clin Biochem* 38:191–196.

Okuyan E, Uslu A, and Ozan Levent M. 2010. Cardiac effects of "mad honey": A case series. *Clin Toxicol (Phila)* 48:528–532.

Omotayo EO, Gurtu S, Sulaiman SA, Ab Wahab MS, Sirajuddin KN, and Salleh MS. 2010. Hypoglycemic and antioxidant effects of honey supplementation in streptozotocin-induced diabetic rats. *Int J Vitam Nutr Res* 80:74–82.

Paul MI, Beiler J, McMonagle A, Shaffer ML, Duda L, and Berlin CM. 2007. Effect of honey, dextromethorphan and no treatment on nocturnal cough and sleep quality for coughing children and their parents. *Arch Pediatr Adolesc Med* 161:1140–1146.

Perticone F, Ceravolo R, Pujia A et al. 2001. Prognostic significance of endothelial dysfunction in hypertensive patients. *Circulation* 104:191–196.

Petrus K, Schwartz H, and Sontag G. 2011. Analysis of flavonoids in honey by HPLC coupled with coulometric electrode array detection and electrospray ionization mass spectrometry. *Anal Bioanal Chem* 400:2555–2563.

Psotová J, Chlopčíková Š, Miketová P, Hrbáč J, and Šimánek V. 2004. Chemoprotective effect of plant phenolics against anthracycline-induced toxicity on rat cardiomyocytes. Part III. Apigenin, baicalelin, kaempherol, luteolin and quercetin. *Phytotherapy Res* 18:516–521.

Pourahmad M and Sobhanian S. 2009. Effect of honey on the common cold. *Arch Med Res* 40:224–225.

Punithavathi VR and Stanely Mainzen Prince P. 2011. The cardioprotective effects of a combination of quercetin and α-tocopherol on isoproterenol-induced myocardial infarcted rats. *J Biochem Mol Toxicol* 25:28–40.

Rajalakhmi D and Narasimham S. 1996. In: *Food Antioxidants*, Madhavi DL, Deshpande SS, Salunkhe DK (Eds.), Marcel Dekker, New York, p 65.

Rakha MK, Nabil ZI, and Hussein AA. 2008. Cardioactive and vasoactive effects of natural wild honey against cardiac malperformance induced by hyperadrenergic activity. *J Med Food* 11(1):91–98.

Ramasamy S, Parthasarathy S, and Harrison DG. 1998. Regulation of endothelial nitric oxide synthase gene expression by oxidized linoleic acid. *J Lipid Res* 39:268–276.

Rimm EB, Katan MB, Ascherio A, Stampfer MJ, and Willett WC. 1996. Relation between intake of flavonoids and risk for coronary heart disease in male health professionals. *Ann Intern Med* 125:384–389.

Rodrigo R, González J, and Paoletto F. 2011. The role of oxidative stress in the pathophysiology of hypertension. *Hypertens Res* 34:431–440.

Roos TU, Heiss EH, Schwaiberger AV, Schachner D, Sroka IM, Oberan T, Vollmar AM, and Dirsch VM. 2011. Caffeic acid phenethyl ester inhibits PDGF-induced proliferation of vascular smooth muscle cells via activation of p38 MAPK, HIF-1α, and heme oxygenase-1. *J Nat Prod* 74:352–356.

Rosenfeld ME. 2000. An overview of the evolution of the atherosclerotic plaque: From fatty streak to plaque rupture and thrombosis. *Z Kardiol* 89(Suppl 7):2–6.

Rossi A, Longo R, Russo A, Borrelli F, and Sautebin L. 2002. The role of the phenethyl ester of caffeic acid (CAPE) in the inhibition of rat lung cyclooxygenase activity by propolis. *Fitoterapia* 73(Suppl 1):S30–S37.

Sánchez MP, Huidobro JF, Mato I, Muniategui S, and Sancho MT. 2001. Evolution of invertase activity in honey over two years. *J Agric Food Chem* 49:416–422.

Shepartz AI and Subers MH. 1964. The glucose oxidase of honey I. Purification and some general properties of the enzyme. *Biochimicaet Biophysica Acta (BBA)-Specialized Section on Enzymological Subjects* 85(2):228–237.

Schulz E, Gori T, and Münzel T. 2011. Oxidative stress and endothelial dysfunction in hypertension. *Hypertens Res* 34:665–673.

Sesso HD, Gaziano JM, Liu S, and Buring JE. 2003. Flavonoid intake and the risk of cardiovascular disease in women. *Am J Clin Nutr* 77:1400–1408.

Shadkam MN, Mozaffari-Khosravi H, and Mozayan MR. 2010. A comparison of the effect of honey, dextromethorphan, and diphenhydramine on nightly cough and sleep quality in children and their parents. *J Altern Complement Med* 16:787–793.

Siddiqui IR. 1970. The sugars of honey. *Adv Carbohyd Chem* 25:285–309.

Song JJ, Cho JG, Hwang SJ, Cho CG, Park SW, and Chae SW. 2008. Inhibitory effect of caffeic acid phenethyl ester (CAPE) on LPS-induced inflammation of human middle ear epithelial cells. *Acta Otolaryngol* 128:1303–1307.

Standridge JB. 2005. Hypertension and atherosclerosis: Clinical implications from the ALLHAT trial. *Curr Atheroscler Rep* 7:132–139.

Steinberg D. 1997. Lewis A. Conner Memorial Lecture. Oxidative modification of LDL and atherogenesis. *Circulation* 95:1062–1071.

Sugiura T, Dohi Y, Takase H, Yamashita S, Tanaka S, and Kimura G. 2011. Increased reactive oxygen metabolites is associated with cardiovascular risk factors and vascular endothelial damage in middle-aged Japanese subjects. *Vasc Health Risk Manag* 7:475–482.

Tomás-Barberán FA, Ferreres F, García-Vignera C, and Tomás-Lorente F. 1993. Flavonoids in honey of different geographical origin. *Z Lebensm Unters For A* 196:38–44.

Tomás-Barberán FA, Martos I, Ferreres FI, Radovic BS, and Anklam E. 2001. HPLC flavonoid profiles as markers for the botanical origin of European unifloral honeys. *J Sci Food Agric* 81:485–496.

Tonks A, Cooper RA, Price AJ, Molan PC, and Jones KP. 2001. Stimulation of TNF-α release in monocytes by honey. *Cytokine* 14:240–242.

Tonks AJ, Dudley E, Porter NG et al. 2007. A 5.8-kDa component of manuka honey stimulates immune cells via TLR4. *J Leukoc Biol* 82:1147–1155.

Tousoulis D, Briasoulis A, Papageorgiou N et al. 2011. Oxidative stress and endothelial function: Therapeutic interventions. *Recent Pat Cardiovasc Drug Discov* 6:103–114.

Vacca M, Degirolamo C, Mariani-Costantini R, Palasciano G, and Moschetta A. 2011. Lipid-sensing nuclear receptors in the pathophysiology and treatment of the metabolic syndrome. *Wiley Interdiscip Rev Syst Biol Med* (in press).

van den Berg AJ, van den Worm E, van Ufford HC, Halkes SB, Hoekstra MJ, and Beukelman CJ. 2008. An *in vitro* examination of the antioxidant and anti-inflammatory properties of buckwheat honey. *J Wound Care* 17:172–174,176–178.

Vanhoutte PM, Shimokawa H, Tang EH, and Feletou M. 2009. Endothelial dysfunction and vascular disease. *Acta Physiol* 196:193–222.

Vinson JA, Su X, Zubik L, and Bose P. 2001. Phenol antioxidant quantity and quality in foods: Fruits. *J Agric Food Chem* 49:5315–5321.

Vita JA. 2005. Polyphenols and cardiovascular disease: Effects on endothelial and platelet function. *Am J Clin Nutr* 81(1 Suppl):292S–297S.

Viuda-Martos M, Ruiz-Navajas Y, Fernández-López J, and V Pérez-Alvarez JA. 2008. Functional properties of honey, propolis, and royal jelly. *Food Sci* 73:R117–R124.

Vlachodimitropoulou E, Sharp PA, and Naftalin RJ. 2011. Quercetin-iron chelates are transported via glucose transporters. *Free Radic Biol Med* 50:934–944.

Vorlova L and Přidal A. 2002. Invertase and diastase activity in honeys of Czech provenience. *Acta univ agric et silvic Mendel Brun* L, No. 5:57–66.

Wahdan HA. 1998. Causes of the antimicrobial activity of honey. *Infection* 26:26–31.

Wang J and Li QX. 2011. Chemical composition, characterization, and differentiation of honey botanical and geographical origins. *Adv Food Nutr Res* 62:89–137.

Watanabe Y, Suzuki O, Haruyama T, and Akaike T. 2003. Interferon-gamma induces reactive oxygen species and endoplasmic reticulum stress at the hepatic apoptosis. *J Cell Biochem* 89:244–253.

White JW and Subers MH. 1964. Studies on honey Inhibine. 4. Destruction of the peroxide accumulation system by light. *J Food Sci* 29:819–828.

Woo KJ, Jeong YJ, Inoue H, Park JW, and Kwon TK. 2005. Chrysin suppresses lipopolysaccharide-induced cyclooxygenase-2 expression through the inhibition of nuclear factor for IL-6 (NF-IL6) DNA-binding activity. *FEBS Lett* 579:705–711.

Woodward G, Kroon P, Cassidy A, and Kay C. 2009. Anthocyanin stability and recovery: Implications for the analysis of clinical and experimental samples. *J Agric Food Chem* 57:5271–5278.

Xu YC, Yeung DKY, Man RYK, and Leung SWS. 2006. Kaempferol enhances endothelium-independent and dependent relaxation in the porcine coronary artery. *Mol Cell Biochem* 287:61–67.

Yaghoobi N, Al-Waili N, Ghayour-Mobarhan M et al. 2008. Natural honey and cardiovascular risk factors; effects on blood glucose, cholesterol, triacylglycerole, CRP, and body weight compared with sucrose. *Sci World J* 8:463–469.

Yochum L, Kushi LH, Meyer K, and Folsom AR. 1999. Dietary flavonoid intake and risk of cardiovascular disease in postmenopausal women. *Am J Epidemiol* 49:943–999.

Yoshizumi M, Kim S, and Kagami S. 1998. Effect of endothelin-1 on extracellular signal-regulated kinase and proliferation of human coronary artery smooth muscle cells. *Br J Pharmacol* 125:1019–1027.

Zhu W, Chen M, Shou Q, Li Y, and Hu F. 2011. Biological activities of Chinese propolis and Brazilian propolis on streptozotocin-induced type 1 diabetes mellitus in rats. *Evid Based Complement Alternat Med* 2011:468529.

# 10 Honey for Diabetic Ulcers

*Hasan Ali Alzahrani*

## CONTENTS

## INTRODUCTION

Diabetes is the global epidemic of the 21st century, claiming as many lives per year as HIV/AIDS. It is the fourth leading cause of global death by disease in developed countries. The number of adults with diabetes in the world was 135 million in 1995 (King et al. 1998). At present, 250 million people are affected by diabetes worldwide and there is no cure. By 2025, the global burden of diabetes is projected to increase to 380 million people, according to the International Diabetes Foundation. Between 2010 and 2030, there will be a 69% increase in the number of adults with diabetes in developing countries and a 20% increase in developed countries (Shaw et al. 2010).

Diabetes mellitus is a serious and complex chronic disease caused by a combination of hereditary and environmental factors. It may lead to devastating short-term and long-term complications, affecting quality of life and health care costs. If left untreated, diabetes may be fatal, and in the long term, complications include heart disease, kidney failure, blindness, and foot problems.

In this chapter, the use of topical honey in treating diabetic foot disorders, particularly in the case of ulceration and wounds, will be discussed in view of the published literature and the author's experience over the past two decades.

## DIABETIC FOOT ULCERS

Diabetic foot ulceration is an important complication of diabetes and represents a major medical, social, and economic problem worldwide (Boulton 2004). Foot ulcers are often difficult to heal due to many extrinsic and intrinsic factors, and for many individuals, this may lead to hospitalization and amputation (Huijberts et al. 2008; Sibbald and Woo 2008). The lifetime risk of developing a diabetic foot ulcer has been estimated to be up to 15% (Reiber 1996) and has been reported to be as high as 25% (Lavery et al. 2003).

Diabetic foot disease and amputations may be reduced considerably only if sufficient attention is paid to the necessary preventative measures (Van Houtum 2005).

A multidisciplinary approach to diabetic foot problems has shown improvements in prevention leading to a reduction in foot ulcers and subsequent amputations (Dargis et al. 1999).

Specialized modern diabetic foot clinics should be therefore multidisciplinary and equipped to coordinate for early diagnosis, off-loading, and preventive care; to perform revascularization procedures; to aggressively treat infections; to locally care for ulcers; and to manage medical comorbidities.

## CAUSES

The diabetic foot ulcers start with trivial physical or mechanical trauma, which is not usually recognized by the diabetic patient, as most diabetics have an underlying diabetic neuropathy (Oguejiofor et al. 2010). In the presence of local risk factors for ulceration, such as peripheral arterial diseases, diabetic peripheral and autonomic neuropathy, and foot deformities, the ulceration usually increases in size and depth. Such deterioration is more in elderly chronically diabetic patients who have poor control on their blood glucose level, particularly those who have associated late complications such as diabetic retinopathy, chronic renal disease, and low immunity. The understanding of the complexity of the pathogenesis is mandatory before any management plan. One should not forget that such a complex clinical presentation needs a holistic multidisciplinary approach when it comes to management (Aydin et al. 2010). Local wound or ulcer care is therefore one step forward in the management, which should take into account the previously mentioned general and local risk factors.

## OUTCOMES OF DIABETIC FOOT ULCERS

Ulcer is the most common presentation in diabetic foot disorders (Boulton et al. 2005). The ultimate endpoint of a diabetic foot ulcer is amputation if not well treated (Unwin 2008). When amputation happens, it is usually associated with significant morbidity (Boulton 2004) and mortality (Badri et al. 2011), in addition to immense social, psychologic, and financial consequences (Dargis et al. 1999; Van Houtum

2005). The eventual outcome of diabetic foot ulceration may be loss of the whole extremity involved (Boulton et al. 2005; Unwin 2008). In the developing world, these fears will influence diabetic patients to try all types of conventional and complementary and alternative medicine (CAM) treatments. Although diabetic foot disorders are manifestations of a systemic metabolic disorder, most patients, and sometimes health professionals, tend to deal with it as a local problem and therefore focus on using local topical agents that may prevent infection or promote healing such as honey (Abdelatif et al. 2008).

## Cost of Diabetic Foot Disorders Care

Diabetes imposes an increasing economic burden on national health care systems worldwide. More prevention efforts are needed to reduce this burden. Meanwhile, the very low expenditures per capita in poor countries indicate that more resources are required to provide basic diabetes care in such settings (Zhang et al. 2010).

The global health expenditure on diabetes is expected to total at least $376 billion in 2010 and $490 billion in 2030. Globally, 12% of the health expenditures and $1330 per person are anticipated to be spent on diabetes in 2010 (Zhang et al. 2010).

Compared with diabetic patients without foot ulcers, the cost of care for patients with a foot ulcer is 5.4 times higher in the year after the first ulcer episode and 2.8 times higher in the second year. Costs for the treatment of the highest-grade ulcers are eight times higher than for treating low-grade ulcers as reported by Driver et al. (2010). Local early, efficient, and affordable wound care products may contribute to better savings in this regard. Honey as a wound care product is expected to be on the top of the list, particularly in developing countries (Alzahrani and Bakhotmah 2010).

## Local Wounds/Ulcers Care

Adequate diabetic foot care includes use of topical agents to treat topical infection and/or prevent potential infection, promote healing, and sometimes debride necrotic tissues. The most frequently used topical antimicrobials in modern wound care practice include iodine- and silver-containing products. In the past, acetic acid, chlorhexidine, hydrogen peroxide, sodium hypochlorite, potassium permanganate, and honey have been used (White et al. 2001). Some of these products seem to be making a return, and other alternatives are being investigated. Oral and intravenous antibiotics do not replace the appropriate local wound care. The search for the best topical agent still continues. Most of the industrialized commercialized products are not affordable for the majority of patients in developing and underdeveloped countries, where patients prefer the use of locally available CAM products (Bakhotmah and Alzahrani 2010).

## Complementary and Alternative Products in Local Foot Care

Self-medication with oral natural preparations and herbs is fairly common across the world, including developed countries, as part of CAM that is used by individuals with and without diabetes (Egede et al. 2002; Garrow and Egede 2006; Modak et al.

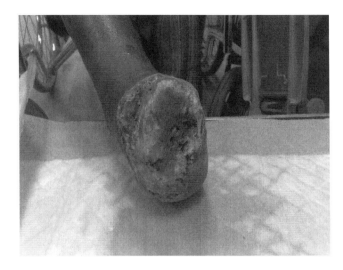

**FIGURE 10.1**   Topical application of natural preparation on a leg of a diabetic foot patient.

2007). For instance in the American general population, estimates of CAM use were not significantly different across selected chronic medical conditions, but diabetes was an independent predictor of CAM use (Garrow and Egede 2006).

Traditional medicines derived from medicinal plants are used by about 60% of the world's population (Modak et al. 2007). In developing countries, people believe that natural products and herbal formulations are preferred due to fewer side effects and lower cost (Modak et al. 2007). However, their preference depends on many factors including cultural background, education, socioeconomic factors, nature of the health problem, and availability of the remedies in the local market (Figure 10.1).

For example, in Saudi Arabia it was found that habits and practices are usually influenced by the Islamic culture and literature, which value the use of CAM as part of spiritual life (Alzahrani and Bakhotmah 2010). Reciting the Holy Quran (62.5%), prescriptions by herb practitioners (43.2%), cautery (12.4%), and cupping (4.4%) were reported (Al-Rowais 2002; Al-Rowais et al. 2010).

Honey was one of the most favored CAM products by diabetic patients for the treatment of diabetic foot disorders in Jeddah, western Saudi Arabia (Alzahrani and Bakhotmah 2010).

## HONEY AS A TOPICAL AGENT IN WOUND CARE

Honey has been appreciated as food since the dawn of history. Its medicinal use is also ancient. The use of honey as a healing agent has witnessed rise and falls in medical history. The Greek physician Hippocrates recommended topical application of honey for infected wounds and ulcers of the lips (Altman 2010). Use of honey was related to history, religions, and cultures. It was widely used to treat various diseases by all religions' followers in the Middle East (i.e., Muslims, Christians, and Jews). Ancient Egyptian and Middle Eastern people also used honey for embalming the dead and wound dressings. For at least 2700 years, honey has been used by humans to treat a variety of ailments through topical application by the Chinese, Romans, Egyptians, and Greeks for medicinal purposes (Altman 2010).

The effectiveness of honey in many of its medical uses is probably due to its antibacterial activity. Honey either kills or inhibits the growth of a wide range of bacterial and fungal species, including many that cause serious infections (Tan et al. 2009; Majtan et al. 2010).

Honey has four main properties that help it to exterminate bacteria: osmosis, high acidity, hydrogen peroxide activity, and a variety of phytochemicals (Sherlock et al. 2010). The honey draws fluid away from the infected wound by osmosis. This may help in killing bacteria because bacteria need liquid in order to grow. Honey is also hygroscopic, meaning it draws moisture out to the environment and thus dehydrates bacteria. The viscosity and the hygroscopic qualities of honey permit its even spread on the wound bed, creating a favorable environment for wound healing (Salomon et al. 2010). Honey is very acidic. Its pH is between 3 and 4. Most types of bacteria thrive at pH levels between 7.2 and 7.4 and cannot survive at levels below pH 4.0. Lowering wound pH can potentially reduce protease activity, increase fibroblast activity, increase oxygen release, and consequently aid in wound healing (Gethin et al. 2008).

Hydrogen peroxide has an extraordinary capacity to stimulate oxidative enzymes (Bang et al. 2003). These activated oxidative enzymes can change the chemical component of other substances (such as viruses, bacteria parasites, viruses, and even some types of tumor cells) without being changed themselves (Brudzynski 2006; Kwakman et al. 2010).

The activated enzyme glucose oxidize enables the honey to produce hydrogen peroxide, an important factor in inhibiting bacterial growth and promoting immune activity. The hydrogen peroxide level in honey, therefore, is a strong predictor of the honey's antibacterial activity (Brudzynski 2006).

Hydrogen peroxide in honey also contributes to its healing properties. Hydrogen peroxide levels produced by diluted honey are much lower than those in 3% hydrogen peroxide solutions (Bang et al. 2003). This low concentration may enable hydrogen peroxide to serve as an intracellular "messenger" that stimulates wound healing without causing oxidative damage from free radicals. Low-level hydrogen peroxide release assists both tissue repair and contributes to the antibacterial activity of honey even on a heavily exuding wound (Allen et al. 1991).

Researchers have identified many other chemicals with antibacterial and antioxidant activity in honey (Tonks et al. 2001). For instance, most types of honey stimulate the release of a variety of cytokines from surrounding tissue cells, mainly monocytes and macrophages (Benhanifia et al. 2011). One of the most important of these cytokines is tumor necrosis factor, a protein that reduces tissue inflammation, induces the destruction of some tumor cells, and activates white blood cells, which is vital to healing (Tonks et al. 2003).

## HOW HONEY WORKS ON WOUNDS?

Honey works on wounds by different mechanisms, which all contribute to better wound healing as it has antibacterial, antiseptic, and anti-inflammatory effects. In addition, it has good physicochemical properties, which enhance healing and make it as a good natural physical barrier (Molan 2006). However, the development of antibiotics during the 20th century marked the decline of many former remedies, including honey, but the emergence of antibiotic-resistant strains of pathogens has led to the need to find alternative treatments (Sherlock et al. 2010). As mentioned above, bacteria cannot live in the presence of honey as the osmotic pressure removes water molecules from bacteria, making them shrivel up and die. Honey placed on a wound creates a physical barrier through which bacteria cannot pass. When honey is diluted with wound exudates, the glucose oxidize contains it becomes active and produces hydrogen peroxide, a powerful antibacterial agent. The overall effect is to reduce infection and to enhance wound healing in burns, ulcers, and other cutaneous wounds (Lusby et al. 2002).

Honey's significant antioxidant content mops up free radicals, which might explain why, in one study, honey dressing prevented partial-thickness burns from converting to full-thickness burns (Allen et al. 1991).

The anti-inflammatory action of honey "provides the most likely explanation for the reduction of hypertrophic scarring observed in wounds that were dressed with honey" (Tonks et al. 2001).

Numerous reported cases were also published on how honey applications either reduce or eliminate wound odor (Molan 2006).

## EVIDENCE FOR USING HONEY AS A TOPICAL AGENT FOR WOUND CARE

Evidence from animal studies and some clinical trials has suggested that honey may accelerate wound healing (Jull et al. 2008). Topical treatment of diabetic foot wound and ulceration is not an exception (Makhdoom et al. 2009).

A hot debate is still ongoing on the availability of enough evidence to recommend using honey topically on all wounds. Moore et al. (2001) conducted a systematic review of six randomized controlled trials and stated, "confidence in a conclusion that honey is a useful treatment for superficial wounds or burns is low. There is biological plausibility." A similar recent Cochrane systematic review found that most studies addressing honey's effect on burns and acute traumatic wounds were of poor quality, which necessitated caution in interpreting their findings. Nevertheless, the review reported that the strongest evidence available supports the use of honey for venous leg ulcers. Jull et al. (2008) identified 19 trials ($n = 2554$) that met the authors' inclusion criteria. They found that, in acute wounds, three trials evaluated the effect of honey in acute lacerations, abrasions, or minor surgical wounds and nine trials evaluated the effect of honey in burns. In chronic wounds, two trials evaluated the effect of honey in venous leg ulcers and one trial in pressure ulcers, infected postoperative wounds, and Fournier's gangrene, respectively. Two trials recruited people with mixed groups of chronic or acute wounds. They also found that the poor quality of most of the trial reports means the results should be interpreted with caution, except in venous leg ulcers. In acute wounds, honey may reduce time to healing compared with some conventional dressings in partial-thickness burns (weighted mean difference, −4.68 days; 95% confidence interval, −4.28 to −5.09 days). All the included burn trials have originated from a single center, which may have impact on replicability. In chronic wounds honey, in addition to compression bandaging, does not significantly increase healing in venous leg ulcers (risk ratio, 1.15; 95% confidence interval, 0.96–1.38). They concluded by stating that there is insufficient evidence to determine the effect of honey compared with other treatments for burns or chronic wound types (Jull et al. 2008).

In contradiction, Dr. Peter Molan, the director of the Waikato Honey Research Unit, University of Auckland, New Zealand (http://www.waikato.ac.nz), has assembled a database of literally hundreds of clinical findings on how honey can be used to treat many wound conditions and the evidence of doing so (Molan 2002). Molan (2006) stated that positive findings on honey in wound care have been reported from 17 randomized controlled trials involving a total of 1965 participants and 5 clinical trials of other forms involving 97 participants treated with honey. The effectiveness of honey in assisting wound healing has also been demonstrated in 16 trials on a total of 533 wounds on experimental animals. There is also a large amount of evidence in the form of case studies that have been reported. Ten publications have reported on multiple cases, totaling 276 cases. There are also 35 reports of single cases. He concluded by stating that these various reports provide a large body of evidence to support honey having the beneficial actions of clearing and preventing wound infection, rapidly debriding wounds, suppressing inflammation, and thus decreasing edema, wound exudate, and hypertrophic scarring, and stimulating the growth of granulation tissue and epithelialization (Molan 2006).

A randomized controlled trial was carried out by leg ulcer specialists at the Aintree University Hospitals in Liverpool, United Kingdom (Robson et al. 2009). A sample of 105 patients was treated with either standardized antibacterial honey (Medihoney) or conventional wound dressings. These results support the proposition that there are clinical benefits from using honey in wound care (Robson et al. 2009).

Millions of people suffer from pressure ulcers each year. Yapucu Güneş and Eşer (2007) conducted a 5-week randomized clinical trial to evaluate the effect of a honey dressing on pressure ulcer healing. The researchers concluded that, by week 5, healing among subjects using a honey dressing was approximately four times the rate of healing in the comparison group. The use of a honey dressing is effective and practical (Yapucu Güneş and Eşer 2007) (Figure 10.2).

The usefulness of honey dressings as an alternative method of managing abdominal surgical wounds was assessed in a 2-year clinical trial at Ramathidodi Hospital, Bangkok, Thailand. The researchers concluded, "We achieved excellent results in all the cases with complete healing within 2 weeks. Honey application is inexpensive, effective and avoids the need to re-suture, which also requires general anesthesia" (Phuapradit and Saropala 1992).

In view of the above, it may be still early to say if there is sufficient evidence to guide clinical practice in wound care in all areas as stated by Jull et al. (2008). However, after Jull et al.'s (2008) review, PubMed cited at least 18 citations of case studies and case reports that discussed the use of honey in various wounds, which may indicate that, in spite of the absence of strong evidence in recommending it as a regular dressing for all wounds, professionals still support the use of honey in infected wounds, particularly the wounds of patients susceptible to methicillin-resistant

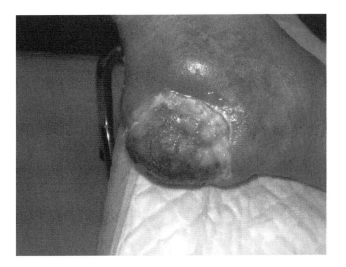

**FIGURE 10.2**    Topical use of honey on a leg of a diabetic foot patient.

*Staphylococcus aureus* and other antibiotic-resistant bacteria. These various reports provide a large body of evidence to support honey having many beneficial actions. The mere absence of evidence does not mean that there is no evidence.

## EVIDENCE FOR USING HONEY IN DIABETIC FOOT ULCERS CARE

There is little scientific evidence published on using honey in treating diabetic foot ulceration and/or wounds. However, Dr. J. Eddy reported treating a refractory diabetic ulcer in a 79-year-old man with topical honey (Eddy and Gideonsen 2005). After 14 months of care costing more than $390,000, which was the cost of five hospitalizations and four surgeries, the ulcers measured 8 × 5 cm and 3 × 3 cm. Deep tissue cultures grew methicillin-resistant *S. aureus*, vancomycin-resistant *Enterococcus*, and *Pseudomonas*. During this time, the patient lost two toes but refused below-the-knee amputation. He was informed by two different surgical teams that without this surgery he would likely die. This opinion was based on the patient's recurrent episodes of heel osteomyelitis and multiple medical complications, including acute renal failure from culture-specific antibiotics. The patient was eventually discharged to his home at his request, after consulting with his family and the hospital's ethics committee. He lost a third toe before consenting to a trial of topical honey. Eddy and Gideonsen (2005) recommended considering topical honey therapy for patients with refractory diabetic foot ulcers.

A few years later, Eddy et al. (2008) hailed the need for a randomized controlled trial about the use of topical honey in treating neuropathic foot ulcers based on the low cost of topical honey therapy with important potential for healing. This therapy can be adopted after a discussion of risks and benefits and in conjunction with standard wound care principles. Researchers advocated publicly funded randomized controlled trials to determine its efficacy (Eddy et al. 2008).

Shukrimi et al. (2008) conducted a comparative study between honey and povidone iodine as a dressing solution for Wagner type II diabetic foot ulcers and concluded that ulcer healing was not significantly different in both study groups. However, they stated that honey dressing is a safe alternative dressing for diabetic foot ulcers.

Moghazy et al. (2010) recently conducted a case series study in which they have applied honey dressing to wounds for 3 months until healing, grafting, or failure of treatment. Changes in grade and stage of wounds, using the University of Texas Diabetic Wound Classification, as well as surface area were recorded weekly. Bacterial load was determined before and after honey dressing. Complete healing was significantly achieved in 43.3% of ulcers. Decrease in size and healthy granulation was significantly observed in another 43.3% of patients. The bacterial load of all ulcers was significantly reduced after the first week of honey dressing. Failure of treatment was observed in 6.7% of ulcers. They concluded by stating that honey is a clinical and cost-effective dressing for diabetic wounds in developing countries. Its omnipresence and concordance with cultural beliefs make it a typical environmentally based method for treating these conditions (Moghazi et al. 2010). A similar trend was reported by Makhdoom et al. (2009) from Pakistan. The medical properties of honey are already described in the management of various chronic wounds and it appears that diabetic foot ulcers are not an exception. In the author's experience, honey has improved outcomes for patients with diabetic foot ulcers. In addition, it reduces the costs associated with long-term care and amputation and reduces the prevalence of drug-resistant organisms that are often fostered by nonhealing diabetic ulcers, and it is not only acceptable by Saudi patients but actually highly requested. It is also readily available with affordable cost for patients in almost all developing countries.

## CONTRAINDICATIONS AND LIMITATIONS FOR TOPICAL USE OF HONEY

Honey as any other wound care product should not be considered as the solo treatment of diabetic foot ulcers, as most of these ulcers are complex ulcers that need a "holistic" approach both generally and locally. For example, a patient with peripheral arterial disease who is presenting with an ischemic ulcer should be assessed and managed by a vascular specialist in addition to a wound care professional. Similarly, one should not expect that honey alone is sufficient to treat underlying osteomyelitis.

Although honey has numerous advantages over other treatments in wound healing, it is important to mention that it has a few disadvantages and some side effects that need mentioning:

1. Honey can become more fluid at higher temperatures. It may liquefy at wound temperature, thus making it less effective.
2. Some patients may experience a temporary stinging sensation in the wound, which is a major cause of discomfort (Eddy et al. 2008).

The availability of commercially produced sterile dressings impregnated with honey (such as Medihoney prepared gels and dressings, Activon Tulle, and Comvita's Apinate dressing) enables health care practitioners to avoid many of these problems.

## Which Honey Should Be Used on Wounds?

At the present time, there is lack of standardization of honey and honey dressings in terms of floral source, the quantity of honey needed for treatment, and the addition of other ingredients (Altman 2010). The results from Dr. Eddy of the University of Wisconsin Medical School show that even some types of common supermarket honey can help heal the most virulent diabetic foot ulcers. That includes microbe-free γ-irradiated raw honey applications from a variety of floral sources that consumers and patients can buy in a drugstore and use easily at home (Eddy and Gideonsen 2005).

On the contrary, the Australian and New Zealand studies indicated that, when it comes to treating wounds, it is best to choose a raw and unfiltered honey with proven antibacterial value (Allen et al. 1991). Other researchers believe that certain types of honey such as manuka honey are superior and attribute that to certain properties (Gethin et al. 2008; Jull et al. 2008; Matjan et al. 2010; Sherlock et al. 2010).

## METHODS OF APPLYING HONEY TO WOUNDS

Bee and honey expert Joe Traynor offered the following simple protocol:

1. Abscesses, cavities, and depressions in the wound should be filled with honey before any dressing is applied.
2. Add 1 ounce of honey to a 4-inch-square dressing pad and apply to the wound.
3. Apply a secondary dry dressing pad on top of the honey pad and then use adhesive tape to hold both dressings in place.
4. Change dressings at least once a day (more frequently if much exudate is produced). Once the wound has stopped producing exudate, dressings can be changed once a week (Traynor 2002).

Dr. Peter Molan offered the following recommendations for dressing wounds with honey, focusing on manuka honey (Molan 2002):

1. Use only honey that has been selected for use in wound care.
2. Use dressings that will hold sufficient honey in place on a wound to get a good therapeutic effect.
3. Ensure that honey is in full contact with the wound bed.
4. If a nonadherent dressing is used between the honey dressing and the wound bed, it must be sufficiently porous to allow the active components of the honey to diffuse through.
5. Ensure that honey dressings extend to cover any area of inflammation surrounding wounds.
6. Use a suitable secondary dressing to prevent leakage of honey.
7. Change the dressings frequently enough to prevent the honey from being washed away or excessively diluted by wound exudates (Molan 2002).

## SUMMARY

In addition to laboratory research on the therapeutic aspects of honey, many hospitals and universities have already undertaken clinical research. The clinical evidence for the use of honey as a safe, effective, and inexpensive wound treatment is accumulating steadily. This chapter features some of the most important studies that discussed the use of topical honey in treating chronic wounds, including those of diabetic patients.

Considering the enormous potential for the use of honey in a clinical setting, it is important that research continue not only on using honeys that are commercially available but also with the locally available honeys. Some of the local products may have equal or sometimes better effectiveness than commercially available therapeutic honey.

Future research should aim to select not only the best honey but also to find out if some of the other CAM products, which are sometimes added to honey in some communities, may potentiate the good effects of honey. Furthermore, information about the common natural preparations will help physicians in outlining interventional plans of diabetic foot disorders. Future research in similar countries should be based on local patients' concepts and practices in dealing with diabetic foot disorders, particularly ulceration. These practices should be taken into consideration when outlining local future health plans for diabetic foot disorders in developing countries.

## REFERENCES

Abdelatif M, Yakoot M, and Etmaan M. 2008. Safety and efficacy of a new honey ointment on diabetic foot ulcers: A prospective pilot study. *J Wound Care* 17(3):108–10.

Allen KL, Molan PC, and Reid GM. 1991. A survey of the antibacterial activity of some New Zealand honeys. *J Pharm Pharmacol* 43(12):817–22.

Al-Rowais N, Al-Faris E, Mohammad AG, Al-Rukban M, and Abdulghani HM. 2010. Traditional healers in Riyadh region: Reasons and health problems for seeking their advice. A household survey. *J Altern Complement Med* 16(2):199–204.

Al-Rowais NA. 2002. Herbal medicine in the treatment of diabetes mellitus. *Saudi Med J* 23(11):1327–31.

Altman N. 2010. *The Honey Prescription: The Amazing Power of Honey as Medicine*. Healing Arts Press, Rochester, Vermont, USA.

Alzahrani HA and Bakhotmah BA. 2010. Top ten natural preparations for the treatment of diabetic foot disorders. *Wound UK* 6(4):1826.

Aydin K, Isildak M, Karakaya J, and Gorlek A. 2010. Change in amputation predictors in diabetic foot disease: Effect of multidisciplinary approach. *Endocrine* 38(1):87–92.

Badri MM, Tashkandi WA, Nawawi A, and Alzahrani HA. 2011. Extremities amputation in King Abdulaziz University Hospital, Jeddah over 5 years. *JKAU-MedSc* 18:13–25.

Bakhotmah BA and Alzahrani HA. 2010. Self-reported use of complementary and alternative medicine (CAM) products in topical treatment of diabetic foot disorders by diabetic patients in Jeddah, Western Saudi Arabia. *BMC Research Notes* 3:254.

Bang LM, Buntting C, and Molan P. 2003. The effect of dilution on the rate of hydrogen peroxide production in honey and its implications for wound healing. *J Altern Complement Med* 9(2):267–73.

Benhanifia MB, Boukraâ L, Hammoudi SM, Sulaiman SA, and Manivannan L. 2010. Recent patents on topical application of honey in wound and burn management. *Recent Pat Inflamm Allergy Drug Discov* 5:81–86.

Boulton AJ. 2004. The diabetic foot: From art to science. The 18th Camillo Golgi lecture. *Diabetologia* 47:1343–53.

Boulton AJ, Vileikyte L, Ragnarson-Tennval G, and Apelqvist J. 2005. The global burden of diabetic foot disease. *Lancet* 366:1719–24.

Brudzynski K. 2006. Effect of hydrogen peroxide on antibacterial activities of Canadian honeys. *Can J Microbiol* 52(12):1228–37.

Dargis V, Pantelejeva O, Jonushaite A, Vileikyte L, and Boulton AJ. 1999. Benefits of a multidisciplinary approach in the management of recurrent diabetic foot ulceration in Lithuania: A prospective study. *Diabetes Care* 22:1428–31.

Driver VR, Fabbi M, Lavery LA, and Gibbons GJ. 2010. The costs of diabetic foot: The economic case for the limb salvage team. *Vasc Surg* 52(3 Suppl):17S–22S.

Eddy JJ, Gideonsen MD, and Mack GP. 2008. Practical considerations of using topical honey for neuropathic diabetic foot ulcers: A review. *WMJ* 107(4):187–90.

Eddy JJ and Gideonsen MD. 2005. Topical honey for diabetic foot ulcers. *J Fam Pract* 54(6):533–5.

Egede LE, Ye X, Zheng D, and Silverstein MD. 2002. The prevalence and pattern of complementary and alternative medicine use in individuals with diabetes. *Diabetes Care* 25(2):324–9.

Facts & Figures. 2010. Did You Know? International Diabetes Federation. Available at http://www.idf.org.

Garrow D and Egede LE. 2006. National patterns and correlates of complementary and alternative medicine use in adults with diabetes. *J Altern Complement Med* 12(9):895–902.

Gethin GT, Cowman S, and Conroy RM. 2008. The impact of Manuka honey dressings on the surface pH of chronic wounds. *Int Wound J* 5(2):185–94.

Huijberts MSP, Schaper NC, and Schalkwijk CG. 2008. Advanced glycation end products and diabetic foot disease. *Diabetes Metab Res Rev* 24:S19–24.

Jull A, Walker N, Parag V, Molan P, and Rodgers A. 2008. Honey as adjuvant leg ulcer therapy trial collaborators. Randomized clinical trial of honey-impregnated dressings for venous leg ulcers. *Br J Surg* 95(2):175–82.

Jull AB, Rodgers A, and Walker N. 2008. Honey as a topical treatment for wounds. *Cochrane Database Syst Rev* 4:CD005083.

King H, Aubert RE, and Herman WH. 1998. Global burden of diabetes, 1995–2025: Prevalence, numerical estimates, and projections. *Diabetes Care* 21(9):1414–31.

Kwakman PH, te Velde AA, de Boer L, Speijer D, Vandenbroucke-Grauls CM, and Zaat SA. 2010. How honey kills bacteria. *FASEBJ* 24(7):2576–82.

Lavery LA, Armastron DG, Wunderlich RP, Tredwell J, and Boulton AJ. 2003. Diabetic foot syndrome: Evaluating the prevalence and incidence of foot pathology in Mexican Americans and non-Hispanic whites from a diabetes disease management cohort. *Diabetes Care* 26:1435–8.

Lusby PE, Coombes A, and Wilkinson JM. 2002. Honey: A potent agent for wound healing? *J Wound Ostomy Continence Nurs* 29(6):295–300.

Majtan J, Majtanova L, Bohova J, and Majtan V. 2010. Honeydew honey as a potent antibacterial agent in eradication of multi-drug resistant *Stenotrophomonas maltophilia* isolates from cancer patients. *Phytother Res* 25:584–7.

Makhdoom A, Khan MS, Lagahari MA, Rahopoto MQ, Tahir SM, and Siddiqui KA. 2009. Management of diabetic foot by natural honey. *J Ayub Med Coll Abbottabad* 21(1):103–5.

Modak M, Dixit P, Londhe J, Ghaskadbi S, Paul A, and Devasagayam T. 2007. Indian herbs and herbal drugs used for the treatment of diabetes. *J Clin Biochem Nutr* 40(3):163–73.

Moghazy AM, Shams ME, Adly OA, Abbas AH, El-Badawy MA, Elsakka DM, Hassan SA, Abdelmohsen WS, Ali OS, and Mohamed BA. 2010. The clinical and cost effectiveness of bee honey dressing in the treatment of diabetic foot ulcers. *Diabetes Res Clin Pract* 89(3):276–81.

Molan PC. 2002. Re-introducing honey in the management of wounds and ulcers—Theory and practice. *Ostomy Wound Management* 48(11):36.

Molan PC. 2004. Clinical usage of honey as a wound dressing: An update. *J Wound Care* 13(9):353–6.

Molan PC. 2006. The evidence supporting the use of honey as a wound dressing. *Int J Low Extrem Wounds* 5(1):40–54.

Moore OA, Smith LA, Campbell F, Seers K, McQuay HJ, and Moore RA. 2001. Systematic review of the use of honey as a wound dressing. *BMC Complement Altern Med* 1:2, Epub.

Oguejiofor OC, Odenigbo CU, and Oguejiotor CB. 2010. Evaluation of the effect of duration of diabetes mellitus on peripheral neuropathy using the United Kingdom screening test scoring system, bio-thesiometry and aesthesiometry. *Niger J Clin Pract* 13(3):240–7.

Phuapradit W and Saropala N. 1992. Topical application of honey in treatment of abdominal wound disruption. *Aust N Z J Obstet Gynaecol* 32(4):381–4.

Reiber GE. 1996. The epidemiology of diabetic foot problems. *Diabet Med* 13:S6–11.

Robson V, Dodd S, and Thomas S. 2009. Standardized antibacterial honey (Medihoney) with standard therapy in wound care: Randomized clinical trial. *J Adv Nurs* 65(3):565–75.

Salomon D, Barouti N, Rosset C, and Whyndham-White C. 2010. Honey: From Noe to wound care. *Rev Med Suisse* 6(246):871–4.

Shaw JE, Sicree RA, and Zimmet PZ. 2010. Global estimates of the prevalence of diabetes for 2010 and 2030. *Diabetes Res Clin Pract* 87(1):4.

Sherlock O, Dolan A, Athman R, Power A, Gethin G, Cowman S, and Humphreys H. 2010. Comparison of the antimicrobial activity of Ulmo honey from Chile and Manuka honey against methicillin-resistant *Staphylococcus aureus*, *Escherichia coli* and *Pseudomonas aeruginosa*. *BMC Complement Altern Med* 2(10):47.

Shukrimi A, Sulaiman AR, Halim AY, and Azril A. 2008. A comparative study between honey and povidone iodine as dressing solution for Wagner type II diabetic foot ulcers. *Med J Malaysia* 63(1):44–6.

Sibbald RG and Woo KY. 2008. The biology of chronic foot ulcers in persons with diabetes. *Diabetes Metab Res Rev* 24:S25–30.

Tan HT, Rahman RA, Gan SH, Halim AS, Hassan SA, Sulaiman SA, and Kirnpal-Kaur B. 2009. The antibacterial properties of Malaysian tualang honey against wound and enteric microorganisms in comparison to manuka honey. *BMC Complement Altern Med* 15(9):34.

Tonks A, Cooper RA, Price AJ, Molan PC, and Jones KP. 2001. Stimulation of TNF-alpha release in monocytes by honey. *Cytokine* 14(4):240–2.

Tonks AJ, Cooper RA, Jones KP, Blair S, Parton J, and Tonks A. 2003. Honey stimulates inflammatory cytokine production from monocytes. *Cytokine* 21:242–7.

Unwin N. 2008. The diabetic foot in the developing world. *Diabetes Metab Res Rev* 24(Suppl 1): S31–2.

Van Houtum WH. 2005. Barriers to the delivery of diabetic foot care. *Lancet* 366:1678–9.

White R, Cooper R, and Kingsley A. 2001. Wound colonization and infection: The role of topical antimicrobials. *Br J Nurs* 10(9):563–78.

Yapucu Güneş U and Eşer I. 2007. Effectiveness of a honey dressing for healing pressure ulcers. *J Wound Ostomy Continence Nurs* 34(2):184–90.

Zhang P, Zhang Z, Brown J et al. 2010. Global healthcare expenditure on diabetes for 2010 and 2030. *Diabetes Res Clin Pract* 87(3):293–301.

# 11 Honey in Pediatrics

*Juraj Majtan*

## CONTENTS

## INTRODUCTION

Honey has been used as a traditional medicine for centuries by different cultures for the treatment of various disorders. It is touted to the public as less toxic and more effective than conventional drugs for various ailments, because it is "natural" and its efficacy is based on knowledge gained over thousands of years. Although one can dispute the theory, primary care clinicians cannot afford to ignore the reality, which is that honey, having potential benefits, is a newly emerging growth industry in the United States, Europe, and Asia.

Honey offers broad-spectrum antimicrobial properties and promotes rapid wound healing (Molan 2006). The *in vitro* antibacterial activity of honey from different floral sources has been intensively studied over the past few decades. It has been found that some types of honey derived from specific floral sources become more potent than others due to the presence of phytochemicals with antibacterial properties (Adams et al. 2008; Mavric et al. 2008; Kwakman et al. 2010). These potent natural honeys, such as manuka and buckwheat honey, are currently being used as medical-grade honeys in clinical applications.

Nowadays, the immunomodulatory effects of honey attract much attention. The fact that honey is able to stimulate human monocytes/macrophages (Tonks et al. 2003, 2007), neutrophils (van den Berg et al. 2008; Fukuda et al. 2009), as well as human fibroblasts and keratinocytes (Majtan et al. 2010) allows the use of honey for treatment of various disorders. Thus, honey becomes a promising therapeutic agent possessing antibacterial activity. On the other hand, most of human randomized clinical studies employing honey as a therapeutic agent were performed in adult populations. Little is known about using honey in the pediatric population.

This chapter aims to review the use of honey in pediatric management. It will focus on the use of honey for treatment of wounds, respiratory tract infections, cough, and gastroenteritis in pediatric populations.

## HONEY IN THE TREATMENT OF PEDIATRIC WOUNDS AND BURNS

It has been assumed that the antibacterial action of honey has its main impact on the healing process of chronic wounds and burns. Honey eliminates pathogens from wounds and provides an appropriate moist environment for proper wound healing. However, the findings of several clinical trials employing honey for treatment of wounds in the adult population are still contradictory. Jull et al. (2008b) established that honey-impregnated dressings did not significantly improve venous ulcer healing or change incidence of infection compared with usual care. On the other hand, two recent clinical trials suggest that healing times and incidence of infection after treatment with honey are reduced compared with conventional treatment, and the results are of clinical significance (Gethin and Cowman 2008; Robson et al. 2009).

The mechanism of honey action has been recently fully elucidated. Honey is a supersaturated solution of sugars with an acidic pH, high osmolarity, and low water content. These characteristics inhibit the growth of microorganisms. Additional antimicrobial activity is generated on dilution by the activation of bee-derived glucose oxidase to produce hydrogen peroxide (Bang et al. 2003). Besides these well-characterized major antibacterial factors in honey, methylglyoxal (Adams et al. 2008; Mavric et al. 2008) and cationic antimicrobial peptide bee defensin-1 (Kwakman et al. 2010), also act as antibacterial substances. However, the concentration of these two distinct factors may vary in honeys depending on floral source and geographical origin. Moreover, honey has a wide range of phytochemicals including polyphenols, a class of natural products possessing a diverse range of pharmacologic properties. Polyphenols including flavonoids and phenolic acids are found in honey as residual secondary metabolites and have been studied for their potential use as botanical and geographical markers and also to explain the antibacterial properties of honey.

To the best of our knowledge, there are only very few clinical studies describing the use of honey for treatment of wounds and/or burns in pediatric patients (Vardi et al. 1998; Ongom et al. 2004; Simon et al. 2006; Smaropoulos et al. 2011). Vardi et al. (1998) reported nine infant case reports with postsurgical open infected wounds that failed to heal with conventional treatment. The majority of these patients had been treated unsuccessfully with local antiseptics and systemic antibiotics for more than 14 days. Systemic intravenous antibiotic treatment was discontinued at the beginning of local honey application in six patients. The wounds were cleaned before application of 5 to 10 mL of commercial, unprocessed, nonpasteurized, and nonirradiated honey. The honey-based dressing was changed twice daily. All infants showed clinical improvement after 5 days of treatment. In addition, all wounds were closed, clean, and sterile after 21 days of treatment. Nonsterile honey was used in the study; however, no adverse reactions were recorded.

Similarly, Simon et al. (2006) observed a positive effect of honey on the healing of chronic wounds in 14 pediatric oncology patients. Due to profound immunosuppression, wound infection can easily spread and act as the source of sepsis in oncology patients. Furthermore, the wound-healing process is impaired in patients receiving chemotherapy. The authors of the study found that the medical-grade manuka honey (MediHoney) was an effective antibacterial agent in elimination of infected wound bacteria from chronic wounds. They suggest using honey also for deep wounds that are usually treated with systemic antibiotics for a long time.

According to recent systematic review (Jull et al. 2008a), the quality of reported trials using honey is variable and evidence to date supports honey only as a treatment of mild-to-moderate superficial and partial-thickness burns. This observation is supported by a very recent study of Smaropoulos et al. (2011) where the honey-based therapy indicated a fast, cost-effective, and patient-friendlier treatment method for pediatric burns and dermal trauma compared with conventional methods. Eight children between the age of 8 months and 13 years were included in the evaluation and were initially treated with povidone iodine 10% solution. Then, the honey-based dressing was applied (L-Mestiran) directly on burn lesions. The mean burn epithelialization time was 19 days. When compared with standard hospital protocol employing povidone iodine in similar lesions, the standard protocol shows delayed wound healing. Overall, 1 month is needed for acceptable wound healing. This might be explained due to the detrimental effects of povidone iodine on human skin cells resulting in delayed wound healing (Wilson et al. 2005; Kataoka et al. 2006). The honey treatment was responsible for the reduction of hospital stay and thus reducing nosocomial infections. In addition, no adverse reactions and allergies such as anaphylaxis or systemic toxicity (e.g., hyperglycemia in diabetic patients) have been reported so far. Nevertheless, it is absolutely necessary to monitor any adverse reactions in clinical trials using honey in the future.

One of the important factors is parent participation in the care of children with burn wounds. Based on parents' feedback during honey-based therapy, the dressing application and removal was a simple process with minimal discomfort for children. No pain was documented in all cases; however, Molan (2006) suggested that honey can sometimes cause pain when applied as a pure honey.

Another study (Ongom et al. 2004) dealing with the management of superficial burn wounds in children using honey was conducted in Uganda. A total of 52 cases with superficial thermal burns were treated with either a mixture of honey and ghee (pure animal fat) in a 1:1 ratio or collagen. Of the 26 patients involved in the honey group, four cases showed bacterial growth. In both groups, similar ability to prevent infection occurred.

In patients treated with honey-ghee dressing, the mean healing duration was 12.3 days, whereas the mean healing duration in the collagen group was 9.9 days. This significant reduction of healing time in the collagen group was probably due to use of diluted 50% pure honey. A 100% pure honey application would probably have resulted in no bacterial growth and also reduced time of healing.

As shown above, honey has been reported to be particularly helpful in wound care of children receiving chemotherapy, in whom the wound-healing process is impaired (Simon et al. 2006). Despite progress in surgical techniques, most pediatric patients face postoperative recurrence of pain. Honey is capable to reduce the pain by reducing both sensitization following inflammation-mediated prostaglandin synthesis and pressure on tissue resulting from edema (Al-Waili and Boni 2003; Kassim et al.

2010). It was shown that oral administration of honey following pediatric tonsillectomy may relieve postoperative pain and may decrease the need of analgesics (Ozlugedik et al. 2006). In addition, honey may also have the positive effect on tissue repair, thus decreasing postoperative pain.

Prospective, randomized, and double-blind pediatric studies should further be conducted to confirm the efficacy of honey either in the treatment of wounds and burns or in reducing the postoperative pain.

## HONEY AND COUGH AND RESPIRATORY TRACT INFECTIONS

Cough is one of the most common symptoms in conjunction with an upper respiratory tract infection (URI) presented by children (Goldsobel and Chipps 2010), resulting in discomfort and substantial health care utilization. Each year, children are usually affected six to eight times by URIs; however, its prevalence in adults is two to three times (Ronald and Gregory 2007). Although they are not associated with disability and mortality, they can cause morbidity and incur medical costs. Most of the symptoms emerge during the first 3 days but are fortunately relieved in a week; the cough, however, may remain for a longer time (Lorber 1996).

Despite the common occurrence of URIs and cough, there are no accepted therapies for these annoying symptoms. The antitussive drugs that are frequently utilized for children are dextromethorphan, codeine, and diphenhydramine. The safety and efficacy of these drugs are controversial (American Academy of Pediatrics Committee on Drugs 1997; Fahey et al. 1998; Chang and Glomb 2006). On the other hand, dextromethorphan and codeine have been reported to decrease cough frequency in children with chronic cough (Matthys et al. 1983; Corrao 1986). Taken together, the cough and cold medicines carry a risk of serious potentially life-threatening adverse events and should be avoided in very young children. Due to this fact, there is urgent need for novel and effective medications either in prevention of URI symptoms or in their treatment.

Nowadays, modern medicine directs attention to natural products as a source of new drugs. In many cultures, alternative natural remedies such as honey are used to treat URI symptoms, including cough (Pfeiffer 2005). Honey is generally believed to be a safe remedy outside of the infant population. Furthermore, honey has been recommended by the World Health Organization for controlling cough and other URI symptoms (Department of Children Adolescent Health and Development 2001). However, the main limitation of honey use among pediatric population is the risk of allergic reactions and botulism.

Due to problems with safety and efficacy of commercially available antitussive drugs, such as dextromethorphan, codeine, and diphenhydramine, honey has recently been used for treatment of cough in pediatric patients (Paul et al. 2007; Shadkam et al. 2010). The results of the study by Paul et al. demonstrated that, in the overall comparison of three treatment groups (placebo, dextromethorphan, and honey), honey was the most effective treatment for all of the outcomes related to cough, child sleep, and parent sleep. Similarly, another recent study (Shadkam et al. 2010) demonstrated that honey, compared with placebo, dextromethorphan, and diphenhydramine groups, had a significantly more effective curing impact on cough frequency, cough severity, and sleep quality of children and their parents. Thus, honey provides a safe and well-tolerated alternative for coughing children.

Although honey seems to be an effective antitussive agent, the mechanism of its action remains unknown. It has been suggested that honey could exert its positive antitussive properties via stimulation of several sensory pathways (Eccles 2006). The sweet taste of honey can readily cause reflex salivation and may also promote secretion of airway mucus (Eccles 2006). In cases of dry unproductive cough, the demulcent effects of a cough medicine may lubricate the pharynx and larynx and help to reduce coughing. The author of the study also suggested that consuming sweet substances causes the production of internal opioids.

The fact that almost all modern cough medicines are formulated as sweet syrup (honey) indicates that the physiologic actions of sweet syrup may contribute to the antitussive activity of the treatment (Eccles 2006).

A cough can also be the result of pneumonia. The therapeutic role of honey against pneumonia in children has been documented by Mannan et al. (2006). In total, 1202 children (2–12 years) suffering from pneumonia were divided into control and test groups. Both groups were given standard treatment, but the test group was also administered with the honey mixture in a dose of 1 to 3 mL twice a day for 3 months. The results of the study showed that the children of the test group were cured early and had no relapse during the next 6 months. Hence, honey can act as a curative as well as preventive agent against pneumonia.

According to a very recent review (Mulholland and Chang 2009), honey has been found to be effective in acute cough, but whether this can be extended to nonacute cough is unknown. The authors of the study identified 381 abstracts, of which 20 potential randomized trials were retrieved. Unfortunately, none of the studies dealt with chronic coughs in children.

## HONEY IN THE TREATMENT OF GASTROENTERITIS IN INFANTS AND CHILDREN

The antibacterial activity of honey against the range of pathogens from chronic wounds is well characterized. Honey is also bactericidal for many enteric bacterial pathogens including *Salmonella* spp. (Badawy et al. 2004; Alnaqdy et al. 2005; Tan et al. 2009), *Shigella* spp. (Tan et al. 2009), *Escherichia coli* (Badawy et al. 2004; Tan et al. 2009), *Campylobacter* spp. (Lin et al. 2009), and *Stenotrophomonas maltophilia* (Tan et al. 2009; Majtan et al. 2011).

Oral administration of honey to treat and protect against gastrointestinal infection such as gastritis, duodenitis, and gastric ulceration caused by bacteria has been reported in adults and children (Haffejee and Moosa 1985; Al Somal et al. 1994).

The efficacy of honey taken orally is greatly affected by dilution in large amounts of body fluids and water from food and drink as well as by a short period of contact with bacterial cells in the gastrointestinal tract. Recently, defensin-1, an antibacterial bee peptide, has been found in honey (Kwakman et al. 2010). It has been suggested that it is one of the major antibacterial factors of honey (Kwakman et al. 2010, 2011). Bee defensin-1 belongs to the widespread family of antimicrobial cationic peptide defensins that function in the innate immune system of most organisms (Raj and Dentino 2002). Insect defensins were documented to be effective against various gram-positive bacteria and some of them also against gram-negative bacteria and fungi (Bulet and Stocklin 2005; Bulet et al. 1999). They almost immediately kill bacterial cells by permeabilization of their cytoplasmic membrane (Cociancich et al. 1993; Otvos 2000; Wong et al. 2007). This fact implies that bee defensin-1 could kill the pathogenic bacteria in the gastrointestinal tract within minutes. Thus, a short period of honey contact with bacterial cells may be satisfactory to kill the enteric pathogens.

Although honey has been used for the treatment of gastrointestinal disorders since ancient times, only two prospective controlled clinical trials using honey for treatment of infantile gastroenteritis (Haffejee and Moosa 1985; Abdulrhman et al. 2010) were identified. In the first study (Haffejee and Moosa 1985), a total of 169 infants and children aged 8 days to 11 years were enrolled into the trial. The patients were randomly assigned to one of two therapeutic groups. The control group received routine management of diarrhea (glucose-based rehydration solution) and the honey group received oral rehydration solution containing 50 mL pure honey per liter instead of glucose. It was found that honey shortened the duration of diarrhea in patients with bacterial gastroenteritis caused by bacteria such as *Salmonella* spp., *Shigella* spp., and *E. coli*. In addition, honey may safely be used as a substitute for glucose in standard rehydration solution containing electrolytes.

In a second study performed by Abdulrhman et al. (2010), 100 infants and children aged 2 months to 7 years with acute gastroenteritis were randomly assigned to two therapeutic groups. The mean ages of patients in the control and honey groups were $1.5 \pm 1.2$ and $1.1 \pm 0.8$ years, respectively. The dose of honey was identical as in the above study (Haffejee and Moosa 1985). Honey was tested for the presence of *Clostridium botulinum* spores. It was found that the frequencies of vomiting and diarrhea were significantly reduced in the honey group compared with the control group. Similarly, the recovery time was significantly shortened after honey ingestion.

Taken together, honey as part of an oral rehydration solution may promote rehydration of the body and speed recovery from vomiting and diarrhea in infants and children.

## LIMITATIONS OF HONEY USE IN PEDIATRIC POPULATIONS

There is a growing body of studies promoting the use of honey as a safe and cost-efficient product in the management of wound care. Honey for therapeutic purposes has to be of "medical grade," which ensures that it has been sterilized and has standardized antibacterial activity. However, in many cultures, raw unsterilized honey has been used for treatment of various disorders.

The potential dangers of honey use for infants are relatively unknown. Infant botulism is most often mentioned as the most dangerous complication of honey use in children under 1 year of age. Honey can also cause dental caries, hyperactivity, nervousness, and insomnia (Ripa 1978; Paul et al. 2007). This bacteria also cause the problem if honey is applied therapeutically but locally (wound botulism).

Infant botulism is an uncommon disease that occurs when ingested spores of *C. botulinum* germinate and produce botulinum neurotoxin in the colon. Ingestion of honey, a recognized vehicle for *C. botulinum* spores, remains the only identified avoidable risk factor for acquiring infant botulism (Arnon et al. 1979; Spika et al. 1989). Approximately 15% of reported cases of infant botulism have been attributed to honey contaminated by spores from *C. botulinum*. This has resulted in warning labels on packing of honey brands in the United Kingdom and in other countries, although there is only little awareness of the potential dangers of the use of honey for children (Kumar et al. 2011). Therapeutic use of honey and also other natural remedies is often based on traditional cultural beliefs. Parents often prefer to use honey because it is perceived as a natural, cheaper, and widely available product.

As mentioned above, only sterilized and standardized honeys can be registered as medical devices and used for therapeutic purposes. The process of honey sterilization is based on γ-irradiation. It has been found that 25 kGy of γ-irradiation is sufficient to achieve sterility (Molan and Allen 1996). Moreover, the antibacterial activity of honey was unaffected after the irradiation process even at double dose (50 kGy). Another study showed that the dose of γ-irradiation for honey samples may be decreased to 18 kGy (Postmes et al. 1995) or 10 kGy (Midgal et al. 2000). The total count of aerobic and anaerobic bacteria and molds decreased in honey by 99% at a dose of 10 kGy.

The elimination of microorganisms in honey increases the quality of honey and its stability during storage. Irradiated honey is free of bacterial spores; thus, it may be therapeutically used for children and infants.

## CONCLUSIONS

A medical-grade honey is a promising remedy that can be safely used in the treatment of wounds, respiratory tract infection symptoms, gastroenteritis, and other disorders in children and infants. The sterilization process of honey by γ-irradiation successfully eliminates microorganisms and bacterial spores and allows the use of honey in pediatric patients. The results of human clinical studies suggest that honey is an effective and cheap natural product, but further research is needed to prove its efficacy in pediatric patients and elucidate the possible mechanisms of action. However, health care practitioners, pediatricians, as well as parents should be aware of the potential risk of allergy in using honey for therapeutic purposes in children.

## ACKNOWLEDGMENTS

This work was supported by the Slovak Research and Development Agency under Contract No. APVV-0115-11. I wish to thank the Ministry of Education of the Slovak Republic and the Slovak Academy of Sciences for supporting my basic research into alternative and complementary medicines.

## REFERENCES

Abdulrhman, M. A., Mekawy, M. A., Awadalla, M. M., and Mohamed, A. H. 2010. Bee honey added to the oral rehydration solution in treatment of gastroenteritis in infants and children. *Journal of Medicinal Food*, 13, 605–609.

Adams, C. J., Boult, C. H., Deadman, B. J. et al. 2008. Isolation by HPLC and characterisation of the bioactive fraction of New Zealand manuka (*Leptospermum scoparium*) honey. *Carbohydrate Research*, 343, 651–659.

Al-Waili, N. S. and Boni, N. S. 2003. Natural honey lowers plasma prostaglandin concentrations in normal individuals. *Journal of Medicinal Food*, 6, 129–133.

Al Somal, N., Coley, K. E., Molan, P. C., and Hancock, B. M. 1994. Susceptibility of *Helicobacter pylori* to the antibacterial activity of manuka honey. *Journal of the Royal Society of Medicine*, 87, 9–12.

Alnaqdy, A., Al-Jabri, A., Al Mahrooqi, Z., Nzeako, B., and Nsanze, H. 2005. Inhibition effect of honey on the adherence of *Salmonella* to intestinal epithelial cells *in vitro*. *International Journal of Food Microbiology*, 103, 347–351.

American Academy of Pediatrics Committee on Drugs. 1997. Use of codeine- and dextromethorphan-containing cough remedies in children. *Pediatrics*, 99, 918–920.

Arnon, S. S., Midura, T. F., Damus, K. et al. 1979. Honey and other environmental risk factors for infant botulism. *Journal of Pediatrics*, 94, 331–336.

Badawy, O. F., Shafii, S. S., Tharwat, E. E., and Kamal, A. M. 2004. Antibacterial activity of bee honey and its therapeutic usefulness against *Escherichia coli* O157:H7 and *Salmonella typhimurium* infection. *Revue Scientifique et Technique*, 23, 1011–1022.

Bang, L. M., Bantting, C., and Molan, P. C. 2003. The effects of dilution rate on hydrogen peroxide production in honey and its implications for wound healing. *Journal of Alternative and Complementary Medicine*, 9, 267–273.

Bulet, P., Hetru, C., Dimarcq, J. L., and Hoffmann, D. 1999. Antimicrobial peptides in insects: Structure and function. *Developmental and Comparative Immunology*, 23, 329–344.

Bulet, P. and Stocklin, R. 2005. Insect antimicrobial peptides: Structures, properties and gene regulation. *Protein and Peptide Letters*, 12, 3–11.

Cociancich, S., Ghazi, A., Hetru, C., Hoffmann, J. A., and Letellier, L. 1993. Insect defensin, an inducible antibacterial peptide, forms voltage-dependent channels in *Micrococcus luteus*. *Journal of Biological Chemistry*, 268, 19239–19245.

Corrao, W. M. 1986. Chronic cough: An approach to management. *Comprehensive Therapy*, 12, 14–19.

Department of Children Adolescent Health and Development. 2001. Cough and cold remedies for the treatment of acute respiratory infections in young children. *World Health Organization, Geneva, Switzerland*.

Eccles, R. 2006. Mechanisms of the placebo effect of sweet cough syrups. *Respiratory Physiology and Neurobiology*, 152, 340–348.

Fahey, T., Stocks, N., and Thomas, T. 1998. Systematic review of the tretament of upper respiratory tract infection. *Archives of Disease in Childhood*, 79, 225–230.

Fukuda, M., Kobayashi, K., Hirono, Y. et al. 2011. Jungle honey enhances immune function and antitumor activity. *Evidence-based Complementary and Alternative Medicine*, 2011, Article ID 908743.

Gethin, G. and Cowman, S. 2008. Manuka honey vs. hydrogel—A prospective, open label, multicentre, randomised contolled trial to compare desloughing efficacy and healing outcomes in venous ulcers. *Journal of Clinical Nursing*, 18, 466–474.

Goldsobel, A. B. and Chipps, B. E. 2010. Cough in the pediatric population. *Journal of Pediatrics*, 156, 352–358.

Haffejee, I. E. and Moosa, A. 1985. Honey in the treatment of infantile gastroenteritis. *British Medical Journal (Clinical Research Ed.)*, 290, 1866–1867.

Chang, A. B. and Glomb, W. B. 2006. Guidelines for evaluating chronic cough in pediatrics: ACCP evidence-based clinical practice guidelines. *Chest*, 129, 260S–283S.

Jull, A., Rodgers, A., and Walker, N. 2008a. Honey as a topical treatment for wounds. *Cohrane Database of Systematics Reviews*, Art. No. CD005083.

Jull, A., Walker, N., Parag, V., Molan, P., and Rodgers, A. 2008b. Randomized clinical trial of honey-impregnated dressings for venous leg ulcers. *British Journal of Surgery*, 95, 175–182.

Kassim, M., Achoui, M., Mansor, M., and Yusoff, K. M. 2010. The inhibitory effects of Gelam honey and its extracts on nitric oxide and prostaglandin E(2) in inflammatory tissues. *Fitoterapia*, 81, 1196–1201.

Kataoka, M., Tsumura, H., Kaku, N., and Torisu, T. 2006. Toxic effects of povidone-iodine on synovial cell and articular cartilage. *Clinical Rheumatology*, 25, 632–638.

Kumar, R., Lorenc, A., Robinson, N., and Blair, M. 2011. Parents' and primary healthcare practitioners' perspectives on the safety of honey and other traditional paediatric healthcare approaches. *Child: Care, Health and Development*, 37, 734–743.

Kwakman, P. H., Te Velde, A. A., De Boer, L. et al. 2010. How honey kills bacteria. *FASEB Journal*, 24, 2576–2582.

Kwakman, P. H., Te Velde, A. A., De Boer, L., Vandenbroucke-Grauls, C. M., and Zaat, S. A. 2011. Two major medicinal honeys have different mechanisms of bactericidal activity. *PLoS ONE*, 6, e17709.

Lin, S. M., Molan, P. C., and Cursons, R. T. 2009. The *in vitro* susceptibility of *Campylobacter spp.* to the antibacterial effect of manuka honey. *European Journal of Clinical Microbiology and Infectious Diseases*, 28, 339–344.

Lorber, B. 1996. The common cold. *Journal of General Internal Medicine*, 11, 229–236.

Majtan, J., Kumar, P., Majtan, T., Walls, A. F., and Klaudiny, J. 2010. Effect of honey and its major royal jelly protein 1 on cytokine and MMP-9 mRNA transcripts in human keratinocytes. *Experimental Dermatology*, 19, e73–e79.

Majtan, J., Majtanova, L., Bohova, J., and Majtan, V. 2011. Honeydew honey as a potent antibacterial agent in eradication of multi-drug resistant *Stenotrophomonas maltophilia* isolates from cancer patients. *Phytotherapy Research*, 25, 584–587.

Mannan, A., Ubaidullah, Ahmad, R., and Khan, R. A. 2006. Therapeutic effect of saffron and honey in pneumonia in children. *Current Pediatric Research*, 10, 25–27.

Matthys, H., Bleicher, B., and Bleicher, U. 1983. Dextromethorphan and codeine: Objective assessment of antitussive activity in patients with chronic cough. *Journal of International Medical Research*, 11, 92–100.

Mavric, E., Wittmann, S., Barth, G., and Henle, T. 2008. Identification and quantification of methylglyoxal as the dominant antibacterial constituent of Manuka (*Leptospermum scoparium*) honeys from New Zealand. *Molecular Nutrition and Food Research*, 52, 483–489.

Midgal, W., Owczarczyk, H. B., Kedzia, B., Holderna-Kedzia, E., and Madajczyk, D. 2000. Microbial decontamination of natural honey by irradiation. *Radiation Physics and Chemistry*, 57, 285–288.

Molan, P. C. 2006. The evidence supporting the use of honey as a wound dressing. *International Journal of Lower Extremity Wounds*, 5, 40–54.

Molan, P. C. and Allen, K. L. 1996. The effect of gamma-irradiation on the antibacterial activity of honey. *Journal of Pharmacy and Pharmacology*, 48, 1206–1209.

Mulholland, S. and Chang, A. B. 2009. Honey and lozenges for children with non-specific cough. *Cohrane Database of Systematics Reviews*, 15, CD007523.

Ongom, P., Kijjambu, S. C., Mutumba, S. K., and Sebbaale, A. K. 2004. Comparision of honey-ghee dressing with collagen dressing in the management of superficial burn wounds in children. *East and Central African Journal of Surgery*, 9, 67–71.

Otvos, L. J. 2000. Antibacterial peptides isolated from insects. *Journal of Peptide Science*, 6, 497–511.

Ozlugedik, S., Genc, S., Unal, A. et al. 2006. Can postoperative pains following tonsillectomy be relieved by honey? A prospective, randomized, placebo controlled preliminary study. *International Journal of Pediatric Otorhinolaryngology*, 70, 1929–1934.

Paul, I. M., Beiler, J., Mcmonagle, A. et al. 2007. Effect of honey, dextromethorphan, and no treatment on nocturnal cough and sleep quality for coughing children and their parents. *Archives of Pediatrics and Adolescent Medicine*, 161, 1140–1146.

Pfeiffer, W. F. 2005. A multicultural approach to the patient who has a common cold. *Pediatrics in Review*, 26, 170–175.

Postmes, T., Van Den Bogaard, A. E., and Hazen, M. 1995. The sterilization of honey with cobalt 60 gamma radiation: A study of honey spiked with spores of *Clostrodium botulinum* and *Bacillus subtilis*. *Experientia*, 51, 986–989.

Raj, P. A. and Dentino, A. R. 2002. Current status of defensins and their role in innate and adaptive immunity. *FEMS Microbiology Letters*, 206, 9–18.

Ripa, L. W. 1978. Nursing habits and dental decay in infants: "Nursing bottle caries." *ASDC Journal of Dentistry for Children*, 45, 274–275.

Robson, V., Dodd, S., and Thomas, S. 2009. Standardized antibacterial honey (Medihoney) with standard therapy in wound care: Randomized clinical trial. *Journal of Advanced Nursing*, 65, 565–575.

Ronald, B. T. and Gregory, F. H. 2007. The common cold. In: Robert, M. K., Richard, E. B., Hall, B. J. and Bonita, F. S. (eds.) *Text Book of Pediatrics*. Philadelphia: Saunders Elsevier.

Shadkam, M. N., Mozaffari-Khosravi, H., and Mozayan, M. R. 2010. A comparison of the effect of honey, dextromethorphan, and diphenhydramine on nightly cough and sleep quality in children and their parents. *Journal of Alternative and Complementary Medicine*, 16, 787–793.

Simon, A., Sofka, K., Wiszniewsky, G. et al. 2006. Wound care with antibacterial honey (Medihoney) in pediatric hematology-oncology. *Supportive Care in Cancer*, 14, 91–97.

Smaropoulos, E., Romeos, S., and Dimitriadou, C. 2011. Honey-based therapy for paediatric burns and dermal trauma compared to standard hospital protocol. *Wounds U.K.*, 7, 33–40.

Spika, J. S., Shaffer, N., Hargrett-Bean, N. et al. 1989. Risk factors for infant botulism in the United States. *American Journal of Diseases of Children*, 143, 828–832.

Tan, H. T., Rahman, R. A., Gan, S. H. et al. 2009. The antibacterial properties of Malaysian tualang honey against wound and enteric microorganisms in comparison to manuka honey. *BMC Complementary and Alternative Medicine*, 9, 34.

Tonks, A. J., Cooper, R. A., Jones, K. P. et al. 2003. Honey stimulates inflammatory cytokine production from monocytes. *Cytokine*, 21, 242–247.

Tonks, A. J., Dudley, E., Porter, N. G. et al. 2007. A 5.8-kDa component of manuka honey stimulates immune cells via TLR4. *Journal of Leukocyte Biology*, 82, 1147–1155.

Van Den Berg, A. J., Van Den Worm, E., Van Ufford, H. C. et al. 2008. An *in vitro* examination of the antioxidant and anti-inflammatory properties of buckwheat honey. *Journal of Wound Care*, 17, 172–178.

Vardi, A., Barzilay, Z., Linder, N. et al. 1998. Local application of honey for treatment of neonatal postoperative wound infection. *Acta Pediatrica*, 87, 429–432.

Wilson, J. R., Mills, J. G., Prather, I. D., and Dimitrijevich, S. D. 2005. A toxicity index of skin and wound cleansers used on *in vitro* fibroblasts and keratinocytes. *Advances in Skin and Wound Care*, 18, 373–378.

Wong, J. H., Xia, L., and Ng, T. B. 2007. A review of defensins of diverse origins. *Current Protein and Peptide Science*, 8, 446–459.

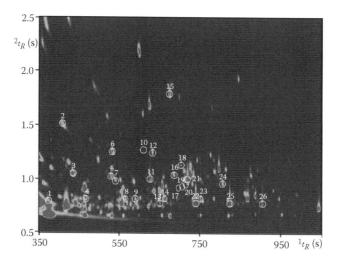

**FIGURE 5.5** HS-SPME-GC×GC-TOF-MS chromatogram of honey volatiles (Corsica sample) with marked markers. For the description of analytes, see Cajka et al. (2009).

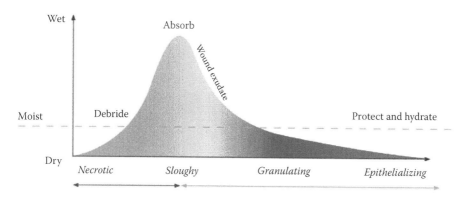

The wound-healing progression model

**FIGURE 7.1** Wound-healing process.

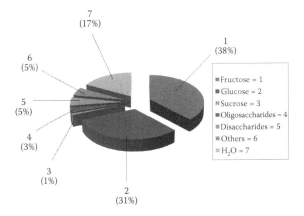

**FIGURE 9.1** General composition of honey.

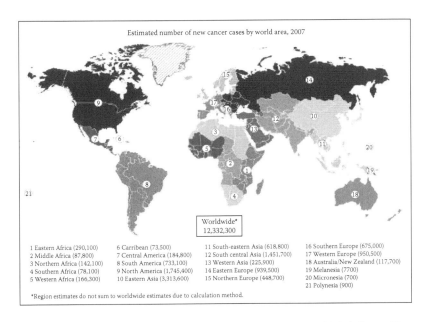

**FIGURE 12.1** Estimated new cancer cases by world areas. (From Global Cancer Facts and Figures 2007.)

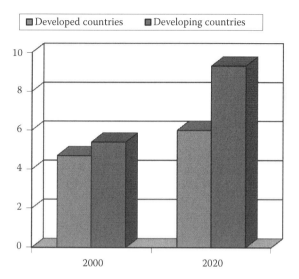

**FIGURE 12.2** Greater number of cases of cancer in developing nations compared with developed nations. (From Ferlay, J. et al. GLOBOCAN 2008 v2.0, Cancer Incidence and Mortality Worldwide: IARC CancerBase No. 10 [internet]. Lyon, France: International Agency for Research on Cancer. Available from: http://globocan.iarc.fr, accessed on 27th September 2013.)

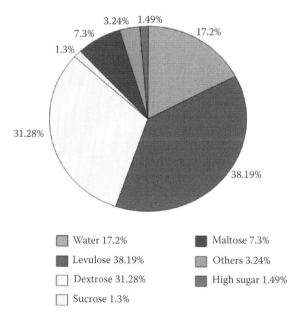

**FIGURE 18.1** Pie chart of honey composition indicating the percentage share of various sugars, water, and other minor constituents. (From Jaganathan, S.K., and Mandal, M., *Journal of Biomedicine and Biotechnology*, 1–13, 2009.)

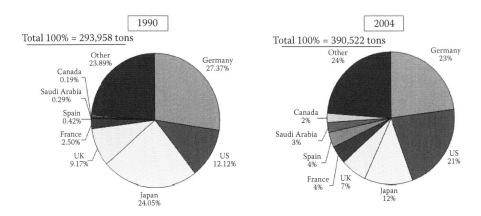

**FIGURE 18.4** Change of share of honey importation from 1993 to 2004. (Courtesy of Food and Agriculture Organization of the United Nations, www.fao.org.)

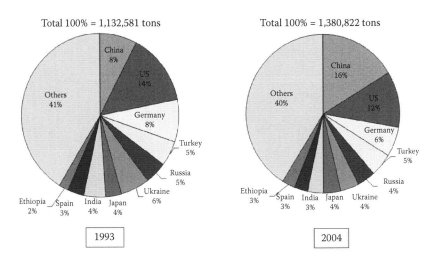

**FIGURE 18.5** Change of share of honey consumption from 1993 to 2004. (Courtesy of Food and Agriculture Organization of the United Nations, www.fao.org.)

# 12 Honey and Cancer
## *A Sustainable Parallel Relationship Particularly for Developing Nations*

*Nor Hayati Othman*

## CONTENTS

## INTRODUCTION

Honey and cancer have a sustainable parallel relationship, although this relationship is yet to be fully explored to date. There are several recently published papers indicating that honey is a natural cancer "vaccine." The cause of cancer is multifactorial and the beneficial effect of honey is multifaceted. This chapter attempts to highlight this parallel relationship.

## CANCER—THE GLOBAL EPIDEMIC

Cancer is a global epidemic. In 2008, it was estimated that there were 12,332,300 cancer cases, of which 5.4 million were in developed countries and 6.7 million were in developing countries (Garcia et al. 2007) (Figure 12.1). Over half of the incident cases occurred in residents of the four World Health Organization (WHO) regions: AFRO (African regions), EMRO (Eastern Mediterranean region), SEARO (South-East Asian region), and WPRO (Western Pacific region). These are countries with large populations with low- and middle-income status (Figure 12.2) (Boyle and Lewin 2008). The world population increased from 6.1 billion in 2000 to 6.7 billion in 2008 (Boyle and Lewin 2008). The increase in population was much more in developing countries than in developed countries. Even if the age-specific rates of cancer remain constant, developing countries would have a higher cancer burden than developed countries.

Cancer trends are showing an upward trends in many developing countries (Lim 2002; Lepage 2008; Yeole 2008) and a mixed pattern in developed countries (Kabir and Lancy 2006; Bouchardy et al. 2008; Westlake and Cooper 2008). By 2050, the cancer burden could reach 24 million cases per year worldwide, with 17 million cases occurring in developing countries (Parkin et al. 2001). Cancers that are associated with diet and lifestyle are seen more in developed countries, while cancers that are due to infections are more in developing countries. In Malaysia, death due to cancer was ranked third (10.11%) after heart disease (14.31%) and septicemia (16.54%) in 2005. According to the WHO, death from cancer is expected to increase to 104% worldwide by 2020, the largest impact being in developing countries in comparison with developed countries (Rastogi et al. 2004).

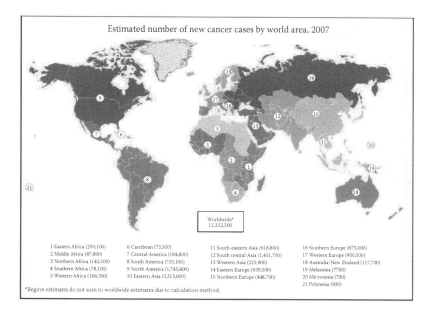

**FIGURE 12.1** **(See color insert.)** Estimated new cancer cases by world areas. (Garcia, M. et al. "Global Cancer: Facts & Figures 2007." *American Cancer Society.*)

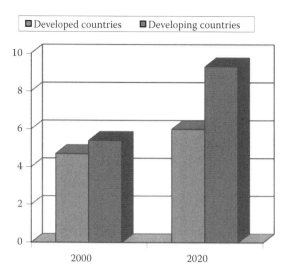

**FIGURE 12.2**   (**See color insert.**) Greater number of cases of cancer in developing nations compared with developed nations. (From Ferlay, J. et al. GLOBOCAN 2008 v2.0, Cancer Incidence and Mortality Worldwide: IARC CancerBase No. 10 [internet]. Lyon, France: International Agency for Research on Cancer. Available from: http://globocan.iarc.fr, accessed on 27th September 2013.)

While the total number of cancer cases is increasing, the trend of certain cancers is changing in developed and developing countries. In developed countries, the trend is declining (Belpomme et al. 2007) because infections by microorganisms are declining and screening facilities are available. In Singapore, there was an average annual increase of 3.6% for breast cancers in women from the 1988 to 1992 period (Seow et al. 1996). In Qatar, there was a 57.1% rise of cancers from 1991 to 2006 (Bener et al. 2008), and in the Netherlands, there was an increase between 1.9% (females) and 3.4% (males) per year for esophageal cancer from 1989 to 2003 (Crane et al. 2007). The cancer trends are also different in developed compared with developing countries. The top 30 cancers of more developed (Figure 12.3a) and less developed nations are different (Figure 12.b) (Boyle and Lewin 2008). While cancers of the prostate, breasts, and colorectal are clearly more prevalent in developed countries than in developing countries, the distinction is not very apparent for cancer of the lung (Figure 12.3), which is as prevalent in more or less developed nations. Except for breast cancers, the top 5 cancers in males and females of developing nations are due to lifestyles or infections (DCP2 2007).

In order to understand the usefulness of honey in cancer, we need to understand the various factors that could cause cancer. Carcinogenesis (understanding how cancer develops) is a multistep process and has multifactorial causes. Development of cancers takes place long after initiation, promotion, and progression steps (Figure 12.4) have taken place. The cellular damage could be by one factor or a multiplicity of these factors. The latter is more frequent.

1. Tobacco use, particularly cigarette smoking
2. Obesity and physical inactivity

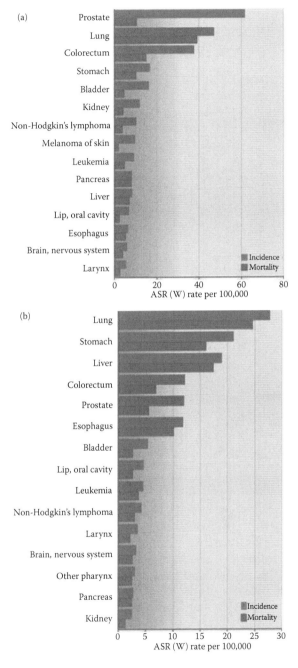

**FIGURE 12.3** (a) Cancers among males seen in more developed nations. (b) Cancers among males seen in less developed nations. (From Ferlay, J. et al. GLOBOCAN 2008 v2.0, Cancer Incidence and Mortality Worldwide: IARC CancerBase No. 10 [internet]. Lyon, France: International Agency for Research on Cancer. Available from: http://globocan.iarc.fr, accessed on 27th September 2013.)

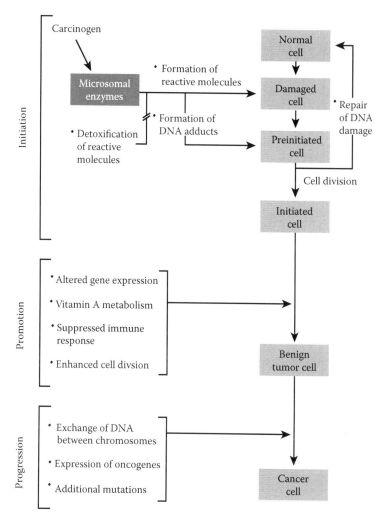

**FIGURE 12.4** Steps in carcinogenesis. *Steps altered by alcohol consumption. (From Garro, A.J. et al. *Alcohol Health & Research World*, *16*(1), 81–86, 1992.)

3. Diabetes, particularly type 2
4. Infections by various microorganisms, particularly bacteria and viruses
5. Low immune status
6. Alcohol consumption
7. Chronic ulcers and wounds

## LIFESTYLE HABITS/DISEASES AS RISKS TO CANCER DEVELOPMENT

Cancer is caused by genetic damage in the genome of a cell. This damage is either inherited or acquired throughout life. The acquired genetic damage is often "self-inflicted" through unhealthy lifestyles. Essentially one-third of cancer is due to tobacco use,

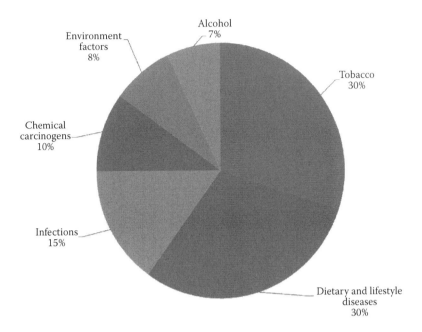

**FIGURE 12.5** Causes of cancer. The unmarked pie segments (in descending order) are chemical carcinogens, environmental factors, and alcohol.

one-third due to dietary and lifestyle factors, and one-fifth due to infections. Other factors include chemical carcinogens, environmental pollutants, and alcohol (Figure 12.5). In developing countries, cancers caused by infections by microorganisms, such as cervical (by human papillomavirus) (Parkin et al. 2008), liver (by hepatitis virus) (Yuen et al. 2009), nasopharynx (by Epstein-Barr virus) (Chou et al. 2008), and stomach (by *Helicobacter pylori*) (Kuniyasu et al. 2003), are more common than in developed countries (DCP2 2007). These are cancers that are theoretically preventable because infections can be eradicated with proper treatment or vaccination, most of which are often not possible or not available in many developing countries due to lack of economic resources.

## Smoking and Tobacco Use

Association of cancer to cigarette smoking is beyond doubt. The prevalence of smoking is higher in developing than in developed countries (Toh 2009). Malaysia is a fast developing nation in Southeast Asia, with an annual GDP of $180,714 million, ranked 36th, ahead of neighboring countries Singapore (ranked 43rd), Philippines (ranked 44th), and Vietnam (ranked 57th) in the GDP listing by the World Bank in 2007 (List of countries by GDP [nominal] by the World Bank 2007). Along with economic progress, Malaysia is seeing a boom in cancer cases related with lifestyle change, such as smoking, obesity, and diabetes. In one state of Kelantan, there was an increment of 20.1% for 1991 to 1996 from the 1987 to 1990 period, 67.4% for 1997 to 2001 from the 1991 to 1996 period, and 143.6% for 2002 to 2007 from the 1997 to 2001 period (Othman et al. 2008).

Smoking is associated with a number of cancers such as larynx, bladder, breast, esophagus, and cervix. While in developed countries the prevalence of smoking is decreasing (Lando et al. 2005), the scenario is the reverse in developing countries. In Kelantan, Malaysia, the prevalence of smoking is 15.6% among primary school children (Norbanee et al. 2006), 33.2% among secondary school children (Shamsuddin and Haris 2000), and 40.6% among secondary school teachers (Naing and Ahmad 2001). Elsewhere in other developing nations, children and adolescents are indulging in this bad habit (Sirichotiratana et al. 2008a,b). Fifty-four percent of smokers in Abu Dhabi started between the age of 10 and 15 years (Bener and Al-Ketbi 1999). The initiation and the influence to start smoking are similar as in developed countries (Warren et al. 2008). Malaysia has increased the price of cigarettes by increasing the import taxation from RM 85/kg in 1990 to RM 216/kg in 2000 (Rahmat et al. 2008); however, the rate of smoking is yet to decline, supporting the idea proposed by Regidor et al. (2007) that increasing the price for cigarettes is not the answer to reduce smoking. Smoking increases the risk of colorectal carcinomas by 43% (Huxley 2007). Ever-smokers were associated with an 8.8-fold increased risk of colorectal cancers (95% confidence interval, 1.7–44.9) when fed on a well-done red meat diet if they have NAT2 and CYP1A2 rapid phenotypes (Le Marchand et al. 2001). No similar association was found in never-smokers (Le Marchand et al. 2001).

## Obesity and Physical Inactivity

The second important risk factor causing acquired genetic damage and thus posing risk for cancer development is obesity. Obese subjects have an approximately 1.5- to 3.5-fold increased risk of developing cancers compared with normal-weight subjects (Pischon et al. 2008). Obesity is associated with a number of cancers (Reeves et al. 2007; Rapp et al. 2008), particularly endometrial (Bjorge et al. 2007; McCourt et al. 2007), breast (Ahn et al. 2007; Dogan et al. 2007), and colorectal (Moghaddam et al. 2007). Adipocytes have the ability to enhance the proliferation of colon cancer cells in vitro (Amemori et al. 2007). The trend of prevalence of overweight/obesity is rising in many developed and developing countries (Low et al. 2009). In a study conducted in 2005 (Nazri et al. 2008) in Kota Bharu District in the state of Kelantan, Malaysia, the overall prevalence of overweight/obesity was 49.1% (Nazri et al. 2008), much higher than the figure reported earlier in 1996 (Jackson et al. 1996). In this community, the rise of cancer is exponential in the period between years 2002 and 2007 (143.6% increment) compared with the previous 5-year period of 1996 to 2001 (Othman et al. 2008).

Obesity is not a social problem but a disease. The greatest risk is for obese persons who are also diabetic, particularly those whose body mass index is above 35 kg/m$^2$. The increase in risk is by 93-fold in women and by 42-fold in men (Jung 1997).

Obesity and physical activity are often reciprocally related. Exercise and physical activity play a role in weight from the prenatal through adolescent time frame (Dugan 2008). In a survey of 11,631 high school students in 1990, those who had low physical activity levels were also found to indulge in many negative lifestyles: cigarette smoking, marijuana use, lower fruit and vegetable consumption, greater television watching, failure to wear a seat belt, and low perception of academic performance (Pate et al. 1996). Participants who were engaged in regular physical activity display more desirable health outcomes across a variety of physical conditions including cancer (Penedo and Dahn 2005).

Smoking and physical activities often have a reciprocal relationship. Nonsmokers are the ones who indulge in physical activities. In Taiwan, those with lower education or income and who are of younger age, smokers, and chewers of betel quid exercised significantly less than their counterparts (Wai et al. 2008). Among Shanghai Chinese, current smokers, particularly heavy smokers, were less active in all kinds of physical activities compared with former smokers and nonsmokers (Lee et al. 2007).

### Diabetes, Particularly Type 2, as Risk of Cancer Development

Obesity is closely related with diabetes (Grandone et al. 2008). A community that has a high prevalence of obesity also has a high prevalence of diabetes (Othman et al. 2008). In Kelantan, Malaysia, the prevalence of diabetes in 1999 was 10.5% and impaired glucose tolerance was 16.5% (Mafauzy et al. 1999). Kelantan is ranked highest in prevalence of diabetes in Malaysia in which the overall national prevalence is 8.3% (Zaini 2000); thus, it was not a surprise to see a rapid rise of cancer prevalence in the state (Othman et al. 2008). The overall prevalence of diabetes mellitus in Kelantan in early 1960s was not known; however, for Malaysia, it was 0.65% (Mustaffa 1990). According to a review on diabetes, the WHO has estimated that, by 2030, there would be 2.48 million diabetics in Malaysia, a jump of 164% from 0.94 million in 2002 (Mafauzy 2006). One of the most common cancers noted in a community that has high diabetics and obesity is colorectal cancer (Yang et al. 2005; Ahmed et al. 2006; Seow et al. 2006; Othman and Zin 2008).

In a study of 138 colorectal cancers seen in Hospital Universiti Sains Malaysia, 47.8% had metabolic diseases, of which 13.8% were diabetes type 2 (Othman and Zin 2008). Those diabetics with colorectal cancer often have distal cancers (Figure 12.6).

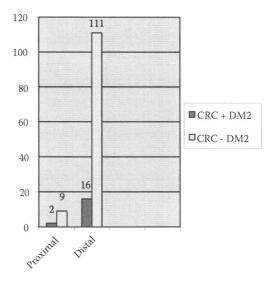

**FIGURE 12.6** Frequency of colorectal cancer cases with or without diabetes mellitus type 2 in Hospital Universiti Sains Malaysia (2001–2006) by anatomic categorization. CRC, colorectal carcinoma; +DM, with diabetes mellitus type 2; –DM, without diabetes type 2. (From Othman, N.H. and A.A. Zin. *Asian Pacific of Journal Prevention*, 9(4), 747–751, 2008.)

## CHRONIC INFECTIONS AS RISK OF CANCER DEVELOPMENT

There are a number of microorganisms that could cause cancer. Common viruses causing cancer (Carrillo-Infante et al. 2007) are Epstein–Barr virus (Siddique et al. 2010) (nasopharyngeal carcinomas), human papillomavirus (cervical cancers and other squamous cancers), and hepatitis B virus (liver cancers). Viruses are oncogenic after a long period of latency (McLaughlin-Drubin and Munger 2008).

Bacteria that have been studied to have associations with cancer are *H. pylori* infections (stomach cancer) (Kuniyasu et al. 2003), *Ureaplasma urealyticum* (prostate cancer) (Hrbacek et al. 2011), and chronic typhoid carrier (gallbladder cancer) (Sharma et al. 2007). Chronic fungi infections have also been studied to be associated with cancer (Hooper et al. 2009). Parasites such as *Schistosoma haematobium* associated with carcinoma of the urinary bladder and liver flukes *Opisthorchis viverrini* and *Clonorchis sinensis* associated with cholangiocarcinoma and hepatocellular carcinoma. There are three main mechanisms by which infections can cause cancer, and they appear to involve initiation as well as promotion of carcinogenesis (Kuper et al. 2001). Persistent infection within the host and induced chronic inflammation are often accompanied by the formation of reactive oxygen and nitrogen species (Kuper et al. 2001). Reactive oxygen and nitrogen species have the potential to damage DNA, proteins, and cell membranes. Chronic inflammation often results in repeated cycles of cell damage and compensating cell proliferation (Cohen et al. 1991). DNA damage promotes the growth of malignant cells. Second, infectious agents may directly transform cells by inserting active oncogenes into the host genome, inhibiting tumor suppressors, or stimulating mitosis (Kuper et al. 2001). Third, infectious agents, such as HIV, may induce immunosuppression (Kuper et al. 2001).

### Low Immune Status as Risk of Cancer Development

#### Cancer and Aging

The most important change that would occur in the world population in the next 50 years is the change in the proportion of elderly people (>65 years): 7% in 2000 to 16% in 2050 (Bray and Moller 2006). According to the Department of Statistics Malaysia, the life expectancy for males and females in Malaysia increases by 0.1 to 0.5 per year annually since 2005 (Key Statistics by Department of Statistics Malaysia 2008) (Key Statistics by Department of Statistics Malaysia 2008). Many cancers are associated with aging; although age per se is not an important determinant of cancer risk, it implies prolonged exposure to carcinogen (Franceschi and La Vecchia 2001). By the year 2050, 27 million people are projected to have cancer. More than half of the estimated number will be residents of developing countries (Bray and Moller 2006). Aging is associated with reduced immune systems.

#### Low Immune Status Due to Chronic Diseases

Patients who have low immune systems are at risk for cancer development. This explains why diabetics are more at risk than nondiabetics to get epithelial cancers such as liver, pancreas, breasts, and endometrium. HIV patients are at risk of developing epithelial and nonepithelial cancers. These persons are also at risk of developing multiple chronic infections implying the multiplicity in cancer genesis. Patients with autoimmune diseases are also at risk of developing cancers such as colorectal carcinomas in ulcerative colitis, Crohn's disease, and cancer of the thyroid in autoimmune thyroiditis.

## Chronic Ulcers and Wounds

Chronic ulcers have a risk to develop cancer. The most common is Marjolin's ulcer (Asuquo et al. 2009) and it is common in developing nations especially in rural areas with poor living conditions (Asuquo et al. 2007). This risk factor is related to chronic infections, as most if not all chronic ulcers are not healing because of persistent infection.

# WHAT IS HONEY?

Honey has been known for centuries for its medicinal and health-promoting properties. It is produced from a complex enzymatic process of nectar and saccharine exudates collected from various kinds of floral sources (Ball 2007). It contains various kinds of phytochemicals with high phenolic and flavonoids content that contribute to its high antioxidant activity (Yao et al. 2003; Iurlina et al. 2009; Pyrzynska and Biesaga 2009). An agent that has strong antioxidant property may have the potential to prevent the development of cancer as free radicals and oxidative stress plays a significant role in inducing the formation of cancers (Valko et al. 2007). Phytochemicals available in honey could be narrowed down into phenolic acids and polyphenols. Variants of polyphenols in honey were reported to have antiproliferative properties against several types of cancer (Jaganathan and Mahitosh 2009).

## HONEY AS A NATURAL IMMUNE BOOSTER

Honey stimulates inflammatory cytokine production from monocytes (Tonks et al. 2003). Manuka, pasture, and jelly bush honey were found to significantly increase tumor necrosis factor-$\alpha$, interleukin-1$\beta$, and interleukin-6 release from Mono Mac 6 cells (and human monocytes) when compared with untreated and artificial honey-treated cells ($P < 0.001$) (Tonks et al. 2003). A 5.8-kDa component of manuka honey was found to stimulate cytokine production from immune cells via TLR4, an important signaling pathway (Tonks et al. 2007). Honey stimulates antibody production during primary and secondary immune responses against thymus-dependent and thymus-independent antigens in mice injected with sheep red blood cells and *Escherichia coli* antigen (Al-Waili and Haq 2004). Consumption of 80 g daily of natural honey for 21 days showed that the prostaglandin levels compared with normal subjects were elevated in patients with AIDS (Al-Waili et al. 2006). Natural honey decreased prostaglandin levels and elevated the NO end product, percentage of lymphocytes, platelet count, and serum protein, albumin, and copper levels. It might be concluded that natural honey decreased prostaglandin levels, elevated NO production, and improved hematologic and biochemical tests in a patient with a long history of AIDS (Al-Waili et al. 2006).

Daily consumption of honey improves one's immune system.

## HONEY AS A NATURAL ANTI-INFLAMMATORY AGENT

In routine everyday life, our cells may be injured by irritants from outside or within our body. Cellular/molecular stress results in inflammatory response, our body's defense mechanism in trying to get rid of the irritants. In general, inflammatory response to any insulting agents is beneficial and protective to us, but at times inflammatory responses are detrimental to health. Honey is a potent anti-inflammatory agent. It provides significant symptom relief of cough in children with an upper respiratory tract infection (Heppermann 2009). It has been shown to be effective in the management of dermatitis and psoriasis vulgaris (Al-Waili 2003). Eight of 10 patients with dermatitis and 5 of 8 patients with psoriasis showed significant improvement after 2 weeks on a honey-based ointment (Al-Waili 2003). Honey at dilutions of up to 1:8 reduced bacterial adherence from $25.6 \pm 6.5$ (control) to $6.7 \pm 3.3$ bacteria per epithelial cell ($P < 0.001$) in vitro (Alnaqdy et al. 2005). Volunteers who chewed "honey leather" showed that there were statistically highly significant reductions in mean plaque scores (0.99 reduced to 0.65; $P = 0.001$) in the manuka honey group compared with the control group, suggesting a potential therapeutic role for honey in the treatment of gingivitis, periodontal disease (English et al. 2004), mouth ulcers, and other problems of oral health (Molan 2001a). It has been shown in experimental mice that honey prevents postoperative peritoneal adhesions (Aysan et al. 2002), probably due to rapid healing of surgical intraperitoneal injuries.

A case study of a patient who had dystrophic epidermolysis bullosa who has been treated with many different dressings and creams for 20 years and healed with honey impregnated dressing in 15 weeks (Hon 2005) illustrates the usefulness of honey as an anti-inflammatory agent.

## HONEY AS A NATURAL ANTIMICROBIAL

Everyday, we are exposed to all kinds of microbial insults from bacteria, viruses, parasites, and fungi. Honey is a potent natural antimicrobial. The most common infection humans get is staphylococcal infection. The bactericidal mechanism is through disturbance in cell division machinery (Henriques et al. 2010). The minimum inhibitory concentration for *Staphylococcus aureus* by *Apis mellifera* honey ranged from 126.23 to 185.70 mg mL$^{-1}$ (Miorin et al. 2003). Honeys are also effective coagulase-negative staphylococci (French et al. 2005). Local application of raw honey on infected wounds reduced signs of acute inflammation (Al-Waili 2004a), thus alleviating symptoms. Antimicrobial activity of honey is stronger in acidic media than in neutral or alkaline media (Al-Waili 2004a). The potency of honey is comparable to some local antibiotics. Honey application into infective conjunctivitis reduced redness, swelling, pus discharge, and time for eradication of bacterial infections (Al-Waili 2004a). When honey is used together with antibiotics, gentamicin enhances anti-*S. aureus* activity by 22% (Al-Jabri et al. 2005).

The effectiveness of honey is best when used at room temperature. Heating honey to 80°C for 1 h decreased antimicrobial activity of both new and stored honey. Storage of honey for 5 years decreased its antimicrobial activity, while ultraviolet light exposure increased its activity against some microorganisms (Al-Waili 2004).

Honey is also effective against common dermatophyte infections such as tinea versicolor (86% response), tinea cruris (78% response), and tinea corporis (75% response) after 4 weeks of topical application three times daily and for patients with tinea faciei with topical application of honey, olive oil, and beeswax mixture (Al-Waili 2004b).

Honey also has been shown to have antiviral properties. In a comparative study, topical application of honey was found to be better than acyclovir treatment on patients with recurrent herpetic lesions (Al-Waili 2004b). Two cases of labial herpes and one case of genital herpes remitted completely with the use of honey, while none with acyclovir treatment (Al-Waili 2004b).

Honey has an antifungal effect against *Candida* species (Irish et al. 2006). Infants suffering from diaper dermatitis improved significantly after topical application of a mixture containing honey, olive oil, and beeswax after 7 days (Al-Waili 2005). When honey is added to bacterial culture medium, the appearance of microbial growth on the culture plates is delayed (Al-Waili et al. 2005).

Mycobacteria did not grow in culture media containing 10% and 20% honey, while it grew in culture media containing 5%, 2.5%, and 1% honey, suggesting that honey could be an ideal antimycobacterial agent (Asadi-Pooya et al. 2003) at certain concentrations.

Honey is also effective in killing hardy bacteria such as *Pseudomonas aeruginosa* and could lead to a new approach in treating refractory chronic rhinosinusitis (Alandejani et al. 2009). Although honey is known to have antibacterial properties, the digestive tract of Slovakian adult honeybees is highly populated by bacteria, fungi, and yeast (Kacaniova et al. 2009). Whether the presence of bacteria in honey has any role in the antibacterial property of honey is yet to be investigated.

Daily consumption of honey reduces risk of chronic infections by microorganisms.

## HONEY AS AN AGENT FOR CONTROLLING OBESITY

Obese individuals are at risk of developing cancer. In a clinical study on 55 overweight or obese patients, the control group (17 subjects) received 70 g of sucrose daily for a maximum of 30 days and patients in the experimental group (38 subjects) received 70 g of natural honey for the same period. Results showed that honey caused a mild reduction in body weight (1.3%) and body fat (1.1%) (Yaghoobi et al. 2008).

## HONEY AS A FIXER FOR CHRONIC ULCERS AND WOUNDS

Increasing numbers of antibiotic-resistant bacteria have made simple wounds become chronic and nonhealing, and as such, honey provides alternative treatment options (Sharp 2009). Honey absorbs exudates released in wounds and devitalized tissue (Cutting 2007). It is effective in recalcitrant surgical wounds (Cooper et al. 2001) and increases the rate of healing by stimulation of angiogenesis, granulation, and epithelialization, making skin grafting unnecessary and giving excellent cosmetic results (Molan 2001b). In a randomized control trial, manuka honey improved wound healing in patients with sloughy venous leg ulcers (Armstrong 2009). Honey was shown to eradicate methicillin-resistant *S. aureus* infection in 70% of chronic venous ulcers (Gethin and Cowman 2008). Honey is acidic, and chronic nonhealing wounds have an elevated alkaline environment. Manuka honey dressings were associated with a statistically significant decrease in wound pH (Gethin et al. 2008). Honey for the purposes of wound management has to be of "medical grade," that is, sterilized by γ-irradiation before it can be registered as a medical device (Evans and Flavin 2008). Available evidence in meta-analysis studies indicates markedly greater efficacy of honey compared with alternative dressings for superficial or partial thickness burns (Wijesinghe et al. 2009). Honey is an inexpensive moist dressing with antibacterial and tissue-healing properties suitable for diabetic feet (Eddy et al. 2008). Intralesional honey application promotes healing and leaves no scar in urethral injuries in experimental animals (Ayyildiz et al. 2007). The average cost of treatment per patient using honey dressing is much cheaper than conventional dressing (Ingle et al. 2006). The honey preparations for wounds and burns are in the form of ointment (Abdelatif et al. 2008) and impregnated gauze (Subrahmanyam 1993) or gel (Wijesinghe et al. 2009).

## HONEY AS A NATURAL CANCER "VACCINE"

Some simple and polyphenols found in honey, namely, caffeic acid, caffeic acid phenyl esters, chrysin, galangin, quercetin, kaempferol, acacetin, pinocembrin, pinobanksin, and apigenin, have evolved as promising pharmacologic agents in the prevention and treatment of cancer (Jaganathan and Mandal 2009). The antioxidant activity of *Trigona carbonaria* honey from Australia was high at 233.96 ± 50.95 µM Trolox equivalents (Oddo et al. 2008). The antioxidant activity of four honey samples from different floral sources showed high antioxidant properties tested by different assay methods (Hegazi and Abd El-Hady 2009). Dark honey had higher phenolic compounds and antioxidant activity than clear honey (Estevinho et al. 2008). The amino acid composition of honey is an indicator of the toxic radical scavenging capacity (Perez et al. 2007).

## HONEY AS A "CANCER THERAPY"

Among these natural products, honey is one of them. Honey contains various vitamins, minerals, and amino acids as well as glucose and fructose and is popular as a natural food. There are wide varieties of honey (manuka honey, pasture honey, jelly bush honey, and jungle honey), and the varieties are due to components of the flower sources (Lusby et al. 2002). In addition, it was reported that oral intake of honey augments antibody production in primary and secondary immune responses against thymus-dependent and thymus-independent antigens (Fukuda et al. 2009). Natural honeys induce interleukin-6 release from Mono Mac 6 cells (Timm et al. 2008). Honey also exhibits antioxidant (Viuda-Martos et al. 2008), chemopreventive (Jaganathan and Mandal 2009), antiatherogenic (Ansorge et al. 2003), immunoregulatory (Ansorge et al. 2003), antimicrobial (Dai et al. 2010), and wound-healing (Boukraa and Sulaiman 2010) properties. Honey may provide the basis for the development of novel therapeutics for patients with cancer and cancer-related tumors. It is investigated that the jungle honey fragments have chemotactic induction for neutrophils and reactive oxygen species, the antitumor factors, proving its antitumor activity and enhancing immunoefficiency (Fukuda et al. 2009). Honey has been shown to have antineoplastic activity in an experimental bladder model in vivo and in vitro (Swellam et al. 2003).

Honey is rich in flavonoids (Gomez-Caravaca et al. 2006; Jaganathan and Mandal 2009). Flavonoids have created a lot of interest among researchers because of their anticancer properties. The mechanisms suggested are rather diverse such as various signaling pathways (Woo et al. 2004), including stimulation of tumor necrosis factor-α release (Tonks et al. 2001), inhibition of cell proliferation, induction of apoptosis (Jaganathan and Mandal 2010), and cell cycle arrest (Pichichero et al. 2010) as well as inhibition of lipoprotein oxidation (Gheldof and Engeseth 2002). Honey is thought to mediate these beneficial effects due to its major components such as chrysin (Woo et al. 2004) and other flavonoids (Jaganathan et al. 2010). These differences are explainable as honeys are of various floral sources and each floral source may exhibit different active compounds. At the same time, honey has about 200 other substances, of which the most predominant are a mixture of sugars (fructose, glucose, maltose, and sucrose) (Aljadi and Kamaruddin 2004). Sugar in itself is carcinogenic (Heuson et al. 1972); thus, those with chronic hyperglycemia are more at risk of developing cancer (Giovannucci et al. 2010). Therefore, the mechanism on how honey has an anticancer effect is an area of great interest recently. The effects of honeys on hormone-dependent cancers such as breast, endometrial, and prostate cancer and tumors remain largely unknown. There is a lot we can learn from nature (Moutsatsou 2007). For example, phytochemicals, such as genistein, lycopene, curcumin, epigallocatechin-gallate, and resveratrol, have been studied to be used for treatment of prostate cancer (Von Low et al. 2007). Phytoestrogens constitute a group of plant-derived isoflavones and flavonoids and honey belongs to plant phytoestrogen (Moutsatsou 2007; Zaid et al. 2010).

With an increasing demand for antioxidant supply in foods, honey had gained vitality since it is rich in phenolic compounds and other antioxidants such as ascorbic acid, amino acids, and proteins (Jaganathan and Mandal 2009). Honey, which has a lower glycemic index and peak incremental index when compared with sucrose, may be used as a sugar substitute in patients with type 1 diabetes mellitus (Abdulrhman et al. 2009) or in those with impaired glucose tolerance or mild diabetes (Agrawal et al. 2007). Natural honey modulates better physiologic glycemic response compared with simulated honey (Ahmad et al. 2008).

## Tualang Honey—Malaysian Natural Jungle Honey

Developing nations are rich with natural jungles where honey could be harvested. Tualang honey (TH) is a jungle honey in Malaysia, produced by bees (*Apis dorsata*) that built their hives high on the Tualang trees (*Kompassia excelsa*) in the tropical rainforest (Kannan et al. 2009; Mohamed et al. 2009; Tan et al. 2009; Erejuwa et al. 2010). Several researchers in Malaysia have established a number of health benefits of TH. A study of TH showed that it has significant anticancer activity against human breast (Fauzi et al. 2011) and cervical cancer cell lines by inducing apoptosis of cancer cells via depolarization of the mitochondrial membrane. TH has also been observed to have an antiproliferative effect on oral squamous cell carcinoma and human osteosarcoma cancer cell lines by inducing early apoptosis (Ghashm et al. 2010). TH modified

7,12-dimethylbenz(a)anthracene (DMBA) (carcinogen)-induced mammary cancers in rats (Erazuliana et al. 2011). TH-treated groups had slower size increment and mean cancer size compared with the control group. The number of cancers developed in TH-treated rats was also fewer than in the control group. Histologic examinations of these experimental cancers showed that the majority of the TH-treated cancers were of grade 2 compared with control in which the majority were of grade 3. TH-treated cancers were seen to have lesser noninvasive components (ductal carcinoma in situ). There was increased apoptosis and reduced angiogenesis in TH-treated DMBA-induced rats. This study showed that TH has a positive modulation effect of chemical carcinogenesis induced by DMBA (Erazuliana et al. 2011).

TH also contains abundant hydroxymethylfurfural compounds that have been shown by others to have antitumor potential (Michail et al. 2007). Honey extracts have polyphenols and phenolic acids (vanillic acid, protocatechic acid, and $p$-hydroxybenzoic acid). These compounds are known to inhibit cancer-related pathways and processes exhibiting significant biological effects in human cancer cells, such as antiestrogenic activity in breast cells (Jenkins et al. 2003) and inhibition of cell viability on prostate cancer (Von Low et al. 2007) and endometrial cancer cells (Michail et al. 2007).

However, there are still areas in anticancer effects of TH that have not been fully understood. Which genes are upregulated or downregulated by TH after chemical carcinogen exposure? What is the mechanism of this change? Why do these genes get altered? In those cases in which honey "succeeded" in converting malignant lesions to benign lesions (Erazuliana et al. 2011), which genes are turned on? Does TH reduce circulating cancer cells in vivo? What are the good proteins that are enhanced by TH and the bad proteins that are alleviated by taking honey? Further research is needed in this area to answer these questions.

## CONCLUSION

Honey, a natural substance, has been shown to have high potential benefits in combating/reducing the causes of cancer. There is now sizeable evidence that honey is a natural immune booster, natural anti-inflammatory agent, natural antimicrobial, natural cancer "vaccine," and natural promoter for healing ulcers and wounds. In addition, it is almost a complete food containing almost all components of nutrients required for the body. For its maximum health effect, it is best taken when fresh, unheated, or stored. Honey and cancer have a sustainable parallel relationship in the setting of developing nations where resources for cancer prevention and treatment are limited.

*Honey bee oh honey bee…*
*You fly free from tree to tree…*
*The sweet stuff from your belly…*
*Not only tastes good but also a remedy…*

*Honey bees oh honey bees…*
*You fly high on tall Tualang trees…*
*Honey, royal jelly and propolis…*
*Have been shown to have anti-cancer properties…*

# REFERENCES

Abdelatif, M., M. Yakoot et al. (2008). "Safety and efficacy of a new honey ointment on diabetic foot ulcers: A prospective pilot study." *J Wound Care* **17**(3): 108–10.

Abdulrhman, M., M. El-Hefnawy et al. (2011). "The glycemic and peak incremental indices of honey, sucrose and glucose in patients with type 1 diabetes mellitus: Effects on C-peptide level-a pilot study." *Acta Diabetol* **48**(2):89–94, doi: 10.1007/s00592-009-0167-7.

Agrawal, O. P., A. Pachauri et al. (2007). "Subjects with impaired glucose tolerance exhibit a high degree of tolerance to honey." *J Med Food* **10**(3): 473–8.

Ahmad, A., M. K. Azim et al. (2008). "Natural honey modulates physiological glycemic response compared to simulated honey and D-glucose." *J Food Sci* **73**(7): H165–7.

Ahmed, R. L., K. H. Schmitz et al. (2006). "The metabolic syndrome and risk of incident colorectal cancer." *Cancer* **107**(1): 28–36.

Ahn, J., A. Schatzkin et al. (2007). "Adiposity, adult weight change, and postmenopausal breast cancer risk." *Arch Intern Med* **167**(19): 2091–102.

Al-Jabri, A. A., S. A. Al-Hosni et al. (2005). "Antibacterial activity of Omani honey alone and in combination with gentamicin." *Saudi Med J* **26**(5): 767–71.

Al-Waili, N. S. (2003). "Topical application of natural honey, beeswax and olive oil mixture for atopic dermatitis or psoriasis: Partially controlled, single-blinded study." *Complement Ther Med* **11**(4): 226–34.

Al-Waili, N. S. (2004a). "An alternative treatment for pityriasis versicolor, tinea cruris, tinea corporis and tinea faciei with topical application of honey, olive oil and beeswax mixture: An open pilot study." *Complement Ther Med* **12**(1): 45–7.

Al-Waili, N. S. (2004b). "Investigating the antimicrobial activity of natural honey and its effects on the pathogenic bacterial infections of surgical wounds and conjunctiva." *J Med Food* **7**(2): 210–22.

Al-Waili, N. S. (2004). "Topical honey application vs. acyclovir for the treatment of recurrent herpes simplex lesions." *Med Sci Monit* **10**(8): MT94–8.

Al-Waili, N. S. (2005). "Clinical and mycological benefits of topical application of honey, olive oil and beeswax in diaper dermatitis." *Clin Microbiol Infect* **11**(2): 160–3.

Al-Waili, N. S., M. Akmal et al. (2005). "The antimicrobial potential of honey from United Arab Emirates on some microbial isolates." *Med Sci Monit* **11**(12): BR433–8.

Al-Waili, N. S., T. N. Al-Waili et al. (2006). "Influence of natural honey on biochemical and hematological variables in AIDS: A case study." *ScientificWorldJournal* **6**: 1985–9.

Al-Waili, N. S. and A. Haq (2004). "Effect of honey on antibody production against thymus-dependent and thymus-independent antigens in primary and secondary immune responses." *J Med Food* **7**(4): 491–4.

Alandejani, T., J. Marsan et al. (2009). "Effectiveness of honey on Staphylococcus aureus and Pseudomonas aeruginosa biofilms." *Otolaryngol Head Neck Surg* **141**(1): 114–8.

Aljadi, A. M. and M. Y. Kamaruddin (2004). "Evaluation of the phenolic contents and antioxidant capacities of two Malaysian floral honeys." *Food Chem* **85**(4): 513–8.

Alnaqdy, A., A. Al-Jabri et al. (2005). "Inhibition effect of honey on the adherence of Salmonella to intestinal epithelial cells *in vitro*." *Int J Food Microbiol* **103**(3): 347–51.

Amemori, S., A. Ootani et al. (2007). "Adipocytes and preadipocytes promote the proliferation of colon cancer cells *in vitro*." *Am J Physiol Gastrointest Liver Physiol* **292**(3): G923–9.

Ansorge, S., D. Reinhold et al. (2003). "Propolis and some of its constituents down-regulate DNA synthesis and inflammatory cytokine production but induce TGF-beta1 production of human immune cells." *Z Naturforsch C* **58**(7–8): 580–9.

Armstrong, D. G. (2009). "Manuka honey improved wound healing in patients with sloughy venous leg ulcers." *Evid Based Med* **14**(5): 148.

Asadi-Pooya, A. A., M. R. Pnjehshahin et al. (2003). "The antimycobacterial effect of honey: An *in vitro* study." *Riv Biol* **96**(3): 491–5.

Asuquo, M., G. Ugare et al. (2007). "Marjolin's ulcer: The importance of surgical management of chronic cutaneous ulcers." *Int J Dermatol* **46 Suppl 2**: 29–32.

Asuquo, M. E., A. M. Udosen et al. (2009). "Cutaneous squamous cell carcinoma in Calabar, southern Nigeria." *Clin Exp Dermatol* **34**(8): 870–3.

Aysan, E., E. Ayar et al. (2002). "The role of intra-peritoneal honey administration in preventing post-operative peritoneal adhesions." *Eur J Obstet Gynecol Reprod Biol* **104**(2): 152–5.

Ayyildiz, A., K. T. Akgul et al. (2007). "Intraurethral honey application for urethral injury: An experimental study." *Int Urol Nephrol* **39**(3): 815–21.

Ball, D. W. (2007). "The chemical composition of honey." *J Chem Educ* **84**(10): 1643–1646.

Belpomme, D., P. Irigaray et al. (2007). "The growing incidence of cancer: Role of lifestyle and screening detection (Review)." *Int J Oncol* **30**(5): 1037–49.

Bener, A. and L. M. Al-Ketbi (1999). "Cigarette smoking habits among high school boys in a developing country." *J R Soc Promot Health* **119**(3): 166–9.

Bener, A., H. Ayub et al. (2008). "Patterns of cancer incidence among the population of qatar: A worldwide comparative study." *Asian Pac J Cancer Prev* **9**(1): 19–24.

Bjorge, T., A. Engeland et al. (2007). "Body size in relation to cancer of the uterine corpus in 1 million Norwegian women." *Int J Cancer* **120**(2): 378–83.

Bouchardy, C., Fioretta, G., Rapiti, E., Verkooijen, H.M., Rapin, C.H., Schmidlin, F. et al. (2008). Recent trends in prostate cancer mortality show a continuous decrease in several countries. *Int J Cancer* **123**(2): 421–9.

Boukraa, L. and S. A. Sulaiman (2010). "Honey use in burn management: Potentials and limitations." *Forsch Komplementmed* **17**(2): 74–80.

Boyle, P. and B. lewin (2008). "World Cancer Report." International Agency for Research on Career (IARC); *WHO-IARC*. ISBN: 9789283204237.

Bray, F. and B. Moller (2006). "Predicting the future burden of cancer." *Nat Rev Cancer* **6**(1): 63–74.

Carrillo-Infante, C., G. Abbadessa et al. (2007). "Viral infections as a cause of cancer (review)." *Int J Oncol* **30**(6): 1521–8.

Chou, J., Y. C. Lin et al. (2008). "Nasopharyngeal carcinoma—Review of the molecular mechanisms of tumorigenesis." *Head Neck* **30**(7): 946–63.

Cohen, S. M., D. T. Purtilo et al. (1991). "Ideas in pathology. Pivotal role of increased cell proliferation in human carcinogenesis." *Mod Pathol* **4**(3): 371–82.

Cooper, R. A., P. C. Molan et al. (2001). "Manuka honey used to heal a recalcitrant surgical wound." *Eur J Clin Microbiol Infect Dis* **20**(10): 758–9.

Crane, L. M., M. Schaapveld et al. (2007). "Oesophageal cancer in the Netherlands: Increasing incidence and mortality but improving survival." *Eur J Cancer* **43**(9): 1445–51.

Cutting, K. F. (2007). "Honey and contemporary wound care: An overview." *Ostomy Wound Manage* **53**(11): 49–54.

Dai, T., Y. Y. Huang et al. (2010). "Topical antimicrobials for burn wound infections." *Recent Pat Antiinfect Drug Discov* **5**(2): 124–51.

DCP2. (2007). "Controlling Cancer in Developing Countries: Prevention and treatment strategies merit further study." Available on www.dcp2.org.

Dogan, S., X. Hu et al. (2007). "Effects of high-fat diet and/or body weight on mammary tumor leptin and apoptosis signaling pathways in MMTV-TGF-alpha mice." *Breast Cancer Res* **9**(6): R91.

Dugan, S. A. (2008). "Exercise for preventing childhood obesity." *Phys Med Rehabil Clin N Am* **19**(2): 205–16, vii, doi:10.1016/j.pmr.2007.11.001.

Eddy, J. J., M. D. Gideonsen et al. (2008). "Practical considerations of using topical honey for neuropathic diabetic foot ulcers: A review." *WMJ* **107**(4): 187–90.

English, H. K., A. R. Pack et al. (2004). "The effects of manuka honey on plaque and gingivitis: A pilot study." *J Int Acad Periodontol* **6**(2): 63–7.

Erazuliana, A. K., S. Siti Amrah et al. (2013). Inhibitory effects of tualang honey on experimental breast cancer in rats: A preliminary study. *Asian Pac J Cancer Prev.* **14**(4): 2249–54.

Erejuwa, O. O., S. A. Sulaiman et al. (2010). "Antioxidant protection of Malaysian tualang honey in pancreas of normal and streptozotocin-induced diabetic rats." *Ann Endocrinol (Paris)* **71**(4): 291–6.

Estevinho, L., A. P. Pereira et al. (2008). "Antioxidant and antimicrobial effects of phenolic compounds extracts of Northeast Portugal honey." *Food Chem Toxicol* **46**(12): 3774–9.

Evans, J. and S. Flavin (2008). "Honey: A guide for healthcare professionals." *Br J Nurs* **17**(15): S24, S26, S28–30.

Fauzi, A. N., M. N. Norazmi et al. (2011). "Tualang honey induces apoptosis and disrupts the mitochondrial membrane potential of human breast and cervical cancer cell lines." *Food Chem Toxicol* **49**(4): 871–8.

Ferlay, J., H. R. Shin et al. (2010). GLOBOCAN 2008 v2.0, Cancer Incidence and Mortality Worldwide: IARC CancerBase No. 10 [internet]. Lyon, France: International Agency for Research on Cancer. Available from: http://globocan.iarc.fr, accessed on 27th September 2013.

Franceschi, S. and C. La Vecchia (2001). "Cancer epidemiology in the elderly." *Crit Rev Oncol Hematol* **39**(3): 219–26.

French, V. M., R. A. Cooper et al. (2005). "The antibacterial activity of honey against coagulase-negative staphylococci." *J Antimicrob Chemother* **56**(1): 228–31.

Fukuda, M., K. Kobayashi et al. (2011). "Jungle honey enhances immune function and antitumor activity." *Evid Based Complement Alternat Med 908743.*

Garcia, M., A. Jemal, E. Ward, M. Center, Y. Hao, R. Siegal et al. (2007). "Global Cancer: Facts & Figures 2007." *American Cancer Society.*

Garro, A. J. et al. (1992). *Alcohol Health & Research World,* **16**(1), 81–86.

Gethin, G. and S. Cowman (2008). "Bacteriological changes in sloughy venous leg ulcers treated with manuka honey or hydrogel: An RCT." *J Wound Care* **17**(6): 241–4, 246–7.

Gethin, G. T., S. Cowman et al. (2008). "The impact of Manuka honey dressings on the surface pH of chronic wounds." *Int Wound J* **5**(2): 185–94.

Ghashm, A. A., N. H. Othman et al. (2010). "Antiproliferative effect of Tualang honey on oral squamous cell carcinoma and osteosarcoma cell lines." *BMC Complement Altern Med* **10**: 49.

Gheldof, N. and N. J. Engeseth (2002). "Antioxidant capacity of honeys from various floral sources based on the determination of oxygen radical absorbance capacity and inhibition of *in vitro* lipoprotein oxidation in human serum samples." *J Agric Food Chem* **50**(10): 3050–5.

Giovannucci, E., D. M. Harlan et al. (2010). "Diabetes and cancer: A consensus report." *CA Cancer J Clin* **60**(4): 207–21.

Gomez-Caravaca, A. M., M. Gomez-Romero et al. (2006). "Advances in the analysis of phenolic compounds in products derived from bees." *J Pharm Biomed Anal* **41**(4): 1220–34.

Grandone, A., A. Amato et al. (2008). "High-normal fasting glucose levels are associated with increased prevalence of impaired glucose tolerance in obese children." *J Endocrinol Invest* **31**(12): 1098–102.

Hegazi, A. G. and F. K. Abd El-Hady (2009). "Influence of honey on the suppression of human Low Density Lipoprotein (LDL) Peroxidation (*In vitro*)." *Evid Based Complement Alternat Med* **6**(1): 113–21.

Henriques, A. F., R. E. Jenkins et al. (2010). "The intracellular effects of manuka honey on Staphylococcus aureus." *Eur J Clin Microbiol Infect Dis* **29**(1):45–50, doi:10.1007/s10096-009-0817-2.

Heppermann, B. (2009). "Towards evidence based emergency medicine: Best BETs from the Manchester Royal Infirmary. Bet 3. Honey for the symptomatic relief of cough in children with upper respiratory tract infections." *Emerg Med J* **26**(7): 522–3.

Heuson, J. C., N. Legros et al. (1972). "Influence of insulin administration on growth of the 7,12-dimethylbenz(a)anthracene-induced mammary carcinoma in intact, oophorectomized, and hypophysectomized rats." *Cancer Res* **32**(2): 233–8.

Hon, J. (2005). "Using honey to heal a chronic wound in a patient with epidermolysis bullosa." *Br J Nurs* **14**(19): S4–5, S8, S10 passim.

Hooper, S. J., M. J. Wilson et al. (2009). "Exploring the link between microorganisms and oral cancer: A systematic review of the literature." *Head Neck* **31**(9): 1228–39.

Hrbacek, J., M. Urban et al. (2011). "Serum antibodies against genitourinary infectious agents in prostate cancer and benign prostate hyperplasia patients: A case-control study." *BMC Cancer* **11**: 53.

Huxley, R. (2007). "The role of lifestyle risk factors on mortality from colorectal cancer in populations of the Asia-Pacific region." *Asian Pac J Cancer Prev* **8**(2): 191–8.

Ingle, R., J. Levin et al. (2006). "Wound healing with honey—A randomised controlled trial." *S Afr Med J* **96**(9): 831–5.

Irish, J., D. A. Carter et al. (2006). "Honey has an antifungal effect against Candida species." *Med Mycol* **44**(3): 289–91.

Iurlina, M. O., A. I. Saiz et al. (2009). "Major flavonoids of Argentinean honeys. Optimisation of the extraction method and analysis of their content in relationship to the geographical source of honeys." *Food Chem* **115**(3): 1141–1149.

Jackson, A., C. Cole et al. (1996). "Obesity in primary care patients in Kelantan, Malaysia: Prevalence, and patients' knowledge and attitudes." *Southeast Asian J Trop Med Public Health* **27**(4): 776–9.

Jaganathan, S. K. and M. Mahitosh (2009). "Antiproliferative effects of honey and of its polyphenols: A review." *J Biomed Biotechnol* **2009**: 13.

Jaganathan, S. K. and M. Mandal (2009). "Antiproliferative effects of honey and of its polyphenols: A review." *J Biomed Biotechnol* **2009**: 830616.

Jaganathan, S. K. and M. Mandal (2010). "Involvement of non-protein thiols, mitochondrial dysfunction, reactive oxygen species and p53 in honey-induced apoptosis." *Invest New Drugs* **28**(5): 624–33.

Jaganathan, S. K., S. M. Mandal et al. (2010). "Studies on the phenolic profiling, anti-oxidant and cytotoxic activity of Indian honey: *In vitro* evaluation." *Nat Prod Res* **24**(14): 1295–306.

Jenkins, D. J., C. W. Kendall et al. (2003). "Soy consumption and phytoestrogens: Effect on serum prostate specific antigen when blood lipids and oxidized low-density lipoprotein are reduced in hyperlipidemic men." *J Urol* **169**(2): 507–11.

Jung, R. T. (1997). "Obesity as a disease." *Br Med Bull* **53**(2): 307–21.

Kabir, Z. and L. Clancy (2006). "Lifestyle-related cancer death rates in Ireland: Decreasing or increasing?" *Ir Med J* **99**(2): 40–2.

Kacaniova, M., S. Pavlicova et al. (2009). "Microbial communities in bees, pollen and honey from Slovakia." *Acta Microbiol Immunol Hung* **56**(3): 285–95.

Kannan, T. P., A. Q. Ali et al. (2009). "Evaluation of Tualang honey as a supplement to fetal bovine serum in cell culture." *Food Chem Toxicol* **47**(7): 1696–702.

Key Statistics by Department of Statistics Malaysia. (2008) Available at http://www.statistics.gov.my/english/frameset_keystats.php.

Kuniyasu, H., Y. Kitadai et al. (2003). "*Helicobacter pylori* infection is closely associated with telomere reduction in gastric mucosa." *Oncology* **65**(3): 275–82.

Kuper, H., H.-O. Adami et al. (2001). "Infections as a major preventable cause of human cancer." *J Intern Med* **249**(S741): 61–74.

Lando, H. A., B. Borrelli et al. (2005). "The landscape in global tobacco control research: A guide to gaining a foothold." *Am J Public Health* **95**(6): 939–45.

Le Marchand, L., J. H. Hankin et al. (2001). "Combined effects of well-done red meat, smoking, and rapid N-acetyltransferase 2 and CYP1A2 phenotypes in increasing colorectal cancer risk." *Cancer Epidemiol Biomarkers Prev* **10**(12): 1259–66.

Lee, S. A., W. H. Xu et al. (2007). "Physical activity patterns and their correlates among Chinese men in Shanghai." *Med Sci Sports Exerc* **39**(10): 1700–7.

Lepage, C., L. Remontet, G. Launoy, B. Tretarre, P. Grosclaude, M. Colonna et al. (2008). "Trends in incidence of digestive cancers in France." *Eur J Cancer Prev*, **17**(1): 13–7.

Lim, G. C. C. (2002). "Overview of cancer in Malaysia." *Jpn J Clin Oncol* **32**: s32–42.

List of countries by GDP (nominal) by the World Bank, sorted by their gross domestic product (GDP). (2007) Available at http://en.wikipedia.org/wiki/List_of_countries_by_GDP_(nominal).

Low, S., M. C. Chin et al. (2009). "Review on epidemic of obesity." *Ann Acad Med Singapore* **38**(1): 57–9.

Lusby, P. E., A. Coombes et al. (2002). "Honey: A potent agent for wound healing?" *J Wound Ostomy Continence Nurs* **29**(6): 295–300.

Mafauzy, M. (2006). "Diabetes mellitus in Malaysia." *Med J Malaysia* **61**(4): 397–8.

Mafauzy, M., N. Mokhtar et al. (1999). "Diabetes mellitus and associated cardiovascular risk factors in north-east Malaysia." *Asia Pac J Public Health* **11**(1): 16–9.

McCourt, C. K., D. G. Mutch et al. (2007). "Body mass index: Relationship to clinical, pathologic and features of microsatellite instability in endometrial cancer." *Gynecol Oncol* **104**(3): 535–9.

McLaughlin-Drubin, M. E. and K. Munger (2008). "Viruses associated with human cancer." *Biochim Biophys Acta* **1782**(3): 127–50.

Michail, K., V. Matzi et al. (2007). "Hydroxymethylfurfural: An enemy or a friendly xenobiotic? A bioanalytical approach." *Anal Bioanal Chem* **387**(8): 2801–14.

Miorin, P. L., N. C. Levy Junior et al. (2003). "Antibacterial activity of honey and propolis from *Apis mellifera* and *Tetragonisca angustula* against *Staphylococcus aureus*." *J Appl Microbiol* **95**(5): 913–20.

Moghaddam, A. A., M. Woodward et al. (2007). "Obesity and risk of colorectal cancer: A meta-analysis of 31 studies with 70,000 events." *Cancer Epidemiol Biomarkers Prev* **16**(12): 2533–47.

Mohamed, M., K. Sirajudeen et al. (2009). "Studies on the antioxidant properties of tualang honey of malaysia." *Afr J Tradit Complement Altern Med* **7**(1): 59–63.

Molan, P. C. (2001a). "The potential of honey to promote oral wellness." *Gen Dent* **49**(6): 584–9.

Molan, P. C. (2001b). "Potential of honey in the treatment of wounds and burns." *Am J Clin Dermatol* **2**(1): 13–9.

Moutsatsou, P. (2007). "The spectrum of phytoestrogens in nature: Our knowledge is expanding." *Hormones (Athens)* **6**(3): 173–93.

Mustaffa, B. E. (1990). "Diabetes in Malaysia: Problems and challenges." *Med J Malaysia* **45**(1): 1–7.

Naing, N. N. and Z. Ahmad (2001). "Factors related to smoking habits of male secondary school teachers." *Southeast Asian J Trop Med Public Health* **32**(2): 434–9.

Nazri, S. M., M. K. Imran et al. (2008). "Prevalence of overweight and self-reported chronic diseases among residents in Pulau Kundur, Kelantan, Malaysia." *Southeast Asian J Trop Med Public Health* **39**(1): 162–7.

Norbanee, T. H., M. N. Norhayati et al. (2006). "Prevalence and factors influencing smoking amongst Malay primary school children in Tumpat, Kelantan." *Southeast Asian J Trop Med Public Health* **37**(1): 230–5.

Oddo, L. P., T. A. Heard et al. (2008). "Composition and antioxidant activity of *Trigona carbonaria* honey from Australia." *J Med Food* **11**(4): 789–94.

Othman, N. H., Z. M. Nor et al. (2008). "Is Kelantan joining the global cancer epidemic?—Experience from hospital Universiti Sains Malaysia; 1987–2007." *Asian Pac J Prev* **9**(3): 473–8.

Othman, N. H. and A. A. Zin (2008). "Association of colorectal carcinoma with metabolic diseases; experience with 138 cases from Kelantan, Malaysia." *Asian Pac J Cancer Prev* **9**(4): 747–51.

Parkin, D. M., M. Almonte et al. (2008). "Burden and trends of type-specific human papillomavirus infections and related diseases in the latin america and Caribbean region." *Vaccine* **26**(Suppl 11): L1–15.

Parkin, D. M., Bray, F. I., and S. S. Devesa (2001). "Cancer burden in the year 2000. The global picture." *Eur J Cancer* **37**(Suppl 8): S4–66.

Pate, R., G. Heath et al. (1996). "Associations between physical activity and other health behaviors in a representative sample of U.S. adolescents." *Am J Public Health* **86**(11): 1577–81.

Penedo, F. J. and J. R. Dahn (2005). "Exercise and well-being: A review of mental and physical health benefits associated with physical activity." *Curr Opin Psychiatry* **18**(2): 189–93.

Perez, R. A., M. T. Iglesias et al. (2007). "Amino acid composition and antioxidant capacity of Spanish honeys." *J Agric Food Chem* **55**(2): 360–5.

Pichichero, E., R. Cicconi et al. (2010). "Acacia honey and chrysin reduce proliferation of melanoma cells through alterations in cell cycle progression." *Int J Oncol* **37**(4): 973–81.

Pischon, T., U. Nothlings et al. (2008). "Obesity and cancer." *Proc Nutr Soc* **67**(2): 128–45.

Pyrzynska, K. and M. Biesaga (2009). "Analysis of phenolic acids and flavonoids in honey." *TrAC Trends Anal Chem* **28**(7): 893–902.

Rahmat, A., R. Ismail et al. (2008). "Policy on Tobacco Industry in Malaysia." Retrieved July 13, 2008, from http://www.tobaccoevidence.net/pdf/sea_activities/actionplan_malaysia.pdf.

Rapp, K., J. Klenk et al. (2008). "Weight change and cancer risk in a cohort of more than 65,000 adults in Austria." *Ann Oncol* **19**(4): 641–8.

Rastogi, T., Hildesheim, A., and R. Sinha (2004). "Opportunities for cancer epidemiology in developing countries." *Nat Rev Cancer* **4**(11): 909–17.

Reeves, G. K., K. Pirie et al. (2007). "Cancer incidence and mortality in relation to body mass index in the Million Women Study: Cohort study." *BMJ* **335**(7630): 1134.

Regidor, E., C. Pascual et al. (2007). "Increasing the price of tobacco: Economically regressive today and probably ineffective tomorrow." *Eur J Cancer Prev* **16**(4): 380–4.

Seow, A., S. W. Duffy et al. (1996). "Breast cancer in Singapore: Trends in incidence 1968–1992." *Int J Epidemiol* **25**(1): 40–5.

Seow, A., J. M. Yuan et al. (2006). "Diabetes mellitus and risk of colorectal cancer in the Singapore Chinese Health Study." *J Natl Cancer Inst* **98**(2): 135–8.

Shamsuddin, K. and M. A. Haris (2000). "Family influence on current smoking habits among secondary school children in Kota Bharu, Kelantan." *Singapore Med J* **41**(4): 167–71.

Sharma, V., V. S. Chauhan et al. (2007). "Role of bile bacteria in gallbladder carcinoma." *Hepatogastroenterology* **54**(78): 1622–5.

Sharp, A. (2009). "Beneficial effects of honey dressings in wound management." *Nurs Stand* **24**(7): 66–8, 70, 72 passim.

Siddique, K., S. Bhandari et al. (2010). "Epstein-Barr virus (EBV) positive anal B cell lymphoma: A case report and review of literature." *Ann R Coll Surg Engl* **92**(3): W7–9.

Sirichotiratana, N., S. Sovann et al. (2008). "Linking data to tobacco control program action among students aged 13–15 in Association of Southeast Asian Nations (ASEAN) member states, 2000–2006." *Tobacco Control* **6**(17): 372–8.

Sirichotiratana, N., C. Techatraisakdi et al. (2008). "Prevalence of smoking and other smoking-related behaviors reported by the Global Youth Tobacco Survey (GYTS) in Thailand." *BMC Public Health* **8**(Suppl 1): S3.

Subrahmanyam, M. (1993). "Honey impregnated gauze versus polyurethane film (OpSite) in the treatment of burns—A prospective randomised study." *Br J Plast Surg* **46**(4): 322–3.

Swellam, T., N. Miyanaga et al. (2003). "Antineoplastic activity of honey in an experimental bladder cancer implantation model: *In vivo* and *in vitro* studies." *Int J Urol* **10**(4): 213–9.

Tan, H. T., R. A. Rahman et al. (2009). "The antibacterial properties of Malaysian tualang honey against wound and enteric microorganisms in comparison to manuka honey." *BMC Complement Altern Med* **9**: 34.

Timm, M., S. Bartelt et al. (2008). "Immunomodulatory effects of honey cannot be distinguished from endotoxin." *Cytokine* **42**(1): 113–20.

Toh, C. K. (2009). "The changing epidemiology of lung cancer." *Methods Mol Biol* **472**: 397–411.

Tonks, A., R. A. Cooper et al. (2001). "Stimulation of TNF-alpha release in monocytes by honey." *Cytokine* **14**(4): 240–2.

Tonks, A. J., R. A. Cooper et al. (2003). "Honey stimulates inflammatory cytokine production from monocytes." *Cytokine* **21**(5): 242–7.

Tonks, A. J., E. Dudley et al. (2007). "A 5.8-kDa component of manuka honey stimulates immune cells via TLR4." *J Leukoc Biol* **82**(5): 1147–55.

Valko, M., D. Leibfritz et al. (2007). "Free radicals and antioxidants in normal physiological functions and human disease." *Int J Biochem Cell Biol* **39**(1): 44–84.

Viuda-Martos, M., Y. Ruiz-Navajas et al. (2008). "Functional properties of honey, propolis, and royal jelly." *J Food Sci* **73**(9): R117–24.

Von Low, E. C., F. G. Perabo et al. (2007). "Review. Facts and fiction of phytotherapy for prostate cancer: A critical assessment of preclinical and clinical data." *In Vivo* **21**(2): 189–204.

Wai, J. P., C. P. Wen et al. (2008). "Assessing physical activity in an Asian country: Low energy expenditure and exercise frequency among adults in Taiwan." *Asia Pac J Clin Nutr* **17**(2): 297–308.

Warren, C. W., N. R. Jones et al. (2008). "Global youth tobacco surveillance, 2000–2007." *MMWR Surveill Summ* **57**(1): 1–28.

Westlake, S. and N. Cooper (2008). "Cancer incidence and mortality: Trends in the United Kingdom and constituent countries, 1993 to 2004." *Health Stat Q* (38): 33–46.

Wijesinghe, M., M. Weatherall et al. (2009). "Honey in the treatment of burns: A systematic review and meta-analysis of its efficacy." *N Z Med J* **122**(1295): 47–60.

Woo, K. J., Y. J. Jeong et al. (2004). "Chrysin-induced apoptosis is mediated through caspase activation and Akt inactivation in U937 leukemia cells." *Biochem Biophys Res Commun* **325**(4): 1215–22.

Yaghoobi, N., N. Al-Waili et al. (2008). "Natural honey and cardiovascular risk factors; effects on blood glucose, cholesterol, triacylglycerole, CRP, and body weight compared with sucrose." *ScientificWorld Journal* **8**: 463–9.

Yang, Y. X., S. Hennessy et al. (2005). "Type 2 diabetes mellitus and the risk of colorectal cancer." *Clin Gastroenterol Hepatol* **3**(6): 587–94.

Yao, L., N. Datta et al. (2003). "Flavonoids, phenolic acids and abscisic acid in Australian and New Zealand Leptospermum honeys." *Food Chem* **81**(2): 159–168.

Yeole, B. B. (2008). "Trends in cancer incidence in female breast, cervix uteri, corpus uteri, and ovary in India." *Asian Pac J Cancer Prev* **9**(1): 119–22.

Yuen, M. F., J. L. Hou et al. (2009). "Hepatocellular carcinoma in the Asia pacific region." *J Gastroenterol Hepatol* **24**(3): 346–53.

Zaid, S. S., S. A. Sulaiman et al. (2010). "The effects of Tualang honey on female reproductive organs, tibia bone and hormonal profile in ovariectomised rats—Animal model for menopause." *BMC Complement Altern Med* **10**: 82.

Zaini, A. (2000). "Where is Malaysia in the midst of the Asian epidemic of diabetes mellitus?" *Diabetes Res Clin Pract* **50**(Suppl 2): S23–8.

# 13 Honey and Bee Products for Animal Health and Wellness

*Laïd Boukraâ*

## CONTENTS

## CONCEPT OF COMPLEMENTARY AND ALTERNATIVE VETERINARY MEDICINE

Complementary and alternative veterinary medicine is an inclusive term that describes treatments, therapies, and modalities that are not accepted as components of mainstream veterinary education or practice but that are performed on animals by some practitioners. Although these treatments and therapies often form part of veterinary postgraduate education, study, and writing, they are generally viewed as alternatives or complementary to more universally accepted treatments. Therefore, performing alternative and complementary veterinary treatments on animals constitutes the practice of veterinary medicine such that these procedures may only be performed by a veterinarian or by a nonveterinarian who is directed and supervised by the veterinarian, within the context of a valid veterinarian–client–patient relationship. The use of apitherapy in animals is not as common as in humans. Ancient civilizations used bee products for animals too, but modern civilization and "education" have seriously lessened our natural instinctive ability and capability.

Having said that, and despite the fact that the modern Western establishment appears to like to relegate apitherapy to the status of "folklore" or "old wives' tales," bee products contain a vast spread of pharmacologically active ingredients and each one has its own unique combination and properties. They are classified in modern herbal medicine according to their spheres of action. Recognized actions include antihelminthic, anticatarrhal, antiemetic, anti-inflammatory, antioxidant, antibacterial, antifungal, antispasmodic, aperient/laxative, aromatic, astringent, diuretic, emollient, expectorant, sedative, stimulant, and tonic.

At present, there is an "Us and Them" division in veterinary medicine between the so-called conventional veterinary surgeons and those who choose to use complementary therapies. Holism is frequently used as a dustbin term that denotes a wholly unscientific, unproven, and, in some people's eyes, totally unethical approach to medicine. There is an assumption by many that a veterinary surgeon who uses complementary therapies and a holistic approach has obliterated the years of training that they received at veterinary school, and suddenly, their qualifications in turn become downgraded. A person previously thought of as an intellectual or at least a colleague is often transformed in the minds of some purely conventional veterinary surgeons into a lost cause of emotions, gullibility, and someone unable to cope with the rigors of modern veterinary practice. This judgment has been exacerbated by the rapid increase in the numbers of laypeople who are advertising themselves as complementary practitioners and who are publishing information to the public via multiple media channels. The level of knowledge across the modalities in the lay sector varies enormously and is largely unregulated. Despite the law putting animal health firmly in the hands of qualified veterinary surgeons, the pet-owning public is using these people to get what they consider to be the medicine of choice. The results are of course as variable as the standards of the practitioners. For many people, the quest for alternative therapies is born out of frustration and dissatisfaction. To label a disease but be no further on in curing it is a powerful source of frustration and questioning. Long-term chronic ill health, in particular, the prevalence of diseases such as arthritis, atopy, and colitis, which are never cured but held at "acceptable" levels only with long-term drug therapy, is another trigger for questioning the validity of our understanding of the disease processes. Long-term medication is a failure not a cure. Then, there is the evaluation of the quality of life for those animals that are on long-term medication (Gerber 1998).

Certainly, for many, the choice between the disease effects and the medication side effects comes out on the side of the medication, but sometimes the line is very fine. Despite all of the frustrations though, many continue working within the limitations of the current system, without questioning. For others, it is the beginning of the search. It is the beginning of opening the mind to accepting that there may be other ways to see a problem and to solve it. Many of the problems that arise from the use of complementary therapies can be traced back to a lack of knowledge and understanding on the part of the person using the therapy. To learn and understand all of the currently available complementary therapies requires an enormous investment from the veterinary surgeon in time and in money. But if they are to be used correctly and safely, they must be used with knowledge and understanding to a level akin to that of our conventional veterinary medicine. Lack of knowledge is not an excuse to do nothing. An inability to come to terms with the philosophy of a modality is also not enough of a reason to stop wanting to push out beyond the limitations of Western conventional medicine. Equally embracing the whole concept of holism and complementary therapies does not give cause to comprehensively reject everything that conventional medicine has taught us (Schoen and Wynn 1998).

We must aim for an integrated system and not an "Us and Them" system. Once we recognize that a living being is more than simply the sum of its parts and is an energy system, we cannot look only to conventional reductionist veterinary diagnostics or therapeutics. A comprehensive holistic approach has to be the way forward. We could finally conclude that there cannot be two kinds of medicines: conventional and alternative. There is only medicine that has been adequately tested and medicine that has not, medicine that works, and medicine that may or may not work (Angell and Kassirer 1998).

## BEE PRODUCTS FOR ANIMAL HEALTH

Apiproducts are animal products; as such, animals may draw profit from honey, propolis, royal jelly, bee venom, and beeswax as humans do. These products are used for animals as medicines and as cosmetics as well. Most animal diseases resemble human diseases due to similarities in physiology and anatomy. Furthermore, both are facing similar threats and pathogens. The sterility and the cleanliness of the hive are good examples for the role of apiproducts in maintaining it in such condition; bees are using their own products to protect the hive against their enemies. Honey is known for its role in wound care for most animals; maybe horses are more exposed to injuries and therefore more treated with honey. Mastitis in dairy cows and goats is an expensive and difficult pathologic condition to treat where honey and propolis extract could be the alternative. Mastitis is an inflammation of the mammary gland that often develops in response to intramammary infection, which is usually due to antibiotic-resistant bacteria. This is an important public health problem in humans in developing countries as well as in domesticated animals used in agriculture worldwide. The mammary gland of dairy cows offers the ideal environment for the optimal action of honey. The standard treatment is the introduction of antibiotics into the teat canal of the infected udder, but milk has to be withheld from use until clear of antibiotic residues. Honey could possibly be suitable for the treatment of mastitis if inserted into the infected udder via the teat canal, as it is harmless to tissues and would leave no undesirable residues in milk. Investigation is carried out to test the potential role of honey to conserve the animal semen. Its antibacterial properties and high energetic content make it a suitable candidate for such a role. Apiproducts are also used for the treatment of other animal illnesses such as gastrointestinal disorders, otitis, sinusitis, ophthalmic, dermatology, and skin care. Royal jelly has been tested for its role in reproductive performance in sheep. The results demonstrate that royal jelly treatment in conjunction with a source of exogenous progesterone can be used to induce estrus and increase first service conception rate in sheep. It may be also effective in improving pregnancy and lambing rates. Many animals can potentially benefit from bee propolis, from pets to livestock. It is often used for promoting the health of the respiratory and immune systems to more specific ailments where its anti-bacterial properties may be helpful. Apiproducts mixed with other natural products are found in the market, honey with essential oils being the most common. Bee products are also used as food supplement for animals, bee pollen being the most used additive in animal foods. It stimulates the growth, decreases the rate of mortality, and prevents morbidity. The present chapter reviews the use of bee products in animal health and their well-being.

## HOW CAN PRODUCTS FROM BEES HELP PROMOTE ANIMAL HEALTH?

### HONEY

### Honey for Wounds

Honey can be applied to wounds to promote healing and be also added to the diet in order to promote health. Its health-promoting properties are thought to assist with the immune system, including animals with seasonal allergies. Many animals, from horses to dogs plus other pets or livestock, can potentially benefit from the health-promoting benefits of honey. A comparative study involving 15 adult, mongrel dogs aged between 4 and 5 years old was used for burn management with honey and silver sulfadiazine. Burn wounds were performed on the dorsolateral region. Reepithelialization was more pronounced in dogs treated with honey, and collagen fibers were also more orderly arranged (Jalali et al. 2007). Conditions often treated with honey, in dogs, cats, horses, and other animals, sometimes in conjunction with other therapies, are laminitis, digestive disturbance, diarrhea, nervousness, arthritis, liver problems (hepatopathy), sinusitis, chronic cough, skin problems, respiratory problems, heart problems, hoof quality (hoof health), and kidney problems.

### Some Clinical Cases with Dogs

Rooster et al. (2008) have discussed the results of the use of honey dressings for dogs with skin wounds on the basis of a few clinical cases. They have reported that the wound healing occurs quickly, that resistant bacterial infections are controlled successfully, that necrotic wounds require less surgical intervention, and that honey dressing and redressing are very well tolerated by patients. Case 1 was a 12-year-old Mechlin Shepherd presented with extensive myiasis. After conventional treatment failed to heal the skin disease, a honey treatment was started; the back was completely covered with compresses drenched in culinary honey starting day 3. When the dog was presented for follow-up 5 days later, all superficial injuries were pretty much completely healed, while the deep injuries had become smaller and were filled with healthy granulation tissue. Case 2 was a 9-year-old shar-pei with a bad general condition and was referred for ulcerations on the tongue and a marked swelling on the medial side of the left heel with discoloration of the skin. The only treatment applied was a daily honey dressing with culinary honey. The necrotic tissue had rapidly demarcated and separated, and on day 7, a clear granulation layer was visible. Six weeks later, the defect was nearly completely healed with relatively little scarring. For case 3, a 9-month-old boxer was burned on all four paws when playfully biting through a plastic bottle of toilet clog remover on the basis of sodium hydroxide in powdered form. The dog showed an exudative, foul smelling necrosis on all four paws with loss of digital and interdigital skin and exposure of tendons and defects in the large foot pads. On the wounds, fluid culinary honey was applied and covered with gauze compresses, and on top of that, a covering bandage was applied. The bandages were changed once daily. In the beginning, the animal showed pain after the application of the honey for a short period of time. After 7 days, the necrotic tissue as well as the odor had disappeared and a clear granulation layer was present at the level of the paws. Case 4 was a 9.5-year-old male intact Berner Sennen that underwent a celiotomy and a marsupialization of the prostate at the treating veterinarian's clinic. The dog recovered well but paprapreputially a fistula developed, probably as a result of the initially performed marsupialization. Despite the initiated intravenous broad-spectrum antibiotic therapy (enrofloxacin combined with amoxicillin), a positive culture with deep sampling in the fistula was obtained 5 days after the last surgical intervention. A resistant *Enterobacter cloacae* was isolated, which was only sensitive to amikacin and chloramphenicol. Because the general condition of the dog was very satisfying, the antibiotic therapy was not adjusted, but a topical therapy was initiated. A 1-mL syringe was filled with therapeutic honey ointment and used as an applicator in order to apply the honey ointment into the fistula as deeply as possible. During the first week, the drainage from the fistula opening was reduced drastically. A new bacteriological study was done at which no *E. cloacae* were observed any more. It took approximately 1 month before the fistulation had disappeared completely.

For as long as people can remember, honey has been applied to wounds in humans and animals for its healing properties. The cases that have been discussed are only a fraction of the numerous applications of honey dressings. Although in theory there is a clear distinction between therapeutic and culinary honey, the authors have clinically established that very favorable results can be obtained with either type of honey. Also, Mathews and Binnington (2002) reported similar findings about the use of unpasteurized honey in pets and a lot of clinical successes have also been described in human medicine after the use of honey from the food chain as wound dressings (Enzlin 2001; Molan 2002a).

In humans, foul smelling wounds are a frustrating problem for the patient (Lee et al. 2007). Although animals are less bothered by this, processes with extensive necrosis will, among others, spread a bothersome odor for the owners. In cases 3 and 4, the deodorizing effect of the treatment with honey ointment was enormously appreciated by the owners.

Myiasis can take on extreme proportions and lead to horrible wounds. The maggots and their breakdown products cause a marked inflammatory cascade (Anderson and Huitson 2004). In case 1, the anti-inflammatory properties of the topical use of honey were already evident at the first change of the dressings. In the small pet sector, there are innumerable indications for the use of honey or honey preparations as wound covering. The cases that have been discussed here show that effectively a very good healing of the wound can be obtained and that such treatment of skin wounds is evidently more efficient and cheaper than other classic ointments and/or wound dressings.

Honey is also used for wound management in horses. According to a research team from the University of Sydney (Australia), a simple application of honey to horses' leg wounds results in smaller wound sizes and faster healing time. Applying a honey gel throughout healing led to 27% faster healing times (Bischofberger et al. 2011).

### Honey for Mastitis Management

Mastitis is an inflammation of the mammary gland that often develops in response to intramammary bacterial infection (Schalm et al. 1971). It is an important public health problem in humans in developing countries as well as in domesticated animals used in agriculture worldwide. It remains one of the most costly diseases in dairy cows (Hortet and Seegers 1998). According to laboratory studies done by Professor Molan from Waikato University (New Zealand), various bacteria that cause mastitis in dairy cattle are sensitive to honey. Honey could possibly be suitable for the treatment of mastitis if inserted into the infected udder via the teat canal, as it is harmless to tissues and would leave no undesirable residues in milk. If the same level of effectiveness is demonstrated when animals with mastitis are treated with honey, then it could be used as a more consumer-friendly alternative to antibiotics (Molan 2002b) (Table 13.1).

---

### TABLE 13.1
### Minimum Inhibitory Concentration of Honeys (% v/v in Nutrient Agar) for Cultures of Various Mastitis-Causing Bacteria Streaked on the Agar Plates

| Bacterial Species | Manuka Honey (%) | Rewarewa Honey (%) | Artificial Honey (%) |
|---|---|---|---|
| *Actinomyces pyogenes* | 1–5 | 1–5 | 5–10 |
| *Klebsiella pneumoniae* | 5–10 | 5–10 | >10 |
| *Nocardia asteroides* | 1–5 | 5–10 | >10 |
| *Staphylococcus aureus* | 1–5 | 1–5 | >10 |
| *Streptococcus agalactiae* | 1–5 | 5–10 | >10 |
| *Streptococcus dysgalactiae* | 1–5 | 5–10 | >10 |
| *Streptococcus uberis* | 1–5 | 5–10 | >10 |

*Source:* Molan, P.C., *Honey as an Antimicrobial Agent*, Waikato Honey Research Unit, University of Waikato. http://honey.bio.waikato.ac.nz/selection.shtml, 2002.

## BEE POLLEN

Bee pollen is different from the pollen we see floating in the air from flowers (and which frequently triggers hay fever); instead, bee pollen is the result of the process that the bees use when collecting pollen from flowers. The bees collect pollen in special pollen baskets and do so in such a way that makes bee pollen a unique product, which cannot successfully be identically replicated artificially. Bee pollen is even thought by some medical experts and scientists to be a complete food and with just adding water would enable your animal to be healthy. Generally, bee pollen is given to animals to help promote immunity, for digestive health, to provide energy, and as an antibiotic. It is also often used to help counter the effects of pollen allergies. Bee pollen can usually be found either in tablet form or as granules. Pollen stimulates ovarian function. The best results were obtained with a pollen supplementation of 2 parts per 100 in the ration and with the substitution of animal proteins with pollen in a proportion of 5 parts per 100; the intensity of ovulation is increased. Parallel to this increase in ovulation, pollen also improves the ability of eggs to withstand the incubation period. Male rabbits given bee pollen had significantly better fertility, sperm quality, and blood profiles.

In addition, baby rabbits given bee pollen had improved body weight and a better survival rating. Furthermore, the use of bee pollen gave the animals increased vitality and improved "powers of reproduction" because of boosted fertility. It has been shown that chickens fed with bee pollen lead to a better development of the small intestine villi from the duodenum, jejunum, and ileum. These findings suggest that bee pollen could promote the early development of the digestive system.

There is some evidence that ingested pollen can protect animals as well as humans against the adverse effects of x-ray radiation treatments (Wang and Zhang 1988; Schmidt and Buchmann 1992). Pollen has been added to diets for domestic animals and laboratory insects resulting in improvements of health, growth, and food conversion rates (Crane 1990; Schmidt and Buchmann 1992). Chickens exhibited improved food conversion efficiency with the addition of only 2.5% pollen to a balanced diet (Costantini and Ricciardelli d'Albore 1971) as did piglets (Schmidt and Buchmann 1992). Beekeepers, too, feed their colonies with pure pollen, pollen supplements, or pollen substitutes, during periods with limited natural pollen sources.

## BEE PROPOLIS

A resinous substance secreted by trees is collected by bees and, using their unique processes, is turned into a substance known as propolis. This is used by the bees to line the hive and so protect it against germs. Beehives are considered to be sterile environments because of the presence of propolis. Propolis has been shown to kill the bee's most ardent bacterial foe, *Penibacillus larvae*, the cause of American foulbrood (Meresta and Meresta 1988). It has been used for thousands of years to help promote the health of both people and animals, often due to its natural antibacterial properties. Many animals can potentially benefit from bee propolis—from pets to livestock. It is often used for promoting the health of the respiratory and immune systems and to more specific ailments where its antibacterial properties may be helpful. It is difficult to determine just what comprises bee propolis. Aside from having a very complex chemical structure with more than 200 components, bee propolis varies in composition depending on where it was obtained, the season in which it was manufactured, and what vegetation, trees, and flowers are nearby. However, bee propolis is often rich in vitamins B-complex, C, and E and pro-vitamin A. It also has minerals such as calcium, iron, zinc, silica, potassium, copper, and magnesium, among other things. Bee propolis has been shown to be a good natural supplement that is great at preventing cancer, inflammation, and tumors even in animals and pets. It hinders the production of T cells, which is responsible for a lot of inflammatory diseases. These properties make it an effective safeguard from cancer, tumors, and inflammation. Bee propolis as a natural medication and is safe for your pets. There are no reported side effects aside from allergies (usually acquired by people who are also allergic to bees and bee stings). Bee propolis can be found as a supplement in tablet form as well as being included in other products—from toothpastes to skincare lotions.

A study was conducted to evaluate the effect of propolis on mortality of fish eggs due to mycosis, to study its efficacy on the fish growth rate, and to analyze the histochemical and ultrastructural characteristics of muscle fibers. The animals fed with propolis showed a more rapid muscular growth compared to a control fed with the standard diet (Velotto et al. 2010).

The effect of propolis extract was also monitored on selected parameters of mineral profiles (calcium, phosphorus, magnesium, potassium, sodium, and chlorides) of Hubbard JV chickens. Propolis preparation caused a significant ($P < 0.05$) decrease in serum phosphorus and magnesium content. Probably, propolis inclusion in the diet increases absorption of phosphorus and magnesium from the blood to the bone and thus decreased the level of these elements in the blood (Petruška et al. 2012). Propolis may also influence immunity of animals. The inclusion of propolis may stimulate IgG and IgM production of laying hens and could be an important factor in immunostimulation of laying hens (Çetin et al. 2010).

The behavior and the productive performance of lambs finished in feedlot receiving diets added with green propolis, brown propolis, or monensin sodium were assessed in a comparative study. The green propolis diet decreased rumination and increased resting time. The diets provided similar feeding rates (g/min). Technically, brown propolis can substitute monensin sodium as a dietary additive for feedlot lambs (Ítavoa et al. 2011).

The effect of propolis intake on the weight gain, the development rate, and productivity of different animals has been studied, the intake being about 10 mL/kg. The following results have been noticed: weight gain, increased rate in development of animals, and productivity and improvement of meat quality (Bogdanov 2011). Table 13.2 summarizes the use of propolis for different ailments in animals.

Water preparation of propolis (WPP) has been used to treat the Cushing's syndrome in dogs (a chronic hyperactivity of the adrenals that could lead to a tumor on either the adrenal or pituitary glands). There is no known cure for Cushing's syndrome and the accepted treatments are expensive and often ultimately ineffective and can have serious to lethal side effects. The treatment consisted of oral administration of WPP at a daily dosage equivalent to 0.4 to 0.5 g of crude propolis per

## TABLE 13.2
### Treatment of Some Animal Pathologic Conditions with Propolis

| Propolis Preparation | Indication |
|---|---|
| Propolis liniment | Mastitis |
| Propolis candles | Gynecologic diseases |
| 0.5% propolis in milk | Gastrointestinal and respiratory diseases |
| 5 mL of 20% ethanol extract | Diarrhea in mild-fed calves |
| 50% aqueous extract | Paratyphoid fever of ducks |
| 5% propolis in fish oil or fat | Wound healing |
| 1%–10% propolis extract preparation (PEP) | Local anesthetic in surgery |
| Propolis extract | Foot-and-mouth disease |
| Propolis extract | Enzootic pneumonia of pigs |
| Propolis extract | Growth stimulation |

*Source:* Bogdanov, S., *Propolis: Biological Properties and Medical Applications. The Propolis Book, Chapter 2*, Bee Product Science, www.bee-hexagon.net, 2011.

kilogram body weight, dosed twice daily, 12 h apart. A potential role of propolis to heal Cushing's syndrome was noticed after 3 months of treatment (Glenn 2006).

Ghisalberti (1979) reports additional weight gains for broiler chickens of up to 20% when 500 ppm propolis was added to their diets. In Japan, the addition of only 30 ppm of propolis to the rations of laying hens increased egg production, food conversion, and hen weight by 5% to 6% (Bonomi et al. 1976).

## ROYAL JELLY

Royal jelly is a secretion produced by worker bees and is effectively a "superfood," as it is only fed to the queen bee. It is also understood to contain several vitamins and amino acids and has been used for hundreds of years to promote health. It is often used to help combat fatigue and to promote skin health. We can usually find it as a supplement in capsule form or in skincare preparations. Researchers from the Jordan University of Science and Technology studied the effects of royal jelly compared with equine human chorionic gonadotropin or placebo on sheep pregnancy rates and lambing rates, which were higher in the groups given royal jelly and eCG than in the placebo groups (Kridli and El-Khetib 2006).

It also has a variety of pharmacologic functions such as maintenance of bone mass, cardiovascular protection, and brain protection. Royal jelly is widely thought to improve menopausal symptoms (Mishima et al. 2005). Aging is usually defined as the progressive loss of function accompanied by decreasing fertility and increasing mortality with advancing age. It is suggested that royal jelly contains growth factors or hormones promoting cell growth (Salazar-Olivo and Paz-González 2005).

*Trans*-2-octenoic acid and *trans*-10-hydroxy-2-decenoic acid were involved in the blood pressure regulation. However, the in vivo hypotensive effect of these unsaturated fatty acids would be in doubt due to their gastrointestinal instability when taken orally (Matsui et al. 2002). It is known that royal jelly can reduce blood plasma levels of cholesterol and triglycerides and arterial cholesterol deposits in rabbits when these disorders were induced experimentally (Nakajin et al. 1982).

Tumor growth inhibition in mice with prophylactic and therapeutic oral administration of royal jelly has been reported. However, inhibition of rapid-growth cancers (leukemia) was insignificant (Tamura et al. 1987). *Trans*-10-hydroxy-2-decenoic acid has been found to have an effect on vascular endothelial growth factor–induced angiogenesis, partly by inhibiting both cell proliferation and migration. Royal jelly also shortened the healing period of desquamated skin lesions, possesses an anti-inflammatory action, and is able to augment wound healing. The healing of skin lesions was accelerated and anti-inflammatory action was shown for rats (Fuji et al. 1990). The royal jelly (3%, 10%, and 30%) ointments significantly improved the recovery from 5-fluorouracil–induced damage in a dose-dependent manner. These results suggest the possibility that the topical application of royal jelly has a healing effect on severe oral mucositis induced by chemotherapy (Suemaru et al. 2008).

The protein fractions in royal jelly have high antioxidative activity and scavenging ability against active oxygen species. Royal jelly seems applicable in both health food and medicine (Nagai and Inoue 2004). It shows some anti-inflammatory activity by decreasing exudation and collagen formation in granulation tissue formation. It also shortens the healing period of desquamated skin lesions. Research results indicate that MRJP3 (major royal jelly protein) can exhibit potent immunoregulatory effects in vitro and in vivo (Okamoto et al. 2003).

Combined with honey, royal jelly has exhibited a synergistic effect against pathogenic bacteria (Boukraa 2008). Therefore, royal jelly, the life prolonging food of queen bees, can help your pet enjoy a longer, healthier life. Because animals have more free radical protection than bees, royal jelly will not multiply a pet's lifespan, but it will minimize the routine cellular stress that causes aging and disease. Eating too much and being overweight can also cause oxidative damage in your pet's body, not to mention extra strain in the joints. We might consider giving royal jelly to our pet for obesity, arthritis, osteoarthritis, congenital joint disorders, or just old age. If you know that your large dog or other pet is predisposed to arthritis and joint problems, early free radical protection from royal jelly may slow the development of the disease and reduce its severity. Despite the fact that royal jelly has many healing properties, its use encounters some limitations. There exist several reports showing acute bronchial asthma, anaphylaxis, and hemorrhagic colitis caused by royal jelly. However, because the precise mechanisms for royal jelly–induced anaphylaxis are still undefined, it is necessary to determine the antigens causing anaphylaxis and to clarify the cross-reactivity with other foods and drugs. The major allergenic proteins of royal jelly are MRJP1 and MRJP2. Royal jelly could prove harmful for those allergic to bee stings, honey, or pollen. Its topical use may cause dermatitis (Shahnaz et al. 2007). Other adverse reactions to royal jelly have included eczema, rhinitis, and bronchospasms, and owners of pregnant animals and nursing mothers are advised to avoid giving royal jelly to the litter. The incidence of allergic side effect in animals that consume royal jelly is unknown; however, it has been suggested that the risk of having an allergy to royal jelly is higher in animals that already have known allergies. The problem of antibiotic residues in royal jelly should also be mentioned. Analyzed samples of royal jelly were found to contain chloramphenicol in many European countries. Chloramphenicol is not authorized for use in the treatment of animals for food production in the European Union and other countries. Chloramphenicol is considered to be a potential genotoxic carcinogen. Medical use of chloramphenicol has also been associated with a rare aplastic anemia, which can be fatal. A minimum required performance limit of 0.3 µg/kg has been set for chloramphenicol (Sabatini et al. 2009). Royal jelly has a limited shelf-life. Refrigeration and freezing delay and reduce the chemical changes. Although freeze-dried jelly is the most stable form of royal jelly, some changes still take place. Refrigeration of royal jelly at 0°C to 5°C is a minimum precaution. Still better is storage, whenever possible, at temperatures below −17°C, which is attainable in most household freezers. The average recommended storage time after production is 18 months under refrigeration. For products stored at −17°C, storage can be extended to 24 months. Freeze-dried royal jelly and royal jelly–based products are generally stored at room temperature, sometimes for several years (Takenaka et al. 1986).

## Bee Venom

Bee venom has been shown to contain several biochemical or pharmacologically active substances including polypeptides (melittin, apamin, and mast cell degranulating peptide), amines (histamine, serotonin, dopamine, and norepinephrine), and enzymes (phospholipase, hyaluronidase, and histidine decarboxylase) (Lariviere and Melzack 1996). These substances were claimed to directly or indirectly express its potency and medical efficacy. Bee venom has been suggested as an effective healing agent for alleviating persistent pain and treating several ailments including different rheumatic disorders involving inflammation and degeneration of connective tissue (different types of arthritis) (Wang et al. 2001), neurologic disorder (Roh et al. 2004), related-immune syndrome (multiple sclerosis) (Hauser et al. 2001), and dermatologic conditions (eczema, psoriasis, and herpesvirus infections) (Kim 2004). Bee venom was revealed to be effective in the reduction of tumors of many different types of malignant diseases (Liu et al. 2002; Oršolić et al. 2003), as it was found to stimulate natural immunity through activation of the pituitary and adrenal glands and to stimulate the body to produce natural cortisone (Nermine and Abeer 2009). Among the several bee products, bee venom has been highlighted due to its chemical composition (Qiu et al. 2011). Bee venom is toxic and contains 88% water and only 0.1 µg dry venom. The dry venom is known to be a very complex mixture of peptides.

Honey VENZTM (a mixture of honey and bee venom from New Zealand) for pets is a proven anti-inflammatory product that eases arthritic pain in pets—including dogs, cats, and even thoroughbred race horses. It is good for pets suffering from joint and muscle diseases and functional disorders of muscles and chords. Bee venom honey provides a very strong pain relief and anti-inflammatory effect. Taken regularly, over time, it helps minimize joint discomfort and helps maintain and restore joint mobility. Concentration of bee venom in a 500 g jar of honey is 20 ppm or 10 mg. This figure equals half a sting (0.5 sting) in a teaspoon of honey.

A study with whole bee venom on arthritic dogs suffering from hip dysplasia (Vick and Brooks 1972) and rats (Dunn 1984) showed that cage activity of the

arthritis dogs increased by as much as 70% following the therapy. The blood monitoring indicated that melittin and apamin stimulated the production of the plasma
cortisol. Although the promise of bee venom application in alternative medicine for
the evidences discussed above, the use of bee venom in conventional medicine has
languished, beyond to the foregoing bee venom therapy is often painful. Human and
animal toxicity cases have been shown for bee venom following bee stings (Sánchez-
Velasco et al. 2005). Understandably, there is no doubt that bee venom and its chemical components are not without cytotoxic effects. Bee stings are indeed unsafe for
allergic people. A recent study revealed that about 1% to 5% of the people worldwide
are hypersensitive to bee venom or other insects such as wasps and hornets (Müller
2011). In addition, it has been referred that both hyaluronidase and phospholipase A2
are the main allergens because they can cause pathogenic reactions in the majority
of patients susceptible to bee venom, where it has been observed that 71% of patients
had specific serum IgE to recombinant hyaluronidase and 78% to recombinant phospholipase A2 (Stuhlmeier 2007).

## RECOMMENDATIONS

It is important to note that, although one bee product may work for one animal, it may
not work for the next. Each animal is different in how it will respond to medication
and other supplementation. It can also take varying amounts of time for your animal
to notice a difference; some animals may respond to a bee supplement or skincare
product within a couple of days, but other animals may take much longer to show
any improvement. Overall, in choosing your bee product, do try to choose a product
that is as pure and natural as possible to help gain the maximum potential from what
are intended to be natural products. As with all complementary therapies, do seek
the advice of your vet first before using a bee product, especially if your animal is on
existing medication; although the products are generally safe, it is always best to be
100% sure there will not be any compatibility issues. Some people and animals may
be allergic to bee products, especially likely with those who are allergic to bee stings.
If in doubt, do consult your veterinary surgeon for advice before using the products.
It may also be helpful to introduce the bee product to your animal's diet gradually,
starting with small amounts; your veterinary surgeon will be able to advise you on
this. Lastly, when you regularly buy your natural bee honey, bee pollen, bee propolis,
or royal jelly, you may find that there are variations in the color of the products. This
should not be anything to worry about, as because of the fact that they are natural
products, it is quite usual for them to have inconsistency in color (e.g., sometimes they
may be light and other times dark) and be not manufactured by nature to be identical.

## REFERENCES

Anderson G and Huitson NR. 2004. Myiasis in pet animals in British Columbia: The potential
    of forensic entomology for determining duration of possible neglect. *Can Vet J*, 45:
    993–998.
Angell M and Kassirer JP. 1998. Alternative medicine—The risks of untested and unregulated
    remedies. *N Engl J Med*, 339(12): 839–841.

Bischofberger AS, Dart CM, Perkins NR, and Dart AJ. 2011. A preliminary study on the effect of manuka honey on second-intention healing of contaminated wounds on the distal aspect of the forelimbs of horses. *Vet Surg*, 40(7): 898–902.

Bogdanov S. Propolis: Biological properties and medical applications. The Propolis Book, Chapter 2. Bee Product Science, www.bee-hexagon.net, retrieved on March 30, 2011.

Bonomi A, Marletto F, and Bianchi M. 1976. Use of propolis in the food of laying hens. *Revista di Avicultura*, 45(4): 43–55.

Boukraa L. 2008. Additive activity of royal jelly and honey against Pseudomonas aeruginosa. *Altern Med Rev*, 13(4): 330–333.

Çetin E, Silici S, Çetin N, and Guçlu BK. 2010. Effects of diets containing different concentrations of propolis on hematological and immunological variables in laying hens. In *Poultry Science*, vol. 89. ISSN 1525-3171. Collins English Dictionary: Fourth Edition 1998, ISBN 0 00 472219 1, Harper Collins Publishers, London, 1703–1708.

Costantini F and Ricciardelli d'Albore G. 1971. Pollen as an additive to the chicken diet. Proc. 23rd International Apicultural Congress, *Apimondia*, 539–542.

Crane E. 1990. *Bees and Beekeeping: Science, Practice and World Resources*. Cornstock Publ., Ithaca, NY, 593 pp.

Dunn JD. 1984. The effect of bee venom on plasma corticosterone levels. *Neuroendocrinol Lett*, 6(5): 273–277.

Enzlin M. 2001. Honing als natuurlijke en superieure wondgenezer. *Verpleegkunde Nieuws*, 11: 1–4.

Fuji A, Kobayashi S, Ishihama S, Yamamoto H, and Tamura T. 1990. Augmentation of wound healing by royal jelly. *Jpn J Pharmacol*, 53(3): 331–337.

Gerber R. *Vibrational Medicine*. Santa Fe, NM, Bear & Company, 1996, ISBN 1-8791-8128-2, 601 pp.

Ghisalberti E. 1979. Propolis: A review. *Bee World*, 60: 59–84.

Glenn P. 2006. *Apimedica*, Athens, Greece.

Hauser RA, Daguio M, Wester D, Hauser M, Kirchman A, and Skinkis C. 2001. Bee-venom therapy for treating multiple sclerosis: A clinical trial. *Altern Complem Ther*, 7: 37–45.

Hortet P and Seegers H. 1998. Loss in milk yield and related composition changes resulting from clinical mastitis in dairy cows. *Prev Vet Med*, 37: 1–20.

Ítavoa CCBF, Moraisa MG, Costab C et al. 2011. Addition of propolis or monensin in the diet: Behavior and productivity of lambs in feedlot. *Anim Feed Sci Technol*, 165(3–4): 161–166.

Jalali FSS, Saifzadeh S, Tajik H, and Farshid AA. 2007. Experimental of repair process of burn-wounds treated with natural honey. *J Anim Vet Adv*, 6(2): 179–184.

Kim CC. 2004. Bee venom treatment without the sting. U.S. Patent 0,081,702, April 29, 2004.

Kridli RT and Al-Khetib SS. 2006. Reproductive responses in ewes treated with eCG or increasing doses of royal jelly. *Anim Reprod Sci*, 92: 75–85.

Lariviere WR and Melzack R. 1996. The bee venom test: A new tonic-pain test. *Pain*, 66: 271–277.

Lee KF, Ennis WJ, and Dunn GP. 2007. Surgical palliative care of advanced wounds. *Am J Hosp Palliat Care*, 24: 154–160.

Liu X, Chen DW, Xie LP, and Rongqing Z. 2002. Effect of honey bee venom on proliferation of K1735M2 mouse melanoma cells *in vitro* and growth of murine B 16 melanomas *in vivo*. *J Pharm Pharmacol*, 54: 1083–1089.

Mathews KA and Binnington AG. 2002. Management of wounds using honey. *Compend Contin Educ Vet Pract*, 24: 53–61.

Matsui T, Yukiyoshi A, Doi S, Sugimoto H, Yamada H, and Matsumoto K. 2002. Gastrointestinal enzyme production of bioactive peptides from royal jelly protein and their antihypertensive ability in SHR. *J Nutri Biochem*, 13(2): 80–86.

Meresta T and Meresta L. 1988. Sensitivity of *Bacillus larvae* to an extract of propolis *in vitro*. *Medycyna Weterynayjna*, 44(3): 169–170.

Mishima S, Suzuki KM, Isohama Y et al. 2005. Royal jelly has estrogenic effects *in vitro* and *in vivo*. *J Ethnopharmacol*, 101(1–3): 215–220.

Molan PC. 2002a. Re-introducing honey in the management of wounds and ulcers: Theory and practice. *Ostomy Wound Manage*, 48: 28–40.

Molan PC. 2002b. *Honey as an Antimicrobial Agent*. Waikato Honey Research Unit, University of Waikato, New Zealand, http://honey.bio.waikato.ac.nz/selection.shtml.

Müller UR. 2011. Hymenoptera venom proteins and peptides for diagnosis and treatment of venom allergic patients. *Inflamm Allergy Drug Targets*, 10: 420–428.

Nagai T and Inoue R. 2004. Preparation and functional properties of water extract and alkaline extract of royal jelly. *Food Chem*, 84: 181–186.

Nakajin S, Okiyama K, Yamashita S, Akiyama Y, and Shinoda M. 1982. Effect of royal jelly on experimental hypercholesterolemia in rabbits. *Shoyakugalku Zasshi*, 36(1): 65–69.

Nermine KMS and Abeer AE. 2009. Immunological effects of honey bee venom in mice with intracerebral candidiasis. *J Med Sci*, 9: 227–233.

Okamoto I, Tanigushi Y, Kunikata T et al. 2003. Major royal jelly protein 3 modulates immune responses *in vitro* and *in vivo*. *Life Sci*, 73(16): 2029–2045.

Oršolić N, Šver L, Verstovšek S, Terzić S, and Bašic I. 2003. Inhibition of mammary carcinoma cell proliferation *in vitro* and tumor growth *in vivo* by bee venom. *Toxicon*, 41: 861–870.

Petruška P, Tušimová E, Kalafová A, Haščík P, Kolesárová A, and Capcarová M. 2012. Effect of propolis in chicken diet on selected parameters of mineral profile. *Biotech Food Sci*, 1(4): 593–600.

Qiu Y, Choo YM, Yoon HJ et al. 2011. Fibrin(ogen)olytic activity of bumble bee venom serine protease. *Toxicol Appl Pharmacol*, 255: 207–213.

Roh DH, Kwon YB, Kim HW et al. 2004. Acupoint stimulation with diluted bee venom (apipuncture) alleviates thermal hyperalgesia in a rodent neuropathic pain model: Involvement of spinal alpha 2-adrenoceptors. *J Pain*, 5: 297–303.

Rooster H, Declercq J and van den Bogaert M. 2008. Honey in wound management: Myth or science? Part 2: Clinical cases in dogs. Faculty of Veterinary Medicine, University of Ghent, Merelbeke, Belgium. *Vlaams Diergeneeskundig Tijdschrift*, 77(2): 75–80.

Sabatini AG, Marcazzan G, Caboni MF, Bogdanov S, and Almeida-Muradian B. 2009. Quality and standardisation of royal jelly. *JAAS*, 1: 1–6.

Salazar-Olivo LA and Paz-González. 2005. Screening of biological activities present in honeybee (*Apis mellifera*) royal jelly. *Toxicol in Vitro*, 19(5): 645–651.

Sánchez-Velasco P, Antón E, Muñoz D et al. 2005. Sensitivity to bee venom antigen phospholipase A2: Association with specific HLA class I and class II alleles and haplotypes in beekeepers and allergic patients. *Hum Immunol*, 66: 818–825.

Schalm OW, Carroll EJ, and Jain NC. 1971. *Bovine Mastitis*. Lea & Febiger, Philadelphia, Pa.

Schmidt JO and Buchmann SL. 1992. Other products of the hive. In *The Hive and the Honeybee*, Graham J.M., ed. Dadant & Sons, Hamilton, IL, 927–988.

Schoen AM Wynn SG. 1998. *Complementary and Alternative Veterinary Medicine*, ISBN 0-8151-7994-4, Mosby Inc., St Louis, MO.

Shahnaz M, Rosmilah M, Geeta P, Dinah R, Meinir J, Abdullah M et al. 2007. Identification of major allergens of royal jelly using 2-dimensional electrophoresis and mass spectrometry analysis. *World Allergy Organ J*, S93; doi: 10.1097/01.

Stuhlmeier KM. 2007. Apis mellifera venom and melittin block neither NF-kappa B-p50-DNA interactions nor the activation of NF-kappa B, instead they activate the transcription of proinflammatory genes and the release of reactive oxygen intermediates. *J Immunol*, 179: 6.

Suemaru K, Cui R, Li B et al. 1986. Changes in quality of royal jelly during storage. *Nippon Shokuhin Kogyo Gakkaishi*, 33(1): 1–7.

Suemaru K, Cui R, Li B, Watanabe S, Okihara K, Hashimoto K, Yamada H, and Araki H. 2008. Topical application of royal jelly has a healing effect for 5-fluorouracil-induced experimental oral mucositis in hamsters. *Methods Find Exp Clin Pharmacol*, 30(2): 103–106.

Tamura T, Fujii A, and Kubiyama N. 1987. Antitumor effects of royal jelly. *Nippon, Yakurigaku Zasshi*, 89(2): 73–80.

Velotto S, Vitale C, Varricchio E, and Crasto A. 2010. Effect of propolis on the fish muscular development and histomorphometrical characteristics. *Acta Vet BRNO*, 79: 543–550; doi: 10.2754/avb201079040543.

Vick JA and Brooks RB. 1972. Pharmacological studies of the major fractions of bee venom. *Amer Bee J*, 112(8): 288–289.

Wang OH, Ahn KB, Lim JK, and Jang HS. 2001. Clinical study on effectiveness of bee venom therapy on degenerative knee arthritis. *J Kor Acu Mox Soc*, 18: 35–47.

Wang BJ and Zhang HJ. 1988. Studies on the chemical constituents of Beijing (China) propolis. *Bull Chinese Materia Medica*, 13(10): 37–38.

# 14 Honey-Based Formulations and Drug Purposes

*Saâd Aissat, Aldina Kesic, Hama Benbarek,*
*and Abdelmalek Meslem*

## CONTENTS

## INTRODUCTION

Although conventional medicine is the mainstream medicine in Western countries, application of traditional medicine is growing worldwide for many reasons, in particular, the side effects or inefficacy of modern drugs. Different types of traditional medicines are widely applied in Asia, Africa, and Latin America to meet primary health care needs. Traditional medicine has maintained its popularity in most regions of the developing world. The application is also rapidly spreading in industrialized countries, where adaptations of traditional medicines are often termed "complementary" or "alternative" (Liu 2011). Honey has been used for thousands of years and used as a folk remedy and is now seeing renewed interest in honey-based wound dressings. Honey is mostly used in the preparation of confection and electuaries and as an adjunct to decoction, pills, and powder (Namdeo et al. 2010). Although the exact mechanism for the beneficial aspects of honey in wound healing is still unknown (Song and Salcido 2011), is it effectively a "worthless but harmless substance" as described by Soffer (1976)? Some clinicians are under the impression that there is little or no evidence to support the use of honey as a wound dressing. This impression is reinforced by it being concluded in systematic reviews that the evidence is not of a high standard. But likewise the evidence for modern wound dressing products is not of a high standard (Molan 2006). For evidence-based medicine to be practiced in wound care, when deciding which product to use to dress a wound, it is necessary to compare the evidence that does exist rather than be influenced by advertising and other forms of sales promotion (Molan 2006). The scientific evidence for using conventional wound care products in pediatric oncology patients is nonexistent, since no prospective randomized studies have been performed in this particular population and no research has been done on the long-term effects of modern conventional treatments such as silver dressings. There has been a report of a silver-coated dressing, which caused raised liver enzymes and an argyria-like syndrome in an adolescent burn patient (Trop et al. 2006). Another common treatment is povidone iodine, which has the advantage of antiseptic properties and is well suited for skin disinfection prior to invasive procedures. However, the antiseptic activity of iodine products is hampered by interactions with the protein content of the wound exudate, and severe adverse effects of systemic absorption of iodine on thyroid function must be considered in infants and toddlers as well as in adult patients with latent hyperthyreosis. In principle, the same problem exists with alcohol-containing antiseptics, since they are almost completely absorbed and need to be metabolized by children, who are treated concomitantly with many other drugs. In contrast, medical honey does not display the problem of systemic absorption and thus can even be utilized in patients with diabetes mellitus without adverse effects on blood glucose levels (Simon et al. 2009). Medihoney is now used in wound care at the Department of Pediatric Oncology, Children's Hospital, University of Bonn, Bonn, Germany (Simon et al. 2006). Currently, information on the use of honey for the treatment of many human diseases can be found in general magazines, beekeeping journals, and natural product leaflets, suggesting a wide variety of unfounded properties. An alternative medicine branch, called apitherapy, has been developed in recent years, offering treatments based on honey and other bee products for many diseases (Alvarez-Suarez et al. 2010).

Being personally convinced that honey is of a great help not only in treating wounds, but that it also offers a myriad of therapeutic properties, the subject of our present chapter does not have the pretension to arbitrate between the pros and cons (see Molan 2006, 2011) but will be based primarily on the formulations commonly used and its application methods. Because of its complex chemical composition as well as various medicinal effects on the body, there are different methods of application of honey.

## HONEY-BASED MIXTURES IN ANCIENT TIMES

It is difficult to know when honey became recognized as a treat. The origins of its earliest dosage forms are lost in the mists of history. The ancient Greeks believed that if bee honey is taken regularly, human life could be prolonged. Early thinkers such as Homer, Pythagoras, Ovid, Democritus, Hippocrates, and Aristotle mentioned that people should eat bee honey to preserve their health and vigor. In classic Greece, laws concerning apiculture were suggested and Plato included honey in his concept of a healthy diet (Skiadas and Lascaratos 2001).

It can safely be assumed that primitive men took parts of plants including leaves, stems, roots, and berries internally for a range of symptoms. A variety of plant and animal products would undoubtedly also have been applied externally to aid the healing of wounds. The vapors of volatile herbs would have been inhaled, and later combinations would have been used, no doubt incorporated into a selection of fats, oils, and honey (Royal Pharmaceutical Society).

In Ayurveda, a 4000-year-old medicine originating from India honey was used for many purposes. According to the Ayurveda classic Ashtanga Hridaya, written about 500 AD, honey can be used against many diseases, for example, healing and cleaning wounds against different internal and external infections (Krishna 2005).

The oldest written record is a prescription written on a clay tablet from Nippur, the religious center of the Sumerians in the Euphrates valley, circa 2000 BC. This prescription states, "Grind to a powder river dust and… (here the words are missing) … then knead it in water and honey and let plain oil and hot cedar oil be spread over it" (Kramer and Levey 1954).

The Egyptian contraceptive tampon, described in the Ebers Papyrus of 1550 BC, was made of lint and soaked in honey and tips from the acacia shrub. The acacia shrub contains gum Arabic, and for the evacuation of the belly: cow's milk; grains; honey; mash, sifted, cooked and take in four portions. Another Ancient Egyptian prescription used powdered crocodile feces, saltpeter and honey, or substitute elephant dung. Allegedly, cotton soaked in honey and lemon juice was still being used as a contraceptive in Egypt in the 1990s (Crane 1990).

The Edwin Smith Papyrus (1700 BC), a fragment of a textbook on trauma surgery from 1800 BC, included the prevention and curing of infection with honey and bread mold, and in wound healing: "Thou shouldst bind [the wound] with fresh meat the first day [and] treat afterwards with grease, honey [and] lint every day until he recovers" (Janick 2010).

Avicenna (980–1037 BC), in his masterpiece in medicine, called the al-Qanun fi'l-tibb or the Canon of Medicine, which was completed in 1025 (Abadi 1976; Jaàfar 2009), noted that fennel juice mixed with honey and wild marjoram is effective in treating the initial stages of cataract. He also recommended honey as a galactogogue (wet nurse) and for constipation during dentition, for aphthous stomatitis, for watery discharge from the ears, for difficulty of breathing, for ulcerated furunculosis (infants) in case of overrepletion, in diet, as a laxative, etc. (Jaàfar 2009).

Hippocrates (460–377 BC) used many herbal treatments; honey and wine were common ingredients mixed with herbal treatment.

If a fistula be already formed, take a stalk of fresh garlic, and having laid the man on his back, and separated his thighs on both sides, push down the stalk as far as it will go, and thereby measure the depth of the fistula. Then, having bruised the root of seseli to a very fine powder, and poured in some water, let it macerate for four days, and, mixing the water with honey, let the patient drink it, fasting, to the amount of three cyathi, and at the same time purge away the ascarides. Those who are left without treatment die. When the fistula has sloughed through, a soft sponge is to be cut into very slender pieces and applied, and then the flowers of copper, roasted, are to be frequently applied with a director; and the sponge smeared with honey is to be introduced with the index finger of the left hand, and pushed forward;

When the gut protrudes and will not remain in its place, scrape the finest and most compact silphium (assafoetida?) into small pieces and apply as a cataplasm, and apply a sternutatory medicine to the nose and provoke sneezing, and having moistened pomegranate rind with hot water, and having powdered alum in white wine, pour it on the gut, then apply rags, bind the thighs together for three days, and let the patient fast, only he may drink sweet wine. If even thus matters do not proceed properly, having mixed vermillion with honey, anoint. (*On Fistulae,* by Hippocrates, translated by Francis Adams.)

Within the *Sumario de la Medicina Romanceada*, the combination of honey and gall was used as a remedy for noctilupia, an eye ailment that blinds the patient only at night. Mondeville's Chirugie noted how eating honey is useful in the digestion of bile (605; 773) and would thus suggest that the ingredients have somewhat of an opposite nature. Honey and bile appear in different recipes used for treating infection (611; 778) and together in the treatment of a concave ulcer: the 1515 translation of the Aromatariorum compendium, Compendio de los Boticarios, notes how honey has a preservative nature [that conserves] all things placed in it, more so than sugar and more so than any other thing in the world. Therefore, honey confections or electuaries last longer than those made with sugar (Guardiola-Griffiths 2011).

Prior to the discovery of the art of distillation by the Arabs in the 9th century, liquid medicinal preparations were often put up as thick, viscous solutions containing honey called electuaries or treacles (Grote and Walker 1946). Electuaries and theriacs include an important proportion of honey in their ingredients; this one has a double purpose. It is an excipient and a curative agent. On one hand, it is used as a preservative, binder, and sweetener, and on the other hand, it is recommended because of its numerous healing virtues. Physicians take care of the selection and preparation conditions to have an optimal quality of the honey and they consider its real nature during its elaboration. Honey will be a perfect food and drug if it was not subjected to the digestion's action after its absorption (Ricordel and Bonmatin 2003).

Theriac, or Venice Treacle, is a medicinal compound first used in Rome as a remedy against poison and then for centuries as a preventative and cure-all. Numerous recipes for it exist: about 70 drugs (herbal, animal, and mineral) were pulverized and reduced with honey to an electuary, a medicated paste prepared with honey or other sweet substance and taken by rubbing on the teeth or gums (Holland 2000). Galen (129–199 AD) prepared his theriac and wrote about various theriac compounds in his books *De Antidotis I*, *De Antidotis II*, and *De Theriaca ad Pisonem*. The basic formula consisted of viper's flesh, opium, honey, wine, cinnamon, and more than 70 ingredients. The final product was supposed to mature for years and was administrated orally as a potion or topically in plasters. Galen claimed that his theriac drew out poisons like a cupping glass and could divide the tissue of an abscess more quickly than a scalpel. The preparation was taken daily by Emperor Marcus Aurelius to protect against poisons and to aid in ensuring good general health (Karaberopoulos et al. 2012).

Galen composed an electuary for a certain child who was suffering from spasm of the entire body: take water mint, rose, cinnamon, mastich, fumaria, spicknel, ameos, arnica, zedoary, cloves, 1 aureus of each; sandalwood, lignum aloes, 1 dram of each; musk, half a dram. Smear all of these drugs with balsam and confect with honey (Brodman 1978).

Between the number of sweet preparations used by the Arabs, confections and electuaries both involved mixing dried and powdered ingredients with syrup or honey and juleps were clear, sweet liquids (Royal Pharmaceutical Society).

In the first compendium of ancient Chinese medicine Shen Nong compiled many years before Christ, and mentioned in a written form for the first time in the West Han Dynasty (206 BC–24 AD), there are many prescriptions and medical indications that contain honey (Ho 1996; Siedentopp 2009).

The wise Solomon praised the virtues of honey in the Old Testament. The Koran says "thy Lord taught the bee to build its cells in hills, on trees and in (men's) habitations … there issues from within their bodies a drink of varying colours, wherein is healing for mankind" (Qur'an 16:68–69).

In the Christian New Testament, Matthew 3:4 (King James Version), John the Baptist is said to have lived for a long period of time in the wilderness on a diet consisting of locusts and wild honey. In Islam, there is an entire Surah in the Qur'an called al-Nahl (the Honey Bee). According to Hadith (the way of life prescribed as normative for Muslims on the basis of the teachings and practices of Muhammad and interpretations of the Koran), Prophet Muhammad strongly recommended honey for healing purposes (Sahih Bukhari volume 7, book 71, numbers 584, 585, 588, and 603).

A formula found in both the *Cambridge Hippiatrica* ([Anon.] 1924, 17.3) and the Geoponica ([Anon.] 1895, Geoponica 16.15) specifies the use of the ash of cannabis combined with honey and "old urine" as a salve for wounds of the lower back (for the "back-biting" of horses and the use of cannabis to control it) (Butrica 2002).

Honey was among most of the ingredients used in a therapeutic purpose. Pedanius Dioscorides the Greek, in his *De Materia Medica* written approximately 2000 years ago, prescribed that lichens, if mixed with honey, are useful against jaundice and heal inflammations of the tongue. There is not enough or satisfactory morphologic evidence in order that one could taxonomically identify the lichen species mentioned in *De Materia Medica*. However, it is obvious that Dioscorides describes a lichen species growing among mosses on rocks (saxicolous), since he mentions mosses in a separate chapter, "II.20. Bryon" (Yavuz 2012).

Moses Maimonides (1135–1204) dealt with specific remedies against the bites of certain animals. These remedies are detailed as follows: Three drams of an extract of leaves of the herba citrine should be imbibed as well as rubbed into the site of the bite. Three drams of the seed of citrus fruits should be imbibed. The colocynth root is an excellent antidote against scorpion sting; its maximum dose is 2 drams. When the root is still fresh, it should be crushed and rubbed into the site of the bite. When dried, it should be crushed, kneaded in vinegar and honey, and applied to the site of the bite for those who either have taken or believe they have taken poison. Emesis should be induced by means of hot water with *Anethum* and much oil followed by fresh milk, butter, and honey, all of which should be vomited. Then, the specific antidote should be administered (Rosner 1968).

Moses is considered to be the original writer on diet, recommending to the Jews, "bread, wine, milk, honey; quadrupeds that divide the hoof, and chew the cud; all the feathered kind, a few only excepted; and fishes that have fins and scales" (Haslam 2007).

Gerard's *Herbal*, a famous English herbal published in 1597, started out with a description of the plant and then went into its uses and qualities. Uses include healing the eyes that hang out; leaves of bramble boiled in water with honey, alum, and a little white wine make an excellent lotion and washing water; and the same decoction fastens the teeth (Janick 2010).

## HONEY-BASED FORMULATION IN MODERN TIMES

Honey has been successfully used in medicine since antiquity and it has continued into present-day folk medicine. However, with the advent of modern medicine, it has been less used, especially in the English-speaking world. Apitherapy (the medical use of honeybee products) has recently become the focus of attention as a form of medicine for treating certain conditions and diseases. The beneficial effects of honey in different disorders have been rediscovered in recent decades on the basis of a series of international scientific studies conducted to investigate the therapeutic properties of this natural product and published on Medline (Pipicelli and Tatti 2009). It is widely available in most communities, and although the mechanism of action of several of its properties remains obscure and needs further investigation, the time has now come for conventional medicine to lift the blinds off this "traditional remedy" and give it its due recognition (Abou El-Soud 2012). The only factor that may limit these effects is inadequate sterility of the preparations, in which case honey may be highly counterproductive.

It is possible therefore to assert that the microorganisms found in honey undergo gradual extinction in honey due to its inhibitory properties. It should be noted, however, that the therapeutic use of honey in everyday clinical practice needs to be validated by relevant guidelines and should only be adopted under medical prescription, in accordance with criteria of efficacy and safety for both patients and health care providers (Pipicelli and Tatti 2009). There are now several brands and types of honey wound-care products available as registered medical devices. Certain types of honey, obtained from particular flowers found in Australia and New Zealand (*Leptospermum* spp.), have been approved for sale as therapeutic honey and used for the treatment of numerous types of skin lesions.

## HONEY IN WOUND CARE

A wound can result from either an external or internal insult. Many acute wounds are caused by external insults, such as mechanical insults, thermal radiation, ultraviolet radiation, or radiation (γ-radiation) therapy. Chronic wounds (leg ulcers, diabetic ulcers, and pressure ulcers), on the other hand, are largely caused by an internal insult in the form of circulatory compromise. Inadequate circulation robs tissue of necessary nutrients and potentiates proinflammatory cytokines, leading to tissue necrosis.

Many topical products are currently being used with the intent of facilitating wound healing. However, there is a lack of strong evidence to support the use of the majority of these products (Nelson and Bradley 2007; Jull et al. 2008; Wasiak et al. 2008; O'Meara et al. 2010; Molan 2011).

Honey is one of the oldest agents used in medicine, mostly applied for wound healing (Majno 1975). The role of honey as a remedy for the treatment of infected wounds is being "rediscovered" by the medical profession, particularly where conventional modern therapeutic agents have failed. Several honey-based wound dressing products for the treatment of wounds, on sale as medical products, are accepted by control authorities and have been introduced in the market for treatment of a wide range of wound infections (Acton 2008). Although several brands and types of honey wound-care products are available as registered medical devices (Table 14.1), there is little promotional advertising of honey products for wound care (Molan 2006).

According to scientific literature and clinical experience, antibacterial honey (Medihoney) seems to fulfill most of the requirements of an ideal antiseptic in wound care. The only open question for Medihoney is the residence time needed to kill bacteria in a colonized wound, which is supposed to be less than 5 min for octenidine or povidone iodine. Theoretical adverse reactions such as anaphylaxis or systemic toxicity (i.e., hyperglycemia in diabetic patients) have not been reported so far (Simon et al. 2006). Kwakman et al. (2011) were able to enhance the bactericidal activity of honey by enrichment with the AMP BP2. BP2-enriched RS (the source for the production of Revamil medical-grade honey) honey had rapid bactericidal activity up to a high dilution against all bacteria tested and had a broader spectrum of bactericidal activity than either agent alone. This offers prospects for the development of clinically applicable honey-based antimicrobials with rapid and broad-range microbicidal activity.

Wet dressings or any form of irrigation moisten the tissues and therefore delay healing. Dry dressings adhere to the surface, causing pain and injuring the granulating surface every time they are changed. Oily dressings prevent the surface secretions from escaping freely and may cause them to spread onto the neighboring skin surfaces and cause undesirable reactions or toxic effects. Since the earlier 1950s, honey has been known as an effective treatment of wounds because it is nonirritating, nontoxic, self-sterile, bactericidal, nutritive, easily applied, and more comfortable than other dressings (Bulman 1955). Honey stimulates the epithelialization growth (Efem 1988; Molan 2002), and its acidic nature is also responsible for released oxygen in high quantities from the hemoglobin in the wound site, which promotes the granulation tissue to repair wounds (Sharp 2009).

An important factor that affects the activity of drugs in formulation is their affinity for the base in which they have been incorporated. Lipophilic drugs are better released from hydrophilic than from lipophilic bases and vehicles. The antibacterial activity of ocimum oil in honey was much higher than in lipophilic Simple Ointment B.P. and petrolatum (Orafidiya et al. 2001, 2002, 2006). However, ocimum oil in honey gave a better structured repaired skin than honey alone (Adesina 2005).

Clinical and mycologic improvement was obtained with the use of a honey mixture containing honey, olive oil, and beeswax (in ratio 1:1:1, v/v/v) in patients with atopic dermatitis, psoriasis, and skin fungal infections (Al-Waili 2003, 2004). In addition, honey mixture was effective in the treatment of diaper dermatitis with or without *Candida albicans* infection (Al-Waili 2005b). A mixture of honey, olive oil, and beeswax is safe and effective in the management of skin diseases. Honey is sticky when applied on the skin. Olive oil is used to decline the viscosity of honey and to facilitate skin application and beeswax is used as a vehicle for preparation of the ointment (Al-Waili 2005a). Furthermore, olive oil and beeswax application could have a beneficial influence on skin lesions (Carbajal et al. 1998; Noa and Mas 1998;

**TABLE 14.1**

**Examples of Manufactured Medical Honeys: Formulations, Indications, Contraindications, and Modes of Action**

| Name of Product and Manufacturer | Description of Product | Indications | Contraindications | Mode of Action |
|---|---|---|---|---|
| HoneySoft— Taureon | Inert acetate dressing impregnated with a medical-grade, multifloral honey (originated from Chile). | Wounds that require bacterial control, e.g., ulcus cruris, diabetic wounds, burns, chronic wounds, acute wounds, infected wounds, oncologic wounds, and pressure sores. | Offer a wound treatment without contraindications, the risk of resistance, allergies, or toxicity, even with long-term use. | Offers the ultimate bacterial control on all levels of wound care; enables the penetration of the biofilm formation that is often seen in persistent wounds; offers an odor neutralizing effect. |
| HoneySoft Skin Recovery Cream— Taureon | 50% of honey (from Chile), cetostearyl alcohol, sorbitol, glycerine, sodium lauryl sulfate, cetiol, and water. | Extremely dry skin, flaky skin, sensitive skin, irritated skin, fungal infections, eczema, and itchy skin. | Do not use if you are allergic to honey, pollen, or one of the other ingredients; do not apply to open wounds. | Unbroken skin, for example, following a process of wound cleansing and healing with HoneySoft honey dressing; can enhance the suppleness and moisture content of the skin. |
| Activon Tulle— Advancis Medical | Knitted viscose mesh dressing impregnated with 100% manuka honey. | Different wound types including leg ulcers, pressure ulcers, malodorous, sloughy, or necrotic wounds; Activon may be used on partial and full-thickness wounds including sloughy wounds, pressure ulcers, surgical wounds, burns, graft sites, and malodorous wounds. | Arterial bleeds and heavily bleeding wounds. Some patients may experience a drawing sensation that can be painful. If pain is of an unacceptable level and cannot be managed by administering an analgesic, the dressing should be removed and its use discontinued. | Creates a moist healing environment and effectively eliminates wound odor while providing antibacterial action; ideally selected for granulating or shallow wounds, it is a good choice when debriding or desloughing small areas of necrotic or sloughy tissue. |

| | | | | |
|---|---|---|---|---|
| Algivon— Advancis Medical | Calcium alginate dressing impregnated with 100% MH. | | | Ideal choice for wetter wounds as the alginate has a small capacity to absorb, meaning the honey is not washed away with exudate therefore staying at the wound site for longer. The dressing is very soft and conformable, ideal for cavities and debriding and desloughing large areas of necrotic and sloughy tissue. |
| Activon— Tube— Advancis Medical | 100% Manuka honey in a tube. | | | Kills harmful bacteria, anti-inflammatory, eliminates odors without masking them, osmotic effect, drawing harmful tissue away from the wound bed, maintains the ideal moist wound healing environment. |
| —Actilite | Nonadherent viscose net dressing coated with 99% manuka honey and 1% manuka oil. | Suitable for all wound types where a primary layer is indicated and an antibacterial effect may be advantageous including cuts, abrasions, burns, surgical wounds, leg ulcers, pressure ulcers, diabetic ulcers, and infected wounds. | | Best suited for granulating or epithelializing wounds. Actilite offers antibacterial protection while promoting the ideal moist wound-healing environment. |
| —Actibalm | White pharmaceutical-grade petroleum jelly and 100% pure Advancis manuka honey. | Replaces moisture for dry, chapped lips, cold sores, good for insect bites, soothing antibacterial honey; helps reduce inflammation. | Do not use if allergic to bee venom. | The petroleum jelly seals the skin preventing bacteria from entering the affected area and keeps the skin's moisture from evaporating, while honey's antibacterial properties can help heal and soothe. |

(continued)

## TABLE 14.1 (Continued)
### Examples of Manufactured Medical Honeys: Formulations, Indications, Contraindications, and Modes of Action

| Name of Product and Manufacturer | Description of Product | Indications | Contraindications | Mode of Action |
|---|---|---|---|---|
| L-Mesitran Ointment—Triticum | Mixture of honey (not manuka), lanolin, cod liver oil, sunflower oil, calendula, aloe vera, zinc oxide, vitamins C and E. | Contaminated, chronic, oncologic, and/or acute wounds; for deep and superficial wounds with low to high exudates. | Known sensitivity to the dressing or any of its components; use with caution in patients with bee venom allergy; not suitable for use on full-thickness burns. | Strong osmotic effect. This prohibits bacterial growth. When the ointment is applied to the wound, it attracts wound fluid from the surrounding tissues. This creates a moist wound-healing environment that in turn stimulates the wound-healing process, facilitates the autolysis of dead tissue, and stimulates the growth of new cells. |
| L-Mesitran Soft—Triticum | Mixture of honey (not manuka) with lanolin, polyethylene glycol, and vitamins C and E. | For dry, necrotic, sloughy, and infected, sensitive wounds (e.g., leg ulcers). | | |
| L-Mesitran Border—Triticum | Combined hydrogel and honey pad on a strong fixation layer. | Ideal for those difficult locations where regular dressings do not adhere or where mobility is key; perfectly suited for those wounds that need to have the exudate locked away and absorbed, preventing maceration and odor; shows its qualities when the dressed wound is in contact with water (e.g., showering and incontinence). The dressing will adhere even in the wettest conditions. | | |

| L-Mesitran Hydro—Triticum | Sheet of acrylic polymer hydrogel containing honey (not manuka). | Primary dressing for the treatment of burns, skin tears, ulcers, and acute trauma wounds. | | The thin honey-hydrogel layer is capable of absorbing five times its own weight of wound fluid and does not disintegrate; promotes a moist healing environment; kills bacteria and fungi and prevents further contamination or growth. |
|---|---|---|---|---|
| L-Mesitran Net—Triticum | Open-weave polyester net impregnated with L-Mesitran Hydro. | | | Allows for passage of wound exudates to be absorbed by a secondary dressing, avoiding maceration of the surrounding tissue. The hydrogel holds the honey in a polymer network and humectant, providing a moist wound environment. The honey will kill bacteria and fungi and will prevent further contamination or growth. |
| Melladerm Plus wound gel—Honey source is Bulgaria—SanoMed | Honey-based wound gel, consists of honey, PEG 4000, propylene glycol, and glycerine. | Superficial chronic wounds—decubitus, ulcus cruris, diabetic foot wounds, fungating wounds (deodorizing and debriding); contaminated acute wounds—surgical wounds (postoperative wounds), traumatic wounds (superficial wounds, cuts), small burn and laser wounds (first and second degree). | No contraindications are known to date. Do not use the gel in cases where the patient is very susceptible to an infection (e.g., in very large burn wounds). Do not use on individuals with a known sensitivity to the gel or its components. | When the gel is in contact with the wound, fluid is extracted from surrounding tissues. The osmotic action creates a moist wound-healing environment and together with honey stimulates the wound-healing process, facilitates autolysis, and promotes epithelial cell migration. Dilutes gradually and can be removed easily with a wound cleanser, if required. SanoSkin Melladerm Plus is a primary wound dressing that can be covered by a secondary dressing. |

*(continued)*

**TABLE 14.1 (Continued)**

**Examples of Manufactured Medical Honeys: Formulations, Indications, Contraindications, and Modes of Action**

| Name of Product and Manufacturer | Description of Product | Indications | Contraindications | Mode of Action |
|---|---|---|---|---|
| Melladerm Plus—Tulle | Gel gauze made of specially woven, wide gauge, nonadherent polyester mesh impregnated with SanoSkin Melladerm PLUS honey gel. | Superficial wounds, abrasions, infected wounds, burns, pressure ulcers, leg ulcers, diabetic foot ulcers, sloughy wounds, necrotic wounds, malodorous wounds, surgical wounds, donor, and recipient graft sites. | Do not use on individuals with a known sensitivity to honey or bee products. | Fluid is extracted from the surrounding tissues. A moist wound-healing environment stimulates the wound healing process, resulting in decreased inflammation (less swelling and pain), facilitating autolysis, and promoting epithelial cell migration. |
| Manuka Health Wound Dressing—Manuka Health NZ | Sheet of hydrogel sheet containing manuka honey. | Abrasions, scrapes, minor cuts, scalds, and burns; surface wounds. | Allergy or sensitivity to honey or bee products; consult your physician if signs of infection occur: redness, swelling, fever, etc. | The low pH maintains a "sour" wound environment, increases the availability of oxygen from hemoglobin, decreases the damaging effects of high protease activity, and creates a medium that is generally inhospitable to bacterial growth. As a supersaturated water solution with natural sugars, and approximately 17% water, this mixture exerts a strong attraction for water then draws fluid from the underlying tissues, bathes the wound bed, and causes autolytic debridement. The dressing conforms to wound bed to maintain a moist wound surface and absorbs large amounts of exudate without drying out the bed. |
| Manuka Health Wound Dressing—Manuka Health NZ | Sheet of hydrogel sheet containing manuka honey. | | | |
| Manuka Health Wound Dressing—Manuka Health NZ | Sheet of hydrogel sheet containing manuka honey. | | | |

| | | | |
|---|---|---|---|
| MANUKAhd—ManukaMed | Superabsorbent polyacrylic fiber dressing pad impregnated with manuka honey, coated with a dry-touch absorbent hydrocolloid. | Pressure ulcers, leg ulcers, trauma wounds, slough and necrotic wounds, surgical wounds, burns, donor site fungating wounds, malodorous wound. | Allergy to honey or bee-venom. Although there is no record of increased blood sugar levels in patients with diabetes due to the use of honey dressings, it is advisable to monitor the levels during use of MANUKAtex. This is a single-use dressing. | Hydrocolloid coating to improve the ease of handling and applying the dressing. The coating combines with exudate to impart a nonadherent property to assist in dressing removal. The dressing will maintain a moist environment conducive to healing. |
| MANUKAtex—ManukaMed | Nonadherent gauze dressing impregnated with manuka honey, coated with a dry-touch absorbent hydrocolloid. | Skin tears, trauma wounds, leg ulcers, pressure ulcers, surgical wounds, burns, slough and necrotic wounds, donor sites. | | |
| MANUKApli—ManukaMed | 100% manuka honey in a tube. | Cuts, abrasions, minor wounds, burns. | If reused, there is a high risk of infection and cross-contamination. | |
| MelMax—Dermasciences | Nonadherent wound dressing impregnated with a mixture of polyhydrated ionogens ointment and buckwheat honey. | Acute wounds (burns, postoperative and trauma wounds), chronic wounds (ulcus cruris, diabetic and pressure ulcers). | MelMax may not be used on patients with a known hypersensitivity to acetate, ionogen formulation, or honey. | Bacteria-regulating agent, reduces the inflammation, stimulates autolytic debridement, enhances cellular growth, reduces the formation of scar tissue, neutralizes odor, strong antioxidant properties. Ionogen formulation (metal ions) regulates the MMP balance in the wound bed by means of pH modulation. Ionogen formulation prevents the typical stalling of chronic wounds during the healing process. |

(continued)

**TABLE 14.1 (Continued)**
**Examples of Manufactured Medical Honeys: Formulations, Indications, Contraindications, and Modes of Action**

| Name of Product and Manufacturer | Description of Product | Indications | Contraindications | Mode of Action |
|---|---|---|---|---|
| Medihoney Antibacterial Medical Honey—Dermasciences | 100% manuka honey. | Wounds with mild to moderate levels of exudate as well as deep cavity wounds. | Should not be used in patients with a known sensitivity to honey, calcium alginate, or sodium alginate. Due to the viscosity (thickness) of Medihoney Wound Gel, it is particularly suited to cavity or deep wounds. However, where gravity may affect, if staying in place an alternative, product may need to be selected such as the Medihoney Gel Sheet or Apinate Dressing. It is contraindicated in very deep wounds or where there is undermining/tracking with sinuses. It is due to the fact that the plant waxes can potentially block sinuses. | As wound fluid enters the dressing, the honey is released while the dressing absorbs and forms a gel. This makes the dressings easy to remove without disrupting the wound bed. This adds to the healing process and inflicts less pain and discomfort on the patient. |
| Medihoney Antibacterial Wound Gel—Dermasciences | 80% manuka honey and 20% plant waxes. | Surface wounds with mild to moderate levels of exudate as well as partial or full thickness wound. | | |
| Medihoney Gel Sheet—Dermasciences | Flexible dressing with manuka honey and sodium alginate. | Suitable for use on mild to moderately exuding wounds and in patients with pressure ulcers presenting with leathery eschar. | | |
| Medihoney Tulle Dressing—Dermasciences | | Lightly exuding wounds with a suspected biofilm; requires a secondary dressing to maintain an optimal moist wound environment; can be used on first and second degree burns, donor sites, and superficial wounds such as abrasions. | | |

| | | | | |
|---|---|---|---|---|
| Medihoney Apinate Dressing—Dermasciences | Calcium alginate dressing impregnated with manuka. | For wounds that require autolytic debridement a suitable secondary absorbent dressing should be used to manage the exudate appropriately; leg ulcers under compression; cavity wounds including pressure ulcers and dehisced surgical wounds. | | |
| Medihoney Barrier Cream—Dermasciences | Manuka honey, coconut oil, German chamomile flower extract, evening primrose oil, aloe vera, and vitamin E. | Protect the skin from breakdown. | | |
| Unadulterated Alpine BIO Honey. Enriched with Royal Jelly and Propolis—Aripmed Armenia | Produced from the nectar of 101 alpine herbs: Sage, Marjoram, Cephalaria, Thyme, Mint, Shepherd's Club, St John's Wort, etc. | Particularly conducive to enhance immune response, boost energy and resilience, as well as used to prevent and treat cardiovascular, gastric, cancerous, and gynecologic conditions and regulate blood circulation and pressure. | For allergics to honey products and patients with dermatitis and Addison's disease. | |
| Active Manuka and Placenta Honey—Oregan New Zealand | Active 30+ manuka honey, sheep placenta extract 30:1; equivalent to 16,600 mg sheep placenta. | Antiaging, enhancing body energy, revitalizing to the skin, boosting the immune system, improving physical vitality, promoting general health. | People who have allergic reactions should seek medical advice before consuming. | Its rich amino acid and gonadotropin play roles of enhancing the energy and nourishing the stomach; resisting allergy placenta contains rich growth factors, hormones and antibodies, and other concentrated nutrients that can rejuvenate organism. |

*(continued)*

**TABLE 14.1 (Continued)**

**Examples of Manufactured Medical Honeys: Formulations, Indications, Contraindications, and Modes of Action**

| Name of Product and Manufacturer | Description of Product | Indications | Contraindications | Mode of Action |
|---|---|---|---|---|
| ActiFlex Honey—Oregan New Zealand[a] | Pure Active 30 manuka honey, freeze-dried bee-venom powder and lucosamine. | Provides relief from joint inflammation and arthritis pain. | | Bee venom stimulates the body to produce cortisol and glucosamine helps with joint restoration and improves cartilage growth. Both are natural anti-inflammatories that helps reduce pain and increase movement in sore joints, while manuka honey is prized for its ability to fortify the body. |
| Manuka and Propolis Honey—Oregan New Zealand[a] | 20% Propolis and 80% organic manuka honey. | Helps to remove toxins and radionuclide out of the body, helps to boost immune system; regulates metabolism; regulates gastric and intestinal function; helps against angina and cold; provides antibacterial effect. | | The well-known properties of bee propolis as a very strong natural antibiotic in combination with antibacterial properties of New Zealand manuka honey make this product very effective for the stimulation of the immune system and help to provide the precedent benefits. |
| Manuka and Royal Jelly Honey | 20% Freeze-dried royal jelly powder and 80% organic manuka honey. | Stimulates physical performance and general health improvement; increases resistance to viral infections. | | The highest quality royal jelly blended with quality organic manuka honey. Royal jelly contains remarkable amounts of vitamins, proteins, lipids, glucides, hormones, enzymes, mineral substances, amino acids, antibacterial and antibiotic components, and specifically factors that act as a biocatalyst in cell regeneration processes and help preserve the youth of the human body. |

| Feropip Syrup—PIP d.o.o.[b] | Honeydew honey, glacial extract of nettle, glacial root extract of parsley, vitamin C, iron (iron fumarate), demineralized water, acidity regulator (citric acid, strawberry flavor). | Reimbursement of iron caused by chronic diseases and infections, increased needs for iron (pregnant women, nursing mothers, children grow and develop, athletes, blood donors), and insufficient intake of iron in the body (vegetarian). | Due to the iron content in this product, it is necessary to abide by the recommended doses. | It contains iron (II) fumarate, which is easily absorbed and well tolerated and does not cause digestive problems. Iron absorption is increased with the addition of vitamin C and honeydew honey, which is exceptionally rich in minerals. |
| Royalpip Pastilles—PIP d.o.o. | Lyophilized royal jelly, vitamin C, honey, fillers, and sweeteners: lactose, sucrose, inverted sugar, lubricant: magnesium stearate, lemon oil (Citrus limonum). | Prevention and therapeutic relief of allergic reactions, restoring psychophysical strength in cases of chronic fatigue and stress, for immune system boost, concentration, and memory. Pastilles help in preventing and easing allergic reactions in natural way by enhancing hyposensibility of body to grass and tree pollen. | Not recommended for persons allergic to bee products and lactose. | Propolis: blocks release of histaminia to ease allergic reactions, stimulates immune system, has anti-inflammatory and anti-oxidant effects. Pollen: in subdosage quantities helps body in developing allergic sensibility to environmental pollen; royal jelly: natural biostimulant and antioxidant, protects organism against free radicals. Honey carries active substances to cells and then reinforces their absorption and exploitability. Honey enhances binding calcium from consumed food. Calcium inhibits histaminia discharge, thus reducing itching at allergic reactions. |
| Royalpip Honey[b]—PIP d.o.o. | Acacia honey, pollen, royal jelly, propolis. | For reduced immunity, fatigue, increased mental and physical activity, persons under stress, preoperative and postoperative treatment, and during and after chemotherapy. | Not recommended for persons allergic to bee products. | Natural biostimulator that boosts the immune system and improves physical condition and vitality, returns strength, strengthens the blood, and increases appetite in children. |

*(continued)*

**TABLE 14.1 (Continued)**

**Examples of Manufactured Medical Honeys: Formulations, Indications, Contraindications, and Modes of Action**

| Name of Product and Manufacturer | Description of Product | Indications | Contraindications | Mode of Action |
|---|---|---|---|---|
| Fortepip Honey[b]—PIP d.o.o. | Sage honey, alcohol-based extract of spruce tips (*Gemmae picea*) propolis. | Eases throat irritation; as an addition to the therapy prescribed by your doctor. | Not recommended for persons allergic to propolis. | Coats the mucous membranes of the pharynx, soothes throat irritation and coughing, facilitates breathing and, thanks to the propolis content, has a mild antibacterial activity. |
| Gastropip Honey[b]—PIP d.o.o. | Chestnut honey, pollen, alcohol extract of anis, alcohol extract of chamomile, alcohol extract of yarrow, propolis. | Relieves stomach ailments. | Not recommended for persons allergic to bee products. | Helps to regulate digestion, for bloating and constipation, regenerates damage to the stomach lining, soothes digestive problems, and stimulates the release of digestive juices. |
| Kardiopip Honey[b]—PIP d.o.o. | Honeydew honey, pollen, royal jelly, alcohol extract of hawthorn, alcohol extract of chestnut. | Particularly recommended as a preventative measure following heart attack or stroke, in combination with regular therapy and in consultation with your physician. | | Positively influence the work of the heart, circulatory system, and regulation of blood pressure; calms arrhythmia and nervous heart, improves blood circulation, and improves sleep and overall health. |

[a] *Oregan New Zealand Limited* is an internationally based healthcare company. The parent company "Evergreen Life Ltd" is located in Auckland, New Zealand.

[b] Commercialized only in Croatia and Bosnia Herzegovina.

Fleischer et al. 1999). In addition, the contents of the mixture might have a synergistic effect.

Any honey should not just be used on just any wound without preparation. This will ensure judicious use of honey for organisms against which it is likely to be active and diminish the possibility of infecting the wounds rather than destroying the microbes. More work may need to be done to give an answer to the conflict (Olaitan et al. 2007). Clinicians need reassurance that any health-related agent is safe and meets its stated therapeutic purpose. Therefore, it is important to emphasize that, although natural in origin, the honey used in wound care should be of a medical-grade standard and not sourced from honey destined for the supermarket shelf. Supermarket-variety honey is attractive because of its low cost and wide availability, but we must be aware that honey is between prime targets for economic adulteration (Fairchild et al. 2003; Moore et al. 2012). In an effort to determine industry opinions on economic adulteration, a mail survey of 14 U.S. honey packers was conducted at the request of the National Honey Board in 1999 (Fairchild 1999). The response rate was 86%. The total volume of honey purchased by survey respondents represented approximately half of the estimated total U.S. honey sales in 1996 to 1998. The survey was not a statistically representative (random) sample; thus, the information generated only represents the experience and opinions of the responding firms. Fifty-eight percent of respondents, representing 88% of respondent volume, reported testing for economic adulteration, while 42% did not test for economic adulteration. The honey sales of those testing for economic adulteration were distributed among product utilization channels as follows: retail sales, 50.2%; food-service sales, including hotel, restaurant, and institutional pack, 13.4%; and bulk sales to the food-ingredient market, 36.4%. All of the responding firms that test for economic adulteration reported using commercial labs, with one firm using both commercial and in-house labs (Fairchild et al. 2003).

Medical-grade honey is filtered, γ-irradiated, and produced under carefully controlled standards of hygiene to ensure that a standardized honey is produced (White et al. 2005). The treatment was not found to affect the physical, biochemical, antibacterial, and organoleptic attributes of honey and the overall quality of honey remains unaltered upon radiation treatment (Saxena et al. 2010). According to Molan and Allen (1996), 25 kGy of γ-irradiation is sufficient to achieve sterility for honey. However, in a study carried out to find the lowest dose of irradiation needed for sterilization, six batches of honey were γ-irradiated with 6, 12, 18, 22, and 25 kGy of cobalt-60. All batches spiked with approximately 10 spores from *Clostridium botulinum* or *Bacillus subtilis* proved to be sterile after irradiation with a dose of 25 kGy (Postmes et al. 1995), but after it was revealed that this article did not study the effects of irradiation on the properties of honey and that ointments made with irradiated honey are rapidly aging and produce gas, which leads to explosive waste of the ointment when the tubes are opened. In a recent study performed by Hussein et al. (2011), it had been shown that antioxidant capacities and total phenolic contents increase with γ-irradiation (cobalt-60 at 25 kGy) in two types of Malaysian honey.

In spite of the high quality of pure bee honey as a topical wound agent compared with other topical wound agents, its use is not without certain problems that may limit its optimal application. Excessive fluid loss through the burn wound may alter the beneficial properties of the honey applied by causing its rapid dilution and accelerating its washout from the wound surface. It is also well known that, within a few hours of application of honey dressing to the wound surface, the dressing will dry off because of evaporative water loss enhanced by body temperature, making subsequent dressing changes difficult and possibly compromising the natural wound-healing process (Osman et al. 2003). In the case of manufactured medical honeys, all dressings must be used in accordance with the manufacturer's instructions. This helps endorse the maxim "do no harm" and ensure that the full benefit of the product is realized. Because of its fluid and viscous nature, honey can be difficult to apply. This is particularly true when profuse exudate is present, diluting the honey. Experience has shown that use of the appropriate honey vehicle, including a secondary dressing, can sometimes circumvent this problem (Bogdanov 2012). If the dressing is inappropriate, the honey may be washed out of the wound by exudate. The clinical benefits of medical honey including antibacterial protection, wound cleaning, and pH modulation may as a result be reduced; the best way to keep the honey in the wound is to soak medical honey into a calcium-alginate or hydrofiber dressing, which forms a gel with the honey as it absorbs the exudate (Simon et al. 2009). The various registered medical honey products available, some in the form of prepared honey-impregnated dressings, have been described (Molan and Betts 2004; Visavadia et al. 2006).

Besides the scientifically based use of honey in wound care in hospitals, it can also be used under home conditions, as it was used for many centuries. Although sterilized honey is only used in hospitals, raw honey can also be used under home conditions without any risk, as no adverse effects have been reported. Indeed, Professor Descotte lectured in several apitherapy conferences that his team has used raw honey routinely for wound care in hundreds of cases in the hospital of Limoge, France (Bogdanov 2009). Complex wounds and wounds of immunocompromised patients should only be treated under professional medical supervision. The additional administration of systemic antibiotics is often necessary in pediatric oncology patients during periods of profound neutropenia ($<0.5 \times 109$/L) (Simon et al. 2006).

With honey, we also need to be aware that it is a natural product and that those characteristics associated with wound healing may be affected by species of bee, geographical location, and botanical origin as well as processing and storage conditions (Moore et al. 2001). It is possible therefore to assert that the microorganisms found in honey undergo gradual extinction in honey due to its inhibitory properties as highlighted earlier in this discourse. It is also recognized that spores are dormant forms of certain microorganisms. The fact that spores cannot transit into vegetative forms and still remain alive in honey persistently is supportive of the inhibitory role of honey on microorganisms. The failure to take into account the large variance in antibacterial potency of different honeys may contribute, in part, to the large discrepancy in results reported between hospitals using honey in similar ways. Is it also possible for the spores in honey to transform to active microorganisms and therefore become pathogenic after honey has been applied to the wounds especially with dilution of the initial high osmolarity and other properties that inhibit microorganisms (Olaitan et al. 2007)? Then, swabs from wounds should be cultured, microorganisms should be isolated, and their sensitivity to honey should be assessed before commencing treatment with honey. This is important not only because of the varying activities of honey but also because of the varying microorganisms they may contain. Any honey should not just be used on just any wound without this preparation. This will ensure judicious use of honey for organisms against which it is likely to be active and diminish the possibility of infecting the wounds rather than destroying the microbes.

Chronic wounds are defined as wounds that have failed to proceed through an orderly and timely reparative process to produce anatomic and functional integrity over a period of 3 months (Mustoe et al. 2006). Most chronic wounds occur in the aged population (older than age 60 years). Although most wounds heal without incident in aged patients, there is a slight, but consistent, decline in wound-healing rates in the elderly. The effect of aging declares itself when a variable such as ischemia or infection is superimposed on an injury. Laboratory studies reveal a decline in molecular processes important for tissue repair in aged fibroblasts and endothelial cells (Galiano and Mustoe 2007).

Chronic pressure ulcers, also known as pressure sores, refer to wounds developed over bony prominences. They are a major health problem, occurring mainly in elderly persons and patients with debilitating illness that renders them immobile. In the attempt to find a cost-effective treatment that decreases pressure ulcers' healing time and severity, a group of researchers in Germany studied 13 men and 7 women. Six of the patients were quadriplegic and 14 were paraplegic. Five patients had grade IV ulcers; 15 patients had grade III ulcers. The study sought to determine the effects of Medihoney—a branded form of *Leptospermum* honey and gelling agents—on bacterial growth, wound size, and stage of healing in pressure ulcers. After 1 week of treatment, all swabs were clear of bacterial growth. Overall, 90% of the patients showed complete wound healing after 1 month, and scars were soft and elastic. No negative effects were noted from the treatment. The scientists recommend that the medical-honey approach to wound care be part of a comprehensive, conservative surgical wound-care concept (Biglari et al. 2011).

The difficulties faced by staff responsible for wound care may be best illustrated by reference to the treatment of leg ulcers, which probably represents one of the greatest challenges faced by community nursing staffs (Robson et al. 2009). A review of 40 patients using honey for venous ulcers showed that Medihoney is a promising agent in the management of leg ulceration. Most patients' pain scores improved, and in many cases, healing was achieved. Furthermore, malodorous wounds were promptly and efficiently deodorized. Dressing removal was easy and was never reported as painful (Dunford and Hanano 2004). Aspects of the patients' existing treatment, such as method of cleansing, dressing-change frequency, and type of compression bandaging/hosiery used, were left unaltered. The only intervention was the application of a honey dressing in place of the previously prescribed primary dressing. Medihoney dressings were prepared by applying the honey to a low-adherent, sterile, contact layer placed on top of a sterile dressing pad. The honey was applied to a depth of approximately 3 mm (roughly 20 g of honey to a $10 \times 10$ cm dressing).

New Zealand manuka (*Leptospermum scoparium*) honey is known to exhibit non-peroxide antibacterial activity caused by the active ingredient methylglyoxal (MGO), which arises by chemical conversion of dihydroxyacetone during honey maturation (Windsor et al. 2012). Since this honey contains high levels of MGO, it has been speculated that patients with diabetes may be at risk due to either the direct negative effect of MGO on cells and components in the wound or the indirect formation of

AGEs, which could impair the wound-healing process (Majtan 2011). However, it is worth noting that certain combinations of compounds result in a decrease in toxic effects, previously noted when compounds were administered singly (Parker et al. 2010).

Biofilms have been identified in association with a number of chronic wound settings including diabetic wounds, venous stasis ulcers, and pressure sores (James et al. 2008; Kirketerp-Møller et al. 2008; Wolcott et al. 2008). In a study to evaluate the contribution of MGO to the biofilmcidal activity of manuka honey (see after) and furthermore determine whether the antibiofilm activity of low-dose honey can be augmented by the addition of exogenous MGO, it was demonstrated that honeys containing at least 0.53 mg/mL MGO had biofilmcidal activity against *Staphylococcus aureus*. Furthermore, the antibiofilm activity of non-MGO honey supplemented with MGO was observed to mimic that of manuka honey (Jervis-Bardy et al. 2011).

Is honey really "the bees' knees" for diabetic foot ulcers, as said by Freeman et al. (2010)? There has been a resurgence of interest in the use of topical honey to treat diabetic foot ulcers, reflecting a growing awareness of the cost and burden of diabetic foot ulcers (Eddy et al. 2008).

The point of product saturation in the wound-dressing market has certainly been reached. Many services now restrict the range of products that are available for staff use. However, it is worthwhile considering additional or alternative wound dressings that may benefit particular wound types that are not catered for within an existing range. It was this situation that led to the introduction of medical-grade honey wound dressing products in an acute, multidisciplinary high-risk foot service (Freeman et al. 2010).

The most common risk associated with honey's use is a burning or stinging sensation due to its low pH. This concern may not be relevant for neuropathic diabetic foot ulcers that result from a lack of sensation (Eddy et al. 2008). According to this author, all patients with adequate blood supply and no evidence of osteomyelitis are candidates for honey therapy.

A prospective study compared the effect of honey dressing for Wagner's grade II diabetic foot ulcers with a controlled dressing group (povidone iodine followed by normal saline). Wound dressing with honey is an option for managing Wagner grade II diabetic foot ulcers with the rate of the wound healing comparable with the use of povidone iodine solution. It decreased wound edema and odor more effectively. Removal of gauze for repeated dressing is easier to perform and is less painful for the patient (Shukrimi et al. 2008).

A study involving 200 patients, performed to compare dressings with honey/normal saline to povidone iodine/hydrogen peroxide in the management of diabetic foot ulcers, showed that the healing and hospital stay times were shorter with honey/normal saline than povidone iodine/hydrogen peroxide (Hammouri 2004).

Five clinicians trialed medical-grade honey wound gel and medical-grade honey alginate on appropriate diabetic foot wounds and completed a simple evaluation form for each application including patient tolerability. Results were as follows: clinician ease of use, clinician overall satisfaction, and patient comfort were rated as "high" in the majority of applications (66%–93%) (Freeman et al. 2010).

At the Department of Pharmaceutics, Faculty of Pharmacy, University of Karachi, and Outpatient Department of Dermatology, honey ointment was prepared with the following ingredients: white soft paraffin 100 g, liquid paraffin (Merck) 100 g, lanolin 100 g, and crude honey (antimicrobial potential) 200 g to make (with white soft paraffin) 1000 g. Tubes and plastic jars were filled under aseptic conditions. For sterilization of the product, an effective dose of γ-radiation was applied. The efficacy of such ointment was evaluated by conducting clinical trials in place of by passing thought clinical trials. A total number of 27 patients (23 skin wound infections and 4 diabetic foot ulcers) were involved in the study. A thin layer of newly formulated honey ointment on gauze was applied two to three times per day. Very significant results (99.15%) of healing were observed in skin wound infection cases with a mean healing time of 5.86 (2–20) days, and 95% diabetic foot ulcers healed with the mean healing time of 20 (8–40) days (Tasleem et al. 2011).

Manuka honey tulle dressings (Activon) have been used in the Maxillofacial Unit, Royal Surrey County Hospital of United Kingdom and have had success in treating recalcitrant surgical wounds within the maxillofacial unit, which had proven to be resistant to antibiotics. An 80-year-old man had a split skin graft harvested from his upper arm in May 2005. The wound was managed until September 2005 with routine wound dressings but remained contaminated with methicillin-resistant *S. aureus* (MRSA). The first manuka honey dressing was applied at the end of September 2005 and the wound had healed 2 weeks later. In another case, a 64-year-old man had a radial forearm flap donor site grafted with an abdominal full-thickness skin graft. The wound was infected with MRSA on removal of the dressings 7 days postoperatively. Treatment consisted of dressings on alternate days with manuka honey tulle and local debridement of eschar when necessary. The wound healed without further complications 5 weeks later (Visavadia et al. 2006).

A 12-year-old patient was submitted to the Pediatric Oncology Department at the Children's Hospital Medical Centre, University of Bonn (Bonn, Germany). Doctors at another hospital had removed an abdominal lymphoma, leaving an open drainage site on his abdomen. On admission, his wound was infected with MRSA. In order to avoid nosocomial spread, the patient was immediately isolated, a difficult situation for the child to comprehend with significant additional costs from the perspective of the hospital. Although the patient was scheduled to receive chemotherapy, treatment could not commence until the infection cleared. The wound was treated with a local antiseptic (octenidine) for 12 days. Since no improvement occurred, we decided to use Australian medical honey (Medihoney). The wound was free of bacteria 2 days later, and the chemotherapy against the underlying illness could be started (Simon et al. 2009).

At the Department of Plastic Reconstructive and Aesthetic, Surgery of the Numune State Hospital (Erzurum, Turkey), medical honey (HoneySoft, MediProf, Moerkapelle, the Netherlands) was used for the fixation of the skin graft in 11 patients who underwent different diagnosis. It has been observed that it has strong adhesive properties for skin graft fixation (Emsen 2007). Skin graft wound area requires dressings to encourage epithelialization; an antimicrobial dressing will be appropriate to reduce the bacterial load. Skin graft sites that produce large amounts of exudate will require an absorbent dressing. It is important to prevent maceration of the remaining skin graft or the surrounding skin. Foam, alginate, or hydrofiber dressings should be used (Beldon 2007). Honey promotes epithelialization and its antimicrobial property is well known; medical honeys contain alginate, hydrofiber, etc., to prevent maceration. The most important factor is that medical honey is a natural material, not synthetic. Comparison of lyophilization and freezing in honey as techniques to preserve cortical bone allografts used to repair experimental femoral defects in domestic adult cats showed that bone grafts preserved in honey or frozen were effective for repairing cortical defects in the femurs of cats compared to autogenous cortical bone grafts (Ferreira 2009).

It has long been understood that microbial species living in natural environments normally associate with surfaces where they collectively secrete sticky materials that help to maintain their position long enough to provide an opportunity to develop complex structured communities or biofilms (Cooper 2010).

Unattached microbes, otherwise known as planktonic cells, are considered to represent the form by which dispersal to other suitable locations is ensured and new biofilms are established (Watnick and Kolter 2000). Now, biofilm diseases are thought to affect more patients than the numbers affected by heart disease and cancer combined (Balaban 2008). Interest in the potential association of biofilms in the pathology of chronic wounds has been demonstrated by the recent influx of publications referencing the keywords "biofilm" and "chronic wound." Increasing evidence from these studies suggests that biofilms may play a role in wound chronicity (Kirketerp-Møller et al. 2011). According to Daniel Rhoads (South Regional Wound Care Centre, Lubbock, Texas, USA), biofilms are the new threat to health and he believes, "the developing world is still struggling with acute infections like cholera and malaria. The developed world has largely overcome these diseases, and is now facing a foe that it does not know how to conquer: chronic bacterial infections associated with biofilms." Alternative antimicrobials have been researched, proposed, and used with some degree of success. Among these naturally occurring compounds are maggots, garlic (Rasmussen et al. 2005), silver (Bjarnsholt et al. 2007), and honey. Despite their popularity and wide use, silver-based modalities are not without complications. Delayed wound healing is often observed. This might be due to the retardation of the sloughing of dead tissue in partial thickness burns. In addition, increased hypertrophic scarring and skin irritation has been described using silver sulfadiazine; black staining of the skin and the possibility of systemic absorption of silver have been reported (Atiyeh et al. 2007; Pham et al. 2007). In fact, it has been reported that honey interferes with the adherence ability of *Pseudomonas aeruginosa* to surfaces, because molecules of fructose (which is the most abundant sugar in honey) competitively block the PA-IIL lectin that mediates adhesion of the bacterium to fucosylated receptors on the membranes of potential host cells (Lerrer et al. 2007). Okhiria et al. (2009) showed that 40% (w/v) manuka honey was necessary to inhibit biofilms of *P. aeruginosa* in vitro and that this inhibition was dependent on exposure time and honey concentration. He concluded that, because the concentration of manuka honey that is incorporated into wounds gels, calcium alginate-impregnated dressings, and tubes of sterile honey ranges between 80% and 100%, it is probable that the topical application of wound-care products containing manuka honey would initially reduce biofilms. When honey is applied to a wound, there can be increased exudation, so it will be important to investigate the rate of dilution in vivo and to determine the concentration at which honey ceases to inhibit biofilms and begins to promote biofilm growth. This will influence recommendations for the frequency of dressing changes (Okhiria et al. 2009).

A study to determine the effectiveness of honey in inhibiting biofilms of *P. aeruginosa* and *S. aureus* grown in the Calgary biofilm device showed that Sidr honey from Yemen and manuka honey from New Zealand were more effective than commonly used antibiotics (Alandejani et al. 2009). The 2006 publication "Bacterial biofilms in chronic rhinosinusitis" reported, "Chronic sinusitis is a prevalent, debilitating condition, and a subpopulation of patients fails to respond to either medical or surgical intervention" (Palmer 2006). A particular bacterium often found in patients with chronic rhinosinusitis, which forms biofilms, is MRSA (Kluytmans et al. 1997). The 2011 publication "Characterization of bacterial community diversity in chronic rhinosinusitis infections using novel culture-independent techniques" reports that *P. aeruginosa* is the most frequently detected species (Stressmann et al. 2011). Nasal lavage with solutions of honey could have a role to play in treating patients with recurrent and persistent infections (Alandejani et al. 2009).

Laboratory investigations into the effect of honey on biofilms indicate that planktonic cells are more susceptible to honey than biofilms. Also, lower concentrations of honey are required to prevent biofilm formation than those required to disrupt established biofilms (Merckoll et al. 2009; Cooper et al. 2011). Manuka honey from New Zealand has been shown to prevent cell division in planktonic cultures of MRSA (Jenkins et al. 2011) and to cause cell surface changes and lysis in *P. aeruginosa* (Henriques et al. 2010). Cultures of planktonic cells of *P. aeruginosa* in suspension tests have been shown to be susceptible to manuka honey at concentrations less than 10% (v/v) (Cooper and Molan 1999; Cooper et al. 2002).

Most bacteria are able to regulate phenotypic characteristics, including virulence factors, as a function of cell density under the control of chemical signal molecules (Gram et al. 2002). Quorum sensing (cell-to-cell communication) can occur within a single bacterial species as well as between multiple species. The quorum sensing-coordinated process is achieved by producing, releasing, and detecting small signal molecules known as autoinducers. Microorganisms can use quorum sensing to coordinate their communal behaviors, biofilm formation, swarming, motility, and production of extracellular polysaccharides. These autoinducer molecules have been identified as oligopeptides in gram-positive bacteria and acylated homoserine lactones in gram-negative bacteria (González and Keshavan 2006; Williams and Williams 2007).

One other interesting attribute of honey is its ability to inhibit quorum sensing. In a survey of 29 unifloral honeys, a pigmented reporter bacterium (*Chromobacter violaceum*) was employed to detect quorum sensing agonists; all the 29 honey samples inhibited the acylated homoserine lactone production, even at the lowest concentration. The anti-quorum sensing activity was concentration dependent, as the inhibition activity increased with increased honey concentration. Among all honeys, chestnut and linden samples had the strongest quorum sensing activity (Truchado et al. 2009).

At the present time, the industrial and clinical implications of biofilms in unwanted situations are far-reaching. The reduced susceptibility of biofilms to antimicrobial agents confounds the use of conventional antibiotics and antiseptics and requires some novel interventions. The ability of honey to interrupt quorum sensing provides a means to prevent and disrupt biofilms in diseased plants and animals, and the eradication of antibiotic-resistant strains from wounds by topical application of honey might also help to reduce cross-infection in health care establishments. There is a need to characterize the active component(s) in honey and to determine how quorum sensing is inhibited. The development of a new class of antimicrobial agent might be possible. Honey no longer seems to be just a quaint folk remedy for wounds that has no place in modern clinical practice (Cooper 2011) (Table 14.2).

## HONEY AS AN INGREDIENT IN MEDICINE AND MEDICINE-LIKE PRODUCTS

Inflammation is widely recognized as a risk factor for gastric *Helicobacter pylori*–associated disease and disruption of this process provides a potential target for intervention. Using an in vitro system, broccoli sprouts, manuka honey, and omega-3 oil, singly and in combination, were screened for their ability to limit *H. pylori*–associated inflammation. Each food significantly attenuated the release of interleukin (IL)-8 by *H. pylori*–infected cells, although the magnitude of this effect was variable. Only broccoli sprouts were able to inhibit IL-8 release in response to tumor necrosis factor (TNF)-$\alpha$. The combination of manuka honey with omega-3 oil failed further to reduce IL-8 levels below those observed with honey alone, but the same concentrations of omega-3 oil and manuka honey independently enhanced the anti-inflammatory effect of the isothiocyanate-rich broccoli sprouts (Keenan et al. 2012).

A study was undertaken to determine the effects of ArmApis BIO Honey (alpine honey produced from the nectar of 101 alpine plants, including Sage, Marjoram, Cephalaria, Thyme, Mint, Shepherd's-club, St John's wort, etc.) enriched with Royal Jelly BIO Honey on the human monocyte activation and release of interferon (IFN)-$\gamma$, TNF-$\alpha$, IL-1$\beta$, and IL-10 and human monocyte-derived dendritic cell activation in vitro. It was reported that the ArmApis BIO Honey enriched with Royal Jelly BIO Honey stimulates innate immune cells and particularly induces human monocyte activation and stimulates cytokine production in vitro. ArmApis BIO Honey stimulates the production of inflammatory cytokines including TNF-$\alpha$ and IL-1$\beta$ as well as induces anti-inflammatory cytokine IL-10 and pleiotropic immunoregulatory cytokine IFN-$\gamma$ production by human monocytes. ArmApis BIO Honey enriched with Royal Jelly BIO Honey induces the upregulation of dendritic cell activation marker CD86 expression on monocyte-derived immature dendritic cells (Table 14.2) (Davtyan 2011).

According to Johnson et al. (2005), thrice weekly application of standardized antibacterial honey (Medihoney) to hemodialysis catheter exit sites was safe, cheap, and effective and that with Medihoney the problem of resistance induction against mupirocin can be circumvented.

Report from the WHO estimates that 46% of the world's 5- to 14-year-old children are anemic; 48% of the world's pregnant women are anemic due to iron deficiency. The iron absorption is carefully regulated to maintain an equilibrium between absorption and body iron. The low absorption and bioavailability with iron salts are the cause of several adverse reactions. Clinical trials were carried out by González and Aznar (2006) using a Cuban formula (Trofin/Biotrofer) with composition about minerals (natural iron), proteins, peptides, and bee honey, which proved its high solubility, absorption, biodisponibility, efficacy, and tolerance.

## TABLE 14.2
## Examples of Clinical Trials with Honey

| Honey Source | Patients | Mode of Application | Results | Reference |
|---|---|---|---|---|
| Medihoney dressings | Forty patients with nonhealing ulcers despite having received compression therapy for at least 12 weeks. | By applying the honey to a low adherent, sterile, contact layer placed on top of a sterile dressing pad. The honey was applied to a depth of approximately 3 mm (roughly 20 g of honey to a 10 × 10 cm dressing). The leg ulcer nurse or community nurse applied the compression bandages. | Ulcers in seven patients healed within the 12-week study period, a significant reduction in wounds in the 20 ulcers of patients who did not heal at end of the study. In contrast, average wound of those who dropped out remained static. Six patients withdrew because of pain and another two because their ulcer deteriorated. | Dunford and Hanano 2004 |
| Medihoney Apinate (calcium alginate) dressings (Derma Sciences) | A 72-year-old diabetic female, with a rapid onset reddened area on the dorsal surface of her foot, which was diagnosed as cellulitis. | The dressings were applied with an absorbent cover dressing and changed daily. | Rapid liquefaction of devitalized tissue was noted within several days of initiating Medihoney and the wound bed was clean within 1 month. | Stephen-Haynes 2011 |
| Medihoney dressing (Woundcare 18+; Comvita) | A group of 108 patients with a history of nonhealing venous were enrolled in a multicenter trial to evaluate the desloughing efficacy and healing outcomes using either active *Leptospermum* honey (ALH) or hydrogel. Each of the patients had failed to respond to standard compression therapy and had greater than 50% slough covering the wound. | Subjects were randomly assigned to a weekly treatment with either ALH (n = 54) or hydrogel (n = 54). The primary wound dressing was covered with hydrocellular foam and all subjects received sustained multilayer compression bandages. Subject withdrawals from the study were consistent with other VU studies. The main reasons for patient withdrawal included VU infections (hydrogel, 22%; ALH, 11%) and patient request (Hydrogel, 5.5%; ALH, 0%). | At the end of week 4, there was a 67% reduction in slough in the ALH group compared to a 53% reduction in slough for those in the hydrogel group. Epithelialization was noted earlier in the ALH group and this difference. At the end of 12 weeks, the percentage of patients healed was higher for ALH patients (44%) than for hydrogel patients (33%). | Gethin and Cowman 2008 |

(continued)

**TABLE 14.2 (Continued)**
**Examples of Clinical Trials with Honey**

| Honey Source | Patients | Mode of Application | Results | Reference |
|---|---|---|---|---|
| Melladerm Plus (SanoMed) | A 76-year-old female developed a venous ulcer on her left leg, but due to excess amounts of exudate, the whole leg became one large wound over a period of 18 months. | Prior to honey gel treatment, the leg was treated by being washed in water twice a week; emollient and compression bandaging were then applied. The honey gel treatment comprised 10 min of rubbing over the entire wound area. | Over 80% of the necrotic tissue was removed at this point with gentle pressure from the handle of disposable forceps. The remaining crust dissolved over the next few days. | Nestjones and Vandeputte 2012 |
| Mesitran Ointment (Aspen Medical) | A 45-year-old woman who presented with an area of redness secondary to a suspected insect bite on the left lower leg pain rating was moderate and the area was described as irritating. | Applied to the area twice daily. | Significant anti-inflammatory effect and within 3 days the irritation resolved quickly. | Stephen-Haynes 2011 |
| Natural unprocessed honey (source not précised) | 38 patients, overweight or obese aged between 20 and 60 years. | 70 g of natural honey dissolved in 250 mL tap water each day for a maximum of 30 days. Subjects in both groups did not undergo a special diet regimen, drug therapy, or change in their lifestyle. | Mild reduction in body weight (1.3%) and body fat (1.1%). Honey reduced total cholesterol (3%), LDL-C (5.8), triacylglycerol (11%), FBG (4.2%), and CRP (3.2%), and increased HDL-C (3.3%) in subjects with normal values, while in patients with elevated variables, honey caused reduction in total cholesterol by 3.3%, LDL-C by 4.3%, triacylglycerol by 19%, and CRP by 3.3% ($p < 0.05$). | Yaghoobi et al. 2009 |

| | | | |
|---|---|---|---|
| L-Mesitran + L-Mesitran Hydro + L-Mesitran Net (Triticum) | 45-year-old female with an open ulcer of 14 years' duration in the left leg, with slough, copious exudation, and an unpleasant odor and an open ulcer, of 8 years' duration at the right leg. | Honey-based ointment (L-Mesitran, Triticum), which was applied. The ointment was used for the first 14 days (daily) and covered with crepe bandages. After 14 days of treatment, a switch was made to a honey hydrogel sheet (L-Mesitran Hydro), with three dressing changes a week. At day 45, because the exudate level was minimal, the authors decided to change the honey dressing again to the mesh (L-Mesitran Net). This was used twice weekly, until the end of treatment when complete healing had been achieved. | After 3.5 months (106 days) of treatment with the honey products, the patient's infected ulcers had successfully healed and she was able to be discharged. Compression therapy with short-stretch bandages (PütterVerband) was initiated at discharge. These were applied at the community health center. The final result was excellent with good cosmetic results. At 6-month follow-up, there was no recurrence of the ulcers and the legs were in a good condition. | Miguéns 2011 |
| Ordinary honey purchased at a supermarket (botanical origin not mentioned) | 79-year-old man with type 2 diabetes mellitus. Informed by two different surgical teams that without below-the-knee amputation he would likely die. | Once-daily, thick applications were smeared on gauze 4 × 4s and placed on the wounds, which were then wrapped. Oral antibiotics and saline dressings were discontinued, but otherwise treatment was unchanged. | Granulation tissue appeared within 2 weeks; in 6–12 months, the ulcers resolved. Two years later, the ulcers have not recurred; the patient ambulates with a walker and reports improved quality-of-life. | Eddy and Gideonsen 2005 |
| Pure honey (botanical origin not mentioned) | 27 patients aged between 8 and 11 years with gastroenteritis. | Patients received either oral fluid only or oral and intravenous fluids depending on the severity of dehydration and the presence of intractable vomiting. Instead of glucose, the oral rehydration solution contained 50 mL pure honey per liter. Antibiotics were given only if there was a concurrent infection. | The mean recovery times for the control (conventional therapy) and honey treated groups were almost identical. One patient in the honey-treated group, but none of the controls, had diarrhea lasting 10 days. | Haffejee and Moosa 1985 |

(continued)

## TABLE 14.2 (Continued)
## Examples of Clinical Trials with Honey

| Honey Source | Patients | Mode of Application | Results | Reference |
|---|---|---|---|---|
| Mixture of Egyptian bee honey and royal jelly | 50 women's age, years (28.6 ± 5.5). Participating couples were recruited between March 2002 and November 2005 at the Specialized Gynecology and Infertility Clinic of Sohag University Hospital and at the first author's private clinic in Asyut. | Mixture of 100 g of Egyptian honey mixed with 3 g of royal jelly and 1 teaspoon of bee bread. Before use, the mixture was diluted at a ratio of 1:1 in a normal saline solution. It was then self-administered intravaginally at each coital act, beginning 1 day after the last menstruation. This protocol was to be repeated over 2 weeks. A plastic piston applicator or a 10 mL syringe was used to insert the mixture, and the insertion was either precoital or postcoital, depending on the couple's preference. | The study had its limitations. It was intended to be a case crossover study, but it is not a clean one because some of the women became pregnant prior to the crossover. Still, it might be considered as an "unadjusted" analysis. In conclusion, the midcycle pericoital intravaginal self-administration of bee honey and royal jelly is a simple, acceptable, and reasonably effective method for the treatment of infertility due to asthenozoospermia. It may be tried before resorting to the more costly assisted reproductive techniques such as intracytoplasmic sperm injection. | Abdelhafiz and Muhamad 2008 |

| Source | Patients | Method | Results | Reference |
|---|---|---|---|---|
| Source not précised | 100 patients with confirmed cutaneous leishmaniasis were selected and randomized into two groups. Ten patients left the study. | Group A was treated with topical honey twice daily along with intralesional injection of glucantime once weekly. Group B were treated with intralesional injection of glucantime alone. | In the glucantime alone-treated group, 32 patients (71.1%) had complete cure, whereas in the group treated with both glucantime and topical honey, 23 patients (51.1%) achieved complete cure. | Nilforoushzadeh et al. 2007 |
| Three Pakistani floral sources (*Trachyspermum copticum*, *Acacia nilotica* species indica, *Zizyphus*) honey | 23 patients (23 males, 1 female), aged between 1.5 and 60 with wide variety of wounds and 4 (3 males, 1 female) aged between 48 and 60 with diabetic ulcers. | Ointment was prepared with the following ingredients: white soft paraffin (Merck) 100 g, liquid paraffin (Merck) 100 g, lanolin (Merck) 100 g, crude honey (antimicrobial potential) 200 g to make (with white soft paraffin) 1000 g. Tubes and plastic jars were filled under aseptic conditions. For sterilization of the product, an effective dose of $\gamma$-radiation was applied. Each wound cleaned with normal saline and then dressed with thin layer gauze. The treatment for wound management was continued two to three times daily until the complete healing. | All the skin wound infection cases healed 99.15% within the mean healing time of 5.86 days. Diabetic foot ulcer cases healed 95% within the mean healing time of 20 days (range 8–40 days). | Tasleem et al. 2011 |

A phytomedicine Syrup (Saratosse, Brazil) composed of several medicinal plants (*Mikania glomerata, Mentha piperita, Eucalyptus globulus*, and *Copaifera multijuga*) along with honey and propolis has been evaluated to treat respiratory diseases. The clinical trial consisted of an open study with 26 adult volunteers of both sexes, who were given an oral dose of 15 mL of Saratosse for 28 consecutive days, four times a day. The laboratory tests included hematologic, biochemical, and serologic analysis. This evaluation was repeated after each week of treatment and 7 days after the last administration. On the whole, the medicine was well tolerated. Some side effects were related, which may or may not be attributed to the phytomedicine. Clinical, electrocardiographic, and laboratory tests did not show any evidence of toxic signs in the organs and systems studied (Tavares et al. 2006).

Studies conducted at some Western universities have reinforced the centuries-old knowledge that honey is effective against arthritis. Taking 1 tablespoon of honey and a half teaspoon of cinnamon powder has helped arthritic patients walk without pain. Honey with lukewarm water and a dash of cinnamon powder made into a paste and applied to the joints also helped in quick recovery (within minutes) from pain. In a recent research done at Copenhagen University, it was found that, when the doctors treated their patients with a mixture of 1 tablespoon honey and a half teaspoon of cinnamon powder before breakfast, within 1 week, of the 200 people so treated, practically 73 patients were totally relieved of pain and, within 1 month, mostly all the patients who could not walk or move around because of arthritis started walking without pain (Kumar et al. 2010).

There is substantial evidence that ghee and honey dressing are at least comparable to other modes of treatment. Honey becomes watery at body temperature, and it is difficult to maintain honey in contact with the wound at a sufficiently high minimum inhibitory concentration without very frequent dressings. However, honey when bound with ghee maintains its viscosity at body temperature and is in longer and stronger contact with the wound surface augmenting its activity. Surprisingly, ghee and honey dressing act as insect repellants, as no insects are found on these dressings. The easy availability and low cost of this treatment make it significant in developing countries (Udwadia 2011).

How can honey help you sleep? Eating honey raises your blood sugar level slightly. This results in a controlled increase in insulin, which then causes the amino acid tryptophan to enter your brain. The tryptophan is converted into serotonin, which promotes relaxation. Finally, in the pineal gland, with the aid of darkness, the serotonin is converted into melatonin, a well-known cure for sleeping disorders. Honey also contains the ideal 1:1 ratio of fructose to glucose making it a superfood for glycogen storage. Sufficient glycogen storage is necessary for restful sleep. When your liver runs out of glycogen at night, your brain starts to trigger stress hormones such as cortisol and adrenalin to convert protein muscle into glucose (http://www.squidoo.com/Honey-honeybees).

With honey and milk, infants suffer less frequently from diarrhea, and their blood contains more hemoglobin compared to those on a diet based on sucrose sweetened milk (Takuma 1955). A study conducted by Al-Jabri (2005) showed that honey alone requires more than 20 h for killing all bacteria (*S. aureus*). However, when milk was combined with honey, 100% killing was achieved in 16 h (Table 14.3).

## TABLE 14.3
### Honey-Based Mixtures in Some Ailments Medicine

| Ailments | Honey-Based Mixtures | References |
|---|---|---|
| As respiratory relief | Dissolve 1 tablespoonful of honey, 1 teaspoon of bee pollen, and some lime juice in a cup of freshly boiled water. | Namdeo et al. 2010 |
| As natural regulatory reaction | Honey taken with pollen either in liquids or as a topping on thick black bread. | |
| Youth elixirs | Combine 2 tablespoons of pollen, 1 teaspoon of chopped ginseng herb and dried orange peel. Take with a spoon. | |
| Sedative | Take 1 teaspoon of honey in warm milk before going to bed. | Kumar et al. 2010 |
| Remedy for colds | Honey can be taken either with warm milk or with lemon juice and radish juice. | |
| Sore throats | Honey in warm milk or water. | |
| Constipation, hyperacidity, and obesity | One spoon of fresh honey mixed with the juice of half a lemon in a glass of lukewarm water taken first thing in the morning. | |
| At the initial stages of tuberculosis | Honey and rose petals when taken in the morning. | |
| Heavy and painful menstrual periods and leucorrhea | Asafetida (Hing) fried in ghee and mixed with 1 tablespoon of honey, taken thrice a day. | |
| Recurrent lesions from labial and genital herpes | Lesions firmly pressed with gauze soaked with honey for 15 min four times a day. | Al-Waili 2004 |
| Hemorrhoids and anal fissure | Honey, olive oil, and beeswax in ratio of 1:1:1 (v/v/v). Apply about a spoon size of the mixture around the skin twice daily. Sleep in a supine position for 15 minutes then apply soft gauze to the anal region to keep the mixture at the site of application. | Al-Waili 2005b |
| Boils and furuncles | Mix liquid honey and flour 1:1. Add a little water and brush the affected area. Cover with gauze and leave it overnight. | Bogdanov 2012 |
| Muscle cramps | Cover the affected area with honey, cover with gauze or cloth, and fix it with adhesive plaster. Cover with a warm wool cloth. Leave for at least 2 h. | |
| Bruises and contusions | Mix honey and olive oil 1:1 and cover with mixture the affected area. Cover with gauze and leave for 4 to 6 h. | |
| Urinary tract infections | Diluted honey is ingested two to three times daily. | Meda et al. 2004 |
| Skin cleansing agent | Honey is applied undiluted overnight to the whole body and then washed off the following morning. | |

*(continued)*

## TABLE 14.3 (Continued)
## Honey-Based Mixtures in Some Ailments Medicine

| Ailments | Honey-Based Mixtures | References |
|---|---|---|
| A revitalizing concentrate | In parts by weight: 4 honey, 1 wheat germ (or wheat extract), 1 pollen extract. Dry yeast (brewers or bakers yeast). 0.1–0.4 royal jelly. | Dany 1988 |
| | 4 honey, 0.5 pollen (or extract), 0.5 yeast (or stimulating plant extract), 0.05–0.5 royal jelly. | |
| For dressing wounds | In parts by weight: honey paste: 10 wax, 3 propolis extract (10% ethanol extract) 2 honey. | Uccusic 1982 |
| Winter solstice cough[a] | Wild cherry bark (4 ounces), white pine bark (3 ounces), Osha root (3 ounces), Elecampane root (3 ounces), balsam root (2 ounces), sweet root (2 ounces), licorice root (2 ounces), Monarda honey (20 ounces), glycerin (10 ounces) | Moore 1995 |
| The Queen of Cramp Remedies[a] | Black haw (*V. runifolium*) (1.5 ounces), cramp bark (*V. opulus*) (1.0 ounce), trillium (dried) (1.0 ounce), OR cotton root bark (1.0 ounce), Dioscorea (wild yam) (1/2 ounce), Skullcap (recently dried) (1/4 ounce), cloves (1.0 ounce), cinnamon bark (3/4 ounce), orange peel (1/2 ounce), glycerin and honey. | |
| Mouth ulcers | *Ficus religiosa* fine powdered bark mixed with honey applied inside the oral cavity for 10–12 days. | Babu and Madhavi 2003 |
| Gouty arthritis | *Ficus religiosa* bark decoction with honey 15 mL twice a day with equal quantities of water. | |
| Cataracts | 2% solution of honey in saline in the eyes for 10 days followed by royal jelly in the eyes. | Solomon and Donnenfeld 2003 |
| | One drop of honey in the eye 4 or 5 times a day. | |
| Plaque, gingivitis, and periodontal disorders | Use honey chewing gum three times a day after meals. | English et al. 2004 |
| Sore and irritated throats | Mix 1 quart of water, 125 g honey, and 25 g alum and gargle. Try it for 2–3 days. | Joshi 2008 |
| Colds, coughs, and ticklish throat | Add 2 tablespoons of honey, a lemon and ginger juice in a cup of hot milk and drink it before going to bed; or add 2 tablespoons of honey, a lemon, and a peg of whisky in hot water and drink it as nightcap; or add 2 tablespoons of honey, one of lemon, 1 tablespoon of glycerine, and if possible 2–3 drops of menthol and eucalyptus oil to a glass of hot water and drink it. | |
| Hay fever, breathing problems, nasal and sinus complaints | Eat raw honey daily. During infection, chew honeycomb for 15 min like chewing gum and throw away whatever remains in your mouth. | |
| Mouth thrush and other infection | Use paste made of 1 part of borax, 1/2 part of glycerine, and 8 parts of honey. | |
| Rough skin and chaps on hands and face | Apply equal parts of honey and glycerine mixed together. | |

## TABLE 14.3 (Continued)
## Honey-Based Mixtures in Some Ailments Medicine

| Ailments | Honey-Based Mixtures | References |
|---|---|---|
| High blood pressure | Before having breakfast, drink a glass of lukewarm water mixed with 2 tablespoons of honey and one of lemon daily. Honey helps to dissolve cholesterol and increases the amount of hemoglobin in the blood. | |
| Abdominal pain | First thing in the morning, every day, take 2 tablespoonfuls of aloe gel together with a little water and honey. | Hirt et al. 2008 |
| Bronchitis, coughs, and colds | Fill a glass jar with peeled and chopped garlic cloves. Slowly pour in honey so that it fills all the gaps between the chopped garlic. Place the jar in a warm place of about 20°C. In 2–4 weeks, the honey will absorb the garlic juice and the garlic will become limp and opaque. Do not filter. Use within 3 months. Take 1 teaspoon of this mixture every few hours. | |
| Cough elixir | Add 250 g of dried and pounded eucalyptus leaves to 5 L of honey wine and stir well. Keep in a warm place, covered but not tightly closed, for 5 days. Filter and store in 1 L airtight bottles in a cool, dark place. Dosage: adults, 1 teaspoon, three times daily; children (1 year and older) 20–40 drops, three times daily. | |
| *Candida* in the mouth | Garlic and honey—Take 1 teaspoonful every few hours. Children may prefer a mixture of 1 teaspoonful of honey with 1 teaspoonful of dried, powdered artemisia leaves. | |
| Chronic metabolism | Mixture of royal jelly, honey, and ginseng, improvements in weight gain and psychological conditions but changes of blood characteristics. | Borgia et al. 1984 |
| Cleansing mask for oily skin (in parts by volume) | 4 Fuller's earth (or substitute), 1 rose water, 1 lemon juice, 2 honey, 15%–10% propolis extract. Should not be mixed until immediately prior to use, since they do not contain preservatives and will spoil rapidly. | Krell 1996 |
| Cracked hooves of animals | A mixture of equal parts melted beeswax and honey. It should be applied after the cracks have been thoroughly cleaned. | |

[a] For more details on preparation, see references.

In a Russian study, the use of a food product (Honey Laminolact) containing milk ferments, amino acids, fruit pectines, and of course honey, was proven effective in protecting the gastrointestinal tract from radiotherapy in women with uterine cancer (Smirnova et al. 2000).

The finding in several studies revealed that honey causes a reduction in blood glucose levels in both normal and diabetic patients and this is an indication that honey has a mechanism, probably an insulin sensitization effect (Al-Waili 2004). Camel milk does seem to contain high levels of insulin or an insulin-like protein, which appears to be able to pass through the stomach without being destroyed (Chaillous et al. 2000). Camel milk either alone or combined with honeybee pollen significantly reduced the hyperglycemia from $217.69 \pm 0.70$ nmol/μL (diabetic untreated rats) to $126.8 \pm 0.68$ and $115.90 \pm 0.60$ nmol/μL, respectively (Hassan and Bayoumi 2010).

In many regions of Europe, people use a combination of honey and pollen to help heal respiratory ailments. Folklorists suggest that you dissolve 1 tablespoonful of honey, 1 teaspoon of bee pollen, and some lime juice in a cup of freshly boiled water. This mixture soothes respiratory distress, sore throat, and other symptoms of lung discomfort (Namdeo et al. 2010).

Honeybee pollen mix (HBM) formulation is claimed to be effective for the treatment of asthma, bronchitis, cancers, peptic ulcers, colitis, various types of infections including hepatitis B, and rheumatism by the herb dealers in northeast Turkey (Küpeli et al. 2010). In a previous study, in vivo antinociceptive, anti-inflammatory, gastroprotective, and antioxidant effects of pure honey and HBM formulation were evaluated comparatively. HBM did not show any significant gastroprotective activity in a single administration at 250 mg/kg dose, whereas a weak activity was observed after 3 days of successive administration at 500 mg/kg dose. On the other hand, HBM displayed significant antinociceptive and anti-inflammatory activities at 500 mg/kg dose orally without inducing any apparent acute toxicity or gastric damage. HBM was also shown to possess potent anti-lipid peroxidant activity at 500 mg/kg dose against an acetaminophen-induced liver necrosis model in mice. On the other hand, pure honey did not exert any remarkable antinociceptive, anti-inflammatory, and gastroprotective activity but showed a potent anti-lipid peroxidant activity (Küpeli et al. 2010).

An evaluation of microbial quality in honey mixture with pollen (2.91% and 3.85%) and also dynamics of microbial groups in honey mixtures with pollen after 14 days of storage at room temperature (approximately 25°C) and in cold storage (8°C) showed, in general, that counts of microorganisms decreased in the honey mixture with pollen compared to raw pollen and these counts increased compared to natural honey raw pollen containing microscopic fungi. Honey mixture with 2.91% pollen after storage (14 days) contained lower microbial counts when compared with the sample analyzed at the beginning; beside sporulating bacteria and filamentous microscopic fungi in samples stored at 8°C, it has recorded growth of anaerobic microorganisms in honey mixture with 3.85% pollen after storage (8°C, 25°C/14 days) (Kňazovická et al. 2011).

To consume tea drink, people like to add lemon and honey in it, which are believed to not only increase its palatability, taste, and aroma but also enhance its health effects. The blended tea, lemon, and honey commonly called "Honey Lemon Tea" may be consumed freshly hot or consumed later after being kept at cool temperatures in a refrigerator or by addition of pieces of ice (Hategekimana et al. 2011).

However, like many other tea-based drinks, honey lemon tea was observed to form cream while it cools down. It is reported that creaming not only has an unattractive appearance but also damages both its taste and color (Yin et al. 2009), with the loss of certain physical attributes or biological activities due to interactions with compounds responsible for sensory attributes (i.e., astringency, aroma, color, and taste) as health-promoting characteristics. Hategekimana et al. (2011) showed that honeys from different sources have contributed to creaming in honey lemon tea depending on their main chemical components. The main components of honey that influence honey lemon tea cream formation are minerals, especially $Ca^{2+}$ and $Mg^{2+}$, proteins, and polyphenols through the complex reactions catalyzed by metal cations. The amounts of proline, total solids, and colloidal matters play a great role in creaming. Honey lemon tea creaming may therefore be reduced by controlling the interaction between proteins and polyphenols (i.e., increasing the solubility of the polyphenols) or by removing $Ca^{2+}$ or $Mg^{2+}$.

To evaluate the clinical and mycologic cure rates of a novel mixture consisting of bee honey and yogurt compared to local antifungal agents for treating patients with vulvovaginal candidiasis during pregnancy, a prospective comparative study that included 129 patients was carried on by Abdelmonem et al. (2012). The clinical cure rate was significantly higher in the study (mixture of bee honey and yogurt) than the control group (local antifungal agents), while the mycologic cure rate was higher in the control than the study group. Both types of therapy were favorably tolerated by most of the patients.

The aromiels are honey and essential oil–based preparations. Honey is a perfect vector for easy penetration of essential oils within the body and their assimilation; these products are characterized by a particularly effective synergy, in which the qualities of plants and honey reinforce and energize each other. They are reserved for curative use only and their preparation requires extreme precision in the choice of essential oils and their dosage. They are taken orally, with the exception of those used to treat skin problems, and are prepared only with chemotyped, essential oils. Combinations of honey and essential oils are endless. According to Domerego (2001), orally, the proportion is 3 to 5 g of essential oil per 100 g of honey, and, for external use, the proportion is 25 drops of essential oil per 100 g of honey.

Propomiel is the combination of honey and propolis in the form of tincture, indicated for the treatment of certain diseases and has been tested successfully on hospital superinfected wounds of burn victims. Indications and usage are similar to aromiels with which they are often used to supplement treatment. Aromiels and propomiels are used in university hospitals of Havana, Calixto Garcia, and FranckPais, and the Finlay Institute (Cuba), and since 2003 in Cameroon and Burkina Faso (Adam 1985). Propoaromiel is a pharmaceutical product containing honey, propolis, and essential oils, with a wide range of biological properties.

Honey may have an unexpected purpose. In Nigeria, natural honey was effectively combined as a composite with Nigerian gum Arabic for core binding for ferrous and nonferrous castings. Both materials are polysaccharides and possess volatile fractions that are easily vaporized off to give clean and accurate cores that do not quickly depreciate during storage. Binders from composites of simple natural materials subjected to simple, easy, and oven-baking processes have been synthesized as alternative materials to hazardous chemical binders that are not only expensive to develop in foundries but are also more difficult to handle and apply, as they require sophisticated skills and equipment (Ademoh 2011).

## HONEY-BASED MIXTURES IN TRADITIONAL PRACTICES

Among several animal sources that have been exploited for medical uses throughout history, honey takes a prominent place. An estimated 80% of the population in much of the developing world rely on traditional systems of medicine, and 70% to 80% of the population in developed countries have used some form of alternative or complementary medicine (WHO 2008). The WHO (2008) defines traditional medicine as "the sum total of knowledge, skills and practices based on the theories, beliefs and experiences indigenous to different cultures that are used to maintain health, as well as to prevent, diagnose, improve or treat physical and mental illnesses." If utilized properly, traditional medicine and health practices can be a source of income as has already been described in countries like India and China, where traditional health practices have evolved and developed over the years and are accepted the world over as established modes of alternative therapies. However, inappropriate use can be harmful and have deleterious effects on health (Khandekar and Al Harby 2006). At some instances, honey and ghee are mixed and served as food. In Ayurveda, Acharya Charaka has quoted that heated honey and honey mixed with equal amount of ghee produce deleterious effects in the body to the extent of causing death (Annapoorani et al. 2010). According to ancient Ayurvedic literature, honey should never be cooked. If cooked, the molecules become a nonhomogenized glue that adheres to mucous membranes and clogs subtle channels, producing toxins. Indeed, after a study on the effect of feeding of honey mixed with ghee and its heated forms on hepatotoxicity, antioxidant enzymes, and lipid profile of rats, Anilakumar et al. (2001) concluded that consumption of honey with equal amounts of ghee and its mixture in the heated forms raises serum alkaline phosphatase, uric acid, hepatic glutathione S-transferase, glucose-6-phosphate dehydrogenase, and γ-glutamyl transpeptidase with an associated increase in serum-conjugated dienes, hydroperoxides, and malondialdehyde. This author also advised against consuming heated honey. However, according to Brudzynski and Miotto (2011), when heating honey, its color is of great importance. Heat treatment of honeys leads to two different outcomes depending on the honey botanical origin: in light-colored honey and medium-colored honey, heat treatment accelerated the formation of melanoidins and increased the antioxidant activity. In the case of dark honey, heat treatment caused a decrease in melanoidins and caused a reduction in the antioxidant activity.

Indigenous people in all ages had some knowledge of plants and, through a systematic trial-and-error approach, applied them to various uses. Thus, the earliest attempts on the use of plants for medicines were based on speculation (Mirutse et al. 2003). The properties of the constituents of composite remedies could be different from those of the mixtures considered as a whole (Farnsworth 1980; Gedif and Hahn 2002). The use of honey in traditional formulations might, in addition to exerting a synergistic effect, also have a direct therapeutic effect on many other concurrent diseases. Traditional but well-studied medicinal systems as the Ayurvedic medicine of India use honey predominantly as a vehicle for faster absorption of various drugs such as herbal extracts (Krell 1996). Most medicines and remedies used in Unani are also used in Ayurveda. The base used in Unani medicine is often honey ("Unani" means "Greek"; Unani medicine originated around 980 AD in Persia). In India, Unani practitioners can practice as qualified doctors (Liu 2011).

Herbs may be processed to change their properties; especially, the meridian affinity of herbs can be changed by baking with honey, wine, vinegar, salt, etc. (Table 14.4). Herbs are used in many forms, and honey takes a place in the most common forms to prepare traditional Chinese medicine herbs; for example, herbal powder is mixed with

## TABLE 14.4
## Honey-Based Mixtures in Ethnomedicine

| Indications | Formulation | Country | Reference |
|---|---|---|---|
| Malaria | Honey and leaves decoction of *Musa sapientum* taken as drink. | Nigeria—Hausa, Yoruba, and Ibo communities | Ene et al. 2010 |
| | Leaves decoction of *Psidium guajava*, pineapple, honey. | | |
| Contraceptive | Mixture infusion of "tamarind" (part nonspecified) with pepper and honey in water, called Konkori Badji. | Mali | Laplante and Soumaoro 1973 |
| Sores, skin diseases, inflammation, and rheumatic joints | Leaves of Calotropisprocera mixed with turmeric, honey, and karanji applied as a paste. | Ethiopia—Bahirdar Zuria District | Muthuswamy and Abay 2009 |
| Constipation | Drink macerate of fruits *Tamarindus indica* L. (tamarind), in water, sweetmeat called bengal prepared of the fruit pulp mixed with honey or lime juice. | Senegal—Wolof communities | Dalziel 1937 |
| Carbuncle | Fruit of *Embelia ribes, Emblica officinalis, Piper longum*, and *Terminalia belerica* are mixed in equal amounts and the crushed form (powder) is added to honey and the concentrated honey is then applied on the infection. | India—Assamese People | Saikia et al. 2006 |
| | Boiled rice of kernel from *Oryza sativa* L. is mixed with curd made of buffalo's milk and is eaten for some days. | | |
| Measles | Flower bud of *Eugenia caryophyllata* is crushed and mixed with honey and the product is orally taken. | | |
| | The dried flower of *Nyctanthes arbor-tristis* is powdered and mixed with honey. The mixture is then orally taken. | | |
| Leprosy | Leaves juice of *Justicia adhatoda* L. is mixed with honey and the mixture is orally taken. | | |
| | Leaves juice from *Tinospora cordifolia* is orally taken with ghee or honey. | | |
| Face becomes fair and beautiful | The juice from the fruit of *Lycopersicon esculentum* Mill. is mixed with honey and the mixture is orally taken. | | |

*(continued)*

## TABLE 14.4 (Continued)
## Honey-Based Mixtures in Ethnomedicine

| Indications | Formulation | Country | Reference |
|---|---|---|---|
| Cancer | Bark powder of *Buchanania lanzan* with cow milk and honey. | India—Chhattisgarh | Jain and Jain 2010 |
| | Leaf juice and bark decoction internally, seeds of *Bambusa* sp. with Shahad (honey). | | |
| Antiepileptic | Adventitious root powder of *Ficus religiosa* mixed with honey. | India—Uttar Pradesh | Singh et al. 2011 |
| Migraine | Leaves and flowers of *Calotropis procera* are crushed and the paste is mixed with honey. | India—Tharus and indigenous people (district Bahraich) | Maliya 2007 |
| Gastric problems, asthma, cough, earache and toothache, migraine hematuria | Leaf juice of *Ficus religiosa* with honey. | Nepal | Kunwar and Bussmann 2006 |
| To tonify body and promote longevity | *Terminalia chebula*, *Terminalia belerica*, *Emblica officinalis*, *Angelica* sp., *Asparagus spinosissimus*, *Polygonatum cirrhifolium*, *Tribulus terrestris*, *Mirabilis himalaica*, clarified butter, milk, and honey (3–4 g daily). | Tibet | Sarong 1986 |
| Inflammation of the stomach and emesis of sour and watery vomitus, cough, hoarseness, difficulty in inhalation | Calcite, *Corydalis* sp., *Terminalia chebula*, *Picrorhiza kurroa*, mineral pitch, honey, *Vladimiria souliei* (2–3 g daily with hot water). | | |
| For all minor disorders specifically for Bad-kan and blood disorders | *Terminalia chebula*, *Picrorhiza kurroa*, calcite, *Corydalis* sp., *Prunus armeniaca*, mineral pitch, honey (2–3 g once in the morning with hot water). | | |
| Febrifuge, inflammation of eye, improves eye sight, watery eyes | *Terminalia chebula*, *Terminalia belerica*, *Emblica officinalis*, iron powder, *Glycyrrhiza glabra*, mineral pitch, *Carthamus tinctorius*, *Carum carvi*, butter, and honey (6 g daily with hot water). | | |

## TABLE 14.4 (Continued)
## Honey-Based Mixtures in Ethnomedicine

| Indications | Formulation | Country | Reference |
|---|---|---|---|
| Specially effective for promoting digestive heat, elixir and tonic, promotes normal flow of urine, tonic for the kidneys and urinary bladder, stops diarrhea, cold parasites, serumal disorders, and arthritis | *Punica granatum*, *Cinnamomum zeylanicum*, *Piper longum*, *Elettaria cardamomum*, *Carthamus tinctorius*, *Malva verticillata*, *Anglica* sp., *Polygonatum cirrhifolium*, *Tribulus terrestris*, *Mirabilis himalaica*, *Asparagus spinosissimus*, honey (2–3 g daily with hot water). | Tibet | Sarong 1986 |
| Cough, influenza | Oral, infusion of *Achyrocline flaccida* with lemon, and abeja's honey. Oral mixture of *Citrus paradisi* with abeja's honey. | Argentina—Nonindigenous population of Misiones | Kujawska et al. 2012 |
| Vaginal mycosis Anuria Urin with blood | Piece of bark of Tusca (*Acacia caven*) is boiled in 1 L of water; it is sweetened with honey and drunk several times a day until the sickness disappears. | Argentina—Zenta River Basin | Hilgert 2001 |
| Postpartum pain | A decoction prepared with a small handful of *Adiantum thalictroides* (or *Adiantopsis chlorophylla*) and about 5 cm of the root of *Cortaderia selloana* are added to the previous recipe. It is drunk cold, sweetened with honey, several times a day for about 10 days. | | |
| Cough | An infusion is prepared with a small handful of the fresh plant (*Geratumconyzoides* L. or *Aloysia citriodora* Palau) in 2 L of water, percolated, sweetened with honey and left to cool. It is drunk as a syrup during sickness. | | |
| Influenza | An infusion is prepared putting three fresh leaves mixed with three leaves of *Persea americana* in 250 mL of boiling water. It is sweetened with honey and drunk several times a day until cure. | | |

*(continued)*

## TABLE 14.4 (Continued)
## Honey-Based Mixtures in Ethnomedicine

| Indications | Formulation | Country | Reference |
|---|---|---|---|
| Common cold, antitussive | Turnip (Brassicaceae *Brassica rapa* var. *rapa* L.) raw slices (root) are chopped, covered in honey or sugar, left overnight, 1 spoon a day. | Bolivia—Cochabamba | Ceuterick et al. 2011 |
| Sore throat, common cold | Extract of fresh leaves of Watercress (Brassicaceae *Nasturtium officinale*), 1/2 cup each meal, with honey or singani, or cataplasm (100 g). | | |
| Antitussive, bronchitis | 1 onion (±100 g), 100 g of carrot, 1 green apple, grated, mixed with 1 L of water, 1/2 spoon of aniseed, boiled until only 1/2 L is left, mixed with honey, 2–3 times a day. | | |
| Stomach ailments, tonic | Leaves of aloe (Liliaceae *Aloe vera* L.) are soaked in water, left overnight and peeled the next day, blended and drunk (potentially mixed with honey). | Peru—Lima | |
| Hepatodepurative, detox remedy | Leaves of aloe (Liliaceae *Aloe vera* L.) are soaked for 3 days in water (peeled), gel blended with honey, drunk or eaten raw (in salads). | | |
| Anti-inflammatory for the bowels | Decoction: 1/2 cup of dried seeds of linseed (Linaceae *Linum*), boiled, and mixed with lemon juice and honey or milk (drunk as a soft drink during the day). | | |
| Health food | Lemon, lime (fresh fruit) mixed with egg shells, left for 4–5 days (the lemon purportedly dissolves the egg shells), blended with honey, 1 cup/day, before or after meals. | | |
| Avoiding pregnancy | *Periplaneta americana* (American cockroach, "barata"): toasted and macerated in order to produce a "powder" to be mixed with honey (of any bee) and ingested. | Brazil—Santa Cruz do Capibaribe | Alves et al. 2008 |
| Sexual weakness, nervous disturbances | By mixing 1 egg of *Anas platyrhynchos*, 1 pinch of cinnamon, 4 spoonfuls of Uruçú honey (a species of stingless bee), and 1 cup of guarana. The drink should be consumed during 9 days. | | |
| Cataract and glaucoma | Honey (stingless bee) as an eyedropper, just one drop of honey. | Northeastern Brazil | Costa-Neto and Oliveira, 2000 |

## TABLE 14.4 (Continued)
## Honey-Based Mixtures in Ethnomedicine

| Indications | Formulation | Country | Reference |
|---|---|---|---|
| Asthma | Decoction of *Cecropia peltata* (leaves), *Ficus carica* (leaves), honey (oral ingestion). | Cuba— Eastern Cuba | Cano and Volpato 2004 |
| | Maceration, juice of *Citrus aurantifolia* (fruit), *Cocos nucifera* (fruit), *Protium cubense* (root/tuber), oil, honey (oral ingestion). | | |
| Catarrh | Juice of *Bidens pilosa* (aerial part), *Solanum torvum* (leaves), oil, and honey (oral ingestion). | | |
| Coolness (frialdad) at the uterus, menstrual irregularity | Decoction of *Bidens pilosa* (aerial part), *Cassia fistula* (fruit), *Cissus sicyoides* (fruit), *Crescentia cujete* (fruit), *Cyrtopodium punctatum* (whole plant), *Phyla scaberrima* (aerial part), *Ruellia tuberosa* (root/tuber), honey (oral ingestion). | | |
| Boils | Triturated *Gossypium arboreum* (seeds), *Jatropha gossypifolia* (leaves), sugar, fat of ram, and honey (topical application). | | |
| Respiratory system affection; skin affection | Aerial parts local application or eaten with honey, wine, or milk. | Spain | San Miguel 2003 |
| Antihelminthic | Leaves of rue eaten in salad; chewed in small quantities with sugar and honey. | Italy | Guarrera et al. 2005; Pieroni and Quave 2005 |
| | Small leaves used very sparely in salad or chewed in small quantities with sugar and honey. | | |
| Tonsillitis, gastric ulcers, duodenal ulcers, pneumonia, urinary tract infections | To rind *Alcea kurdica* add 1 teaspoon (4–5 whole flowers, white flowers preferred) to 1 cup warm water, drink 1 cup/day (2 cups/day, if sickness remains: add 1 teaspoon honey and 1 teaspoon lemon juice). | Iraq— Kurdistan | Mati and de Boer 2011 |
| Diabetes, antihypertensive | Mix *Nigella sativa* seeds with honey 1:1 or 2:1, eat 1 teaspoon/day. | | |

*(continued)*

**TABLE 14.4 (Continued)**
**Honey-Based Mixtures in Ethnomedicine**

| Indications | Formulation | Country | Reference |
|---|---|---|---|
| Cancer, pneumonia, tonsillitis, hyperlipidemia, blood circulation | Mix *Nigella sativa* seeds 1:2 with honey, eat 1 teaspoon in morning and 1 teaspoon at night. | | |
| Emmenagogue | Break 1 seed of *Entada gigas*, mix the inside with 1 cup honey, eat 1 tablespoon/day. | | |
| Infertility (both sexes), expectorant | Grind 1 teaspoon of inula helenium, mix with 1 tablespoon honey, divide into three portions, eat 1 portion each time, three times a day. | | |

*Note:* Bad-kan (http://en.wikipedia.org/wiki/Traditional_Tibetan_medicine-cite_note-The_Basic_Tantra-5) is characterized by the quantitative and qualitative characteristics of cold and is the source of many functions such as aspects of digestion, the maintenance of our physical structure, joint health, and mental stability.

honey or starch to form pills/tablets for easy carriage and usage; sometimes, pills are used because the particular herb is not suitable for decoction (ChingKwei Kang 2010).

Each country has its own set of medical knowledge based on the local culture and past experience. As a result, medical concepts and understandings can vary significantly from one country to the next. For instance, a traditional Chinese medicine for the heart will not treat the same conditions as heart medications in conventional medicine. This is because the term "heart" in traditional Chinese medicine does not only mean the physical organ "heart" but also includes some functions conventional medicine would attribute to the brain. This is an example of how simple cultural misunderstanding can easily occur.

When medicines are being traded in foreign markets, the domestic consumers often wrongly apply their own medical concepts and understanding to the imported traditional medicines, often resulting in misunderstanding and subsequent misuse (WHO 2004).

An unusual application of honey described in Burkina Faso was its use as a means of confirming the presence of measles during its early stages. Honey is reportedly massaged onto the eruptions that then, in the case of measles, become more pronounced on the following day. Continued application of honey is performed until total disappearance of the eruptions occurs (Meda et al. 2004).

Therapeutic intake of honey is often done for treating gastroenteric disorders and respiratory infections by ingesting honey before sleep and again the following morning 2 to 3 h before eating (Meda et al. 2004).

The use of complementary and alternative medicine products in the topical treatment of diabetic foot disorders is fairly common among Saudi diabetic patients. Honey headed the list as a solo topical preparation or in combination with other herbs, namely, black seeds and myrrh (Bakhotmah and Alzahrani 2010).

## CONCLUSION

In order to justify its acceptability in the modern system of medicine, the quality assessment and standardization of honey formulations is essential in order to assess the quality of drugs. Medical devices containing honey are manufactured under the same regulations as any other "medical device." Different manufacturers set their own quality standards for the antibacterial activity of honey, so users should be informed what these are. Despite the number of published papers during the past decades concerning therapeutic properties of honey and honey-based formulations, according to Chepulis (2008), a nutrition scientist who has been researching the health benefits of honey and other natural foods for nearly 15 years, research (including drug companies and food manufacturers) often "sits" on the findings and results. There is generally a commercial aspect to science research these days, and very little is done purely for interest's sake alone. It seems that natural and complementary therapies are pushed aside by pharmaceutical companies. Prescribed drugs are more convenient for patients and physicians, although natural products might offer an alternative in the treatment of many different diseases. In resource-limited countries, conventional medications are often not affordable or not available; consequently, natural products are the medication of choice (Schnitzler 2012) because these drugs are easily available at low cost and are safe, and people have faith in them.

## REFERENCES

Abadi, N.M. 1976. *History of Medicine in Iran*. 1st edition. Tehran University Publication, Tehran.

Abdelhafiz, A.T. and Muhamad, J.A. 2008. Midcycle pericoital intravaginal bee honey and royal jelly for male factor infertility. *Int J Gynecol Obstet*. 101:146–149.

Abdelmonem, A.M., Rasheed, S.M., and Mohamed, A.S. 2012. Bee-honey and yogurt: A novel mixture for treating patients with vulvovaginal candidiasis during pregnancy. *Arch Gynecol Obstet*. 286:109–114.

Abou El-Soud, N.H. 2012. Honey between traditional uses and recent medicine. *Maced J Med Sci*. 5:205–214.

Acton, C. 2008. Medihoney: A complete wound bed preparation product. *Br J Nurs*. 17:S44, S46–S48.

Adam, F. 1985. *Les croisements et l'apiculture de demain*. SNA, Paris, 127p.

Ademoh, N.A. 2011. Natural honey and Nigerian gum arabic as composite binder for expendable foundry cores. *AU JT*. 15:70–76.

Adesina, S.K. 2005. Effect of formulation variables on the wound healing properties of the leaf essential oil of *Ocimum gratissimum* Linn. Lamiaceae. M.Sc. (Pharmaceutics) Thesis, Obafemi Awolowo University, Ile-Ife, Nigeria.

Alandejani, T., Marsan, J., Ferris, W., Slinger, R., and Chan, F. 2009. Effectiveness of honey on Staphylococcus aureus and *Pseudomonas aeruginosa* biofilms. *Otolaryng Head Neck*. 139:107–111.

Alvarez-Suarez, J.M., Tulipani, S., Romandini, S., Bertoli, E., and Battino, M. 2004. *J Wound Care*. 13:193–197.

Alves, R.R.N., Lima, H.N., Tavares, M.C., Souto, W.M.S., Barboza, R.R.D., and Vasconcellos, A. 2008. Animal-based remedies as complementary medicines in Santa Cruz do Capibaribe, Brazil. *BMC Complement Altern Med*. 8:2008.

Al-Jabri, A.A. 2005. Honey, milk and antibiotics. *Afr J Biotech*. 4:1580–1587.

Al-Waili, N.S. 2003. Topical application of natural honey, beeswax, and olive oil mixture to treat patients with atopic dermatitis or psoriasis: Partially controlled study. *Complement Ther Med.* 11:225–234.

Al-Waili, N.S. 2005a. Mixture of honey, beeswax, and olive oil inhibits growth of Staphylococcus aureus and *Candida albicans. Arch Med Res.* 36:10–13.

Al-Waili, N.S. 2005b. Clinical and mycological benefits of topical application of honey, olive oil and beeswax in diaper dermatitis. *Clin Microbiol Infect.* 11:160–163.

Al-Wali, N.S. 2004. Natural honey lowers plasma glucose, C-reactive protein, homocysteine, and blood lipids in healthy, diabetic, and hyperlipidemic subjects: Comparison with dextrose and sucrose. *J Med Food.* 7:100–107.

Annapoorani, A., Anilakumar, K.R., Khanum, F., Anjaneya Murthy, N., and Baw, A.S. 2010. Studies on the physicochemical characteristics of heated honey, honey mixed with *ghee* and their food consumption pattern by rats. *Ayu.* 31:141–146.

Anilakumar, K.R., Annapoorani, A., Khanum, F., Murthy, N.A., and Bawa, A.S. 2001. Effect of feeding of honey mixed with ghee and its heated forms on hepatotoxicity, antioxidant enzymes and lipid profile of rats. *IJRAP.* 2:1758–1762.

Anilakumar, K.R., Khanum, F., Anjaneya Murthy, N., and Baw, A.S. 2010. Studies on the physicochemical characteristics of heated honey, honey mixed with ghee and their food consumption pattern by rats. *Ayu.* 31:141–146.

Atiyeh, B.S., Costagliola, M., Hayek, S.N., and Dibo, S.A. 2007. Effect of silver on burn wound infection control and healing: Review of the literature. *Burns.* 33:139–148.

Babu, S.S. and Madhavi, M. 2003. *Green Remedy.* Pustak Mahal, Delhi, India, 138–139.

Bakhotmah, B.A. and Alzahrani, H.A. 2010. Self-reported use of complementary and alternative medicine (CAM) products in topical treatment of diabetic foot disorders by diabetic patients in Jeddah, Western Saudi Arabia. *BMC Res Notes.* 3:254.

Balaban, N. 2008. Preface. In Balaban, N., ed. *Control of Biofilm Infections by Signal Manipulations.* Springer-Verlag, Berlin. ISBN 978-3-540-73853-7.

Beldon, P. 2007. What you need to know about skin grafts and donor site wounds. *Wound Essentials.* 2:149–155.

Biglari, B., Linden, P.H., Simon, A., Aytac, S., Gerner, H.J., and Moghaddam, A. 2011. Use of Medihoney as a non-surgical therapy for chronic pressure ulcers in patients with spinal cord injury. *Spinal Cord.* 50:165–169.

Bjarnsholt, T., Kirketerp-Moller, K., Kristiansen, S., Phipps, R., Nielsen, A.K., Jensen, P.O., Hoiby, N., and Givskov, M. 2007. Silver against Pseudomonas aeruginosa biofilms. *APMIS.* 115:921–928.

Bogdanov, S. 2012. Honey in Medicine. Bee Product Science, www.bee-hexagon.net 15.

Bogdanov, S. 2009. External Applications of Honey. Book of Honey, Chapter 8, p1. Bee Product Science. Available at www.bee-hexagon.net.

Borgia, M., Sepe, N., Brancato, V., Simone, P., Costa, G., and Borgia, R. 1984. Efficacia e tollerabilita' di un preparato a base di miele, pappa reale e ginseng in un gruppo di pazienti affette da tubercolosi cronica. *Clinica Dietologica.* 11:443–447.

Brodman, E. 1978. Pediatrics in an Eighteenth Century Remedy Book. *Bull Med Libr Assoc.* 66:36–46.

Brudzynski, K. and Miotto, D. 2011. The recognition of high molecular weight melanoidins as the main components responsible for radical-scavenging capacity of unheated and heat-treated Canadian honeys. *Food Chem.* 125:570–575.

Bulman, M.W. 1955. Honey as a surgical dressing. *Middlesex Hosp J.* 55:188–189.

Butrica, J.L. 2002. The medical use of cannabis among the Greeks and Romans. *J Cannabis Therapeut.* 2:51–70.

Cano, J.H. and Volpato, G. 2004. Herbal mixtures in the traditional medicine of Eastern Cuba. *J Ethnopharmacol.* 90:293–316.

Carbajal, D., Molina, V., Valdes, S., and Arruzazbala, L. 1998. Anti-inflammatory activity of D-002: An active product isolated from beeswax. *Prostaglandins Leukot Essent Fatty Acids*. 59:235–238.

Ceuterick, M., Vandebroek, I., and Pieroni, A. 2011. Resilience of Andean urban ethnobotanies: A comparison of medicinal plant use among Bolivian and Peruvian migrants in the United Kingdom and in their countries of origin. *J Ethnopharmacol*. 136:27–54.

Chaillous, L., Lefevre, H., Thivolet, C., Boitard, C., Lahlou, N., Atlan-Gepner, C., Bouhanick, B., Mogenet, A., Nicolino, M., Carel, J.C., Lecomte, P., Marechaud, R., Bougneres, P., Charbonnel, B., and Sai, P. 2000. Oral insulin administration and residual beta-cell function in recent-onset type 1 diabetes: A multicentre randomised controlled trial. Diabetes Insuline Orale group. *Lancet*. 356:545–549.

Chepulis, L. 2008. *Healing Honey: A Natural Remedy for Better Health and Wellness*. Brown Walker Press, Boca Raton, FL.

ChingKwei Kang, M. 2010. Foundations of Traditional Chinese Medicine Chapter 1 Herbs. Available at http://site.magiherbs.com/FTCMEssentials.pdf.

Cooper, R. 2010. Biofilms and wounds: Much ado about nothing? *Wounds U.K.* 6:84–90.

Cooper, R. 2011. Is the use of honey for the treatment of biofilms a sticky subject? *The Magazine of the Society for Applied Microbiology*. 12:30–32.

Cooper, R., Halas, E., and Molan, P.C. 2002. The efficacy of honey in inhibiting strains of *Pseudomonas aeruginosa* from infected burns. *J Burn Care Rehabil*. 23(6):366–370.

Cooper, R., Jenkins, L., and Rowlands, R. 2011. Inhibition of biofilms through the use of manuka honey. *Wounds U.K.* 7:24–32.

Cooper, R. and Molan, P. 1999. The use of honey as an antiseptic in managing *Pseudomonas* infection. *J Wound Care*. 8:161–164.

Costa-Neto, E.M. and Oliveira, M.V.M. 2000. Cockroach is good for asthma: Zootherapeutic practices in Northeastern Brazil. *Hum Ecol Rev*. 7:41–51.

Crane, E. 1990. *Bees and Beekeeping: Science, Practice and World Resources*. Cornstock Publ., Ithaca, NY, 593 pp.

Dalziel, J.M. 1937. *The Useful Plants of West Tropical Africa*. Crown Agents for the Colonies, London, 612 pp.

Dany, B. 1988. *Selbstgemachtes aus Bienenprodukten*. Ehrenwirth Verlag, Munchen, 174 pp.

Davtyan, T.K. 2011. ArmApis BIO Honey and enriched with Royal Jelly BIO Honey induce human monocytes activation and stimulate cytokine production in vitro. Available at http://www.arpimed.am/e/novelties/arm-apis.

Domerego, R. 2001. Ces abeilles qui nous guérissent, Ed. J.-C. Lattès. 205 pp. ISBN: 2-01-235496-3.

Dunford, C.E. and Hanano, R. 2004. Acceptability to patients of a honey dressing for non-healing venous leg ulcers. *J Wound Care*. 13:193–197.

Eddy, J.J. and Gideonsen, M.D. 2005. Topical honey for diabetic foot ulcers. *J Fam Pract*. 54:533–535.

Eddy, J.J., Gideonsen, M.D., and Mack, G.P. 2008. Practical considerations of using topical honey for neuropathie diabetic foot ulcers: A review. *WMJ*. 107:187–190.

Efem, S.E. 1988. Clinical observation on wound healing properties of honey. *Br J Surg*. 75:679–681.

Emsen, I.M. 2007. A different and safe method of split thickness skin graft fixation: Medical honey application. *Burns*. 33:782–787.

Ene, A.C., Atawodi, S.E., Ameh, D.A., Kwanashie, H.O., and Agomo, P.U. 2010. Locally used plants for malaria therapy amongst the *Hausa*, *Yoruba* and *Ibo* communities in Maiduguri, Northeastern Nigeria. *IJTK*. 9:486–490.

English, H.K., Pack, A.R., and Molan, P.C. 2004. The effects of manuka honey on plaque and gingivitis: A pilot study. *J Int Acad Periodontol*. 6:63–67.

Farnsworth, N.R. 1980. The development of pharmacological and chemical research for application to traditional medicine in developing countries. *J Ethnopharmacol.* 2:173–171.

Fairchild, G.F. 1999. Estimated impacts of economic adulteration on the U.S. honey industry, "unplubished research report prepared for the National honey Board," 54 pp.

Fairchild, G.F., Nichols, J.P., and Capps, O., Jr. 2003. Observations on economic adulteration of high-value food products: The honey case. *JFDRS.* 34:38–45.

Ferreira, M.P. 2009. Comparison of lyophilization, freezin or honey as techniques to preserve allogenous cortical bone grafts used to repair experimental femoral bone defects in domestic cats. *Acta Scien Vet.* 37:211–212.

Fleischer, A., Feldman, S., and Krowchuk, D. 1999. Use of the potent corticosteroid agents and clotrimazole-betamethasone dipropionate (Lotrisone) by pediatricians. *Clin Ther.* 21:1725–1731.

Freeman, A., May, K., and Wraight, P. 2010. Honey: The bees' knees for diabetic foot ulcers? *Wound Practice Res.* 18:144–147.

Galiano, R.D. and Mustoe, T.A. 2007. *Grabb & Smith's Plastic Surgery.* 6th edition, Philadelphia. by Charles H. Thorne, Chap 3: Wound care, 929 pp.

Gedif, T. and Hahn, H.J. 2002. Herbalists in Addis Ababa and Butajira, Central Ethiopia: Mode of service delivery and traditional pharmaceutical practice. *Ethiop J Health Dev.* 16:191–197.

Gethin, G. and Cowman, S. 2008. Manuka honey vs. hydrogel-a prospective, open label, multicentre, randomised controlled trial to compare delsoughing efficacy and healing outcomes in venous ulcers. *J Clin Nurs.* 18:466–474.

González, H.R. and Aznar, G.E. 2006. P086-iron absortion and efficacy the antianemics iron hem trofin/biotrofer and different formulations. *Rev Cubana Farm.* 40(Suplemento Especial):72.

González, J.E. and Keshavan, N.D. 2006. Messing with bacterial quorum sensing. *Microbiol Mol Biol Rev.* 70:859–875.

Gram, L., Grossart, H.P., Schlingloff, A., and Kiørboe, T. 2002. Possible quorum sensing in marine snow green tea infusions processed from different parts of new shoots. *Food Chem.* 114:665–670.

Grote, I.W. and Walker, P. 1946. Studies in the preservation of liquid pharmaceutical preparations: The relationship between the preservative action of alcohol and the solid content of the preparation. *J Am Pharm Assoc Am Pharm Assoc.* 35:182–187.

Guardiola-Griffiths, C. 2011. Medieval mean girls: On sexual rivalry and the uses of cosmetics in La Celestina. *EHumanista.* 19:172–192.

Guarrera, P.M., Forti, G., and Marignoli, S. 2005. Ethnobotanical and ethnomedicinal uses of plants in the district of Acquapendente (Latium, Central Italy). *J Ethnopharmacol.* 76:11–34.

Haffejee, I.E. and Moosa, A. 1985. Honey in the treatment of infantile gastroenteritis. *Brit Med J.* 290:1866–1867.

Hammouri, S. 2004. The role of honey in the management of diabetic foot ulcers. *JRMS.* 11:20–22.

Haslam, D. 2007. Obesity: A medical history. *Obes Rev.* 8(Suppl 1):31–36.

Hassan, A.I. and Bayoumi, M.M. 2010. Efficiency of camel milk and honey bee in alleviation of diabetes. *Nature Sci.* 8:333–341.

Hategekimana, J., Ma, J.G., Li, Y., Karangwa, E., Tang, W., and Zhong, F. 2011. Effect of chemical composition of honey on cream formation in honey lemon tea. *Adv J Food Sci Technol.* 3:16–22.

Henriques, A.F., Jenkins, R.E., Burton, N.F., and Cooper, R.A. 2010. The effect of manuka honey on the structure of Pseudomonas aeruginosa. *Eur J Clin Microbiol Infect Dis.* 30:167–171.

Hilgert, N.I. 2001. Plants used in home medicine in the Zenta River basin, Northwest Argentina. *J Ethnopharmacol.* 76:11–34.

Hippocrates. On Fistulae. Translated by Francis Adams. Provided by The Internet Classics Archive. Available at http://classics.mit.edu//Hippocrates/fistulae.html.

Hirt, H.M., Lindsey, K., and Balagizi, I. 2008. *Natural Medicine in the Tropics IV.* AIDS and Natural Medicine. A resource book for carers of AIDS patients anamed. 3rd Edition.

Ho, N.K. 1996. Traditional Chinese medicine and treatment of neonatal jaundice. *Singapore Med J.* 37:645–651.

Holland, B.K. 2000. Treatments for bubonic plague: Reports from seventeenth century British epidemics. *J R Soc Med.* 93:322–324.

Hussein, S.Z., Yusoff, K.M., Makpol, S., and Yusof, Y.A. 2011. Antioxidant capacities and total phenolic contents increase with gamma irradiation in two types of Malaysian honey. *Molecules.* 16:6378–6395.

Jaàfar, H. 2009. Avicenna, Al-Qanun fi'l-tibb. I. Al-Hilal. Beirut: Dar-ol-Behar; 2009. In: M. Nejabat, M. et al. 2012. Avicenna and cataracts: A new analysis of contributions to diagnosis and treatment from the canon. *Iran Red Crescent Med J.* 14:265–270.

James, G.A., Swogger, E., Wolcott, R., Pulcini, E., Secor, P., Sestrich, J., Costerton, J.W., and Stewart, P.S. 2008. Biofilms in chronic wounds. *Wound Repair Regen.* 16:37–44.

Janick, J. 2010. Healing, health, and horticulture: Introduction to the workshop. *Hortscience.* 45:1584–1695.

Jenkins, R., Burton, N., and Cooper, R. 2011. Effect of manuka honey on the expression of universal stress protein A in meticillin-resistant *Staphylococcus aureus*. *Int J Antimicrob Agents.* 37:373–376.

Jervis-Bardy, J., Foreman, A., Bray, S., Tan, L., and Wormald, P.-J. 2011. Methylglyoxal-infused honey mimics the anti-*Staphylococcus aureus* biofilm activity of manuka honey: Potential implication in chronic rhinosinusitis. *Laryngoscope.* 121:1104–1107.

Johnson, D.W., van Eps, C., Mudge, D.W., Wiggins, K.J., Armstrong, K., Hawley, C.M., Campbell, S.B., Isbel, N.M., Nimmo, G.R., and Gibbs, H. 2005. Randomized, controlled trial of topical exit-site application of honey (Medihoney) versus mupirocin for the prevention of catheter-associated infections in hemodialysis patients. *J Am Soc Nephrol.* 16:1456–1462.

Jull, A.B., Rodgers, A., and Walker, N. 2008. Honey as a topical treatment for wounds. *Cochrane Database Syst Rev.* 4:CD005083.

Karaberopoulos, D., Karamanou, M., and Androutsos, G. 2012. The art of medicine. *Theriac Antiquity Pers.* 379:1942–1943.

Keenan, J.I., Salm, N., Wallace, A.J., and Hampton, M.B. 2012. Using food to reduce *H. Pylori* associated inflammation. *Phytother Res.* 26:1620–1625.

Khandekar, R. and Harmy, A.l. 2006. National register of the blind: A tool for health programme management. *EMRO Health J.* 12 WHO fact sheet no. 282.

Kirketerp-Møller, K., Jensen, P.Ø., Fazli, M., Madsen, K.G., Pedersen, J., Moser, C., Tolker-Nielsen, T.H., Høiby, N., Givskov, M., and Bjarnsholt, T. 2008. Distribution, organization, and ecology of bacteria in chronic wounds. *J Clin Microbiol.* 46:2717–2722.

Kirketerp-Møller, K., Zulkowski, K., and James, G. 2001. Chronic wound colonization, infection, and biofilms, in *Biofilm Infections*, T. Bjarnsholt, P.Ø. Jensen, C. Moser, and N. Høiby, Eds., pp. 11–24, Springer, New York.

Kluytmans, J., van Belkum, A., and Verbrugh, H. 1997. Nasal carriage of *Staphylococcus aureus*: Epidemiology, underlying mechanisms, and associated risks. *Clin Microbiol Rev.* 10:505–520.

Kňazovická, V., Kačániová, M., Dovičičová, M., Melich, M., and Kadási-Horáková, M. 2011. Microbial quality of honey mixture with pollen. *Potravinarstvo.* 5:27–32.

Kramer, S.N. 1954. An older pharmacopoeia. *J Am Med Ass.* 155(1): 26. In: Jones, R. 2001. Honey and healing through the ages. Munn, P. and Jones, R. eds. *Honey and Healing*, pp. 1–4, International Bee Research Association, Cardiff, U.K.

Krell, R. 1996. Value-added products from beekeeping. FAO Agricultural Services Bulletin No. 124; Chapter 2.

Krishna, R. 2005. Therapeutic uses of Honey in Ayurveda. Available at http://www.boloji.com/index.cfm?md=Content&sd=Articles&ArticleID=1117.

Kujawska, M., Zamudio, F., and Hilgert, N.I. 2012. Honey-based mixtures used in home medicine by nonindigenous population of misiones, Argentina. *Evid Based Complement Alternat Med.* 2012: Article ID 579350, 15 pages doi:10.1155/2012/579350.

Kumar, K.P.S., Bhowmik, D., Chiranjib, B., and Chandira, M.R. 2010. Medicinal uses and health benefits of honey: An overview. *J Chem Pharm Res.* 2:385–395.

Kunwar, R.M. and Bussmann, R.W. 2006. *Ficus* (Fig) species in Nepal: A review of diversity and indigenous uses. *Lyonia* 11:85–97.

Küpeli, A.E., Orhan, D.D., Gürbüz, I., and Yesilada, E. 2010. In vivo activity assessment of a "honey-bee pollen mix" formulation. *Pharm Biol.* 48:253–259.

Kwakman, P.H.S., de Boer, L., Ruyter-Spira, C.P., Creemers-Molenaar, T., Helsper, J.P.F.G., Vandenbroucke-Grauls, C.M.J.E., Zaat, S.A.J., and te Velde, A.A. 2011. Medical-grade honey enriched with antimicrobial peptides has enhanced activity against antibiotic-resistant pathogens. *Eur J Clin Microbiol Infect Dis.* 30:251–257.

Laplante, A. and Soumaoro, B. 1973. Planning' traditionnel au mali [Traditional "planning" in Mali]. Education sexuelle en Afrique tropicale [Proceedings of an Inter-African Seminar, Bamako, April 16–25, 1973]. Ottawa, International Development Research Centre, pp. 54–60.

Lerrer, B., Zinger-Yosovich, K.D., Avrahami, B., and Gilboa-Garber, N. 2007. Honey and royal jelly, like human milk, abrogate lectin-dependent infection preceding *Pseudomonas aeruginosa* adhesion. *ISME J.* 1:149–155.

Liu, W.J.H. 2011. Traditional herbal medicine research methods. Liu, W.J.H., ed. *Chap 1 Introduction to Traditional Herbal Medicines and Their Study.* John Wiley & Sons, Inc., Hoboken, NJ, 488 pp.

Majno, G. 1975. *Man and Wound in the Ancient World.* Harvard University Press, Cambridge, MA, 571 pp.

Majtan, J. 2011. Methylglyoxal-a potential risk factor of manuka honey in healing of diabetic ulcers. *Evid Based Complement Alternat Med.* Article ID 295494, 5 pp.

Maliya, S.D. 2007. Traditional fruit and leaf therapy among Tharus and indigenous people of district. Bahraich, India. *Ethnobotany.* 19:131–133.

Mati, E. and de Boer, H. 2011. Ethnobotany and trade of medicinal plants in the Qaysari Market, Kurdish Autonomous Region, Iraq. *J Ethnopharmacol.* 133:490–510.

Meda, A., Lamien, C.E., Millogo, J., Romito, R., and Nacoulma, O.G. 2004. Therapeutic uses of honey and honeybee larvae in central Burkina Faso. *J Ethnopharmacol.* 95:103–107.

Merckoll, P., Jonassen, T.O., Vad, M.E., Jeansson, S.L., and Melby, K.K. 2009. Bacteria, biofilm and honey: A study of the effects of the honey on 'planktonic' and biofilm-embedded wound bacteria. *Scand J Infect Dis.* 41:341–347.

Miguéns, C. 2011. Use of honey products in lower limb lymphoedema and recalcitrant wounds. *Wounds U.K.* 7:88–90.

Mirutse, G., Zemeda, A., Thomas, E., and Zerihum, W. 2003. An ethnobotanical study of medicinal plants used by the Zay people in Ethiopia. *J Ethnopharmacol.* 85:43–52.

Molan, P.C. 2002. Reintroducing honey in the management of wounds and ulcers-theory and practice. *Ostomy Wound Manage.* 48:28–40.

Molan, P.C. 2006. The evidence supporting the use of honey as a wound dressing. *Low Extrem Wounds.* 5:40–54.

Molan, P.C. 2011. The evidence and the rationale for the use of honey as wound dressing. *Wound Practice Res.* 19:204–220.

Molan, P.C. and Allen, K.L. 1996. The effect of gamma-irradiation on the antibacterial activity of honey. *J Pharm Pharmacol.* 48:1206–1209.

Molan, P.C. and Betts, J.A. 2004. Clinical usage of honey as a wound dressing: An update. *J Wound Care*. 13(9):353–356.

Moore, M. 1995. Herb formulas for clinic and home. Southwest School of Botanical Medicine; www.swsbm.com.

Moore, J.C., Spink, J., and Lipp, M. 2012. Development and application of a database of food ingredient fraud and economically motivated adulteration from 1980 to 2010. *J Food Sci*. 77:18–26.

Moore, O.A., Smith, L.A., Campbell, F., Seers, K., McQuay, H.J., and Moore, R.A. 2001. Systematic review of the use of honey as a wound dressing. *BMC Complement Altern Med*. 1:2.

Mustoe, T.A., O'Shaughnessy, K., and Kloeters, O. 2006. Chronic wound pathogenesis and current treatment strategies: A unifying hypothesis. *J Plast Reconstr Surg*. 117:35–41.

Muthuswamy, R. and Abay, S.M. 2009. Ethnomedicinal survey of folk drugs used in Bahirdar Zuria district, Northwestern Ethiopia. *IJTK*. 8:281–284.

Namdeo, K.P., Shekhar, V., Bodakhe, S.H., Shrivastava, S.K., and Dangi, J.S. 2010. Chemical investigations of honey: A multiactive component of herbal therapeutic agent. *IJRAP*. 1:85–89.

Nelson, E.A. and Bradley, M.D. 2007. Dressings and topical agents for arterial leg ulcers. *Cochrane Database Syst Rev*. 7:CD001836.

Nestjones, D. and Vandeputte, J. 2012. Clinical evaluation of Melladerm® Plus: A honey-based wound gel. *Wounds U.K.* 8:106–112.

Nilforoushzadeh, M.A., Jaffary, F., Moradi, S., Derakhshan, R., and Haftbaradaran, E. 2007. Effect of topical honey application along with intralesional injection of glucantime in the treatment of cutaneous leishmaniasis. *BMC Complement Altern Med*. 7:13.

Noa, M. and Mas, R. 1998. Effect of D-002 on the pre-ulcerative phase of carrageenan-induced colonic ulceration in the guinea pig. *J Pharma Pharmacol*. 50:549–553.

O'Meara, S., Al-Kurdi, D., Ologun, Y., and Ovington, L.G. 2010. Antibiotics and antiseptics for venous leg ulcers. *Cochrane Database Syst Rev*. 1:CD003557.

Okhiria, O.A., Henriques, A.F.M., Burton, N.F., Peters, A., and Cooper, R.A. 2009. Honey modulates biofilms of Pseudomonas aeruginosa in a time and dose dependent manner. *JAAS*. 1:6–10.

Olaitan, P.B., Adeleke, O.E., and Ola, I.O. 2007. Honey: A reservoir for microorganisms and an inhibitory agent for microbes. *Afr Health Sci*. 7:159–165.

Orafidiya, L.O., Adesina Jr., S.K., Igbeneghu, O.A., Akinkunmi, E.O., Adetogun, G.E., and Salau, A.O. 2006. The effect of honey and surfactant type on the antibacterial properties of the leaf essential oil of *Ocimum gratissimum* Linn. against common wound-infecting organisms. *Int J Aroma*. 16:57–62

Orafidiya, L.O., Agbani, E.O., Oyedele, A.O., Babalola, O.O., and Onayemi, O. 2002. Preliminary clinical tests on topical preparations of *Ocimum gratissimum* Linn. leaf essential oil for the treatment of Acne vulgaris. *Clin Drug Invest*. 22:313–319.

Orafidiya, L.O., Oyedele, A.O., Shittu, A.O., and Elujoba, A. 2001. Theformulation of an effective topical antibacterial product containing *Ocimum gratissimum* leaf essential oil. *Inter J Pharmaceutics*. 224:177–183.

Osman, O.F., Mansour, I.S., and El-Hakim, S. 2003. Honey compound for wound care: A preliminary report. *Ann Burns Fire Disasters*. 16:131–134.

Palmer, J. 2006. Bacterial biofilms in chronic rhinosinusitis. *Ann Otol Rhinol Laryngol* Suppl. 196:35–39.

Parker, T.L., Miller, S.A., Myers, L.E., Miguez, F.E., and Engeseth, N.J. 2010. evaluation of synergistic antioxidant potential of complex mixtures using oxygen radical absorbance capacity (ORAC) and electron paramagnetic resonance (EPR). *J Agric Food Chem*. 58:209–217.

Pham, C., Greenwood, J., Cleland, H., Woodruff, P., and Maddern, G. 2007. Bioengineered skin substitutes for the management of burns: A systematic review. *Burns*. 33:946–957.

Pieroni, A. and Quave, C.L. 2005. Traditional pharmacopoeias and medicines among Albanians and Italians in southern Italy: A comparison. *J Ethnopharmacol.* 101:258–270.

Pipicelli, G. and Tatti, P. 2009. Therapeutic properties of honey. *Health.* 1:281–283.

Postmes, T., Van Den Boogaard, A.E., and Hazen, M. 1995. The sterilization of honey with cobalt 60 gamma radiation: A study of honey spiked with spores of *Clostridium botulinium* and *Bacilus subtilus. Experientia, Basel.* 51:986–989.

Rasmussen, T.B., Bjarnsholt, T., Skindersoe, M.E., Hentzer, M., Kristoffersen, P., Kote, M., Nielsen, J., Eberl, L., and Givskov, M. 2005. Screening for quorum sensing inhibitors (QSI) by use of a novel genetic system, the QSI selector. *J Bacteriol.* 187:1799–1814.

Ricordel, J. and Bonmatin, J.-M. 2003. [Honey's virtues in theriacs]. *Rev Hist Pharm (Paris)* 51:21–28.

Ritesh Jain, R. and Jain, S.K. 2010. Traditional medicinal plants as anticancer agents from Chhattishgarh, India: An overview. *Int J Phytomed.* 2:186–196.

Robson, V., Dodd, S., and Thomas, S. 2009. Standardized antibacterial honey (Medihoney) with standard therapy in wound care: Randomized clinical trial. *J Adv Nurs.* 65:565–575.

Rosner, R. 1968. Moses Maimonides' Treatise on Poisons. *JAMA.* 205:914–916.

Royal Pharmaceutical Society. The evolution of pharmacy Theme D, Level 1. www.rpharms.com/./d1-the-origins-of-dosage-2.

Sahih Bukhari vol. 7, book 71, number 584, 585, 588, and 603.

Saikia, A.P., Ryakala, V.K., Sharma, P., Goswami, P., and Bora, U. 2006. Ethnobotany of medicinal plants used by Assamese people for various skin ailments and cosmetics. *J Ethnopharmacol.* 106:149–157.

San Miguel, E. 2003. *Ruta* spp. (*Ruta* L., Rutaceae) in traditional Spain: Frequency and distribution of its medicinal and symbolic applications. *Econ Bot.* 57:231–244.

Sarong, T.J. 1986. *Handbook of Traditional Tibetan Drugs. Their Nomenclature, Composition, Use, and Dosage.* Tibetan Medical Publications, Kalimpong, 101 pp.

Saxena, S., Gautam, S., and Sharma, A. 2010. Microbial decontamination of honey of Indian origin using gamma radiation and its biochemical and organoleptic properties. *J Food Sci.* 75:M19–M27.

Schnitzler, P. 2012. Clinical Studies. *Medicinal Aromatic Plants.* 1:1.

Sharp, A. 2009. Beneficial effects of honey dressings in wound management. *Nurs Stand J.* 24:66–74.

Shukrimi, A., Sulaiman, A.R., Halim, A.Y., and Azril, A. 2008. A comparative study between honey and povidone iodine as dressing solution for Wagner type II diabetic foot ulcers. *Med J Malaysia.* 63:44–46.

Siedentopp, W. 2009. Honey: Effective against inflammation, cough and hoarseness. *Dt Ztschr f Akup.* 52:57–60.

Simon, A., Sofka, K., Wieszniewsky, G., and Blaser, G. 2006. Antibacterial honey (Medihoney®) for wound care of immunocompromised pediatric oncology patients. *GMS Krankenhaushyg Interdiszip.* 1:Doc18.

Simon, A., Traynor, K., Santos, K., Blaser, G., Bode, U., and Molan, P. 2009. Medical honey for wound care—Still the 'latest resort'? *Evid Based Complement Alternat Med.* 6:165–173.

Singh, D., Singh, B., and Goel, R.K. 2011. Traditional uses, phytochemistry and pharmacology of Ficus religiosa: A review *J Ethnopharmacol* 134:565–583.

Skiadas, P.K. and Lascaratos, J.G. 2001. Dietetics in ancient Greek philosophy: Plato's concepts of healthy diet. *Eur J Clin Nutr.* 55:532–537.

Smirnova, I.I., Filatova, E.I., Suvorov, A.N., and Bylinskaia, E.N. 2000. The use of therapeutic/prophylactic dragee "honey laminolact" in radiotherapy of uterine tumors. *Vopr Onkol.* 46:748–750.

Soffer, A. 1976. Editorial: Chihuahuas and laetrile, chelation therapy, and honey from Boulder, Colo. *Arch Intern Med.* 136:865–866.

Solomon, R. and Donnenfeld, E.D. 2003. Recent advances and future frontiers in treating age-related cataracts. *JAMA*. 290:248–251.

Song, J.J. and Salcido, R. 2011. Use of honey in wound care: An update. *Adv Skin Wound Care*. 24:40–44.

Stephen-Haynes, J. 2011. Achieving clinical outcomes: The use of honey. *Wound Essent*. 6:14–19.

Stressmann, F.A., Rogers, G.B., Chan, S.W., Howarth, P.H., Harries, P.G., Bruce, K.D., and Salib, R.J. 2011. Characterization of bacterial community diversity in chronic rhino-sinusitis infections using novel culture-independent techniques. *Am J Rhinol Allergy*. 25:e133–e140.

Surndra Raj Joshi, S.R. 2008. Honey in Nepal. Approach, Strategy and Intervention for Subsector Promotion. German Technical Cooperation/Private Sector Promotion-Rural Finance Nepal (GTZ/PSP-RUFIN) ISBN 978-9937-2-0713-3.

Takuma, D.T. 1955. Honig bei der Aufzucht von Säuglingen. *Monatsschrift Kinderheilkunde*. 103:160–161.

Tasleem, S., Naqvi, S.B.S., Khan, S.A., and Hashimi, K. 2011. Honey ointment': A natural remedy of skin wound infections. *J Ayub Med Coll Abbottabad*. 23:26–31.

Tavares, J.P., Martins, I.L., Vieira, A.S., Lima, F.A.V., Bezerra, F.A.F., Moraes, M.O., and Moraes, M.E.A. 2006. Clinical toxicology study of a phytomedicine syrup composed of plants, honey, and propolis. *Rev bras Farmacogn*. 16:350–356.

Trop, M., Novak, M., Rodl, S., Hellbom, B., Kroell, W., and Goessler, W. 2006. Silver-coated dressing acticoat caused raised liver enzymes and argyria-like symptoms in burn patient. *J Trauma*. 60:648–652.

Truchado, P., Gil-Izquierdo, A., Tomas-Barberan, F., and Allende, A. 2009. Inhibition by chestnut honey of N-Acetyl-L-homoserine lactones and biofilm formation in *Erwinia carotovora*, *Yersinia entercolitica*, and *Aeromonas hydrophila*. *J Agric Food Chem*. 57:11186–11193.

Uccusic, P. 1982. *(Doctor Bee: Bee Products, their Curative Power and Application in Medicine.) Doktor Biene: Bienenprodukte—Ihre Heilkraft und Anwendung in der Heilkunst*. Ariston Verlag, Genf, Switzerland, 2nd edition in 1983, 198 pp.

Udwadia, T.E. 2011. Ghee and honey dressing for infected wounds. *Indian J Surg*. 3:278–283.

Visavadia, B.G., Honeysett, J., and Danford, M.H. 2006. Manuka honey dressing: An effective treatment for chronic wound infections. *Br J Oral Maxillofac Surg*. 46:55–56.

Wasiak, J., Cleland, H., and Campbell, F. 2008. Dressings for superficial and partial thickness burns. *Cochrane Database Syst Rev*. 4:CD002106.

Watnick, P. and Kolter, R. 2000. Biofilm, city of microbes. *J Bacteriol*. 182:2675–2679.

White, R., Molan, P., and Copper, R., eds. 2005. *Honey: A Modern Wound Management Product*. Wounds U.K., Aberdeen, 160 pp.

WHO. 2004. World Health Organization Guidelines on Developing Consumer Information on Proper Use of Traditional, Complementary and Alternative Medicine. *apps.who.int/medicinedocs/pdf*.

WHO. 2008. Traditional Medicine. Retrieved October, 2012, from http://www.who.int/mediacentre/factsheets/fs134/en/index.html.

Williams, P. and Williams, P. 2007. Quorum sensing, communication and cross-kingdom sig-nalling in the bacterial world. *Microbiology*. 153:3923–3938.

Windsor, S., Pappalardo, M., Brooks, P., Williams, S., and Manley-Harris, M. 2012. A con-venient new analysis of dihydroxyacetone and methylglyoxal applied to Australian *Leptospermum* honeys. *Journal of Pharmacognosy and Phytotherapy*. 4:6–11.

Wolcott, R.D., Rhoads, D.D., and Dowd, S.E. 2008. Biofilms and chronic wound inflamma-tion. *J Wound Care*. 17:333–341.

Yaghoobi, N., Al-Waili, N., Ghayour-Mobarhan, M., Parizadeh, S.M.R., Abasalti, Z., Yaghoobi, Z., Yaghoobi, F., Esmaeili, H., Kazemi-Bajestani, S.M.R., Aghasizadeh, R., Saloom, K.Y., and Ferns, G.A.A. 2009. Natural honey and cardiovascular risk factors: Effects on blood glucose, cholesterol, triacylglycerole, CRP, and body weight compared with sucrose. *Scientific World Journal*. 20:463–469.

Yavuz, Y. M. 2012. Lichens mentioned by Pedanios Dioscorides. *Ethno Med*. 6(2):103–109.

Yin, J., Xu, Y., Yuan, H., Luo, L., and Qian, X. 2009. Cream formation and main chemical components of green tea infusions processed from different parts of new shoots. *Food Chem*. 114:665–670.

# 15 Modern Methods of Analysis Applied to Honey

*Satyajit D. Sarker and Lutfun Nahar*

## CONTENTS

## INTRODUCTION

Honey is a sweet natural product produced by bees (*Apis mellifera*) from the nectar of flowers, honeydew, or both. Transformation of nectar or honeydew into honey takes place in bees and afterward in honeycombs. The process is comprised of water evaporation and a number of biochemical changes (e.g., enzymatic decomposition of sucrose to glucose and fructose). Although the main nutritional and health relevant components of honey are carbohydrates, predominantly D-fructose and D-glucose (~95% of honey solids) (Figure 15.1), and about 25 different oligosaccharides (e.g., raffinose), it also contains various other natural compounds (e.g., proteins, enzymes, amino acids, minerals, trace elements, vitamins, aroma compounds, flavonoids, and other polyphenols) as well as various contaminants from the soil or environment

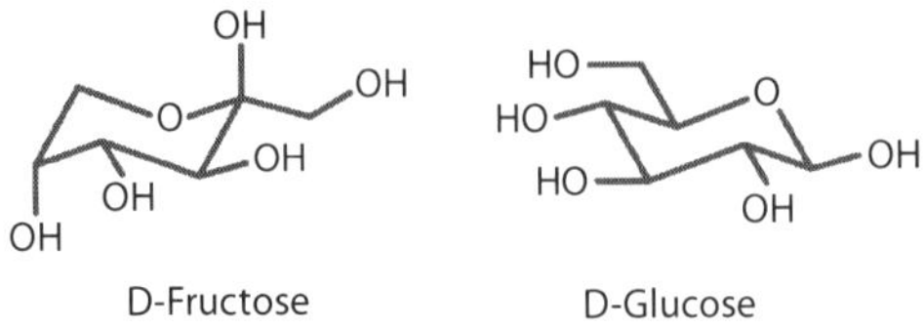

D-Fructose                    D-Glucose

**FIGURE 15.1** Major sugars in honey.

(e.g., pesticides and antibiotics). The ratio of D-glucose to D-fructose is characteristic of certain types of honey. In mixed-flower honeys, they are present in almost
equal proportions. On the other hand, unifloral honeys contain appreciably more
D-fructose than D-glucose.

The chemical composition of honey is very much dependent on the floral origin of
the nectar collected by bees. Although honey from a single floral source has higher
commercial value, it is often marketed as mixed-flower honey with blended flavors
and tastes. It is not only the nectar source that dictates the quality and content of
honey, but maturity, production methods, climate, process, and storage conditions
may also have significant influence. With regards to the quality of honey, in addition to the chemical composition, other important aspects of honey are its sensory
properties (e.g., taste, aroma, odor, and dissolution in tamate). The level of purity of
honey is of paramount importance not only for consumers but also for its commercial value. Over the years, various methods have been developed and validated to
assess the various purity criteria, quality, and chemical composition of honey. Honey
quality criteria were initially specified in a European Directive and in the *Codex
Alimentarius* Standard (Council Directive 1974; Crane 1975; *Codex Alimentarius*
1993). Later, the quality criteria were revised along with the available methods of
analysis (Swiss Food Manual 1995; Bogdanov et al. 1997), and most of those methods of analysis of the quality of honey revolve around determination of moisture content, mineral content (ash), electrical conductivity, acidity, 5-hydroxymethylfurfural
(5-HMF) content, diastase activity, invertase activity, sugar content, water-insoluble
solids content, proline content, and specific rotation. Analytical methods applied
to honey mainly deal with different topics such as determination of botanical or
geographical origin, quality control according to the current standards, and detection of adulteration or residues such as pesticides, pharmaceutical residues, antibiotics (e.g., tetracyclines, sulfonamides, aminoglycosides, and chloramphenicol), and
heavy metals.

This chapter will, however, focus only on the modern methods of analysis that
have been introduced and applied for honey analysis over the past few years.

## MODERN METHODS OF ANALYSIS

Whatever may be the analytical method applied to honey analysis, in most cases, it
requires a pretreatment step or preparation of honey samples for analysis. Preparation
of honey samples may include preliminary treatment, extraction of analytes or

enrichment, and cleanup before it can be subjected to a particular analytical technique (e.g., gas chromatography [GC] or high-performance liquid chromatography [HPLC]) for determination of target analytes. The most popular method for sample extraction or cleanup of a honey sample is either solvent extraction or solid-phase extraction (SPE) (Reid and Sarker 2012), which significantly reduces interferences of the matrix and thus increases the sensitivity of the subsequent analytical techniques. It also allows lower limits of quantification and renders unequivocal identification and confirmation of analytes. In recent years, to improve the efficiency of extraction and cleanup of honey samples even further, solid-phase microextraction (SPME), which is a rapid, solvent-free, and easy-to-use extraction method, and stir-bar sorptive extraction have been introduced. For a complex matrix like honey, often a combination of different extraction and cleanup techniques is employed. A summary of various modern extraction and cleanup methods for honey analysis has been presented by Kujawski and Namiesnik (2008). Although several methods of analysis of honey have been reported over the years, and a number of review articles focusing on the analysis of particular type of compounds or contaminants (e.g., antibiotics [Barganska et al. 2011], flavonoids, and phenolics [Pyrzynska and Biesaga 2009]) are available, some of the recent advances in techniques and methodologies applied to honey analysis have been depicted below under individual subheadings.

## Amperometric Method

Amperometry is a detection technique for analyzing ions in a solution based on electric current or changes in electric current. Such an amperometric method, employing separation on the CarboPac PA1 column with sodium hydroxide eluent and homemade electrochemical detector cells that contain copper working electrodes, for quantification of the main sugar components of honey and nectars was reported (Nagy et al. 2010). In this method, the electrochemical oxidation of the sugar components was studied, estimating the number of electrons exchanged and the rate constant of the electrode reaction.

A sensor for catalase amperometric detection in association with flow injection analysis (FIA) and a tubular reactor containing Amberlite IRA-743 resin was developed for the analysis of honey (Franchini et al. 2011). Catalase was quantitatively determined in 14 samples of Brazilian commercial honeys. FIA is an approach to chemical analysis, which is accomplished by injecting a plug of a sample into a non-segmented continuous flowing carrier stream of a suitable liquid. The injected sample forms a zone, which is then transported toward a detector that continuously records the changes in absorbance, electrode potential, or other physical parameter originating from the passage of the sample material through the flow cell. The principle of this technique is similar to that of segmented flow analysis, but no air is injected into the sample or reagent streams. A differential amperometric method coupled with a FIA and a tubular reactor containing the ascorbate oxidase enzyme immobilized on amberlite IRA-743 was used for the specific determination and quantification of ascorbic acid (Figure 15.2) in honey samples (Da Silva et al. 2012). The working electrode was a gold electrode modified by electrochemical deposition of palladium.

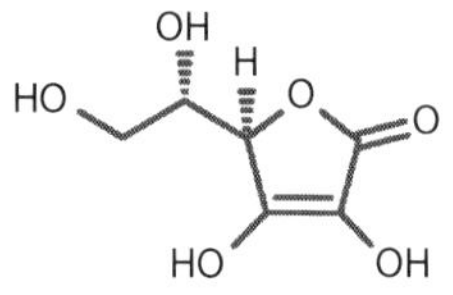

Ascorbic acid (vitamin C)

**FIGURE 15.2**   Structure of ascorbic acid.

## Capillary Electrophoresis

A capillary zone electrophoresis technique, coupled with a diode array detector, was employed for the determination of 26 phenolic compounds in honey samples with different floral origins (e.g., from heather, lavender, acacia, rape, sunflower, rosemary, citrus, rhododendron, thyme, chestnut tree, and calluna) (Andrade et al. 1997). Compounds were separated on a fused-silica column (50 cm × 50 mm) using 100 mM sodium borate buffer (pH 0.5)–20% methanol. A correlation was observed between the phenolics profiles and the botanical origin of the honey. For example, thyme honey was characterized by the presence of rosmarinic acid, heather honey by ellagic acid, citrus honey by hesperetin, and lavender honey by naringenin.

An analytical method based on capillary zone electrophoresis coupled to electro-spray ionization–mass spectrometry (CE-ESI-MS) was reported for the identification and simultaneous quantification of several endocrine-disrupting chemicals (e.g., 2,4-dichlorophenol, 2,4,5-trichlorophenol, pentachlorophenol, bisphenol-A, 4-*tert*-butylphenol, and 4-*tert*-butylbenzoic acid) in honey (Rodriguez-Gonzalo et al. 2009) (Figure 15.3). A new sample treatment based on the combined use of a restricted access material and a polymeric sorbent for SPE was also used.

Programmed nebulizing-gas pressure (PNP) could be used as a simple strategy for the separation of anions by CE-ESI-MS. Dominguez-Alvarez et al. (2012) described the application of the PNP approach to the quantitative analysis of pollutants in real samples by CE-ESI-MS for the first time, especially in relation to the determination of several endocrine disruptors (Figure 15.3) in honey samples. With the application of the PNP approach to CE-ESI-MS, the limits of detection could be achieved in the 14 ng/g range with a simple liquid–liquid procedure without any further cleanup step. It was suggested that this analytical method might be used in the control analysis of the above pollutants in honey. The analysis of inorganic cations in honey samples by CE with the indirect ultraviolet (UV) detection technique was reported by Shi et al. (2012). A fast CE method involving direct injection mode for determination and quantitative analysis of sugars (e.g., fructose, glucose, and sucrose) in honey samples was demonstrated by Rizello et al. (2012a). The samples were previously dissolved in deionized water and filtered with no other sample treatment.

Micellar electrokinetic chromatography (MEKC) is a modification of classic CE, where the samples are separated by differential partitioning between micelles (pseudo-stationary phase) and a surrounding aqueous buffer solution (mobile phase) (Rizvi et al. 2011). The basic setup and detection methods used for MEKC are very

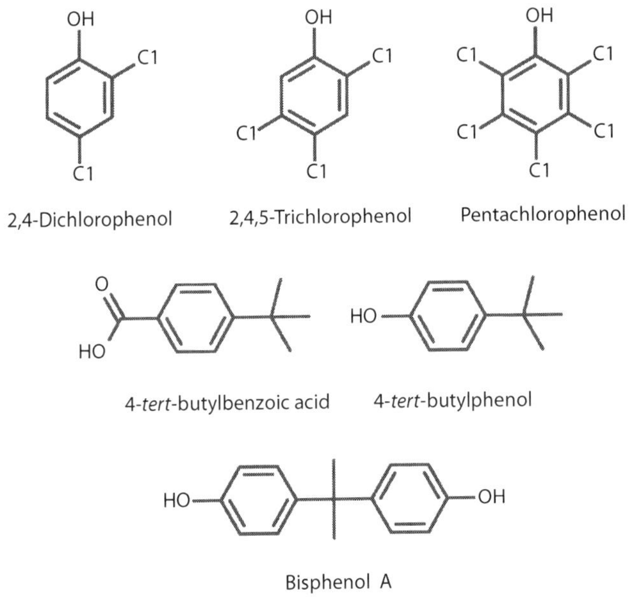

2,4-Dichlorophenol        2,4,5-Trichlorophenol        Pentachlorophenol

4-tert-butylbenzoic acid        4-tert-butylphenol

Bisphenol A

**FIGURE 15.3** Endocrine-disrupting chemicals in honey.

similar to those used in CE. The difference is that the solution contains a surfactant at a concentration that is greater than the critical micelle concentration. Above this concentration, surfactant monomers are in equilibrium with micelles. Teixido et al. (2011) reported a MEKC analytical method for the analysis of 5-HMF (Figure 15.4) content in honey. A phosphate buffer (75 mM) solution at pH 8.0 containing sodium dodecyl sulfate (100 mM) was used as a background electrolyte, and the separation was performed by applying +25 kV in a 50 μm I.D. uncoated fused-silica capillary. The procedure was validated by comparing the results with those obtained with liquid chromatography–tandem MS (LC-MS/MS). A similar MEKC was successfully employed in the determination of 5-HMF in honey samples using caffeine as the internal standard (IS) (Rizello et al. 2012b). Under optimal CE conditions, separation of the investigated substance was achieved in less than 0.7 min. It was proposed that the analytical performance of this method could make it suitable for implementation in food laboratories for the routine determination of 5-HMF in honey samples.

5-Hydroxymethylfurfural

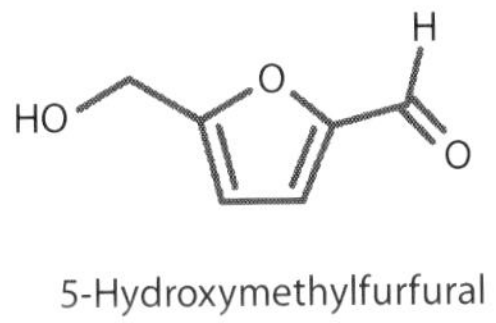

**FIGURE 15.4** Structure of 5-HMF.

## GAS CHROMATOGRAPHY

GC-MS detection has been the method of choice for analyzing volatile and non-volatile components of honey samples (Dobrinas et al. 2008; Aliferis et al. 2010; Ceballos et al. 2010; Daher and Gulacar 2010; Ruiz-Matute et al. 2010; Kadar et al. 2011; Wiest et al. 2011; Manyl-Loh et al. 2012; Sarker and Nahar 2012). While GC-MS is more suitable for volatile components, and less so for the nonvolatiles, it can, however, be used successfully to characterize complex mixtures of carbohydrates up to degree of polymerization and with the derivatization of the sample being a prior requirement. Trimethylsilyloximes are adequate for derivatization, since reducing sugars produces only two different derivatives corresponding to the syn (*E*) and anti (*Z*) isomers. Various MS detectors can be coupled to a GC for the analysis of honey, but the electron impact ionization MS detector has been found to be the most popular one. For example, an electron impact ionization MS detector was used with GC for the determination of multiple pesticide residues in commercial honey (Zhen et al. 2006).

The analysis of the volatile fraction of unifloral Greek thyme honey by means of ultrasound-assisted extraction followed by GC-MS was reported (Alissandrakis et al. 2009). The volatile profile of thistle honey was determined by the headspace-SPME (HS-SPME) and GC-MS (Bianchi et al. 2011) and compared with a dynamic headspace extraction method. This method was proven to be effective in recognizing the botanical origin of thistle honey. Similar methodology was also applied for the analysis of volatile components of *Asphodelus microcarpus* honey (Jerkovic et al. 2011a). HS-SPME combined with ultrasonic solvent extraction and GC-MS was applied for the analysis of honey from *Prunus mahaleb* using coumarin and vomifoliol as nonspecific biomarkers (Jerkovic et al. 2011b) (Figure 15.5). A similar method was also previously described by Fontana et al. (2010) for the determination of organophosphate pesticides in honey samples, where a coacervative microextraction ultrasound-assisted back-extraction technique was applied for the first time for extracting and preconcentrating organophosphate pesticides from honey samples prior to GC-MS analysis. The extraction/preconcentration technique was supported on the micellar organized medium based on nonionic surfactant. To enable coupling with GC, it was necessary to back-extract the analytes into hexane.

Tsimeli et al. (2008) developed a rapid and sensitive method, HS-SPME (with a 100-mm film thickness polydimethylsiloxane fiber) coupled with GC-MS (in selected ion monitoring mode), for the determination of 1,2-dibromoethane,

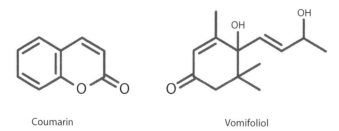

Coumarin                                                                    Vomifoliol

**FIGURE 15.5**  Nonspecific biomarkers used in the analysis of *Prunus* honey.

1,4-dichlorobenzene, and naphthalene residues in honey samples. GC has also been used for the determination of phenolic compounds, especially phenolic acids in honey (Pyrzynska and Biesaga 2009), with limited success.

A new, straightforward, and rapid single-drop microextraction procedure combined with GC-MS was developed, validated, and applied for the determination of multiclass pesticide residues in honey samples (Tsiropoulos and Amvrazi 2011). Later, Zacharis et al. (2012) demonstrated a simple but efficient dispersive liquid–liquid microextraction method, involving acetonitrile, chloroform, and aqueous honey, for the determination of organochlorine pesticide residues in various honey samples collected in Greece. The pesticide residues were separated by the GC and identified by electron capture detector (ECD) or ion trap MS (GC-IT/MS). The possible effect of several parameters on the extraction efficiency, for example, type and volume of organic extraction solvent, type and volume of disperser solvent, sample pH, ionic strength, extraction time, and centrifugation speed, was extensively studied. Mean recoveries were found to be within the range of 75% to 119% and the precision was better than 20% in both methodologies (i.e., GC-ECD and GC-IT/MS). GC-MS together with principal component analysis (PCA) was utilized for analyzing the volatiles, extracted by HS-SPME, to identify the origins of various honeys (Yang et al. 2012). The volatile fraction of Corsican *Erica arborea* flowers mainly contained octen-3-ol, (*E*)-β-ocimene, and (*Z*)-β-ocimene, while 4-propylanisol, *p*-anisaldehyde, benzaldehyde, and 3-furaldehyde (Figure 15.6) were found to be the major volatile components of Corsican honeys.

A similar method involving HS-SPME and GC for the analysis of organotin compounds in honey is also available (Campillo et al. 2012).

Cuevas-Glory et al. (2012) reported a similar floral classification method of honeys by PCA and HS-SPME/GC-MS. The volatile compounds of the honey obtained by HS-SPME/GC-MS with polydimethylsiloxane/divinylbenzene fibers were used in this study. The PCA discriminated between the three types of honeys and established the fact that this method could be used to successfully differentiate samples from different unifloral honeys according to their volatile composition. An application of a comprehensive 2D-GC coupled to time-of-flight MS for the analysis of complex mixtures of disaccharides in honey samples was demonstrated (Brokl et al. 2010). This 2D-GC method allowed the separation of most of the usual honey disaccharides as well as several others previously not reported. Soria et al. (2009) reported

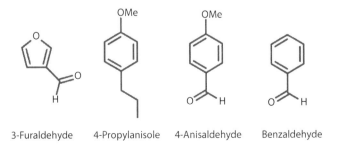

3-Furaldehyde    4-Propylanisole    4-Anisaldehyde    Benzaldehyde

**FIGURE 15.6**   Major volatile components of Corsican honey.

a purge-and-trap GC-MS analytical method for the analysis of volatile components in honey. It was concluded that this method could be used for a better selection of IS in quantitative analysis of volatiles in honey.

It is not only the GC-MS but also the GC olfactometry that could be used for the analysis of honey. Analysis of odor-active compounds from the honey of *Avicennia germinans* by SPME combined with GC-MS and GC olfactometry has recently been reported (Pino 2012). Similarly, GC coupled with a nitrogen–phosphorous detector (NPD) was used for the analysis of pesticides in honey (Chienthavorn et al. 2012). Polydivinylbenzene and silica monolithic materials were synthesized in capillaries. They were then used to adsorb nine organonitrogen pesticides extracted from honey. After adsorption, each monolith capillary was connected to a GC-NPD system. The detection limits of the pesticides determined by the GC-NPD and GC-MS ranged between 0.36–1.75 and 0.13–0.25 ng/g, respectively.

A GC-flame ionization detector method was successfully employed in the sugar analysis for authenticity evaluation of 15 natural and 1 artificial honey samples from the Lithuanian market (Kaskoniene et al. 2011). The composition of carbohydrates in the tested samples of natural honey appeared to be quite similar, except for trisaccharides, which were absent in the samples. A GC coupled with a flame photometric detector was used for the determination of organophosphorous compounds in honey samples (Amendola et al. 2011). A combination of SPME and GC coupled with microwave-induced plasma atomic emission detection was applied for the determination of 16 pesticides of different chemical families (e.g., organochlorines, organophosphorous, and pyrethrins) in honey (Campillo et al. 2006).

## High-Performance Liquid Chromatography

HPLC (Latif and Sarker 2012; Sarker and Nahar 2012) coupled with various detection techniques, mass spectroscopy being the most popular detector, has been used for a long time for the analysis of honey (Afredsson et al. 2005; Hammel et al. 2008; Truchado et al. 2009, 2011; Kelly et al. 2010; Ouchemoukh et al. 2010; Wang et al. 2010a; Arena et al. 2011; Blasco et al. 2011; Kempf et al. 2011; Kujawski and Namiesnik 2011; Tanner and Czerwenka 2011; Uchiyama et al. 2011; Wiest et al. 2011; Zhang et al. 2011; Bohm et al. 2012; Kanda et al. 2012; Ramanauskiene et al. 2012; Sakamoto et al. 2012; Tolgyesi et al. 2012). Among the LC-MS techniques, an HPLC coupled to MS/MS (LC-MS/MS) has revolutionized trace analysis in complex samples. It has replaced many older techniques, where less selective detectors (e.g., fluorescence, electrochemical, or single-stage MS) were used. The much higher selectivity offers less matrix-related signals and improved signal-to-noise ratio. For example, a sensitive and precise analytical method using hollow fiber renewal liquid membrane extraction technology followed by LC-MS/MS was developed for determination of five sulfonamides in honey samples (Bedendo et al. 2010). However, the coupling of various other detectors to HPLC, such as UV detection (Granja et al. 2008; Li et al. 2008), photo-diode array detection, or fluorescence detection (Bernal et al. 2009; Peres et al. 2010; Tsai et al. 2010; Escuredo et al. 2012; Soto et al. 2012), has also been applied for the analysis of honey. A new method for the simultaneous analysis at trace level of sulfonamides (e.g., sulfaguanidine, sulfanilamide, sulfacetamide, sulfathiazole, sulfapyridine, sulfachloropyridazine, sulfamerazine, sulfameter, sulfamethazine, sulfadoxine, sulfadiazine, sulfamonomethoxine, and sulfadimethoxine) in honey was reported (Bernal et al. 2009) (Figure 15.7). Methanol was used in the sample treatment step to avoid the emulsion formation and to break the N-glycosidic bond between sugars and sulfonamides. The determination was performed by HPLC in gradient elution mode, with fluorescence detection after the online precolumn derivatization with fluorescamine.

It is not only the phenolics or antibiotics that can be analyzed by an HPLC method; water-soluble vitamins, such as vitamins B2, B3, B5, B9, and C (Figure 15.8), have recently been simultaneously analyzed by a fully validated simple and fast reverse-phase HPLC method (Ciulu et al. 2011). The method provided low detection and quantification limits, good linearity in a large concentration interval, good precision, and the absence of any bias. The HPLC equipment from Perkin-Elmer comprised a Series 200 binary pump, a sampling valve, a 20-µL sample loop, and a Series 200 UV-visible variable wavelength detector. Separation was performed on an Alltima $C_{18}$ reverse-phase silica column (250 × 4.6 mm, 5 µm particle size) fitted with a guard cartridge packed with the same stationary phase.

An HPLC method coupled with diode array detection and fluorescence detection and ESI-MS (LC-DAD-FLD and LC-ESI-MS) was developed for the determination of residues of resveratrol in honey (Soto et al. 2012). This procedure also incorporated pretreatment by SPE on polymeric cartridges for the isolation of geometric isomers of resveratrol and piceid from diluted honey samples (Figure 15.9).

Dispersive liquid–liquid microextraction coupled with HPLC with a variable wavelength detector was applied for the extraction and detection of chloramphenicol and thiamphenicol in honey (Chen et al. 2009) (Figure 15.10). Liang et al. (2009)

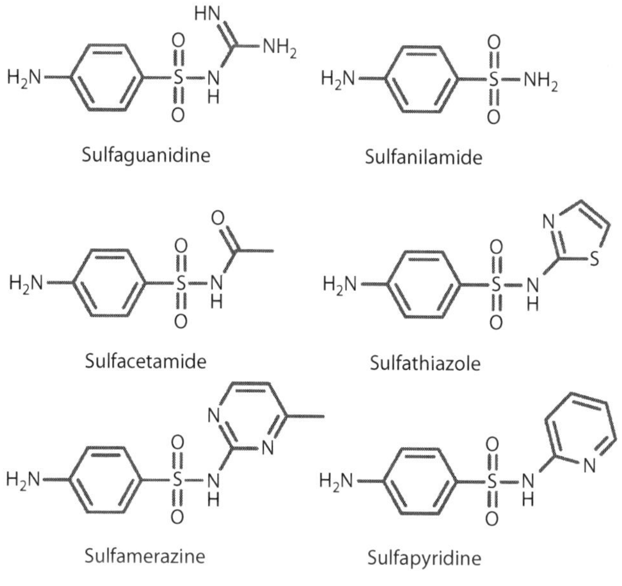

FIGURE 15.7 Examples of some sulfonamides found in honey as contaminants.

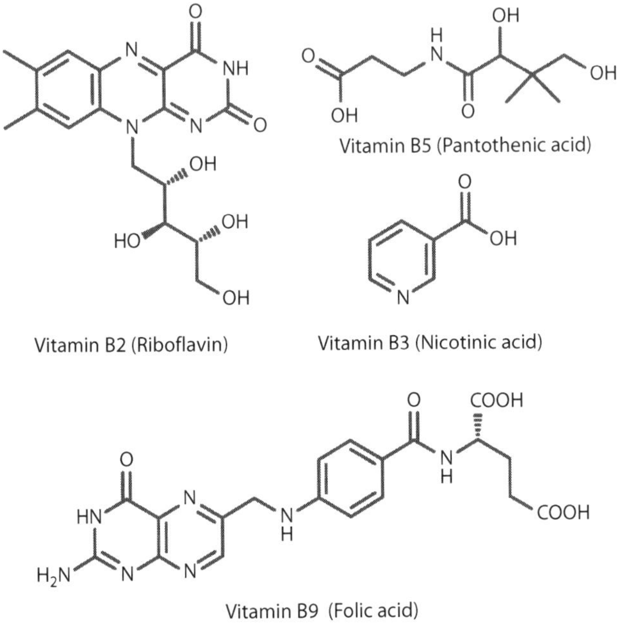

**FIGURE 15.8** Various vitamin Bs generally found in honey.

developed a sensitive and accurate HPLC method using an electrochemical detection technique for the simultaneous determination of four phenolic components (e.g., caffeic acid, *p*-coumaric acid, ferulic acid, and hesperetin) in Chinese citrus honey (Figure 15.9). In most of such HPLC analyses, usually a reverse-phase silica $C_{18}$ column is used, and a mobile phase generally is composed of water, acetonitrile, or methanol.

Hyphenated HPLC techniques are particularly useful for the analysis of phenolics, including flavonoids in honey (Biesaga and Pyrzynska 2009; Pyrzynska and Biesaga 2009; Truchado et al. 2009, 2011). Bertoncelj et al. (2011) determined the flavonoid profiles of seven different types of Slovenian honey. SPE was used to extract flavonoids from the honey samples, and the extracted flavonoids were analyzed by LC-DAD-ESI-MS. It was observed that the honey samples had similar,

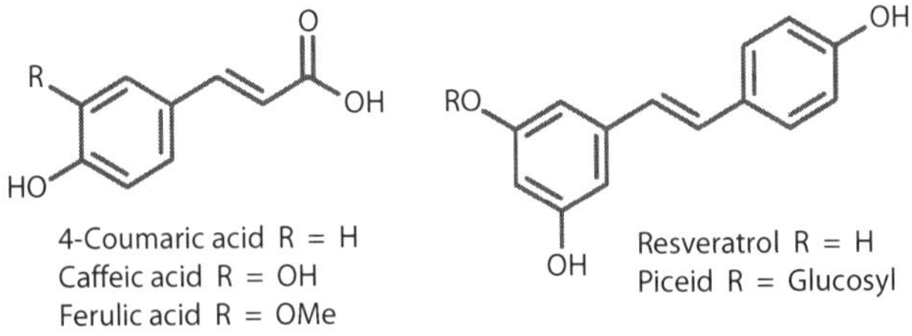

**FIGURE 15.9** Cinnamic acid derivatives, resveratrol and piceid, from honey.

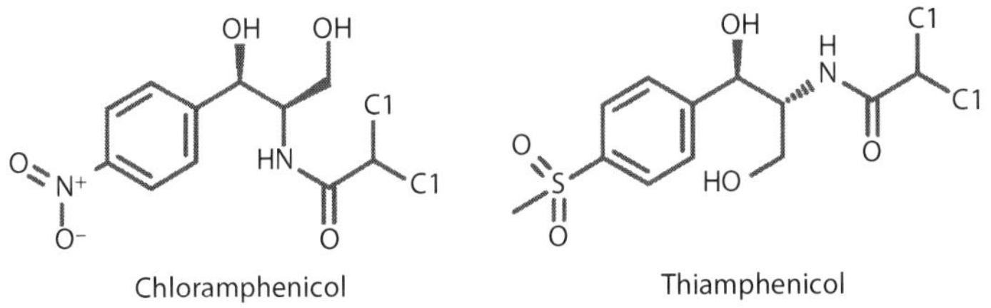

**Chloramphenicol**

**Thiamphenicol**

**FIGURE 15.10** Two common antibiotic contaminants in honey.

but quantitatively different, flavonoid profiles. Myricetin, luteolin, quercetin, naringenin, apigenin, kaempferol, pinocembrin, chrysin, and galangin were identified using reference standards, while pinobanksin was tentatively identified through its retention time, UV, and MS/MS data (Figure 15.11). In another similar study involving flavonoids in honey, an HPLC coupled with coulometric electrode array detection and ESI-MS was used (Petrus et al. 2011).

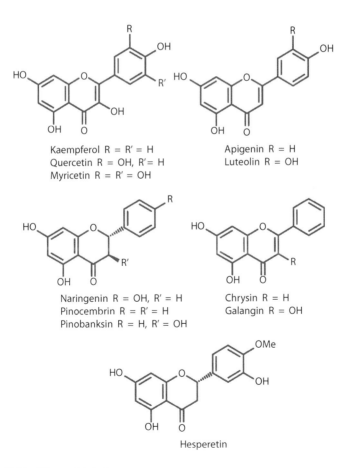

Kaempferol R = R′ = H
Quercetin R = OH, R′ = H
Myricetin R = R′ = OH

Apigenin R = H
Luteolin R = OH

Naringenin R = OH, R′ = H
Pinocembrin R = R′ = H
Pinobanksin R = H, R′ = OH

Chrysin R = H
Galangin R = OH

Hesperetin

**FIGURE 15.11** Flavonoids in honey.

The recent advances in technology and electronics have incorporated the modern version of the HPLC technique, ultra-HPLC (UHPLC or UPLC). UPLC takes advantage of technological advances made in particle-chemistry performance, system optimization, detector design, data processing, and control. It uses 1.5 to 2 µm particles, narrower analytical columns, and instrumentation that operates at higher pressures than those used in HPLC. UPLC offers remarkable enhancement in resolution, sensitivity, and speed of analysis. For example, the same separation on reverse-phase HPLC with a run time of more than 20 min can be achieved in under 3 min by UPLC. While various detection techniques can be coupled to UPLC, the coupling with a mass spectrometer has been probably the most popular of all couplings when it comes to analyzing honey (Alechaga et al. 2012). A UPLC-MS method was developed and validated for the simultaneous determination of different veterinary drug residues (e.g., macrolides, tetracyclines, quinolones, and sulfonamides) in 16 honey samples (Vidal et al. 2009). Honey samples were dissolved in $Na_2EDTA$, and veterinary residues were extracted from the supernatant by SPE using OASIS HLB cartridges and analyzed by UPLC-MS/MS. An ESI-MS in positive mode was used. Gomez-Perez et al. (2012) have established a database for the simultaneous analysis of more than 350 pesticides and veterinary drugs (including antibiotics) using UPLC coupled to high-resolution Orbitrap MS (UHPLC-Orbitrap-MS) and utilized this technique for the qualitative and quantitative analysis of 26 real honey samples. A similar hyphenated technique UPLC-PDA-MS/MS was employed for the classification and characterization of manuka honeys based on phenolic compounds and methylglyoxal (Oelschlaegel et al. 2012). Several samples of manuka honeys were analyzed by UPLC-PDA-MS/MS after SPE and differentiated successfully. For example, the honeys of one group had typically high concentrations of 4-hydroxybenzoic acid, dehydrovomifoliol, and benzoic acid, while another group displayed high amounts of kojic acid and 2-methoxybenzoic acid intensities.

Comparison between an HPLC technique and a spectrophotometric White method, in relation to determination of extremely low levels of 5-HMF in natural unifloral honey and honeydew samples, was reported by Truzzi et al. (2012). It was concluded that the HPLC method was more appropriate for the determination of HMF in honey in the range 14 mg/kg because of its greater precision.

### IMMUNOASSAY

A simple and sensitive biotin-avidin-mediated immunoassay (enzyme-linked immunosorbent assay [ELISA]) was developed by Jeon and Paeng (2008) for the quantitative determination of tetracycline residues in honey samples without any cleanup or extraction. Centi et al. (2010) described an electrochemical immunoassay based on the use of eight-electrode carbon screen-printed arrays as transducers coupled with magnetic beads coated with protein A for the determination of antimicrobial sulfonamides (Figure 15.7) in honey samples. Screen-printed eight-electrode arrays were utilized as transducers because of the possibility to repeat multiple analysis and to test different samples simultaneously. Alkaline phosphatase was used as an enzyme label and differential pulse voltammetry as a fast electrochemical technique. It was suggested that short incubation times (25 min) and the fast electrochemical measurement (10 s) could make these systems a possible alternative to classic ELISA tests.

An enzyme immunoassay technique for the detection of sulfamethoxypyridazine in 24 honey samples was developed using rabbit polyclonal antibodies raised against *N*-sulfonyl-4-aminobutyric acid, which contains a structural group characteristic of sulfonamides (Tafintseva et al. 2010). Under the optimized conditions, the sulfamethoxypyridazine detection limit was found to be 0.05 ng/mL.

Thongchai et al. (2010) reported a microflow chemiluminescence system for determination of chloramphenicol (Figure 15.10) in honey with preconcentration using a molecularly imprinted polymer, while Wutz et al. (2011) introduced, for the first time, a simultaneous identification and quantification method for antibiotic derivatives in honey chemiluminescence multianalyte chip immunoassays. An electrochemiluminescence inhibition method was developed for quantitative determination of tetracyclines in honey samples (Guo and Gai 2011). O'Mahony et al. (2011) developed a multiplexing chemiluminescence biochip array screening assay for simultaneous determination of four nitrofuran metabolites in honey. An immunologic characterization of honey proteins was established by Hayashi et al. (2011).

A lateral flow immunoassay was developed in the competitive reaction format and applied to analyze sulfathiazole residues in honey samples (Guillen et al. 2011). A hapten conjugate and goat anti-rabbit antiserum as capture and control reagent, respectively, were dispensed on the nitrocellulose membrane. Polyclonal antiserum against sulfathiazole was conjugated to colloidal gold nanoparticles and utilized as the detection reagent.

## INDUCTIVELY COUPLED PLASMA-MASS SPECTROMETRY

Inductively coupled plasma-MS (ICP-MS) is a type of MS technique that is capable of detecting metals and various nonmetals at extremely low concentrations (parts per trillion). This is achieved by ionizing the sample with ICP followed by the use of a mass spectrometer to separate and quantify those ions. Compared with conventional atomic absorption techniques, ICP-MS has greater speed, precision, and sensitivity. Such technique was employed for simultaneous determination of several elements in honey samples (Chudzinska and Baralkiewicz 2010, 2011; Chudzinska et al. 2012). The detection limits of all elements studied demonstrated the suitability of this method for routine analyses of honey. Chudzinska and Baralkiewicz (2010) demonstrated that the analysis of quality and quantity of honey elements could be used successfully to define the origins of honey samples by using an ICP-MS technique for simultaneous determination of elements. Chemometric methods could then be used to classify honey samples according to mineral content. In that study, the mineral content of 55 honey samples, representing three different types of honey (honeydew, buckwheat, and rape honey) from different areas in Poland, was assessed. A total of 13 elements (e.g., Al, B, Ba, Ca, Cd, Cu, K, Mg, Mn, Na, Ni, Pb, and Zn) were determined using ICP-MS. It was shown that the analysis of quality and quantity of honey elements could be used to define honey origin by using ICP-MS as a technique for simultaneous determination of elements.

## INFRARED SPECTROSCOPY

Infrared (IR) spectroscopic technique has been reported for the analysis of honey samples. Quantitative analysis of glucose, fructose, sucrose, and maltose in honey samples from different geographic origins was done using Fourier transform IR (FTIR) spectroscopy and chemometrics (Wang et al. 2010b). It was shown that FTIR could be used to qualitatively and quantitatively determine the presence of glucose, fructose, sucrose, and maltose in multiple regional honey samples.

Attenuated total reflectance-FTIR spectroscopy, along with multivariate analysis, was applied for the analysis of honey intentionally adulterated with standard sugar solutions (e.g., glucose, fructose, and sucrose) and also with cheap syrups (e.g., corn, inverted, and cane sugar) (Rios-Corripio et al. 2012). PCA method on pure and adulterated (0%–100%) honey samples afforded the determination of the type of adulterant. A similar method together with multivariate analytical tools was also used earlier by Gallardo-Velazquez et al. (2009) to the quantification of adulterants in Mexican honeys.

In recent years, near-IR (NIR) spectroscopy has gained widespread acceptance in many different applications by virtue of its many advantages over other analytical techniques, the most salient of these being its ability to record spectra from solid and liquid samples with no prior treatments. It has traditionally been used for the quantification of food compositional parameters but can equally be useful for the determination of complex quality properties such as texture and sensory attributes. The NIR region spans the wavelength range 780 to 2500 nm, where absorption bands correspond mainly to overtones and combinations of fundamental vibrations. The development of instrumentation, measurement techniques, and chemometric applications that are relevant to the food industry have been widely reviewed (Woodcock et al. 2008). NIR coupled with chemometrics methods was used to detect adulteration of various honey samples (Zhu et al. 2010). A similar method was applied for the prediction of soluble solids content and moisture in honey (Li et al. 2010), jaggery syrup in honey (Mishra et al. 2010), and fructose and glucose determination in honey samples (Tu et al. 2010a). A couple of reviews on quality control of honey using IR spectroscopy have recently been published (Tu et al. 2010b; Cozzolino et al. 2011).

## INSTRUMENTAL NEUTRONIC ACTIVATION ANALYSIS

The concentrations of 14 trace elements (Br, Ce, Co, Cr, Cs, Eu, Fe, La, Rb, Sb, Sc, Sm, Th, and Zn) were quantified in 120 samples of multifloral Argentine honeys, using instrumental neutronic activation analysis (NAA) (Pellerano et al. 2012), which is a nuclear process. Generally, NAA allows discrete sampling of elements, as it disregards the chemical form of a sample and focuses solely on its nucleus. The method is based on neutron activation and therefore needs a source of neutrons. The sample is bombarded with neutrons, causing the elements to form radioactive isotopes. The radioactive emissions and radioactive decay paths for each element are well documented, and this information allows analysis of spectra of the emissions of the radioactive sample and determination of the concentrations of the elements within it. In the NAA of multifloral Argentine honeys, the elemental composition was used in multivariate statistical analysis to discriminate the honeys according to geographical origin. It was suggested that NAA-based element analysis could provide a good prospect for discriminating honeys by regions, even if the element composition is not usually dependent on the year of harvest.

## ION CHROMATOGRAPHY

High molecular weight oligosaccharides of 9 sugar syrups and 25 honey samples were analyzed by high-performance anion-exchange chromatography pulsed amperometric detection (PAD) (Morales et al. 2008). Samples were previously treated with activated charcoal to remove monosaccharides and disaccharides. Dionex DX-300 equipment containing a gradient pump and an eluent degas module was used, and the separation of carbohydrates was carried out on a CarboPac PA 100 guard column (4 × 50 mm) and a CarboPac PA-100 anion-exchange column (4 × 250 mm) with an injection volume of 20 μL and a flow rate of 0.7 mL/min. Carbohydrates (oligosaccharides) were detected by PAD (Concorde Waters) with a gold working electrode and a hydrogen reference electrode using triple pulsed amperometry. Oligosaccharides were eluted by a gradient prepared from 1 M sodium acetate (eluent A), deionized water (eluent B), and 1 M sodium hydroxide (eluent C). Eluent C was kept constant (10%) during the whole process and eluent A changed from 3% to 10% at 30 min and increased to 20% at 70 min. This proportion was kept constant until 85 min, when it recovered the initial conditions. Oligosaccharides were quantified by an external standard method using a mixture containing from maltotriose to maltoheptose.

The qualitative and quantitative analyses of various organic acids were performed by ion chromatography with an electrochemical detector (Daniele et al. 2012). The $^{13}C/^{12}C$ isotopic ratios of the honeys, and of the organic acids extracted from them with an anion exchange resin, were determined by isotope ratio MS. It was observed that the isotopic ratios of honeys and of their acids were strongly linked. This method was successful to differentiate honeys from seven botanical origins, based on organic acid analysis, and could be used as an alternative method to the protein-based White method, using organic acids as new IS instead of proteins.

## NUCLEAR MAGNETIC RESONANCE AND CHEMOMETRIC METHODS

1D and 2D nuclear magnetic resonance (NMR) coupled with multivariate statistical analysis were employed for the detection of honey adulteration by sugar (Bertelli et al. 2010). The best discriminant model was observed with 1D NMR spectra, while 2D NMR also displayed reasonable results. It was, however, concluded that $^1H$ NMR sequence was preferable because of its simplicity and short experimental time. As a $^1H$ NMR spectrum provides a fingerprint for each honey type, showing many characteristic peaks in all spectral regions, $^1H$ NMR together with chemometrics for the determination of the botanical origins of honey samples, based on nonvolatile organic components, was reported by Schievano et al. (2010). A combination of HPLC-DAD-ESI-MS and NMR techniques (1D $^1H$ NMR and diffusion-ordered spectroscopy NMR) was applied for the determination of quinoline alkaloids in honey samples (Beretta et al. 2009). Just before this, the same authors reported an SPE procedure coupled with [1]H NMR, with chemometric analysis, to see reliable markers of the botanical origin of honey (Beretta et al. 2008). The [1]H NMR spectra were obtained at 300 K on a Varian Mercury VX300 or Bruker DRX500 NMR spectrometer equipped with a 5-mm z-gradient inverse broad-band probe, and trimethylsilane was used as the IS. Proton–proton connectivities were confirmed through a correlation spectroscopy.

Boffo et al. (2012) evaluated the potential of NMR spectroscopy, particularly the [1]H NMR technique, in combination with chemometric methods to differentiate honeys concerning the nectar employed in its production, and were able to discriminate the honeys produced in the state of Sao Paulo. Principal component analysis (PCA) and high content analysis were employed to analyze [1]H NMR resulting in the natural clustering of the samples. Honeys from different botanical origins were successfully identified by the presence of certain characteristic components; wildflower honeys were characterized by higher concentration of phenylalanine and tyrosine, and *Citrus* honeys displayed higher amounts of sucrose than other compounds. Chemometric methods were also applied to classify honey according to mineral content, and cluster analysis was used to show three clear clusters corresponding to three different geographical origins (Yucel and Sultanoglu 2012).

An NMR-based metabolomic approach that uses O2PLS-DA multivariate data analysis to identify the botanical origin of honey was described by Sachievano et al. (2012). It was demonstrated that NMR could be used in the analysis of honey for botanical origin discrimination and biomarker discovery using a metabolomic approach. The NMR spectra of 353 chloroform extracts of selected honey samples, both of monofloral and polyfloral origins, were analyzed to detect possible markers of their floral origin. The high precision of the classification obtained indicated that this approach could be applied for introducing generally applicable metabolomic tools to discriminate the origin of honey samples.

## POTENTIOMETRIC ANALYSIS

A simple and rapid potentiometric method for determination of honey diastase activity was reported by Sakac and Sak-Bosnar (2012). Potentiometric analysis involves measurement of the potential of electrochemical cells without drawing appreciable current. A platinum redox sensor was utilized to quantify the amount of free triiodide released from a starch triiodide complex after starch hydrolysis by honey diastase. This method was compared with classic Schade and commercial Phalebas procedures revealing good correlations with both methods.

The effectiveness of a potentiometric electronic tongue, made of various metals and metallic compounds, was applied for the differentiation of honey samples at three different states: raw, liquefied, and pasteurized (Escriche et al. 2012). PCA and a neural network established that potentiometric electrodes could be useful for classifying honey by its botanical origin. A noticeable correlation between the electronic tongue and the physicochemical parameters was observed. However, the correlation with volatile compounds was much weaker.

## Size Exclusion Chromatography

Size exclusion chromatography (SEC) coupled with UV diode array detection technique and aided by chemometric techniques involving PCA and hierarchical cluster analysis has recently been used for rapid simultaneous profiling of minor components of 32 diluted honeys of different botanical and geographical origins (Beretta et al. 2012). A single chromatographic run was found to be adequate to provide a fast profile of high molecular weight components (e.g., proteins and enzymes), intermediate (e.g., terpenoid glycosides), and low molecular weight compounds (e.g., kynurenic acid). The method can be summarized as follows. Native honey (1 g) was dissolved in mobile phase (10 mL final volume) and 10 μL of the clear solution obtained after centrifugation to remove solid particles was analyzed by SEC-UV-DAD. Chromatographic runs were carried out using a Varian LC-940 analytical/ semipreparative HPLC system (Varian, Turin, Italy) equipped with a binary pump system, an autosampler, a fraction collector, and a UV-DAD detector operating (200–400 nm) at wavelengths of 220 and 280 nm. Analytical separations were performed on a TSKgel G2000 SWXL column (300 × 7.8 mm and 250 × 21.2 mm). The solvent system was NaCl (8.5 g/L)/$NaH_2PO_4$ (2 g/L)/$Na_2HPO_4$ (1 g/L) (pH 6.7), and the flow rate is 1.2 mL/min. Analyses were carried out from 0 to 20 min of the chromatographic run to ensure that analytes with the lowest molecular weight were eluted. The chromatographic reproducibility was found to be highly satisfactory with the average relative standard deviations of the peak retention times less than 0.3%.

SEC was also used to characterize α-amylase from Mandarin orange honey (Nagai et al. 2012). DEAE-Toyopearl 650M, CM-Toyopearl 650M, and Toyopearl HW-55F size exclusion columns were used. The molecular weight of the purified enzyme was approximately 58 kDa as estimated by Toyopearl HW-55F gel chromatography and sodium dodecyl sulfate-polyacrylamide gel electrophoresis.

## UV-Visible Absorption Spectroscopy

A simple method incorporating UV–visible absorption spectroscopy and chemometrics analysis was applied for the identification of authentic and adulterated honeys (Ou et al. 2011). D-fructose and D-glucose, at a mass ratio typical of honey composition (1.2:1.0), were used for the preparation of adulterant solution and added to individual honeys at levels of 5%, 10%, 15%, and 20%. Absorption spectra of authentic and adulterated honeys were obtained (220–750 nm). The absorbance values of the best sensitive band (250–400 nm) were selected to establish models by PCA in combination with back-propagation artificial neural network. The scores of optimal principal components were used as the input vectors of the model. The correct identification rate was 100% for both the calibration and prediction sets.

## Miscellaneous

A spectroscopic fingerprint technique, FT-Raman spectroscopy, combined with chemometric tools was used as a rapid and reliable method for the discrimination of honey samples based on their sources (Pierna et al. 2011). Also different chemometric models were constructed to discriminate between Corsican honeys and honey coming from other regions in France, Italy, Austria, Germany, and Ireland in accordance to their FT-Raman spectra. Several Romanian honey samples of multifloral origins were analyzed for their fructose corn syrup adulteration content by coupling an isotope ratio mass spectrometer to an elemental analyzer (Dordai et al. 2011). Exploiting the difference in stable carbon isotope ratio (delta C-13) between honey and its protein fraction, the adulteration with small extent (minimum of 7%) could be successfully evaluated.

Micellar liquid chromatography (MLC) together with the chemometrics approach was applied for the optimization of separation of flavonoid markers (e.g., quercetin hesperetin and chrysin) in honey (Hadjmohammadi et al. 2010). This method incorporated SPE of honey using a reverse-phase silica $C_{18}$ cartridge followed by the separation and quantification of flavonoids by MLC.

An electrochemical method for the determination of organophosphorous compounds, azinphos-methyl and parathion-methyl, in honey samples was developed by adsorptive stripping differential pulse voltammetry at hanging mercury working electrode (Tsiafoulis and Nanos 2010). It was observed that, compared with available chromatographic methods for the determination of pesticide residues, the sample preparation of the voltammetric method was minimal and the instrumentation was simpler.

Total reflection x-ray spectrometry in combination with chemometric methods was applied for the determination of the botanical origin of Slovenian honey based on mineral contents (Necemer et al. 2009). By employing PCA and regularized discriminant analysis, it was established that, from all of the measured elements, only the four characteristic key elements (Cl, K, Mn, and Rb) could be used to best discriminate the types of honey samples.

## CONCLUSION

Among all the modern approaches, HPLC and GC coupled with various pretreatments and detection techniques are the most commonly used techniques for the analysis of the chemical components in honey. Significant method developments and advances as outlined earlier have taken place involving these analytical tools, especially in relation to honey analysis.

## REFERENCES

Alechaga E, Moyano E, and Galceran MT (2012). Ultra-high performance liquid chromatography-tandem mass spectrometry for the analysis of phenicol drugs and florfenicolamine in foods. *Analyst* **137**, 2486–2494.

Afredsson G, Branzell C, Granelli K, and Lundstrom A (2005). Simple and rapid screening and confirmation of tetracyclines in honey and egg by a dipstick test and LC-MS/MS. *Analytica Chimica Acta* **529**, 47–51.

Aliferis KA, Tarantilis PA, Harizanis PC, and Alissandrakis E (2010). Botanical discrimination and classification of honey samples applying gas chromatography/mass spectrometry fingerprinting of headspace volatile compounds. *Food Chemistry* **121**, 856–862.

Alissandrakis E, Tarantilis PA, Pappas C, Harizanis PC, and Polissiou M (2009). Ultrasound-assisted extraction gas chromatography-mass spectrometry analysis of volatile compounds in unifloral thyme honey from Greece. *European Food Research and Technology* **229**, 365–373.

Amendola G, Pelosi P, and Dommarco R (2011). Solid-phase extraction for multi-residue analysis of pesticides in honey. *Journal of Environmental Science and Health Part B—Pesticides Food Contaminants and Agricultural Wastes* **46**, 24–34.

Andrade P, Ferreres F, Gil MI, and Tomas-Barberian FA (1997). Determination of phenolic compounds in honeys with different floral origin by capillary zone electrophoresis. *Food Chemistry* **60**, 74–84.

Arena E, Ballistreri G, Tomaselli F, and Fallico B (2011). Survey of 1,2-dicarbonyl compounds in commercial honey of different floral origin. *Journal of Food Science* **76**, C1203–C1210.

Barganska Z, Slebioda M, and Namiesnik J (2011). Determination of antibiotic resifues in honey. *Trends in Analytical Chemistry* **30**, 1035–1041.

Bedendo GC, Jardim ICSF, and Carasek E (2010). A simple hollow fiber renewal liquid membrane extraction method for analysis of sulfonamides in honey samples with determination by liquid chromatography-tandem mass spectrometry. *Journal of Chromatography A* **1217**, 6449–6454.

Bernal J, Nozal MJ, Jimenez JJ, Martin MT, and Sanz E (2009). A new and simple method to determine trace levels of sulfonamides in honey by high performance liquid chromatography with fluorescence detection. *Journal of Chromatography A* **1216**, 7275–7280.

Beretta G, Caneva E, Regazzoni L, Bakhtyari NG, and Facino RM (2008). A solid-phase extraction procedure coupled to [1]H NMR, with chemometric analysis, to seek reliable markers of the botanical origin of honey. *Analytica Chimica Acta* **620**, 176–182.

Beretta G, Artali R, Caneva E, Orlandini S, Centini M, and Facino RM (2009). Quinoline alkaloids in honey: Further analytical (HPLC-DAD-ESI-MS, multidimensional diffusion-ordered NMR spectroscopy), theoretical and chemometric studies. *Journal of Pharmaceutical and Biomedical Analysis* **50**, 432–439.

Beretta G, Fermo P, and Facino RM (2012). Simple and rapid simultaneous profiling of minor components of honey by size exclusion chromatography (SEC) coupled to ultraviolet diode array detection (UV-DAD), combined with chemometric methods. *Journal of Pharmaceutical and Biomedical Analysis* **58**, 193–199.

Bertelli D, Lolli M, Papotti G, Bortolotti L, Serra G, and Plessi M (2010). Detection of honey adulteration by sugar syrups using one-dimensional and two-dimensional high-resolution nuclear magnetic resonance. *Journal of Agricultural and Food Chemistry* **58**, 8495–8501.

Bertoncelj J, Polak T, Kropf U, Korosec M, and Golob T (2011). LC-DAD-ESI/MS analysis of flavonoids and abscisic acid with chemometric approach for the classification of Slovenian honey. *Food Chemistry* **127**, 206–302.

Bianchi F, Mangia A, Mattarozzi M, and Musci M (2011). Characterization of the volatile profile of thistle honey using headspace solid-phase microextraction and gas chromatography-mass spectrometry. *Food Chemistry* **129**, 1030–1036.

Biesaga K and Pyrzynska M (2009). Liquid chromatography/tandem mass spectrometry studies of the phenolic compounds in honey. *Journal of Chromatography A* **1216**, 6620–6626.

Blasco C, Vazquez-Roig P, Onghena M, Masia A, and Pico Y (2011). Analysis of insecticides in honey by liquid chromatography-ion trap-mass spectrometry: Comparison of different extraction procedures. *Journal of Chromatography A* **1218**, 4892–4901.

Boffo EF, Tavares LA, Tobias ACT, Ferreira MMC, and Ferreira AG (2012). Identification of components of Brazilian honey by H-1 NMR and classification of its botanical origin by chemometric methods. *LWT-Food Science and Technology* **49**, 55–63.

Bogdanov S, Martin P, and Lullmann C (1997). Harmonised methods of the European Honey Commission. *Apidolgie* (extra issue), 1–59.

Bohm DA, Stachel CS, and Gowik P (2012). Confirmatory method for the determination of streptomycin and dihydrostreptomycin in honey by LC-MS/MS. *Food Additives and Contaminants, Part A—Chemistry and Analysis Control Exposure & Risk Assessment* **29**, 189–196.

Brokl M, Soria AC, Ruiz-Matute AI, Sanz ML, and Ramos L (2010). Separation of disaccharides by comprehensive two-dimensional gas chromatography-time-of-flight mass spectrometry. Application to honey analysis. *Journal of Agricultural and Food Chemistry* **58**, 11561–11567.

Campillo N, Penalver R, Aguinaga N, and Hernandez-Cordoba M (2006). Solid-phase microextraction and gas chromatography with atomic emission detection for multiresidue determination of pesticides in honey. *Analytica Chimica Acta* **562**, 9–15.

Campillo N, Vinas P, Penalver R, Cacho JI, and Hernandez-Cordoba M (2012). Solid-phase microextraction followed by gas chromatography for the speciation of organotin compounds in honey and wine samples: A comparison of atomic emission and mass spectrometry detectors. *Journal of Food Composition and Analysis* **25**, 66–73.

Ceballos L, Pino JA, Quijano-Celis CE, and Dago A (2010). Optimization of a HS-SPME/GC-MS method for determination of volatile compounds in some Cuban unifloral honeys. *Journal of Food Quality* **33**, 507–628.

Centi S, Stoica AI, Laschi S, and Mascini M (2010). Development of an electrochemical immunoassay based on the use of an eight-electrodes screen-printed array coupled with magnetic beads for the detection of antimicrobial sulfonamides in honey. *Electroanalysis* **22**, 1881–1888.

Chen H, Chen H, Ying J, Huang J, and Liao L (2009). Dispersive liquid-liquid microextraction followed by high-performance liquid chromatography as an efficient and sensitive technique for simultaneous determination of chloramphenicol and thiamphenicol in honey. *Analytica Chimica Acta* **632**, 80–85.

Chienthavorn O, Dararuang K, Sasook A, and Ramnut N (2012). Purge and trap with monolithic sorbent for gas chromatographic analysis of pesticides in honey. *Analytical and Bioanalytical Chemistry* **402**, 955–964.

Chudzinska M and Baralkiewicz D (2010). Estimation of honey authenticity by multielements characteristics using inductively coupled plasma-mass spectrometry (ICP-MS) combined with chemometrics. *Food and Chemical Toxicology* **48**, 284–290.

Chudzinska M and Baralkiewicz D (2011). Application of ICP-MS method of determination of 15 elements in honey with chemometric approach for the verification of their authenticity. *Food and Chemical Toxicology* **49**, 2741–2749.

Chudzinska M, Debska, A and Baralkiewicz D (2012). Method validation for determination of 13 elements in honey samples by ICP-MS. *Accreditation and Quality Assurance* **17**, 65–73.

Ciulu M, Solinas S, Floris I, Panzanelli A, Pilo MI, Piu PC, Spano N, and Sanna G (2011). RP-HPLC determination of water-soluble vitamins in honey. *Talanta* **83**, 924–929.

*Codex Alimentarius* Standards for Honey (1993). Ref Nr CL 1993/14-SH FAO and WHO, Rome.

Council Directive of July 22, 1974 on the harmonization of the laws of the member states relating to honey, 74/409/EEC (1974). *Official Journal of the European Communities* **L221**, 14.

Cozzolino D, Corbella E, and Smyth HE (2011). Quality control of honey using infrared spectroscopy: A review. *Applied Spetroscopy Reviews* **46**, 523–538.

Crane E (1975). *A Book of Honey*. Oxford University Press, Oxford, New York, Toronto, Melbourne.

Cuevas-Glory L, Ortiz-Vazquez E, Pino JA, and Sauri-Duch E (2012). Floral classification of Yucatan Peninsula honeys by PCA & HS-SPME/GC-MS of volatile compounds. *International Journal of Food Science and Technology* **47**, 1378–1383.

Daniele G, Maitre D, and Casabianca H (2012). Identification, quantification and carbon stable isotopes determinations of organic acids in monofloral honeys. A powerful tool for botanical and authenticity control. *Rapid Communications in Mass Spectrometry* **26**, 1993–1998.

Da Silva VL, Cerqueira MRF, Lowinsohn D, Matos MAC, and Matos RC (2012). Amperometric detection of ascorbic acid in honey using ascorbate oxidase immobilised on amberlite IRA-743. *Food Chemistry* **133**, 1050–1054.

Daher S and Gulacar FO (2010). Identification of new aromatic compounds in the New Zealand Manuka honey by gas chromatography-mass spectrometry. *E-Journal of Chemistry* **7**, S7–S14.

Dobrinas S, Birghila S, and Coatu V (2008). Assessment of polycyclic aromatic hydrocarbons in honey and propolis produced from various flowering trees and plants in Romania. *Journal of Food Composition and Analysis* **21**, 71–77.

Dominguez-Alvarez J, Rodriguez-Gonzalo E, Hernanadez-Mendez J, and Carabias-Martinez R (2012). Programed nebulizing-gas pressure mode for quantitative capillary electrophoresis-mass spectrometry analysis of endocrine disruptors in honey. *Electrophoresis* **33**, 2374–2381.

Dordai E, Magdas DA, Cuna SM, Cristea G, Futo I, Vodila G, and Mirel V (2011). Detection of some Romanian honey types adulteration using stable isotope methodology. *Studia Universitatis Babes-Bolyai Chemia* **56**, 167–163.

Escriche I, Kadar M, Domenech E, and Gil-Sanchez L (2012). A potentiometric electronic tongue for the discrimination of honey according to the botanical origin. Comparison with traditional methodologies: Physicochemical parameters and volatile profile. *Journal of Food Engineering* **109**, 449–456.

Escuredo O, Silva LR, Valentao P, Seijo MC, and Andrade PB (2012). Assessing Rubus honey value: Pollen and phenolic compounds content and antibacterial capacity. *Food Chemistry* **130**, 671–678.

Fontana AR, Camargo AB, and Altamirano JC (2010). Coacervative microextraction ultrasound-assisted back-extraction technique for determination of organophosphates pesticides in honey samples by gas chromatography-mass spectrometry. *Journal of Chromatography A* **1217**, 6334–6341.

Franchini RAD, Matos MAC, and Matos RC (2011). Amperometric determination of catalase in Brazilian commercial honeys. *Analytical Letters* **44**, 232–240.

Gallardo-Velazquez T, Osorio-Revilla G, Loa MZ, and Rivera-Espinoza Y (2009). Application of FTIR-HATR spectroscopy and multivariate analysis to the quantification of adulterants in Mexican honeys. *Food Research International* **42**, 313–318.

Gomez-Perez ML, Plaza-Bolanos P, Romero-Gonzalez R, Martinez-Vidal JL, and Garrido-Frenich A (2012). Comprehensive qualitative and quantitative determination of pesticides and veterinary drugs in honey using liquid chromatography-Orbitrap high resolution mass spectrometry. *Journal of Chromatography A* **1248**, 130–138.

Granja RHMM, Nino AMM, Rabone F, and Salerno AG (2008). A reliable high-performance liquid chromatography with ultraviolet detection for the determination of sulphonamide in honey. *Analytica Chimica Acta* **613**, 116–119.

Guillen I, Gabaldon JA, Nunez-Delicado E, Puchades R, Maquieira A, and Morais S (2011). Detection of sulphathiazole in honey samples using a lateral flow immunoassay. *Food Chemistry* **129**, 624–629.

Guo ZY and Gai PP (2011). Development of an ultrasensitive electrochemiluminescence inhibition method for the determination of tetracyclines. *Analytica Chimica Acta* **688**, 197–202.

Hadjmohammadi MR, Saman S, and Nazari SJ (2010). Separation optimization of quercetin, hesperetin and chrysin in honey by micellar liquid chromatography and experimental design. *Journal of Separation Science* **33**, 2144–2151.

Hammel Y-A, Mohamed R, Gremaud E, LeBreton M-H, and Guy PA (2008). Multi-screening approach to monitor and quantify 42 antibiotic residues in honey by liquid chromatography–tandem mass spectrometry. *Journal of Chromatography A* **1177**, 58–76.

Hayashi T, Takamatsu N, Nakashima T, and Arita T (2011). Immunological characterization of honey proteins and identification of MRJP 1 as an IgE-binding protein. *Bioscience, Biotechnology and Biochemistry* **75**, 556–560.

Jeon M and Paeng IR (2008). Quantitative detection of tetracycline residues in honey by a simple sensitive immunoassay. *Analytical Chimica Acta* **626**, 180–185.

Jerkovic I, Tuberoso CIG, Kasum A, and Marijanovic Z (2011a). Volatile Compounds of *Asphodelus microcarpus* SALZM. et VIV. Honey Obtained by HS-SPME and USE Analyzed by GC/MS. *Chemistry and Biodiversity* **8**, 587–598.

Jerkovic I, Marijanovic Z and Staver MM (2011b). Screening of natural organic volatiles from *Prunus mahaleb* L. honey: Coumarin and vomifoliol as nonspecific biomarkers. *Molecules* **16**, 2507–2518.

Kadar M, Juan-Borras M, Carot JM, Domenech E, and Escriche I (2011). Volatile fraction composition and physicochemical parameters as tools for the differentiation of lemon blossom honey and orange blossom honey. *Journal of the Science of Food and Agriculture* **91**, 2768–2776.

Kanda M, Sasamoto T, Takeba K, Hayashi H, Kusano T, Matsushima Y, Nakajima T, Kanai S, and Takano I (2012). Rapid determination of nitroimidazole residues in honey by liquid chromatography/tandem mass spectrometry. *Journal of AOAC International* **95**, 923–031.

Kaskoniene V, Venskutonis PR, and Ceksteryte V (2011). Sugar analysis for authenticity evaluation of honey in Lithuanian market. *Acta Alimentaria* **40**, 205–216.

Kelly MT, Blaise A, and Larroque M (2010). Rapid automated high performance liquid chromatography method for simultaneous determination of amino acids and biogenic amines in wine, fruit and honey. *Journal of Chromatography A* **1217**, 7385–7392.

Kempf M, Wittig M, Reinhard A, von der Ohe K, Blacquiere T, Raezke KP, Michel R, Schreier P, and Beuerle T (2011). Pyrrolizidine alkaloids in honey: Comparison of analytical methods. *Food Additives and Contaminants Part A—Chemistry Analysis Control Exposure and Risk Management* **28**, 332–347.

Kujawski MW and Namiesnik J (2008). Challenges in preparing honey samples for chromatographic determination of contaminants and trace residues. *Trends in Analytical Chemistry* **27**, 785–793.

Kujawski MW and Namiesnik J (2011). Levels of 13 multi-class pesticide residues in Polish honeys determined by LC-ESI-MS/MS. *Food Control* **22**, 914–919.

Latif Z and Sarker SD (2012). In *Natural Products Isolation* (editors: S. D. Sarker and L. Nahar), 3rd edition, Isolation of natural products by preparative high performance liquid chromatography (prep-HPLC), Humana Press/Springer-Verlag, New Jersey, 255–276.

Li J, Chen L, Wang X, Jin H, Ding L, Zhang K, and Zhang H (2008). Determination of tetracyclines residues in honey by on-line solid-phase extraction high-performance liquid chromatography. *Talanta* **75**, 1245–1252.

Li SF, Zhang X, Shan Y, and Li ZH (2010). Prediction analysis of soluble solids content and moisture in honey by near infrared spectroscopy. *Spectroscopy and Spectral Analysis* **30**, 2377–2380.

Liang Y, Cao W, Chen W-J, Xiao X-H, and Zheng J-B (2009). Simultaneous determination of four phenolic components in citrus honey by high performance liquid chromatography using electrochemical detection. *Food Chemistry* **114**, 1537–1541.

Manyl-Loh CE, Clarke AM, and Ndip RN (2012). Detection of phytoconstituents in column fractions of n-hexane extract of Goldcrest honey exhibiting anti-*Helicobacter pylori* activity. *Archives of Medical Research* **43**, 197–204.

Mishra S, Kamboj U, Kaur H, and Kapur P (2010). Detection of jaggery syrup in honey using near-infrared spectroscopy. *International Journal of Food Sciences and Nutrition* **61**, 306–315.

Morales V, Corzo N, and Sanz ML (2008). HPAEC-PAD oligosaccharide analysis to detect adulterations of honey with sugar syrup. *Food Chemistry* **107**, 922–928.

Nagy L, Batai R, Nagy G, and Nagy G (2010). Application of copper electrode based ampero-metric detector cell for LV analysis of main sugar component of honey and nectar. *Analytical Letters* **43**, 1411–1425.

Nagai T, Inoue R, Suzuki N, Tanoue Y, and Kai N (2012). Characterization of α-amylase from mandarin orange honey. *Journal of Apicultural Research* **51**, 3–9.

Necemer M, Kosir IJ, Kump P, Kropf U, Jamnik M, Bertoncelj J, Ogrinc N, and Golob T (2009). Application of total reflection X-ray spectrometry in combination with chemo-metric methods for determination of the botanical origin of Slovenian honey. *Journal of Agriculture and Food Chemistry* **57**, 4409–4414.

Oelschlaegel S, Gruner M, Wang PN, Boettcher A, Koelling-Speer I, and Speer K (2012). Classification and characterization of Manuka honeys based on phenolic compounds and methylglyoxal. *Journal of Agriculture and Food Chemistry* **60**, 72229–72237.

O'Mahony J, Moloney M, McConnell RI, Benchikh El O, Lowry P, Furey A, and Danaher M (2011). Simultaneous detection of four nitrofuran metabolites in honey using a multi-plexing biochip screening assay. *Biosensors and Bioelectronics* **26**, 4076–4081.

Ou WJ, Meng YY, Zhang XY, and Kong M (2011). Application of UV-visible absorption spec-troscopy and principal components-back propagation artificial neural network to iden-tification of authentic and adulterated honeys. *Chinese Journal of Analytical Chemistry* **39**, 1104–1108.

Ouchemoukh S, Schweitzer P, Bey MB, Djoudad-Kadji H, and Louaileche H (2010). HPLC sugar profiles of Algerian honeys. *Food Chemistry* **121**, 561–568.

Pellerano RG, Unates MA, Cantarelli MA, Camina JM, and Marchevsky EJ (2012). Analysis of trace elements in multifloral Argentine honeys and their classification according to provenance. *Food Chemistry* **134**, 578–582.

Peres GT, Rath S, and Reyes FGR (2010). A HPLC with fluorescence detection method for the determination of tetracyclines residues and evaluation of their stability in honey. *Food Control* **21**, 620–625.

Petrus K, Schwartz H, and Sontag G (2011). Analysis of flavonoids in honey by HPLC cou-pled with coulometric electrode array detection and electrospray ionization mass spec-trometry. *Analytical and Bioanalytical Chemistry* **400**, 2555–2563.

Pierna JAF, Abbas O, Dardenne P, and Baeten V (2011). Discrimination of Corsican honey by FT-Raman spectroscopy and chemometrics. *Biotechnologie Agronomie Societe et Environment* **15**, 75–84.

Pyrzynska K and Biesaga M (2009). Analysis of phenolic acids and flavonoids in honey. *Trends in Analytical Chemistry* **28**, 893–903.

Ramanauskiene K, Stelmakiene A, Briedis V, Ivanauskas L, and Jakstas V (2012). The quan-titative analysis of biologically active compounds in Lithuanian honey. *Food Chemistry* **132**, 1544–1548.

Reid RG and Sarker SD (2012). In *Natural Products Isolation* (editors: S. D. Sarker and L. Nahar), 3rd edition, Isolation of natural products by low pressure chromatography, Humana Press/Springer-Verlag, New Jersey, 155–188.

Rios-Corripio MA, Rojas-Lopez M, and Delgado-Macuil R (2012). Analysis of adulteration in honey with standard sugar solutions and syrups using attenuated total reflectance-Fourier transform infrared spectroscopy and multivariate methods. *CYTA-Journal of Food* **10**, 119–122.

Rizelio VM, Tenfen L, da Silveira R, Gonzaga LV, Costa ACO, and Fett R (2012a). Development of a fast capillary electrophoresis method for determination of carbohydrates in honey samples. *Talanta* **93**, 62–66.

Rizello VM, Gonzaga LV, Borges GDC, Micke GA, Fett R, and Costa ACO (2012b). Development of a fast MECK method for determination of 5-HMF in honey samples. *Food Chemistry* **133**, 1640–1645.

Rizvi SAA, Do DP, and Saleh AM (2011). Fundamentals of micellar electrokinetic chromatography (MEKC). *European Journal of Chemistry* **2**, 276–281.

Rodriguez-Gonzalo E, Dominguez-Alvarez J, Garcia-Gomez D, Garcia-Jimenez MG, and Carabias-Martinez R (2009). Determination of endocrine disruptors in honey by CZE-MS using restricted access materials for matrix cleanup. *Electrophoresis* **31**, 2279–2288.

Ruiz-Matute AL, Brokl M, Soria AC, Sanz ML, and Martinez-Castro I (2010). Gas chromatographic-mass spectrometric characterisation of tri- and tetrasaccharides in honey. *Food Chemistry* **120**, 637–642.

Sachievano E, Stocchero M, Morelato E, Facchin C, and Mammi S (2012). An NMR-based metabolomic approach to identify the botanical origin of honey. *Metabolomics* **8**, 679–690.

Sakac N and Sak-Bosnar M (2012). A rapid method for the determination of honey diastase activity. *Talanta* **93**, 135–138.

Sakamoto Y, Saito K, Hata R, Katabami S, Iwasaki Y, Ito R, and Nakazawa H (2012). Determination of chloramphenicol in honey and royal jelly by clean-up with molecular imprinted polymer. *Bunseki Kagaku* **61**, 383–389.

Sarker SD and Nahar L (2012). In *Natural Products Isolation* (editors: S. D. Sarker and L. Nahar), 3rd edition, Hyphenated techniques and their applications in natural products analysis, Humana Press/Springer-Verlag, New Jersey, 301–340.

Schievano E, Peggion E, and Mammi S (2010). H-1 Nuclear magnetic resonance spectra of chloroform extracts of honey for chemometric determination of its botanical origin. *Journal of Agricultural and Food Chemistry* **58**, 57–65.

Shi M, Gao QY, Feng JM, and Lu YC (2012). Analysis of inorganic cations in honeys by capillary zone electrophoresis with indirect UV detection. *Journal of Chromatographic Science* **50**, 547–552.

Soria AC, Martinez-Castro I, and Sanz J (2009). Study of the precision in the purge-and-trap–gas chromatography–mass spectrometry analysis of volatile compounds in honey. *Journal of Chromatography A* **1216**, 3300–3304.

Soto ME, Bernal J, Martin MT, Higes M, Bernal JL, and Nozal MJ (2012). Liquid chromatographic determination of resveratrol and piceid isomers in honey. *Food Analytical Methods* **5**, 162–171.

Swiss Food Manual (1995). *Schweizerisches Lebensmittelbuch, Chapter 23A: Honey*. Eidg., Drucksachen und Materialzentralle, Bern.

Tafintseva IY, Zherdev AV, Eremin SA, and Dzantiev BB (2010). Enzyme immunoassay for determination of sulfamethoxypyridazine in honey. *Applied Biochemistry and Microbiology* **46**, 216–220.

Tanner G and Czerwenka C (2011). LC-MS/MS analysis of neonicotinoid insectisides in honey: Methodology and residue findings in Austrian honeys. *Journal of Agriculture and Food Chemistry* **59**, 12271–12277.

Teixido E, Nunez O, Santos FJ, and Galceran MT (2011). 5-Hydroxymethylfurfural content in foodstuffs determined by micellar electrokinetic chromatography. *Food Chemistry* **126**, 1902–1908.

Thongchai W, Liawruangath B, Liawruangrath S, and Greenway GM (2010). A microflow chemiluminescence system for determination of chloramphenicol in honey with preconcentration using a molecularly imprinted polymer. *Talanta* **82**, 560–566.

Tolgyesi A, Sharma VK, Fekete S, Fekete J, Simon A, and Farkas S (2012). Development of a rapid method for the determination and confirmation of nitroimidazoles in six matrices by fast liquid chromatography-tandem mass spectrometry. *Journal of Pharmaceutical and Biomedical Analysis* **64–65**, 40–48.

Truchado P, Ferreres F, and Tomas-Barberan F (2009). Liquid chromatography-tandem mass spectrometry reveals the widespread occurrence of flavonoid glycosides in honey, and their potential as floral origin markers. *Journal of Chromatography A* **1216**, 7241–7248.

Truchado P, Vit P, Ferreres F, and Tomas-Barberan F (2011). Liquid chromatography tandem mass spectrometry analysis allows the simultaneous characterization of C-glycosyl and O-glycosyl flavonoids in stingless bee honeys. *Journal of Chromatography A* **1218**, 7601–7607.

Truzzi C, Annibaldi A, Illuminati S, Finale C, Rossetti M, and Scarponi G (2012). Determination of very low levels of 5-(hydroxymethyl)-2-furaldehyde (HMF) in natural honey: Comparison between the HPLC technique and the spectrophotometric White method. *Journal of Food Science* **77**, C784–C790.

Tsai W-H, Chuang H-Y, Chen H-H, Wu Y-W, Cheng S-H, and Huang T-C (2010). Application of sugaring-out extraction for the determination of sulfonamides in honey by high-performance liquid chromatography with fluorescence detection. *Journal of Chromatography A* **1217**, 7812–7815.

Tsiafoulis CG and Nanos CG (2010). Determination of azinphos-methyl and parathion-methyl in honey by stripping voltammetry. *Electrochimica Acta* **56**, 566–574.

Tsimeli K, Triantis TM, Dimotikali D, and Hiskia A (2008). Development of a rapid and sensitive method for the simultaneous determination of 1,2-dibromoethane, 1,4-dichlorobenzene and naphthalene residues in honey using HS-SPME coupled with GC-MS. *Analytica Chimica Acta* **617**, 64–71.

Tsiropoulos NG and Amvrazi EG (2011). Determination of pesticide residues in honey by single-drop microextraction and gas chromatography. *Journal of AOAC International* **94**, 634–644.

Tu ZH, Zhu DZ, Ji BP, Meng CY, Wang LG, and Qing ZS (2010a). Difference analysis and optimization study for determination of fructose and glucose by near infrared spectroscopy. *Chinese Journal of Analytical Chemistry* **38**, 45–50.

Tu ZH, Zhu DZ, Ji BP, Meng CY, Wang LG, and Qing ZS (2010b). Progress in quality analysis of honey by infrared spectroscopy. *Spectroscopy and Spectral Analysis* **30**, 2971–2975.

Uchiyama K, Kondo M, Yokochi R, Takeuchi Y, Yamamoto A, and Inoue Y (2011). Derivative spectrum chromatographic method for the determination of trimethoprim in honey samples using an on-line solid-phase extraction technique. *Journal of Separation Science* **34**, 1525–1530.

Vidal JLM, Aguilera-Luiz MD, Romero-Gonzalez R, and Frenich AG (2009). Multiclass analysis of antibiotic residues in honey by ultraperformance liquid chromatography-tandem mass spectrometry. *Journal of Agricultural and Food Chemistry* **57**, 1760–1767.

Wang Y, You J, Ren R, Xia Y, Gao S, Zhang H, and Yu A (2010a). Determination of triazines in honey by dispersive liquid–liquid microextraction high-performance liquid chromatography. *Journal of Chromatography A* **1217**, 4241–4246.

Wang J, Kliks MM, Jun S, Jackson M and Li QX (2010b). Rapid analysis of glucose, fructose, sucrose, and maltose in honeys from different geographic regions using fourier transform infrared spectroscopy and multivariate analysis. *Journal of Food Science* **75**, C208–C214.

Wiest L, Bulete A, Giroud B, Fratta C, Amic S, Lambert O, Pouliquen H, and Arnaudguilhem C (2011). Multi-residue analysis of 80 environmental contaminants in honeys, honeybees and pollens by one extraction procedure followed by liquid and gas chromatography coupled with mass spectrometric detection. *Journal of Chromatography A* **1218**, 5743–5756.

Woodcock T, Downey G, and O'Donnell CP (2008). Better quality food and beverages: The role of near infrared spectroscopy. *Journal of Near Infrared Spectroscopy* **16**, 1–29.

Wutz K, Niessner R, and Seidel M (2011). Simultaneous determination of four different antibiotic residues in honey by chemiluminescence multianalyte chip immunoassays. *Microchimica Acta* **173**, 1–9.

Yang Y, Battesti MJ, Paolini J, Muselli A, Tomi P, and Costa J (2012). Melissopalynological origin determination and volatile composition analysis of Corsican "*Erica arborea* spring maquis" honeys. *Food Chemistry* **134**, 37–47.

Yucel Y and Sultanoglu P (2012). Determination of industrial pollution effects on citrus honeys with chemometric approach. *Food Chemistry* **135**, 170–178.

Zacharis CK, Rotsias I, Zachariadis PG, and Zotos A (2012). Dispersive liquid-liquid microextraction for the determination of organochlorine pesticides residues in honey by gas chromatography-electron capture and ion trap mass spectrometric detection. *Food Chemistry* **134**, 1665–1672.

Zhang J, Gao H, Peng B, Li S, and Zhou Z (2011). Comparison of the performance of conventional, temperature-controlled, and ultrasound-assisted ionic liquid dispersive liquid–liquid microextraction combined with high-performance liquid chromatography in analyzing pyrethroid pesticides in honey samples. *Journal of Chromatography A* **1218**, 6621–6629.

Zhen J, Zhuguang L, Meiyu C, Yu M, Jun T, Yulan F, Jiachen W, Zhaobin C, and Fengzhang T (2006). Determination of multiple pesticide residues in honey using gas chromatography-electron impact ionisation-mass spectrometry. *Chinese Journal of Chromatography* **24**, 440–446.

Zhu XR, Li SF, Shan Y, Zhang ZY, Li GY, Su DL, and Liu F (2010). Detection of adulterants such as sweetener materials in honey using near-infrared spectroscopy and chemometrics. *Journal of Food Engineering* **101**, 92–97.

# 16 Mad Honey
## *The Reality*

*Abdülkadir Gunduz and Faik Ahmet Ayaz*

## CONTENTS

## INTRODUCTION

### Brief History of the Secret Weapon in Honey

The use of mad honey as a "biological weapon" was first described in 401 BC by the Athenian historian and army commander Xenophon (430–355 BC) (Gökçöl 1998; Gunduz et al. 2011). Xenophon fought the Makrons in what is now the Turkish province of Trabzon, and in his "Anabasis, The March of the Ten Thousand," he describes in detail the effects on soldiers subjected to mad honey poisoning (Gökçöl 1998; Okmen 2004; Gunduz et al. 2011). These are the keys to the use of the terms "mad honey," "hidden poison," or "the first biological weapon."

The substance now called "mad honey" subdued a number of troops from different empires in the past, some examples of which are given in the present chapter. For the most part, there was nothing in it which the victims found particularly untoward, but the swarms of bees in the neighborhood were very numerous, and the soldiers who ate the honey all suffered temporary insanity, vomiting, and diarrhea. And not one was able to remain standing. Those who had only eaten a little behaved as if they were drunk, while those who had eaten a great deal seemed crazy or even, in some cases, to be dying. So they lay there in great numbers as though the army had suffered a defeat, and "great despondency" prevailed. On the next day, however, no one had died, and at approximately the same hour as they had eaten the honey, they began to return to their senses; on the third or fourth day they got up, as if they were recovering from a drugging. From there they marched two stages, seven parasangs, and reached the sea at Trapezus, a Greek city on the Euxine Sea, a colony of the Sinopeans in the territory of Colchis (Gökçöl 1998; Gunduz et al. 2011).

Mad honey was also used in Northern Anatolia by King Mithridates Eupator of Pontus against the armies of Pompey in 97 BC. On the advice of his chief counselor, the Greek Kateuas, Mithridates placed combs full of mad honey in the path of the advancing Romans and staged a strategic withdrawal. The Romans who ate the honey from these combs collapsed with fatigue and were easily overcome (Gökçöl 1998; Gunduz et al. 2011).

These similar incidents described by ancient historians involving two different periods are particularly noteworthy. This foodstuff known as mad honey was clearly used for a very different purpose than it is today. Instead of being a food product, it may be described as a weapon of war in ancient times. A similar incident involving this honey has also come down from Russian history. When the Russian Queen Olga defeated rebel forces by a ruse and captured the city of Kiev, they accepted several tons of honey liquor from her allies. We do not know whether this liquor was reinforced with mad honey, but sources report that 5000 Russians were put to the sword as they lay in a stupor on the ground (Pekman 2005).

In the chapter on plants of his famous *Historia Naturalis*, the renowned ancient naturalist Pliny the Elder, who lived between 79 and 23 BC, provides the following information about mad honey: "Indeed the food of bees is so important that it may be poisonous for us. In Pontus, in Herakleia, different honeys from the same bees may be fatal after a few years. However, the authorities had not stated the type of flowers from which these honeys were made. There is a specific plant that meets these conditions and has proven to be a fatal danger, especially for goats and pack animals, and is called 'aegolcthron.' It makes interaction with poisonous toxins sprouting in the steppes during spring rains. A poisonous honey has a dense appearance and an abnormally red color and its smell causes sneezing. However, it is much more effective than a similar good quality honey" (Isik 2001; Ozturk 2005; Gunduz et al. 2011).

In the next part of his work, he refers to the effects of this honey and means of treating it. Elsewhere in his work, he makes the following statements about this honey: "In the country of Sanni, located in the same Pontus region, there is another type of honey causing madness. It is called 'maenomenon.' This poison is generally related with the *Rhododendron* flower, which exists in thick forests. Although people here pay tax to the Romans for it, they cannot gain any income from this honey because of its very harmful effects" (Rechham 1949, cited in Pereira 1971). Pereira (1971), who carried out a voyage of exploration in the Northeast Anatolia region in 1970, mentioned economic activities in his discussion of Trabzon. He stated that beekeeping was widespread in the region and referred to the honey on sale there being known as "Thousand and One Flower Plateau Honey" (Pereira 1971).

Cases of mad honey poisoning were reported from Europe and North America in the 18th century (Ozhan et al. 2004). Speaking at the Annual Meeting of the American Pharmacists Union in 1896, Kebler added cases of mad honey poisoning in the United States to the literature (Kebler 1896). That publication described eight cases of poisoning in Princeton, New Jersey. Kebler (1896) discussed previous research as well as his own cases. According to Kebler (1896), Barton (2007; cited in Gunduz et al. 2008) was the first American to report the effects of mad honey poisoning. He first read his findings at a meeting of American philosophers in 1794, subsequently publishing these in 1802. The case concerned involved a 54-year-old woman who ingested mad honey. She presented to hospital with a burning forehead, facial pallor, breathlessness, and an irregular pulse. The patient suffered visual hallucinations and subsequently lost mental awareness. She had generalized tonic-clonic seizure for a short period of time. The patient regained mental awareness after vomiting, following which lingual numbness and mydriasis remained (Kebler 1896). He also referred to Coleman's (1853) study (cited in Kebler 1896), in which one fatality (out of 14 cases affected) in New Jersey and 3 (out of 20 affected) in Branchville, South Carolina, were from mad honey poisoning. Additionally, Plugge and de Zaayer (1889) conducted a study of several plants from the Ericaceae family in 1891 and discovered that the active substance in these was andromedotoxin (grayanotoxin or GTX) (Kebler 1896). Kebler's (1896) paper, republished in the *British Medical Journal* in 1999, concerned typical mad honey poisoning (BMJ 1999). Dioscorides, Diodorus of Sicily, and Aristotle also refer to the intoxicating effects of honey collected from a species of *Rhododendron* in Harakleia Pontika at certain times of the year. J.P. Tournefort described poisonous honey as originating from *Azalea pontica* and said, "whoever eats of this honey becomes intoxicated" (Kebler 1896).

## RHODODENDRON L.

The name *Rhododendron* derives from the Greek "rhodo," rose, and "dendron" tree. *Rhododendron* is a genus characterized by shrubs and small to (rarely) large trees, the smallest species growing up to 10 to 100 cm in length and the largest, *Rhododendron giganteum*, reported to grow more than 30 m (Cox and Kenneth 1997; Nelson et al. 2007; Michie et al. 2011). The leaves are spirally arranged; leaf size can range from 1 to 2 cm to more than 50 cm and exceptionally to 100 cm in *Rhododendron sinogrande* AGM. They may be either evergreen or deciduous. In some species, the underside of the leaves is covered with scales (lepidote) or hairs (indumentum). Some of the best-known species are noted for their many clusters of large flowers (Cox and Kenneth 1997).

*Rhododendron* is primarily a northern hemisphere genus, extending from North America across Europe and Asia to Japan and from the extreme north to the equator. In a recent paper, the exact number of *Rhododendron* species described is put at about 800 to 850, found naturally across four continents, divided into eight subgenera (Nelson et al. 2007; Michie et al. 2011).

## WHERE DOES THE POISONING AGENT/COMPOUND OF MAD HONEY COME FROM?

Not all rhododendrons produce GTXs. Some well-known examples of the species that are active in the production of these compounds are cited here: *Rhododendron ponticum*, called the "Common *Rhododendron*" or "Pontic *Rhododendron*," is the most common species in the world and is especially native to southern Europe and southwest Asia (Stevens 1978; Nelson et al. 2007; Michie et al. 2011) (Figures 16.1 and 16.2). It contains a considerable amount of these neurotoxic compounds, called GTXs (Figures 16.3 and 16.4).

In most countries of the northern hemisphere, these compounds and related intoxication have been reported for a number of *Rhododendron* species. *R. ponticum* is found from the Balkans to the Caucasus, including the Black Sea region of Turkey, *Rhododendron mucronulatum* extends from Korea, and *Rhododendron decorum* Franchet is found in China. In the western United States, toxic *Rhododendron* species are the western azalea (*R. occidentale* [Torr. & A. Gray] A. Gray), the California rosebay (*R. macrophyllum* D. Don ex G. Don.), and *R. albiflorum* Hook. In addition to *Rhododendron*, in the eastern part of North America, the mountain laurel (*Kalmia*

FIGURE 16.1   Purple *Rhododendron ponticum*, also known as the "mountain rose."

**FIGURE 16.2** Yellow *Rhododendron flavum*, also known as *R. luteum*.

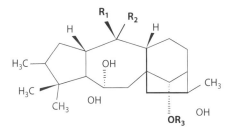

| GTXs | R₁ | R₂ | R₁ = R₂ | R₃ |
|------|------|------|---------|------|
| GTX1 | –OH | –CH₃ | – | _H |
| GTX2 | – | – | =CH₂ | _H |
| GTX3 | –OH | –CH₃ | – | –Ac |

**FIGURE 16.3** General structure of GTX.

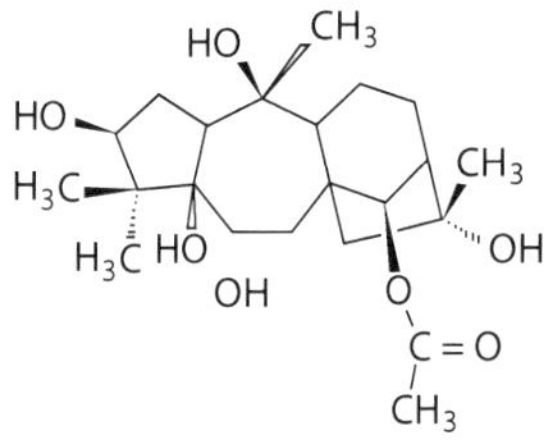

Grayanotoxin I

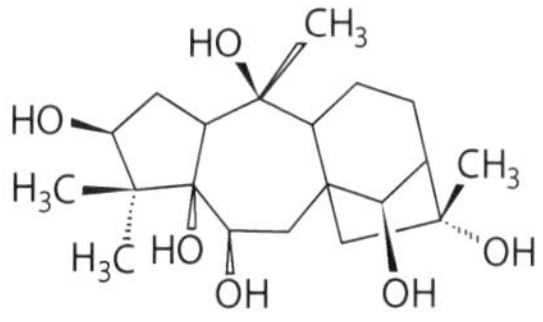

Grayanotoxin III

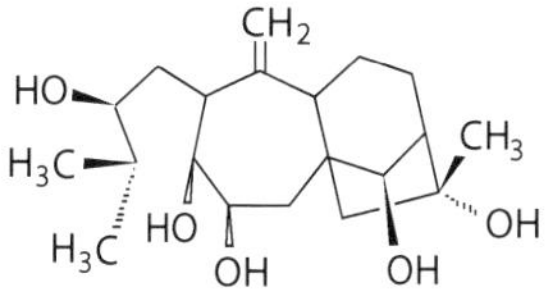

Grayanotoxin II

**FIGURE 16.4**  General structure of GTX types.

*latifolia* L.) and sheep laurel (*Kalmia angustifolia* L.) are also well-known sources of GTX (Stevens 1978; Gunduz et al. 2008) (see Table 16.1).

Turkey heads the list of those countries reporting cases and complaints of mad honey intoxication. The genus *Rhododendron* is represented by six species in *Flora of Turkey* (Stevens 1978). In Turkey, the common toxic *Rhododendron* species are *Rhododendron luteum* L. and *R. ponticum* L. Collected mad honey in Turkey is represented by the abundance of pollens of *R. ponticum* (Figure 16.1).

The question "where does the poisoning agent/compound of mad honey come from?" used to be a very difficult one to answer. The structure could only be determined using heavy and tiring early chromatographic equipment. Today, however, increasingly sensitive and powerful analytical instruments involved in the separation, characterization, and structure determination processes using developed liquid chromatography–tandem mass spectrometry (LC-MS/MS), LC time-of-flight MS (LC-TOF-MS), etc., have revealed that the answer is "*Rhododendron* toxins." The principle toxins in *Rhododendron* species, commonly found in the nectar, flowers, pollens, and leaves of the plants belonging to the family Ericaceae (Table 16.1), are GTXs (Figures 16.1 and 16.2) (Stevens 1978; Adler 2000; Gunduz et al. 2008; Michie et al. 2011).

We now know that the honey guide is on the upper corolla lobe (sometimes also on the adjacent lobes) and that pollinating bees collect pollen on their undersides from the declinate stamens as they land on the lower lobes in the flowers. When bees ingest nectar containing GTX, the GTX, along with the other components of nectar, is included in the honey that is subsequently produced. The complex sugars in nectar are enzymatically broken down into glucose and fructose in the bees' second stomach (Gunduz et al. 2008). The honey is then secreted by the bees into the honeycomb. Here, forced evaporation removes water from the honey and concentrates all its components (Gunduz et al. 2008).

## TABLE 16.1
### Sources of GTXs in Some Plant Species Toxic to Humans

| Species | Family | GTX[a] (andromedotoxin) |
| --- | --- | --- |
| *Agauria* (DC.) Hook. f. | Ericaceae | |
| *Agauria* spp. (DC.) Hooker | Ericaceae | + |
| *Andromeda* (Pieris) L. | Ericaceae | |
| *P. japonica* (Thunb.) D. Don | Ericaceae | +++ |
| *P. floribunda* (Pursh ex Simms) Benth. & Hook. | Ericaceae | +++ |
| *P. formosa* (Wall) D. Don | Ericaceae | +++ |
| *Rhododendron* L. | Ericaceae | |
| *R. luteum* Sweet | Ericaceae | ++ |
| *R. ponticum* L. | Ericaceae | ++ |
| *R. occidentale* | Ericaceae | + |
| *R. macrophyllum* | Ericaceae | + |
| *R. albiflorum* | Ericaceae | + |
| *R. maximum* L. | Ericaceae | ++ |
| *R. japonicum* (Gray) Suringar | Ericaceae | + |
| *R. catawbiense* Michx. | Ericaceae | +++ |
| *R. decorum* Fr. | Ericaceae | ++ |
| *R. mucronulatum* | Ericaceae | ++ |
| *Kalmia* L. | Ericaceae | |
| *K. latifolia* L. | Ericaceae | + |
| *K. angustifolia* L. | Ericaceae | +++ |
| *K. polifolia* Wangenh var. microphylla (Hook.) Calder & Roy Taylor | Ericaceae | +++ |
| *Pernettya* Gaud. | Ericaceae | |
| *P. coriaceae* Klotzsch | Ericaceae | ++ |

[a] Possible presence level (spectroscopic and chromatographic data): +, less; ++, mid; +++, high.

## ARE THERE ANY OTHER NECTAR-PRODUCING PLANTS APART FROM *RHODODENDRON* SPECIES?

GTX-producing plants are not unique to Turkey. The toxic species of Ericaceae are widespread in North America, Europe, and Asia (Cox and Kenneth 1997; Gunduz et al. 2008). However, the toxic Ericaceae may not be present in sufficient density outside Turkey to produce toxic honey. Sources of honey toxic to humans are given in Table 16.1. The number of *Rhododendron* spp. growing on the hills and mountains of eastern Turkey is quite impressive. These *Rhododendron* species threaten other commercial plant species (Esen et al. 2006).

## IS IT TRUE THAT THE MORE DIVERSE THE BEES, THE MORE DIVERSE MAD HONEYS ARE PRODUCED?

It seems unlikely that the distribution of GTX/mad honey poisoning can be attributed to a difference between Turkish honey bees and those in the rest of the world, because historically the honey bees of Europe and North America have also been capable of producing mad honey (Kebler 1896; BMJ 1999; Gunduz et al. 2006, 2007a,b, 2009; Lampe 1988).

## STRUCTURE OF GTX

In general, GTXs I–III (Figures 16.3 and 16.4) (sometimes written 1–3) are stable, resistant to temperatures of 300°C, and known as nonvolatile diterpenes or a unique class of toxic diterpenoids (polyhydroxylated cyclic hydrocarbon) that are water- and lipid-soluble (Wong et al. 2002). GTXs have also been described as andromedotoxin, acetylandremodol, and rhodotoxin in the past (Wong et al. 2002; Terai et al. 2003; Hough et al. 2010; Holstege et al. 2001; Michie et al. 2011).

Figure 16.1 shows the general structure of GTXs with their substituents, and Figure 16.2 shows the common characteristic structure of types of GTXs. There are 18 known forms of GTXs (Wong et al. 2002; Michie et al. 2011), but the principle variants are GTX I and III. GTX II is also present in *Rhododendron*, but in much lower amounts (Wong et al. 2002; Hough et al. 2010; Michie et al. 2011), and is less toxic than GTX I and III (Gunduz et al. 2006, 2008; Koca and Koca 2007). GTXs are also found in other plants of the family Ericaceae, including those in the species *Aguaria*, *Pieris*, *Kalmia*, *Pernettya*, etc. (Tallent et al. 1957; Power et al. 1991; Adler 2000). The reported toxicities of GTXs vary among the species of the genera (see Table 16.1). Wong et al. (2002) and Scott et al. (1971) reported that GTX II is much less toxic than GTX I and III, with far larger quantities being required for a lethal dose when injected in mice. In contrast, Hikino et al. (1976; cited in Holstege et al. 2001; Hough et al. 2010; Michie et al. 2011) reported that GTX I, II, and IV are the most toxic to mice. GTXs other than GTX I–III are also present in *Rhododendron* species, but to our knowledge, no reports have reported the presence of GTXs other than GTX I–III in *Rhododendron*. The levels of the different GTXs have been found to change seasonally in plants, with summer leaves of *Leucothoe grayana* containing different amounts of GTXs to spring leaves (Miyajima and Takei 1936; cited in Scott et al. 1971). GTX concentrations have also been found to vary within types of rhododendron material (Wong et al. 2002; Michie et al. 2011).

### BIOSYNTHESIS, DERIVATIVES, INTERMEDIATES, AND TOXICITY OF GTXS

GTXs are characterized by the A-nor-B-homo-kaurene skeleton, an unusual tetracyclic carbon framework, and by the dense arrangement of hydroxyl groups (Kan et al. 1994). Total synthesis of (-)-GTX III originating from (*R*)-2-(benzyloxy)propionaldehyde was successfully accomplished by Kan et al. (1994). Some 60 grayanotoxane compounds, GTXs, 1,5-seco- GTXs, and their glucosides, have been isolated from various ericaceous plants. GTXs are the toxic components of these compounds (Terai et al. 2002; Michie et al. 2011).

Katakawa et al. (2004) determined that GTX derivatives control the degree of bioactivity of GTXs, as this is dependent on the position and type of a C(10) functional group. One of the key intermediates to convert GTX I, GTX III, and others into their C(10) isomers is 10,20-epoxy-GTX II (Katakawa et al. 2004). However, no information is available on the toxicity of this derivative.

GTX III has been found to transform to 9α- and 9β-hydroxy-GTX derivatives (Terai et al. 2002; Michie et al. 2011). The toxicity of these derivatives has been compared with natural GTX II, and the transformation of GTX II to 9α- and 9β-hydroxy GTX derivatives was found to result in a slight or marked reduction of acute toxicity, respectively (Terai et al. 2002; Michie et al. 2011). GTX III has also been found to transform to 10-epi-GTX III following a series of oxidation and reduction steps. The dosage level of acute toxicity of 10-epi-GTX III was estimated at about half that of natural GTX III (Terai et al. 1996; Michie et al. 2011). There has been very little research into the toxicity of GTX-related compounds. However, the limited research conducted to date indicates that the toxicity of these compounds is lower than that of the GTXs themselves (e.g., Terai et al. 1996, 2002; Michie et al. 2011). No reports are available concerning the natural degradation of GTXs to these derivatives. In our opinion, long-term storage and exposure to daylight or industrial processing may reduce the toxic effect of the honey.

## DETERMINATION OF GTXs

The GTX can be isolated from the suspect commodity by typical extraction procedures for naturally occurring terpenes, especially using methods valid for the lower terpenes. GTXs, as diterpenes that are composed of four isoprene units having a molecular formula as $C_{20}H_{32}$ derived from geranylgeranyl pyrophosphate, are less volatile than the sesquiterpenes and require the use of some chromatographic techniques during identification. At the beginning of separation, paper electrophoresis and thin-layer chromatography (TLC) are preferred for class separations. Gas chromatography (GC) and slightly different gas LC (GLC) are often required due to the compounds' instability (they oxidize or decompose easily) on heating and having low vapor pressure during analysis. The compounds therefore require derivatization before GC or GLC analysis. Further identification is largely based on infrared (IR), nuclear magnetic resonance (NMR), and MS. In recent years, developed LC-MS/MS and LC-TOF-MS techniques, etc., are also in use in the detection of the GTXs in biological, plant, and honey samples (Harborne 1998; Holstege et al. 2001; Terai et al. 2003; Gunduz et al. 2008; Hough et al. 2010).

## ISOLATION OF GTXs FROM HONEY AND PLANT PARTS

Having discussed the history and structure of GTXs, it is now worth describing their extraction, isolation, and identification from honey and plants. For the extraction of GTX from mad honey, the method described by Scott et al. (1971) have commonly been adopted by various research groups with slight modifications (Karakaya 1977; Onat et al. 1991a,b; Sutlupinar et al. 1993). In that method, an aliquot (50 g) of honey is homogenized with 10 or 100 mL of aqueous methanol (water–methanol, 3:1 or 1:1, v/v), adjusted to pH 6.5 with the addition of 0.1 N NaOH solution, and then extracted with chloroform varying between 5 × 30 and 3 × 100 mL in volume. The organic phases are collected and exposed to evaporation either by leaving the material to decant at 50°C or using a rotary vacuum evaporator.

Various authors have used different amounts of plant parts with different solvents, either alone or in aqueous solutions. For the extraction of GTX from leaf or flower samples, Karakaya (1977) and Sutlupinar et al. (1993) adhered to Scott et al.'s (1971) method, with slight modifications. Each group used 5 g of *R. ponticum* and *R. flavum* leaves and flowers in a Soxhlet apparatus with aqueous methanol (water–methanol, 1:1, v/v) over 40 h. The homogenate was filtered, decanted, or evaporated using a rotary evaporator at 50°C. The residue was then dissolved in a mixture of aqueous methanol (water–methanol, 3:1, v/v) and extracted with differing amounts of chloroform, $5 \times 30$ mL, etc. The organic phases were then combined, dried with anhydrous sodium sulfate, and evaporated to dryness under vacuum.

After preparing appropriate standards of GTXs (GTX I–III) in a chloroform–acetone (9:1, v/v) mixture at a 0.9% concentration, the research groups ran both honey and plant samples on a Silicagel G plate (0.250 mm) in a mobile phase in an appropriate mixture of toluene, ethyl acetate, and formic acid (5:4:1, v/v/v). The developed TLC plates were visualized using Godin's reactive solution (cited in Scott et al. 1971).

Zhang et al. (2005) used 10 kg of powdered *R. decorum* extracted with 95% ethanol three times. After evaporation, the residue was dissolved in water and extracted successively with petroleum ether, chloroform, and ethyl acetate. Following evaporation, the ethanol phase was applied to a silica gel column and eluted with chloroform containing an increasing amount of methanol. Repeating column chromatography yielded GTXs. In that yield, with the aid of IR, $^1$H and $^{13}$C NMR, electrospray ionization (ESI)-MS, and high-resolution ESI-MS, they identified a new minor 1,5-seco-5-oxo-GTX known as XX1 (1), together with three known GTXs, GTX I, IV, and VIII.

Hough et al. (2010) used 0.2 g of dried and ground *R. ponticum* leaves extracted by shaking over 48 h in 10 mL methanol. The extracts were diluted 1:10 with methanol and analyzed directly by LC-TOF-MS using a mobile phase of 1% (v/v) acetic acid in water (A) and 1% (v/v) acetic acid in methanol (B) at a flow rate of 0.3 mL min$^{-1}$ with an initial gradient condition of 70% A and 30% B. This was changed in a linear fashion to 90% B over 15 min, returning to the initial conditions over 1 min and then held for 9 min for column equilibration. After mass spectral analysis, LC-TOF-MS analysis yielded GTX I–III, respectively.

A rapid LC-MS/MS method was developed by Holstege et al. (2001) for the quantitative determination of GTX I–III in biological samples such as rumen contents, feces, and urine. The GTXs were extracted from solid samples with methanol. The methanol extract was diluted with water and cleansed using a reverse-phase, solid-phase extraction column. High-performance liquid chromatography (HPLC) separation was performed by reverse-phase HPLC using a gradient of water and methanol containing 1% acetic acid. Determination was by positive ion ESI and ion trap MS/MS. GTX I quantitation was based on fragmentation of the sodium adduct ion at $m/z$ 435 to a product ion at $m/z$ 375. GTX II and III were quantitated on the basis of fragmentation of the ion at $m/z$ 335 to the product ion at $m/z$ 299. The diagnostic usefulness of the method was tested by analyzing samples submitted to the veterinary toxicology laboratory.

## *RHODODENDRON* TOXICITY

GTX poisoning most commonly results from the ingestion of GTX-contaminated honey, although it may result from the ingestion of the leaves, flowers, and nectar of rhododendrons. The toxicity of these plants is often attributed to the GTXs contained in their nectar, flowers, pollens, and leaves. The honey guide is on the upper corolla lobe (sometimes also on the adjacent lobes), and pollinating bees collect pollen on their undersides from the declinate stamens as they alight on the lower lobes (Stevens 1978). Symptoms of GTX poisoning are predominantly neurologic and cardiac (Nelson et al. 2007). GTXs act on animal tissue by exerting a specific stimulatory action on membrane permeability to sodium ions in various tissues (Seyama et al. 1988; Yuki et al. 2001). GTXs are not fatal to humans in normal doses (Wong et al. 2002), although they can cause death in livestock (Black 1991).

Every winter, the Scottish Agricultural College veterinary laboratories diagnose outbreaks of rhododendron poisoning in out-wintered cattle and sheep (Michie et al. 2011). This is typically caused by animals grazing on fresh leaves or by clippings being put in a field with grazing livestock (Hosie 2006). Vomiting caused by GTXs in the ingested leaves leads to inhalation pneumonia, which kills the animal (Hosie 2006). *Rhododendron* is also poisonous to livestock (Forsyth 1968). Honey made from *Rhododendron* pollen (known as "mad honey") is also poisonous and has the potential to cause death if untreated (Koca and Koca 2007). Scott et al. (1971) determined that the mean $LD_{50}$ (the lethal dose that leads to the death of 50% of a group of test animals) values for standard GTXs injected intraperitoneally into male albino mice (weighing 21–44 g) were 1.28 mg/kg for GTX I and 0.908 mg/kg for GTX III. No toxic effects were noted for GTX II, although the maximum dose used was 4 mg/kg. Terai et al. (2002) found that all mice (body weight 27.1–32.4 g) injected intraperitoneally with 10 and 100 mg/kg GTX II died within 15 and 5 min of dosing, respectively. No studies are available concerning the acute toxicity of GTXs to livestock.

*Rhododendron* species other than *R. ponticum* are also toxic to a range of animals (Tokarnia et al. 1996; ShaoChen et al. 1997; Brown 2006).

## CLASSIC WORKS ON HONEY POISONING IN EUROPE AND NORTH AMERICA

GTX/mad honey poisoning was well documented in 19th century Europe and North America. A 1999 issue of the *British Medical Journal* reprinted an 1899 *British Medical Journal* article that described a typical case of mad honey poisoning (BMJ 1999). Cases from the United States and Germany were discussed in the same article (Gunduz et al. 2008).

In 1896, Kebler reviewed honey poisoning in the United States (Kebler 1896). That review may have been precipitated by the eight cases of honey intoxication, which occurred in Princeton, New Jersey, during the preceding year. He also reported earlier studies in his article. According to Kebler (1896), Barton (2007 cited in Gunduz et al. 2008) was the first American to report the effects of honey intoxication. He first read his findings at a meeting of the American Philosophical Society in 1794 and later published his work in 1802. Kebler (1896) also quoted the 1853 work of Coleman, who had a series of 14 patients from New Jersey with honey intoxication. One patient from that series died. There was another series in Branchville, South Carolina. In that series, there were 23 patients poisoned, of whom 3 patients died. By 1891, Plugge and de Zaayer (1889) had examined a number of plants from the Ericaceae family and had isolated andromedotoxin in many of them. Andromedotoxin was later shown to be identical to GTX (Tallent et al. 1957).

## MODERN EXPERIMENTAL WORK

### ROLE OF THE CENTRAL NERVOUS SYSTEM, VAGUS NERVE, AND MUSCARINIC RECEPTORS IN GTX POISONING

In animal studies, Onat et al. (1991a) determined that the respiratory and cardiac effects of GTX occur within the central nervous system rather than at a peripheral site (Table 16.2). They evaluated the doses of GTX needed to produce bradycardia and respiratory depression in rats. Very small doses delivered intracerebroventricularly yielded the same physiologic effects as much larger doses delivered intraperitoneally (Onat et al. 1991a). In the same study, they discovered that bilateral vagotomy eliminated the bradycardic effect of GTX. They concluded that the bradycardic effect of GTX is mediated peripherally by the vagus nerve. In another rat study, Onat et al. (1991b) observed that atropine, a nonspecific antimuscarinic agent, improved both GTX-induced bradycardia and respiratory depression (Table 16.2). AF-DX 116 is a selective M2 muscarinic receptor antagonist. When they administered AF-DX 116 to GTX-poisoned rats, bradycardia was eliminated, but respiratory depression was unaffected. They concluded that M2 muscarinic receptors are involved in the cardiotoxicity of GTX.

### EFFECTS OF GTX AT THE CELLULAR LEVEL

The toxic cellular effect of GTXs works on the sodium channel. The work of a number of researchers was summarized by Maejima et al. (2003). They stated that GTX performs three actions on the voltage-dependent sodium channel. First, GTX binds to the voltage-dependent sodium channel in its open state. Second, the modified sodium channel is unable to inactivate. Third, the activation potential of the modified sodium channel is shifted in the direction of hyperpolarization.

### EFFECTS ON RENAL AND HEPATIC TISSUE AND GLUCOSE METABOLISM

High doses of GTX I administered to rats caused proteinuria and hematuria but no histologic changes in renal parenchyma in one study. Transaminases were elevated in that study, and there were also significant changes in hepatic central vein dilation, congestion, focal necrosis, and inflammatory cell infiltration in the hepatic portal triad and parenchyma (Ascioglu et al. 2000). In experimentally induced diabetes, GTX served to normalize blood sugar in diabetic rats (Oztasan et al. 2005).

## TABLE 16.2
## Summary Report of Cardiac Dysrhythmia of 269 Patients and 21 Case Series Exposed to Mad Honey Poisoning

| References | NP | NsB | NSR | SB | NR | 2AVB | AVB | WPW | Asystole | SH |
|---|---|---|---|---|---|---|---|---|---|---|
| Yavuz et al. (1991) | 7 | | | 7 | | | | | | Kd |
| Onat et al. (1991a,b) | 2 | | | 2 | | | | | | Kd |
| Gossinger et al. (1983) | 2 | 2 | | | | | | | | Tr |
| Biberoglu et al. (1988) | 16 | | 8 | 5 | 1 | | 1 | 1 | | Kd |
| Sutlupinar et al. (1993) | 11 | 11 | | | | | | | | Kd |
| Von Malottki and Wiechmann (1996) | 1 | | | 1 | | | | | | Tr |
| Dilber et al. (2002) | 1 | | | 1 | | | | | | Kd |
| Ozhan et al. (2004) | 19 | | | 15 | | | 4 | | | Kd |
| Kumral et al. (2005) | 1 | | | | | | 1 | | | Kd |
| Gunduz et al. (2006) | 8 | | | 4 | 3 | | 1 | | | Kd |
| Dursunoglu et al. (2007) | 1 | | | | | | 1 | | | Kd |
| Gunduz et al. (2007b) | 1 | | | | | | | | 1 | Kd |
| Choo et al. (2008) | 1 | | | | 1 | | | | | Korea |
| Aliyev et al. (2009) | 1 | | | | 1 | | | | | Kd |
| Gunduz et al. (2009) | 47 | | 3 | 37 | 6 | | 1 | | | Kd |
| Bostan et al. (2010) | 33 | | 3 | 30 | | | | | | Kd |
| Weiss et al. (2010) | 1 | | | | | 1 | | | | Tr |
| Hancı et al. (2010) | 72 | 71 | | | | | 1 | | | Tr |
| Okuyan et al. (2010) | 42 | | | 18 | 9 | | 15 | | | Tr |

*(continued)*

**TABLE 16.2 (Continued)**
**Summary Report of Cardiac Dysrhythmia of 269 Patients and 21 Case Series Exposed to Mad Honey Poisoning**

| References | NP | NsB | NSR | SB | NR | 2AVB | AVB | WPW | Asystole | SH |
|---|---|---|---|---|---|---|---|---|---|---|
| Yorgun et al. (2010) | 1 | | | 1 | | | | | | Kd |
| Baltacı et al. (2011) | 1 | | | | | | 1 | | | Kd |
| Total (%) | 269 | 84 (31) | 14 (5) | 121 (44) | 21 (8) | 1 (0.3) | 26 (10) | 1 (0.3) | 1 (0.3) | |

*Sources:* Tallent, W.H. et al., *Journal of the American Chemical Society*, 79, 4549–4554, 1957; Adler, L.S., Oikos, 91, 409–420, 2000; Zhang, H.P. et al., *Journal of Asian Natural Products Research*, 87–90, 2005; Wang, W.G. et al., *Journal of Asian Natural Products Research*, 70–75, 2010.

*Note:* 2AVB, second-degree Av block; AVB, complete AV block; Kd, Karadeniz (the Black Sea region); NP, number of patients; NsB, nonspecific bradycardia; NSR, normal sinus rhythm; NR, nodal rhythm; SB, sinus bradycardia; SH, source of honey; Tr, Turkey; WPW, Wolff–Parkinson–White.

## CLINICAL COURSE OF GTX/MAD HONEY INTOXICATION—A CHOLINERGIC TOXIDROME

The manifestations of GTX/mad honey intoxication do not take the form of a classic cholinergic toxidrome but may be regarded as a cholinergic toxidrome.

Xenophon provides a vivid picture of the clinical manifestations of mad honey intoxication: "Here, generally speaking, there was nothing to excite their wonderment, but the numbers of bee-hives were indeed astonishing, and so were certain properties of the honey. The effect upon the soldiers who tasted the combs was, that they all went for the nonce quite off their heads, and suffered from vomiting and diarrhea, with a total inability to stand steady on their legs. A small dose produced a condition not unlike violent drunkenness, a large one an attack very like a fit of madness, and some dropped down, apparently at death's door. So they lay, hundreds of them, as if there had been a great defeat, a prey to the cruelest despondency. But the next day, none had died; and almost at the same hour of the day at which they had eaten they recovered their senses, and on the third or fourth day got on their legs again like convalescents after a severe course of medical treatment." (Gunduz et al. 2008, 2011).

Several less poetic clinical references to mad honey intoxication from the literature are summarized in Figure 16.5 (Yavuz et al. 1991; Yilmaz et al. 2006). A larger series of mad honey intoxication-related findings obtained from 66 patients was tabulated by Yilmaz et al. (2006) (Figure 16.5).

Mad honey poisoning exhibits many, but not all, of the symptoms and signs of a cholinergic toxidrome and responds to atropine, as do other cholinergic toxidromes (Mokhlesi et al. 2003). Significant hypotension (mean systolic blood pressures of 70 mmHg) and bradycardia (mean pulse rate of 48 beats/minute) are the most frequent manifestations in studies (Yilmaz et al. 2006). These occurred over 90% of the

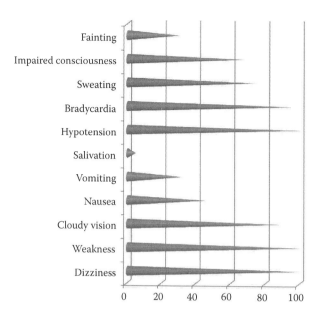

**FIGURE 16.5** Major signs and symptoms of GTX in mad honey poisoning (%) in the literature.

time (Yavuz et al. 1991). Diaphoresis, dizziness, and altered mental status were the next most frequent symptoms, occurring about 70% of the time (Yavuz et al. 1991; Yilmaz et al. 2006). Syncope occurred about 30% of the time. Visual symptoms of blurred vision and/or diplopia were reported 20% to 80% of the time (Yavuz et al. 1991; Yilmaz et al. 2006). Salivation was reported 14% of the time in one series (Yilmaz et al. 2006). Lacrimation, urination, bronchorrhea, and meiosis have not been reported (Biberoglu et al. 1988; Yavuz et al. 1991; Gunduz et al. 2006; Yilmaz et al. 2006) (Table 16.2).

Cardiac dysrhythmias were prominent in the 21 different series and case reports summarized in Table 16.2 (Gossinger et al. 1983; Biberoglu et al. 1988; Onat et al. 1991a,b; Yavuz et al. 1991; Sutlupinar et al. 1993; Von Malottki and Wiechmann 1996; Dilber et al. 2002; Ozhan et al. 2004; Ergun et al. 2005; Gunduz et al. 2006, 2007, 2009; Dursunoglu et al. 2007; Choo et al. 2008; Aliyev et al. 2009; Bostan et al. 2010; Hancı et al. 2010; Okuyan et al. 2010; Weiss et al. 2010; Baltacı et al. 2011). Either a nonspecific bradyarrhythmia or a sinus bradycardia was reported in about 75% of cases. Heart blocks of varying degrees were present in 25% of patients. Nodal rhythms were present in 11% of patients, while 8.7% had a complete heart block and 2.9% had a second-degree heart block. One (1.45%) patient had asystole. Another patient was reported as having Wolff–Parkinson–White syndrome, which is most likely unrelated to the intoxication.

Although reported in animal studies (Terai et al. 2003; Oztasan et al. 2005), clinically significant alterations in blood glucose and renal and hepatic toxicity have neither been reported nor specifically studied in human case series (Yavuz et al. 1991; Gunduz et al. 2006; Yilmaz et al. 2006).

## Toxic Dose and Duration of Illness

The amount of honey needed to produce toxicity is quite small. The average amount of ingested honey in one report was $13.45 \pm 5.39$ (5–30 g). Symptoms began 1.5 to 3 h after ingesting honey (Yilmaz et al. 2006). Several sources report that honey has an unusual sharp, pungent taste (Dilber et al. 2002; Gunduz et al. 2006).

In untreated cases of severe intoxication, the worst signs and symptoms last about 24 h. By the end of that time, the patient is alert and vital signs are normal. Complete recovery may take several more days (Kebler 1896; Gunduz et al. 2008). The exact duration of symptoms has not been accurately documented, but one investigator was able to safely discharge mild cases of mad honey poisoning after 2 to 6 h of cardiac monitoring (Gunduz et al. 2006). The modern medical literature contains no detailed studies of the duration of individual signs and symptoms in severe GTX/mad honey poisoning.

## Treatment

As with other cholinergic toxidromes, treatment with atropine can be life-saving (Yavuz et al. 1991; Yilmaz et al. 2006; Gunduz et al. 2006, 2007).

Although symptoms and signs can be alarming, and sometimes life threatening, usual supportive care with electrocardiographic monitoring, normal saline infusion, and intravenous atropine resulted in no fatalities in 66 cases of mad honey intoxication (Yilmaz et al. 2006). A number of cases of compete heart block have been recorded (Biberoglu et al. 1988; Ozhan et al. 2004; Ergun et al. 2005; Gunduz et al. 2006) (Table 16.2). One patient required a temporary transvenous pacemaker because of complete heart block (Lampe 1988). Another developed asystole, which was treated successfully with atropine (Gunduz et al. 2007, 2008). In those unusual cases when atropine and intravenous saline are not adequate, Advanced Cardiac Life Support bradyarrhythmia protocols should be considered.

## Mortality

In a classic case series from the 1800s, when intravenous atropine and normal saline were not available, there was a significant mortality rate: 1 of 14 (7%) in one series and 3 of 23 (13%) in another (Kebler 1896). There have been no modern reports of fatality from mad honey poisoning (Ozhan et al. 2004; Yilmaz et al. 2006; Gunduz et al. 2007) (Table 16.2).

## A Change in the Geographic Distribution of Human GTX/Mad Honey Poisoning

The classic medical literature contains a number of reports of toxic honey in North America and Europe (Kebler 1896; BMJ 1999). However, all cases of mad honey poisoning in the modern medical literature are from the ingestion of honey produced in Turkey. There are rare anecdotal reports of mild to moderate illness, which possibly represent GTX/mad honey poisoning on contemporary North American beekeeping and personal Web sites (Burton 2007, cited in Gunduz et al. 2008). Why did this change in the geographic distribution of GTX/mad honey poisoning occur?

## CULTURAL FACTORS

Cultural factors are certainly involved. Some Turkish beekeepers purposely harvest mad honey for use as an alternative health product. In Turkey, mad honey used in small quantities is widely believed to promote general health; it is used as a pain reliever and for the treatment of abdominal pain and dyspepsia and is also regarded as a sexual stimulant (Gunduz et al. 2006). Since the amount of GTX in mad honey is variable, accidental poisoning from intentional ingestion may occur (Scott et al. 1971).

## HONEY PRODUCTION TECHNIQUES AND THE STABILITY OF MAD HONEY

Although there are large commercial honey packers in Turkey, a significant propor-tion of honey in the northeast of Turkey is sold by individual beekeepers in local mar-kets. This may occur more frequently in Turkey than in North America or Europe. Individual beekeepers' hives produce honey from flowers in a 5-$km^2$ surrounding area (Dilber et al. 2002). If the area in which the hives are located contains a large number of toxic *Rhododendron* species, toxic honey may result. Commercial honey packers receive honey from thousands of hives; therefore, toxin originating from any one hive is diluted (Lampe 1988). Additionally, the honey sold by an individual beekeeper is often unprocessed. Commercially processed honey is usually heated to retard later crystallization and to kill yeast spores (Gunduz et al. 2008). Since GTX may be heat-liable, commercial thermal processing of honey may destroy GTX in it (Koca and Koca 2007). The amount of heat needed to destroy GTX is not known with certainty.

## CLINICAL APPLICATION

Physicians in North America and Europe will encounter this condition only very rarely and should ask whether the patient has been to Turkey or ingested honey from that part of the world. Clinicians should consider the possibility of GTX/mad honey poisoning in patients presenting with prominent bradycardia, heart blocks, hypoten-sion, and altered mental status. Very likely nausea, vomiting, diaphoresis, dizziness, and prostration will be present. Syncope, blurred vision, and diplopia are possible. Lacrimation, urination, bronchorrhea, bronchospasm, and meiosis, if present, would suggest an alternative diagnosis, as these signs have not been reported with GTX/mad honey poisoning.

# CONCLUSION

In a patient suspected of having GTX poisoning, the practitioner should obtain a detailed dietary history. The consumption of unprocessed honey produced and sold by a single beekeeper in an area with an unusually high concentration of toxic Ericaceae would be confirmatory. A sample of the suspected toxic honey should be saved for later identification of the pollen and toxin. Documented diagnosis of GTX/ mad honey poisoning in Europe or North America from locally produced honey would be reportable.

## ACKNOWLEDGMENT

This study was financially supported by the Research Fund of Karadeniz Technical University (Project No. 2008.114.002.10).

## REFERENCES

Adler LS. 2000. The ecologic significance of toxic nectar. *Oikos* 91: 409–420.
Aliyev F, Türkoğlu C, and Celiker C. 2009. Nodal rhythm and ventricular parasystole: An unusual electrocardiographic presentation of mad honey poisoning. *Clin Cardiol* 32: 52–54.
Ascioglu M, Ozesmi C, Dogan P, and Ozturk F. 2000. Effects of acute grayanotoxin-1 administration on hepatic and renal functions in rats. *Turk J Med Sci* 30: 23–27.
Baltacı D, Öztürk S, Kahraman K, Kandiş H, and Sarıtaş A. 2011. Mad-honey intoxication leading to severe arrhythmia. *Clin Exp Invest* 2: 216–218.
Biberoglu S, Biberoglu K, and Komsuoglu B. 1988. Mad honey. *J Am Med Assoc* 259: 1943.
Black DH. 1991. *Rhododendron* poisoning in sheep. *Vet Rec* 128: 363–364.
BMJ. 1999. British Medical Journal. ABC of Complementary Medicine: Herbal medicine. One hundred years ego-honey poisoning (bmj 1899; ii: 674) 319: 1422.
Bostan M, Bostan H, Kaya AO, Bilir O, Satiroglu O, Kazdal H, Karadag Z, and Bozkurt E. 2010. Clinical events in mad honey poisoning: A single centre experience. *Bull Environ Contam Toxicol* 84: 19–22.
Brown S. 2006. *Rhododendron* species intoxication in Testudo species tortoises. *Vet Rec* 158: 742.
Choo YK, Kang HY, and Lim SH. 2008. Cardiac problems in mad-honey intoxication. *Circ J* 7: 1210–1211.
Cox PA and Kenneth NEC. 1997. *The Encyclopedia of Rhododendron Species*. Glendoick Publishing, Scotland. ISBN: 0-9530533-0-X.
Dilber E, Kalyoncu M, Yarifi N, and Okten A. 2002. A case of mad honey poisoning presenting with convulsion: Intoxication instead of alternative therapy. *Turk J Med Sci* 32: 361–362.
Dursunoglu D, Gur S, and Semiz E. 2007. A case with complete atrioventricular block related to mad honey intoxication. *Ann Emerg Med* 4: 484–485.
Ergun K, Tufekcioglu O, and Aras D, Korkmaz S, Pehlivan S. 2005. A rare cause of atrioventricular block: Mad honey intoxication. *Int J Cardiol* 99: 347–348.
Esen D, Yildiz O, Kulac S, and Sarginci M. 2006. Controlling *Rhododendron* spp. in the Turkish Black Sea Region. *Forestry* 79: 177–184.
Forsyth A. 1968. *British Poisonous Plants, Bulletin Number 161*. 2nd edition. Ministry of Agriculture, Fisheries and Food, Her Majesty's Stationery Office, London.
Gökçöl T. 1998. *Anabasis (Onbinlerin Dönüşü)*. Ksenophon. Sosyal Yayınları/Dünya Klasikleri/Kültür Dizisi, İstanbul, 265. ISBN: 9786051271705.
Gossinger H, Hruby K, Pohl A, Davogg S, Sutterlutti G, and Mathis G. 1983. Poisoning with andromedotoxin-containing honey. *Deutsch Med Wochenschr* 108: 1555–1558.
Gunduz A, Turedi S, Uzun H, and Topbas M. 2006. Mad honey poisoning. *Am J Emerg Med* 24: 595–598.
Gunduz A, Bostan H, Turedi S, Nuhoglu I, and Patan T. 2007a. Wild flowers and mad honey. *Wilderness Environ Med* 18: 69–71.

Gunduz A, Durmus I, Turedi S, Nuhoglu I, and Ozturk S. 2007b. Mad honey poisoning-related asystole. *Emerg Med J* 24: 592–593.

Gunduz A, Russel MR, Turedi S, and Ayaz FA. 2008. Clinical review of grayanotoxin/mad honey poisoning past and present. *Clin Toxicol* 46: 437–442.

Gunduz A, Meriçé ES, Baydin A, Topbaş M, Uzun H, Türedi S, and Kalkan A. 2009. Does mad honey poisoning require hospital admission? *Am J Emerg Med* 27: 424–427.

Gunduz A, Turedi S, and Oksuz H. 2011. Lessons from history. The honey, the poison, the weapon. *Wilderness Environ Med* 22: 182–184.

Hancı V, Bilir S, Kırtaç N, Akkız S, Yurtlu S, andÖzkoçak Turan I. 2010. Zonguldak Bölgesinde Deli Bal Zehirlenmesi: Yetmiş İki Olgunun Analizi. *Türk Anest Rean Der* 38: 278–284.

Harborne JB. 1998. *Phytochemical Methods: A Guide to Modern Techniques of Plant Analysis.* 3rd edition. Chapman & Hall, London, 302.

Hikino H, Ohta T, Ogura M, Ohizumi Y, Konno C, and Takemoto T. 1976. Structure-activity relationship of Ericaceous toxins on acute toxicity in mice. *Toxicol Appl Pharmacol* 35: 305–310.

Holstege DM, Puschner B, and Le T. 2001. Determination of grayanotoxins in biological samples by LC-MS/MS. *J Agric Food Chem* 49: 1648–1651.

Hosie B. 2006. Look out for *Rhododendron* and yew. SAC Sheep & Beef News: December, 3.

Hough LR, Crews C, White D, Driffield M, Campbell CD, and Maltin C. 2010. Degradation of yew, ragwort and rhododendron toxins during composting. *Sci Total Environ* 408: 4128–4137.

Isik A. 2001. *Antik Kaynaklarda Karadeniz Bölgesi.* Turkish Historical Society, Ankara, Turkey, pp. 109–110.

Kan T, Hosokawa S, Nara S, Oikawa M, Ito S, Matsuda F, and Shirahama H. 1994. Total synthesis of (-)-grayanotoxin III. *J Org Chem* 59: 5532–5534.

Karakaya AE. 1977. Grayanotoxin content of the toxic honey and research on its relation with the *Rhododendron* species. *J Fac Pharm Ankara* 7: 111–115.

Katakawa J, Tetsumi T, Sakaguchi K, Katai M, and Terai T. 2004. Crystal and molecular structure of 10,20-epoxy-grayanotoxin-II. *J Chem Crystallogr* 34: 311–315.

Kebler LF. 1896. Poisonous honey. *Am Pharm Assoc Proc* 44: 167–174.

Koca I, Koca AF. 2007. Poisoning by mad honey: A brief review. *Food Chem Toxicol* 45: 1315–1318.

Kumral E, Tufekcioglu O, Aras D, Korkmaz S, and Pehlivan S. 2005. A rare cause of atrioventricular block: Mad honey intoxication. *Int J Cardiol* 18: 347–348.

Lampe KF. 1988. *Rhododendrons*, mountain laurel, and mad honey. *J Am Med Assoc* 259: 2009.

Maejima H, Kinoshita E, Seyama I, and Yamaoka K. 2003. Distinct site regulating grayanotoxin binding and unbinding to D4S6 of Nav1.4 sodium channel as revealed by improved estimation of toxin sensitivity. *J Biol Chem* 278: 9464–9471.

Michie D, Litterick A, and Crews C. 2011. The influence of outdoor windrow composting on the concentration of grayanotoxins in *Rhododendron* leaves. *Compost Sci Util* 19: 44–51.

Miyajima S and Takei S. 1936. Grayanotoxins, the active constituents of the leaves of *Leucothea grayana* II. *J Agric Cem Soc Japan* 12: 497.

Mokhlesi B, Leiken JB, Murray P, and Corbridge TC. 2003. Adult toxicology in critical care: Part I: General approach to the intoxicated patient. *Chest* 123: 577–592.

Nelson LS, Shih RD, and Balick MJ. 2007. *Section 5: Individual Plants. Handbook of Poisonous & Injurious Plants.* 2nd edition. Springer Science, NY, 55–306.

Onat F, Yegen BC, Lawrence R, Oktay A, and Oktay S. 1991a. Site of action of grayanotoxins in mad honey in rats. *J Appl Toxicol* 11: 199–201.

Onat FY, Yegen BC, Lawrence R, Oktay A, and Oktay S. 1991b. Mad honey poisoning in man and rat. *Rev Environ Health* 9: 3–9.

Okuyan E, Uslu A, and Ozan Levent M. 2010. Cardiac effects of "mad honey": A case series. *Clin Toxic (Phila)* 6: 528–532.

Okmen M. 2004. *Herodot Tarihi*. 2nd edition. Kültür Yayınları, Istanbul, Turkey, 120.

Ozhan H, Akdemir R, Yazici M, Gunduz H, Duran S, and Uyan C. 2004. Cardiac emergencies caused by honey ingestion: A single centre experience. *Emerg Med J* 21: 742–744.

Oztasan N, Altinkaynak K, Akcay FA, Gocer F, and Dane S. 2005. Effect of mad honey on blood glucose and lipid levels in rat with streptozocin induced diabetes. *Turk J Vet Anim Sci* 29: 1093–1096.

Ozturk O. 2005. Karadeniz: Ansiklopedik Sözlük, I/II, Istanbul, Heyamola Yayincilik, Turkey.

Pekman A. 2005. *Strabon Geopraphika, Antik Anadolu Coğrafyası*. Arkeoloji ve Sanat Yayinlari, Istanbul, Turkey, 18. ISBN: 975-7538-20-5.

Pereira M. 1971. *East of Trebizond*. Bles, London, England, 243.

Plugge PC and de Zaayer HG. 1889. Andromedotoxin. *Am J Pharm* 61(7).

Power SB, O'Donnell PG, Quirk, EG. 1991. *Pieris* poisoning in sheep. *Vet Rec* 128: 599–600.

Seyama I, Yamada K, Kato R, Masutani T, and Hamada M. 1988. Grayanotoxin opens Na channels from inside the squid axonal membrane. *Biophys J* 53: 271–274.

Scott PM, Coldwell BB, and Wiberg GS. 1971. Grayanotoxins. Occurrence and analysis in honey and a comparison of toxicities in mice. *Food Cosmet Toxicol* 9: 179–184.

ShaoChen Z, YanSheng Y, and Tuan L. 1997. Isolation of toxin of *Rhododendron molle* and its toxicity and killing efficiency against rodents. *Chin J Vector Biol Control* 8: 89–91.

Sutlupinar N, Mat A, and Satganoglu Y. 1993. Poisoning by toxic honey in Turkey. *Arch Toxicol* 67: 148–150.

Stevens PF. 1978. *Rhododendron* L. In: Davis PF ed. *Flora of Turkey and the East Aegean Islands*. Edinburgh University Press, Edinburgh 6: 90–94.

Tallent WH, Riethof ML, and Horning EC. 1957. Studies on the occurrence and structure of acetylandromedol (Andromedotoxin). *J Am Chem Soc* 79: 4549–4554.

Terai T, Sato M, Narama I, Matuura T, Katakawa J, and Tetsumi M. 1996. Transformation of grayanotoxin-III to 10-epi-grayanotoxin-III. Its X-ray crystallographic analysis and acute toxicity in mice. *Chem Pharm Bull* 44: 1245–1247.

Terai T, Osakabe K, Katai M, Sakaguchi K, Narama I, Matsuura T, Katakawa J, and Tetsumi T. 2002. Preparation of 9-hydroxy grayanotoxin derivatives and their acute toxicity in mice. *Chem Pharm Bull* 51: 351–353.

Terai T, Osakabe K, Katai M, Sakaguchi K, Narama I, Matsuura T, Katakawa J, and Tetsumi T. 2003. Preparation of 9-hydroxy grayanotoxin derivatives and their acute toxicity in mice. *Chem Pharm Bull (Tokyo)* 51: 351–353.

Tokarnia CH, Armien, AG, Peixoto PV, Barbosa JD, Brito MF, and Döbereiner J. 1996. Experimental study on the toxicity of some ornamental plants in cattle. *Pesq Agropec Bras* 16: 5–20.

Wang WG, Li HM, Li HZ, Wu ZY, and Li RT. 2010. New grayanol diterpenoid and new phenolic glucoside from the flowers of *Pieris fromosa*. *J Asian Nat Prod Res* 70–75.

Weiss TW, Smetana P, Nurnberg M, and Huber K. 2010. The honey man—second degree heart block after honey intoxication. *Int J Cardiol* 25: 142:e6–e7.

Wong J, Youde E, Dickinson B, and Hale M. 2002. *Report on the Rhododendron Feasibility Study*. School of Agricultural and Forest Science, University of Wales, Bangor, U.K.

Von Malottki K and Wiechmann HW. 1996. Acute life-threatening bradycardia: Food poisoning by Turkish wild honey. *Dtsch Med Wochenschr* 121: 936–938.

Yavuz H, Ozel A, Akkus I, and Erkul I. 1991. Honey poisoning in Turkey. *Lancet* 337: 789–790.

Yilmaz O, Eser M, Sahiner A, Altintop L, and Yesildag O. 2006. Hypotension, bradycardia and syncope caused by honey poisoning. *Resuscitation* 68: 405–408.

Yuki T, Yamaoka K, Yakehiro M, and Seyama I. 2001. State-dependent action of grayanotoxin-I on Na channels in frog ventricular myocytes. *J Physiol* 534: 777–790.

Yorgun H, Ülgen A, and Aytemir K. 2010. A rare cause of junctional rhythm causing syncope: Mad honey intoxication. *J Emerg Med* 39: 656–658.

Zhang HP, Wang HB, Wang LQ, Bao GH, and Qin GW. 2005. A new 1,5-seco grayanotoxin from *Rhododendron decorum. J Asian Nat Prod Res* 87–90.

# 17 Honey as a Nutrient

## Md. Ibrahim Khalil and Nadia Alam

## CONTENTS

## CHEMICAL COMPOSITION OF HONEY

The composition of honey is relatively variable and primarily depends on the floral source; however, certain external factors also play a role, such as seasons, environmental factors, and processing. Honey contains at least 181 substances (Chow 2002). It is a complex, nutritious food and its main ingredients are simple sugars (or monosaccharides) such as glucose, fructose, and galactose and complex sugars (oligosaccharides) such as sucrose (Figure 17.1). The biggest components are glucose and fructose. Honey is mainly a watery solution of two invert sugars, dextrose (glucose or grape sugar) and levulose (fructose or fruit sugar), in nearly equal proportions. The terms dextrose and levulose originated from the use of the two prefixes, dexter (right) and levis (left), because the former turns the polarized light to the right and the latter to the left. We may call these two invert sugars natural or simple sugars because they are readily absorbed by the bloodstream without requiring the assistance of the salivary, gastric, or intestinal secretions to accomplish the process of inversion. Cane and some other artificial sugars must first be inverted into simple sugars before they are assimilated (Chow 2002; Terrab et al. 2003; Dimitrova et al. 2007).

In addition to the two invert sugars, honey contains aromatic volatile oils, which bestow its flavor, mineral elements (sodium, potassium, calcium, magnesium, iron, copper, phosphorus, etc.), some protein, various enzymes, vitamins, and coloring matter. It can be concluded that the contribution of honey to the recommended daily intake (RDI) is small. However, its importance with respect to nutrition lies in its manifold physiologic

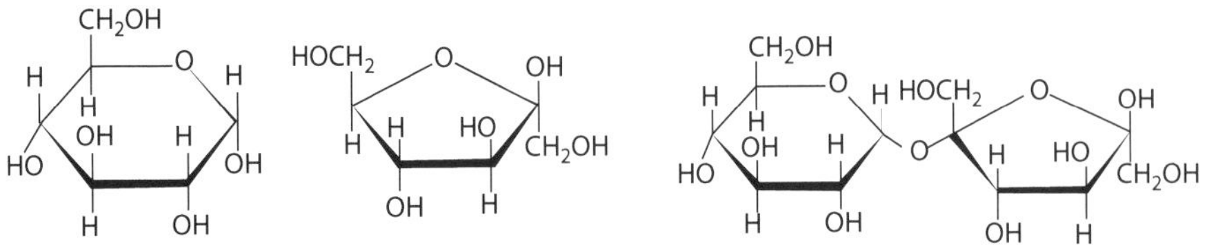

**FIGURE 17.1** Chemical structure of monosaccharides and disaccharides in honey.

effects (Pérez 2002; Persano-Oddo and Piro 2004). It should be noted that the composition of honey depends greatly on its botanical origin, a fact that has seldom been considered in nutritional and physiologic studies (Persano-Oddo and Piro 2004).

## CARBOHYDRATE COMPOSITION

Honey is mainly made up of carbohydrates, which constitute about 95% of its dry weight. It is a highly complex mixture of sugars, most of which are in the immediately digestible form in the small intestine. In addition to those named in Table 17.1,

## TABLE 17.1
## Average Composition in Honey (Data in g/100 g)

| Component | Average (%) |
| --- | --- |
| Water | 17.2 |
| Fructose | 38.19 |
| Glucose | 31.28 |
| Sucrose | 1.31 |
| Disaccharides, calculated as maltose | 7.31 |
| Melezitose | <0.1 |
| Erlose | 0.8 |
| Higher sugars | 1.5 |
| Total sugars | 79.7 |
| Free acid as gluconic | 0.43 |
| Lactone as gluconolactone | 0.14 |
| Total acid as gluconic | 0.57 |
| Ash | 0.169 |
| Nitrogen | 0.041 |
| Minerals | 0.2 |
| Amino acids and proteins | 0.3 |
| Acids | 0.5 |
| pH value | 3.9 |

*Sources:* Chow, J., *Journal of Renal Nutrition*, 12, 76–86, 2002; Pérez, R.A., *Journal of Agricultural and Food Chemistry*, 50, 2633–2637, 2002; Terrab, A. et al., *European Food Research and Technology*, 218, 88–95, 2003; Bogdanov, S. et al., *Journal of the American College of Nutrition*, 27, 677–689, 2008.

the following constituents have also been identified in honey: isomaltose, nigerose, turanose, maltulose, kojibiose, α,β-trehalose, gentiobiose, laminaribiose, maltotriose, 1-kestose, panose, isomaltosyl glucose, erlose, isomaltosyltriose, theanderose, centose, isopanose, isomaltosyltetraose, and isomaltosylpentaose (Jeffrey and Echazarreta 1996). Table 17.2 summarizes different disaccharides and trisaccharides reported by

## TABLE 17.2
## Disaccharides and Trisaccharides Reported in Honey

| Trivial Nomenclature | Systematic Nomenclature |
|---|---|
| *Disaccharide* | |
| Cellobiose[a] | O-β-D-glucopyranosyl-(1→4)-D-glucopyranose |
| Gentiobiose[a] | O-β-D-glucopyranosyl-(1→6)-D-glucopyranose |
| Isomaltose[a] | O-α-D-glucopyranosyl-(1→6)-D-glucopyranose |
| Isomaltulose[b] | O-α-D-glucopyranosyl-(1→6)-D-fructofuranose |
| Kojibiose[c] | O-α-D-glucopyranosyl-(1→2)-D-glucopyranose |
| Laminaribiose[d] | O-β-D-glucopyranosyl-(1→3)-D-glucopyranose |
| Leucrose[b] | O-α-D-glucopyranosyl-(1→5)-D-fructofuranose |
| Maltose[c] | O-α-D-glucopyranosyl-(1→4)-D-glucopyranose |
| Maltulose[a] | O-α-D-glucopyranosyl-(1→4)-D-fructose |
| Melibiose[b] | O-α-D-galactopyranosyl-(1→6)-D-glucopyranose |
| Neo-trehalose[d] | O-α-D-glucopyranosyl-β-D-glucopyranoside |
| Nigerose[a] | O-α-D-glucopyranosyl-(1→3)-D-glucopyranose |
| Palatinose[a] | O-α-D-glucopyranosyl-(1→6)-D-fructose |
| Saccharose[c] | O-α-D-glucopyranosyl-β-D-fructofuranoside |
| Turanose[c] | O-α-D-glucopyranosyl-(1→3)-D-fructose |
| *Trisaccharide* | |
| Kestose[b] | O-α-D-glucopyranosyl-(1→4)-O-α-D-glucopyranosyl-(1→2)-D-glucopyranose |
| 1-Kestose[b] | O-α-D-glucopyranosyl-(1→2)-β-D-fructofuranosyl-(1→2)-β-D-fructofuranoside |
| Erlose[c] | O-α-D-glucopyranosyl-(1→4)-α-D-glucopyranosyl-β-D-fructofuranoside |
| Isomaltotriose[a] | O-α-D-glucopyranosyl-(1→6)-O-α-D-glucopyranosyl-(1→6)-D-glucopyranose |
| Isopanose[a] | O-α-D-glucopyranosyl-(1→4)-O-α-glucopyranosyl-(1→6)-D-glucopyranose |
| Laminaritriose[b] | O-β-D-glucopyranosyl-(1→3)-O-β-D-glucopyranosyl-(1→3)-D-glucopyranose |
| Maltotriose[a] | O-α-D-glucopyranosyl-(1→4)-O-α-D-glucopyranosyl-(1→4)-D-glucopyranose |
| Melezitose[a] | O-α-D-glucopyranosyl-(1→3)-O-β-D-fructofuranosyl-(2→1)-α-D-glucopyranoside |
| Panose[a] | O-α-D-glucopyranosyl-(1→6)-O-α-D-glucopyranosyl-(1→4)-D-glucopyranose |
| Raffinose[a] | O-α-D-galactopyranosyl-(1→6)-O-α-D-glucopyranosyl-β-D-fructofuranoside |
| Teanderose[a] | O-α-D-glucopyranosyl-(1→6)-α-D-glucopyranosyl-β-D-fructofuranoside |

*Source:* Moreira, R.F.A. and De Maria, C.A.B., *Quim Nova, 24*, 516–525, 2001.

[a] Minority.
[b] Not confirmed.
[c] Majority.
[d] Traces.

Moreira and De Maria (2001). Many of these sugars are not found in nectar but are formed during the ripening and storage effects of bee enzymes and the acids of honey (Jeffrey and Echazarreta 1996). In the process of digestion after honey intake, the principal carbohydrates, fructose and glucose, are quickly transported into the blood and can be utilized for energy requirements by the human body. A daily dose of 20 g honey will cover about 3% of the required daily energy (Bogdanov et al. 2008).

## PROTEINS, ENZYMES, AND AMINO ACIDS

Honey contains roughly 0.5% proteins, mainly enzymes and free amino acids. Protein content has been reported in honey from different floral sources, where high protein contents were considered as more than 1000 μg/g (Azeredo et al. 2003). Nevertheless, the contribution of that fraction to human protein intake is low. The three main honey enzymes are diastase (amylase), decomposing starch or glycogen into smaller sugar units; invertase (sucrase, α-glucosidase), decomposing sucrose into fructose and glucose; and glucose oxidase, producing hydrogen peroxide and gluconic acid from glucose (Bogdanov et al. 2008). Amino acids in honey account for 1% (w/w). The amount of total free amino acids in honey corresponds to between 10 and 200 mg/100 g, with proline as their major contributor, corresponding to about 50% of the total free amino acids (Iglesias et al. 2004). Besides proline, there are 26 amino acids in honeys, their relative proportions depending on its origin (nectar or honeydew). Since pollen is the main source of honey amino acids, the amino acid profile of a honey could be characteristic of its botanical origin. The main amino acids identified in honey from different botanical and geographical origin are glutamic acid (Glu), aspartic acid (Asp), asparagine (Asn) serine (Ser), glutamine (Gln), histidine (His), glycine (Gly), threonine (Thr), β-alanine (β-Ala), arginine (Arg), α-alanine (α-Ala), γ-aminobutyric acid (GABA), proline (Pro), tyrosine (Tyr), valine (Val), ammonium ion $\left( NH_4^+ \right)$, methionine (Met), cysteine (Cys), isoleucine (Ile), leucine (Leu), tryptophan (Trp), phenylalanine (Phe), ornithine (Orn), and lysine (Lys) (Hermosín et al. 2003; Iglesias et al. 2004; González-Paramás et al. 2006; Pérez et al. 2007).

## VITAMINS, MINERALS, AND TRACE COMPOUNDS

It is known that different trace elements and mineral concentrations in honey depend on its botanical and geological origin (Bengsch 1992). Trace elements play a key role in the biomedical activities associated with this food, as these elements have a multitude of known and unknown biological functions. For this reason, the concentrations of minerals and trace elements in honey were investigated.

Different trace elements (Al, Ba, Sr, Bi, Cd, Hg, Pb, Sn, Te, Tl, W, Sb, Cr, Ni, Ti, V, Co, and Mo) and minerals (P, S, Ca, Mg, K, Na, Zn, Fe, Cu, and Mn) were systematically investigated in botanically and geologically defined honey (Conti 2000; Stocker et al. 2005). The vitamin content in honey is low. Vitamins such as thiamine ($B_1$), riboflavin ($B_2$), niacin ($B_3$), pantothenic acid ($B_5$), pyridoxine ($B_6$), folic acid ($B_9$), ascorbic acid (C), and phylloquinone (K) are reported in honey, but in general the amount of vitamins and minerals is little and the contribution of honey to the RDI of the different trace substances is tiny (Bogdanov et al. 2008).

## Polyphenols

In recent years, there has been growing interest in functional foods, that is, foods that can provide not only basic nutritional and energetic requirements but also additional physiologic benefits (Goldberg 1996). The functionality of a food is usually related to some of the ingredients that it contains, and at present, consumers prefer these ingredients to have a natural rather than synthetic origin. Thus, they are commonly extracted from plants, food byproducts, and other natural sources (Herrero et al. 2005). Among the functional ingredients, the group most widely studied is the family of antioxidants. Traditionally, these kinds of compounds has played an important role in food science and technology because of their usefulness in preserving foodstuffs against oxidative degradation (Madhavi et al. 1996). Interest in antioxidant compounds has increased nowadays in the light of recent evidence regarding the important role of antioxidants in human health. In fact, several preventative effects against different diseases such as cancer, coronary diseases, inflammatory disorders, neurologic degeneration, and aging, have been related to the consumption of antioxidants (Madhavi et al. 1996; Wollgast and Anklam 2002).

Honey has been used as a food since the earliest times. Only in recent years, however, has evidence emerged of its antioxidant capacity (FAO 1996). It is also used as a food preservative (Cherbuliez and Domerego 2003; Meda et al. 2005), preventing deteriorative oxidation reactions in foods, such as lipid oxidation in meat (Anthony et al. 2000; McKibben and Engeseth 2002) and the enzymatic browning of fruits and vegetables (McLellan et al. 1995; Chen et al. 2000). Antioxidants specifically retard deterioration, rancidity, or discoloration due to oxidation caused by light, heat, and some metals. Nevertheless, the antioxidant activity of honey varies greatly depending on the floral source (Frankel et al. 1998; Gheldof and Engeseth 2002) and external factors, such as the season and environment, and finally its processing.

Honey is reported to contain at least 181 substances (White 1975) and is considered as part of traditional medicine. Apitherapy has recently become the focus of attention as a form of folk and preventive medicine for treating certain conditions and diseases as well as promoting overall health and well-being (Inoue et al. 2005). Honey has a wide range of minor constituents, many of which are known to have antioxidant properties (Vit et al. 1997; Antony et al. 2000). These include flavonoids and phenolic acids (Cherchi et al. 1994), certain enzymes (glucose oxidase and catalase) (White 1975), ascorbic acid (White 1975), Maillard reaction products (White 1975), carotenoid-like substances (Tan et al. 1989), organic acids (Cherchi et al. 1994), and amino acids and proteins (White 1978). The natural antioxidants, especially flavonoids, exhibit a wide range of biological effects, including antibacterial, anti-inflammatory, antiallergic, antithrombotic, and vasodilatory actions (Cook and Sammon 1996).

Phenolic compounds or polyphenols are one of the most important groups of compounds occurring in plants, where they are widely distributed, comprising at least 8000 different known structures (Bravo 1998). Polyphenols are also products of the secondary metabolism of plants. These compounds are reported to exhibit anticarcinogenic, anti-inflammatory, antiatherogenic, antithrombotic, immunomodulating, and analgesic activities, among others, and exert these functions as antioxidants (Salah et al. 1995; Catapano 1997; Vinson et al. 1998). In general, phenolic compounds can be divided into at least 10 types depending on their basic structure: (1) simple phenols, (2) phenolic acids, (3) coumarins, (4) isocoumarins, (5) naphthoquinones, (6) xanthones, (7) stilbenes, (8) anthraquinones, (9) flavonoids, and (10) lignins. Flavonoids constitute the most important polyphenolic class, with more than 5000 compounds already described (Wollgast and Anklam 2002).

## ANTIOXIDANT PROPERTIES OF HONEY

Polyphenols are the most abundant antioxidants in the diet. Their total dietary intake could be as high as 1 g/day, which is much higher than that of all other classes of phytochemicals and known dietary antioxidants. For perspective, this is ~10 times higher than the intake of vitamin C and 100 times higher than the intakes of vitamin E and carotenoids (Scalbert and Williamson 2000; Manach et al. 2004). Current evidence strongly supports a contribution of polyphenols to the prevention of cardiovascular diseases, cancers, and osteoporosis and suggests a role in the prevention of neurodegenerative diseases and diabetes mellitus (Scalbert et al. 2005). Polyphenols clearly improve the status of different oxidative stress biomarkers (Williamson and Manach 2005). Significant progress has been made in the field of cardiovascular diseases, and today, it is well established that some polyphenols, administered as supplements or with food, do improve health status, as indicated by several biomarkers closely associated with cardiovascular risk (Keen et al. 2005; Sies et al. 2005; Vita 2005).

A considerable body of literature supports a role for oxidative stress in the pathogenesis of age-related human diseases and a contribution of dietary polyphenols to their prevention. The complex relationships between antioxidant status and disease are still poorly understood and have been studied intensively. For many years, polyphenols and other antioxidants were thought to protect cell constituents against oxidative damage through scavenging of free radicals. However, this concept now appears to be an oversimplified view of their mode of action (Azzi et al. 2004). More likely, cells respond to polyphenols mainly through direct interactions with receptors or enzymes involved in signal transduction, which may result in modification of the redox status of the cell and may trigger a series of redox-dependent reactions (Forman et al. 2002; Halliwell et al. 2005; Moskaug et al. 2005). Both antioxidant and prooxidant effects of polyphenols have been described, with contrasting effects on cell physiologic processes. As antioxidants, polyphenols may improve cell survival; as prooxidants, they may induce apoptosis and prevent tumor growth (Lambert et al. 2005). However, the biological effects of polyphenols may extend well beyond the modulation of oxidative stress. The current evidence for protective effects of polyphenols against diseases has generated new expectations for improvements in health, with great interest from the food and nutritional supplement industry regarding promotion and development of polyphenol-rich products (Mennen et al. 2005). Some hazards associated with the consumption of polyphenols are documented, but evaluation among humans is still very limited. Last, we should not forget that many polyphenols have a taste and/or a color (Lesschaeve and Noble 2005); food must be not only good for health but also acceptable to consumers. Integration of the results of past and future experiments in various disciplines, including biochemistry, cell biology, physiology, pathophysiology, epidemiology, and food chemistry, will be needed to identify the most effective polyphenols and to determine the optimal levels of intake for better health.

Although free radicals of oxygen are a natural byproduct of metabolism within the organism, they cause cellular damage and break down the structure of DNA. Exactly, these processes cause premature aging. Antioxidants bind these dangerous molecules, preventing their harmful effects. Unlike synthetic compounds, honey represents a natural product that does not carry side effects that can be harmful to health. Among the compounds found in honey, vitamin C, catalase enzymes, peroxidase, glucose oxidase, and phenol compounds have antioxidant properties. Honey also contains flavonoids and carotenoids. High levels of these indicators ensure a high level of antioxidants in honey. Antioxidant properties of honey act as an antidepressant during high emotional, physical, and intellectual stress. (As an example, Bashkir honey is supplied to the international space station and is part of the daily meals of astronauts, and it is also used by deep-sea divers.)

Various polyphenols are reported in honey (Table 17.3). Some of the polyphenols of honey such as caffeic acid, caffeic acid phenyl ester, chrysin, galangin, quercetin, acacetin, kaempferol, pinocembrin, pinobanksin, and apigenin have evolved as promising pharmacologic agents in the treatment of cancer (Jaganathan and Mandal 2009).

"Gram for gram, antioxidants in buckwheat honey equal to those of fruits and vegetables," said Dr. May Berenbaum, head of the University of Illinois Entomology Department. "It packs the antioxidant power of vitamin C in a tomato." Researchers at the University of Illinois–Champaign/Urbana have identified the antioxidant values of 14 unifloral honeys. The antioxidant components of honey were compared with an ascorbic acid standard. The water-soluble antioxidant content of the honey samples varied more than 20-fold, from a high value of $4.32 \times 10^{-3}$ Eq for Illinois buckwheat honey to a low value of $21.3 \times 10^{-5}$ Eq for California button sage honey (National Honey Board 2011).

Research showed a correlation between color and antioxidant capacity, with the darker honeys providing the highest levels of antioxidants. With antioxidant levels reaching $4.32 \times 10^{-3}$ mEq, honey rivals those levels found in tomatoes ($2.83 \times 10^{-3}$ mEq) and sweet corn ($1.36 \times 10^{-3}$ mEq). Although honey by itself may not serve as a major source of dietary antioxidants, it demonstrates the potential for honey to play a role in providing antioxidants in a highly palatable form. Due to honey's pleasing taste, it may be more readily consumed by individuals reluctant to ingest plant-derived antioxidants. Certainly, compared with sucrose, which has no antioxidant value, honey can be a flavorful, supplementary source of antioxidants (Jaganathan and Mandal 2009). Honey has been found to contain significant antioxidant compounds including glucose oxidase, catalase, ascorbic acid, flavonoids, phenolic acids, carotenoid derivatives, organic acids, Maillard reaction products, amino acids, and proteins (Frankel et al. 1998; Fahey and Stephenson 2002; Aljadi and Kamaruddin 2004; Beretta et al. 2005; Inoue et al. 2005; Blasa et al. 2006; Nagy et al. 2006; Pérez et al. 2007). The antioxidative activity of honey polyphenols can be measured in vitro by comparing the oxygen radical absorbance capacity (ORAC) with the total phenolic concentration (Table 17.4).

**TABLE 17.3**
**Some Important Polyphenols Identified in Different Honey Samples**

| Observations | Polyphenols | References |
|---|---|---|
| Honeys from various floral sources | p-Hydroxybenzoic acid, vanillic acid, syringic acid, p-coumaric acid, cis-trans-abscisic acid, cinnamic acid, pinobanksin, quercetin, pinocembrin, kaempferol, chrysin, galangin | Gheldof and Engeseth 2002 |
| Apis mellifera and Melipona spp. honeys | Ellagic acid, myricetin, chalcone, glycoside, quercetin, luteolin, 8-methoxikaempferol, kaempferol, apigenin, isorhamnetin, pinocembrin, chrysin, genkwanin, tectochrysin | Vit et al. 1997 |
| Leptospermum honeys | Myricetin, tricetin, quercetin, luteolin, quercetin 3-methyl ether, kaempferol, kaempferol 8-methyl ether, pinocembrin, quercetin 3,3-dimethyl ether, isorhamnetin, chrysin, pinobanksin, tectochrysin | Yao et al. 2003 |
| European unifloral honeys | Pinocembrin, pinobanksin, chrysin, galangin, tecthochrysin, quercetin, kaempferol, 8-methoxykaempferol, caffeic acid, p-coumaric acid, cis-trans-abscisic acid, ferulic acid, apigenin, quercetin 3,7-dimethyl ether, quercetin 3,3-dimethyl ether, hesperetin | Tomás-Barberán et al. 2001 |
| Heather honey | Trans-trans-abscisic acid, cis-trans-abscisic acid, pinobanksin, pinocembrin, chrysin, galangin | Ferreres et al. 1996 |
| Rosemary honey | Pinobanksin, quercetin, luteolin, 8-methoxykaempferol, kaempferol, apigenin, isorhamnetin, pinocembrin, chrysin, galangin, tectochrysin | Gil et al. 1995 |
| Stingless bee honey | Quercetin glycoside, luteolin glycoside, 8-methoxykaempferol glycoside, kaempferol glycoside, quercetin, luteolin, methylated luteolin, 8-methoxykaempferol, isorhamnetin, genkwanin | Vit et al. 1997 |
| Manuka honey | Caffeic acid, phenyllactic acid, methyl syringate, cinnamic acid, pinobanksin, pinocembrin, chrysin, galangin | Weston et al. 1999 |
| Eucalyptus honey | Myricetin, tricetin, quercetin, luteolin, kaempferol | Martos et al. 2000a |
| Eucalyptus Australian honeys | Myricetin, tricetin, quercetin, luteolin, quercetin 3-methyl ether, kaempferol, pinobanksin, pinocembrin, chrysin | Martos et al. 2000b |
| Eucalyptus honey | Gallic acid, chlorogenic acid, caffeic acid, p-coumaric acid, o-coumaric acid, ferulic acid, ellagic acid, abscisic acid | Yao et al. 2004c |
| Melaleuca, Guioa, Lophostemon, Banksia and Helianthus honeys | Myricetin, tricetin, quercetin, luteolin, quercetin 3-methyl ether, kaempferol, 8-methoxy kaempferol, pinocembrin, quercetin 3,3-dimethyl ether, isorhamnetin, chrysin, pinobanksin, genkwanin | Yao et al. 2004b |

*(continued)*

| Honey | Compounds | Reference |
|---|---|---|
| Australian *Eucalyptus* honeys | Myricetin, tricetin, quercetin, luteolin, quercetin 3-methyl ether, kaempferol, kaempferol 8-methyl ether, pinocembrin, quercetin 3,3-dimethyl ether, isorhamnetin, chrysin, pinobanksin | Yao et al. 2004a |
| *Melaleuca, Guioa, Lophostemon, Banksia* and *Helianthus* honeys | Gallic acid, chlorogenic acid, coumaric acid, ferulic acid, ellagic acid, syringic acid | Yao et al. 2003 |
| Phenolic acids in Malaysian honey | Gallic acid, caffeic acid, ferulic acid, benzoic acid, cinnamic acid | Aljadi and Yusoff 2003 |
| Manuka honey | Methyl syringate | Inoue et al. 2005 |
| Flavonoids in Tunisian honeys | Ellagic acid, pinobanksin, hesperetin, quercetin, luteolin, 3-methylquercetin, 8-methoxykaempferol, kaempferol, apigenin, isorhamnetin, pinocembrin, phenylethyl caffeate, pinobanksin 3-acetate, dimethylallyl caffeate, quercetin 3,7-dimethyl ether, chrysin, galangin, galangin 3-methyl ether, myricetin 3,7,4,5-methyl ether, pinocembrin 7-methylether, tecthochrysin | Martos et al. 1997 |
| Strawberry-tree honey | Homogentisic acid | Cabras et al. 1999 |
| Acacia honeys | p-Hydroxybenzoic acid, vanillic acid, p-coumaric acid, ferulic acid, t-cinnamic acid, abscisic acid, pinobanksin, apigenin, kaempferol, pinocembrin, chrysin, acacetin | Marghitas et al. 2010 |
| Czech honey | 3′,4′-Dihydroxyflavones, flavonols, chrysin, ferulic acid, 3,4-dimethoxycinnamic acid, 4-hydroxybenzaldehyde, vanillin, syringaldehyde, salicylic acid, syringic acid, dihydrochrysin, tectochrysin, and galangin | Lachman et al. 2010 |
| *Robinia, Castanea, Eucalyptus, Helianthus, Erica arborea* | Apigenin, chrysin, hesperetin, kaempferol, myricetin, naringenin, pinocembrin, quercetin, p-coumaric acid, ellagic, ferulic, gallic, homogentisic, mandelic | Pulcini et al. 2006 |
| Lime-tree, rape, heather, buckwheat | Gallic acid, chlorogenic acid, caffeic acid, ferulic acid, benzoic acid, rutin, quercetin, kaempferol, isorhamnetin, pinocembrin, chrysin, galangin | Kaškonienė et al. 2009 |
| Malaysian honey | Gallic acid, chlorogenic acid, caffeic acid, p-coumaric acid, ferulic acid, ellagic acid, myricetin, quercetin, hesperetin, luteolin, kaempferol, chrysin | Kassim et al. 2010 |

**TABLE 17.4**
**Antioxidative Activity (ORAC) and Total Phenol Content of Different Unifloral Honeys**

| Honey Type | ORAC μmol TE/g | Total Phenolics GAE mg/kg |
|---|---|---|
| Buckwheat Illinois | 16.95 ± 0.76 | 796 ± 32 |
| Buckwheat | 9.81 ± 0.34 | ND |
| Buckwheat New York | 9.75 ± 0.48 | 456 ± 55 |
| Buckwheat | 9.34 ± 0.57 | ND |
| Buckwheat | 9.17 ± 0.63 | ND |
| Buckwheat | 7.47 ± 0.27 | ND |
| Soy (2000) | 9.49 ± 0.29 | ND |
| Soy (1996) | 8.34 ± 0.51 | 269 ± 22 |
| Hawaiian Christmas berry | 8.87 ± 0.33 | 250 ± 56 |
| Clover (January 2000) | 6.53 ± 0.70 | ND |
| Clover (July 2000) | 6.05 ± 1.00 | 128 ± 11 |
| Tupelo | 6.48 ± 0.37 | 183 ± 9 |
| Fireweed | 3.09 ± 0.27 | 62 ± 6 |
| Acacia | 3.00 ± 0.16 | 46 ± 2 |

*Source:*   Gheldof, N. and Engeseth, N.J., *Journal of Agricultural and Food Chemistry*, 50, 3050–3055, 2002.

*Note:*   GAE, gallic acid equivalent; ND, not determined; TE, Trolox equivalent.

There is a significant correlation between the antioxidant activity, the phenolic content of honey, and the inhibition of the in vitro lipoprotein oxidation of human serum (Gheldof et al. 2003). Furthermore, in a lipid peroxidation model system, buckwheat honey showed a similar antioxidant activity as 1 mM α-tocopherol (Nagy et al. 2006). The influence of honey ingestion on the antioxidative capacity of plasma was tested in two studies (Schramm et al. 2003). In the first one, the trial persons were given maize syrup or buckwheat honeys with a different antioxidant capacity in a dose of 1.5 g/kg body weight. In comparison with the sugar control, honey caused an increase in both the antioxidant and reducing serum capacities.

In the second study, humans received a diet supplemented with a daily honey serving of 1.2 g/kg body weight. Honey increased the body antioxidant agents: blood vitamin C concentration by 47%, β-carotene by 3%, uric acid by 12%, and glutathione reductase by 7% (Al-Waili 2003). It should be borne in mind that the antioxidant activity depends on the botanical origin of honey and varies to a great extent in honeys from different botanical sources (Baltrušaitytė et al. 2007; Kücük et al. 2007; Vela et al. 2007).

The Departments of Nutrition and Internal Medicine at the University of California and National Honey Board showed that free radicals and reactive oxygen species (ROS) have been implicated in contributing to the processes of aging and disease (Schramm et al. 2003). Humans protect themselves from these damaging compounds, in part, by absorbing antioxidants from high antioxidant foods. This report describes the effects of consuming 1.5 g/kg body weight of corn syrup or buckwheat honey on the antioxidant and reducing capacities of plasma in healthy human adults. The corn syrup treatment contained 0.21 mg of phenolic antioxidants per gram, and the two buckwheat honey treatments contained 0.79 and 1.71 mg of phenolic antioxidants per gram. Following consumption of the two honey treatments, plasma total phenolic content increased ($P < 0.05$) as did plasma antioxidant and reducing capacities ($P < 0.05$). These data support the concept that phenolic antioxidants from processed honey are bioavailable and that they increase the antioxidant activity of plasma. It can be speculated that these compounds may augment defenses against oxidative stress and that they might be able to protect humans from oxidative stress.

Given that the average sweetener intake by humans is estimated to be in excess of 70 kg/year, the substitution of honey in some foods for traditional sweeteners could result in an enhanced antioxidant defense system in healthy adults (Schramm et al. 2003). Beretta et al. demonstrated the protective activity of a honey of multifloral origin, standardized for total antioxidant power and analytically profiled (high-performance liquid chromatography–mass spectrometry) in antioxidants, in a cultured endothelial cell line (EA.hy926) subjected to oxidative stress. Cumene hydroperoxide (CuOOH) was used as a free radical promoter. Native honey (1% w/v [pH 7.4], $10^6$ cells) showed strong quenching activity against lipophilic cumoxyl and cumoperoxyl radicals, with significant suppression/prevention of cell damage, complete inhibition of cell membrane oxidation and intracellular ROS production, and recovery of intracellular GSH. Experiments with endothelial cells fortified with the isolated fraction from native honey enriched in antioxidants, exposed to peroxyl radicals from 1,1-diphenyl-2-picrylhydrazyl (10 mM) and to hydrogen peroxide (50–100 μM), indicated that phenolic acids and flavonoids were the main causes of the protective effect. They suggested that, through the synergistic action of its antioxidants, honey by reducing and removing ROS may lower the risks and effects of acute and chronic free radical induced pathologies in vivo (Beretta et al. 2007).

## AROMA COMPOUNDS IN HONEY

Honey is a sweet and flavorful product produced by bees from nectar and/or honeydew; it has been consumed since the ancient times as a high nutritive value food distinguished by its characteristic aroma and pleasant sweet taste. In general, the aroma of honey is formed by volatile compounds, which may come from the nectar or honeydew collected by bees; consequently, it may largely depend on the plant of honey origin. Additionally, flavor constituents may be formed by the honeybee as well as during thermal processing and/or storage of honey (Bonvehí and Coll 2003; Soria et al. 2003). Honey flavor is an important quality for its application in the food industry and also a selection criterion for consumer's choice. More than 500 different compounds have been identified in the volatile flavor fraction of honey originated from different floral types (Bentivenga et al. 2004; Bogdanov et al. 2008).

The aroma of the majority of honeys depends on the group of constituents. For instance, furfuryl mercaptan, benzyl alcohol, d-octalactone, c-decalactone, euge-nol, benzoic acid, isovaleric acid, phenylethyl alcohol, and 2-methoxyphenol were reported to be particularly important impact volatile compounds for Brazilian caju honey (Moreira et al. 2002). It is worth mentioning that phenylethyl alcohol is well known in the perfume industry as possessing floral, spicy, and herblike odor. This compound was reported as an important aroma compound in lime honey (Moreira et al. 2002). However, Radovic et al. (2001) found phenylethyl alcohol only in two lime honey samples of the four analyzed and concluded that the authenticity of such hon-eys may be confirmed by the presence of one of the following substances: 2-methyl-furan, α-terpinene, α-pinene oxide, bicyclo-3,2,1-octane-2,3-bis(methylene), methyl isopropyl benzene, aromatic hydrocarbon, 3-cyclohexen-1-ol-5-methylene-6-isopro-pylene, and 4-methylacetophenone (Kaškoniene et al. 2008).

In the literature, five sources of aroma compounds identified in honey have been considered: (1) plant constituents, (2) transformation of plant constituents by the honeybee, (3) direct generation of aroma compounds by the honeybee, (4) generation of aroma compounds by thermal processing of honey, and (5) flavor contaminants in honey (Steeg and Montag 1987, 1988; Rowland et al. 1995). Apparently, honey flavor is the result of a complex combination of trace components. It is well known that the Maillard or nonenzymatic browning reaction has a great impact on food flavors. The effect of different amino acids on the formation of Maillard-related aroma compounds has been widely studied in various products (e.g., roasted cocoa beans), generating methylpyrazines to compare them with flavor index development (Serra Bonvehı and Ventura Coll 2000). This parameter constitutes an important factor that can ultimately be used to successfully identify fresh and processed honeys. The processing steps gen-erally require exposure to heat, contributing to quality loss (Serra Bonvehı et al. 2000). Heating honey at temperatures as low as 50°C leads to the formation of new volatile flavor compounds, and the peak areas of many compounds vary significantly as a result of different heating conditions (Visser et al. 1988). Tan et al. (1989) have also suggested that most of the compounds on which assessment of floral sources of New Zealand fresh honeys is based seem to originate from nectar without identified methylpyrazines. Not all individual components identified should be regarded as significant aroma com-pounds, since the aroma impact depends on the odor concentration and odor inten-sity. Aroma compounds are present in honey at very low concentrations as complex mixtures of volatile components of different functionality and relatively low molecular weight. Table 17.5 shows important volatile compounds found in some unifloral honeys.

## TOXICITY OF HONEY

Plant toxins may, in some cases, be transferred to honey produced from their nec-tar. Honeys produced by bees feeding on flowers of the Ericaceae and Solanaceae families are known to have caused toxicity (Palmer-Jones 1965; White 1975), as have those produced from tansy ragwort (*Senecio jacobaea* L.) (Deinzer et al. 1977). Some substances that are toxic to humans have no effect on bees. If bees obtain their nectar from certain flowers, the resulting honey can be psychoactive, or even toxic, to humans but innocuous to bees and their larvae (Mayor 1995).

## TABLE 17.5
## Important Volatile Compounds Found in Some Unifloral Honeys

| Honey Source | Compound | References |
|---|---|---|
| Chestnut | 3-Aminoacetophenone, 2-methyldihydrofuranone, α-methylbenzyl alcohol, 3-hexen-1-ol, dimethylstyrene | Radovic et al. 2001 |
| | Acetophenone, 1-phenylethanol, and 2-acetophenone | Verzera et al. 2001; Piasenzotto et al. 2003 |
| | 2-Methylcyclopentanol, diethylphenol | Guidotti and Vitali 1998 |
| | 1-Phenylethanol, 2-aminoacetophenone | Guyot et al. 1998 |
| | styrene | Radovic et al. 2001 |
| *Erica arborea* | Benzoic acid, decanoic acid, high levels of cinnamic acid, isophorone, 4-(3-oxobut-1-enylidene)-3,5,5-trimethylcyclohexen-2-en-1-one Shikimate-pathway derivatives: 4-methoxybenzaldehyde, 4-methoxybenzoic acid, and methyl vanillate | Guyot et al. 1999 |
| *Calluna vulgaris* | Phenylacetic acid, dehydrovomifoliol, (4-(3-oxo-1-butynyl)-3,5,5-trimethylcyclohexen-2-en-1-one), higher levels of 3,5,5-trimethylcyclohexene derivatives | Guyot et al. 1999 |
| Heather | 1-Penten-3-ol, 4-methylbicyclo-octan-1-ol, phenyl acetaldehyde | Radovic et al. 2001 |
| | Isophorone | Soria et al. 2003 |
| | | Tan et al. 1989 |
| Eucalyptus | Acetoin, aldimethyldisulfide, dimethyltrisulfide, alkane, nonane | Bouseta et al. 1992 |
| | Acetoin | Graddon et al. 1979; Radovic et al. 2001 |
| | | Verzera et al. 2001; Pérez 2002; Piasenzotto et al. 2003 |
| | Nonanol, nonanal, nonanoic acid | Guidotti and Vitali 1998; Verzera et al. 2001; Piasenzotto et al. 2003 |

*(continued)*

**TABLE 17.5 (Continued)**

**Important Volatile Compounds Found in Some Unifloral Honeys**

| Honey Source | Compound | References |
|---|---|---|
| Lime | 2-Pentanone, acetoin, furfural, 4-methylacetophenone, methyl isopropyl benzene, dimethylstyrene | Radovic et al. 2001 |
| | Ethylmethylphenol, estragole, carvacrol | Bouseta et al. 1992 |
| | Ethylmethylphenol, carvacrol, estragole | Guyot et al. 1998 |
| Citrus | 1-p-Menthen-9-al | Alissandrakis et al. 2005a,b, 2007 |
| | Lilac aldehyde | Piasenzotto et al. 2003; Soria et al. 2003, Alissandrakis et al. 2005a,b, 2007 |
| | Limonene diol, hotrienol | Guidotti and Vitali 1998; Verzera et al. 2001 |
| | Methyl anthranilate | Guidotti and Vitali 1998; Verzera et al. 2001; Piasenzotto et al. 2003; Soria et al. 2003; Alissandrakis et al. 2005a,b, 2007 |
| Lavender | Ethanol, 2-methyl-1-propanol, 3-methyl-1-butanol, 3-methyl-3-buten-1-ol, 3,7-dimethyl-1,5,7-octatrien-3-ol (hotrienol), furfuryl alcohol, hexanal, heptanal, 1-hexanol, furfural, phenyl acetaldehyde, benzaldehyde | Radovic et al. 2001 |
| | Hexanal, heptanal, ethyl propionate | Bouseta et al. 1992 |
| Acacia | Acetone, furfural, benzaldehyde | Radovic et al. 2001 |
| Rosemary | Acetone, 2-pentanone, benzaldehyde, 4-oxoisophorone | Radovic et al. 2001 |

| Honey | Compounds | Reference |
|---|---|---|
| Haze | Methyl-p-anisaldehyde, trimethoxybenzene, 5-hydroxy-2-methyl-4H-pyran-4-one, lilac aldehyde isomer A, lilac aldehyde isomer B, lilac aldehyde isomer C, lilac aldehyde isomer D | Shimoda et al. 1996 |
| Cashew | Trans-linalool oxide, nonacosane | Moreira et al. 2002 |
| Marmeleiro | Linalool, linalool acetate | |
| Sardinian strawberry tree | α-Isophoron, β-isophoron, 4-oxo-isophoron | |
| Cambara honey | Benzaldehyde, 2,3-methylbutanoic acid, benzonitrile, 2-phenylethanol, phenyl-1,2-propanedione, benzoic and phenylacetic acid | Moreira and De Maria 2005 |
| Alfalfa | Benzene acetaldehyde, nonanal, and 2-ethoxyphenol | Baroni et al. 2006 |
| Carob | Nonanal and octanal | |
| White clover | 2-H-1-Benzopyran-2-one | |
| Sunflower | 2-Methoxyphenol | |
| Calder | 1-Octanol | |
| Dalmatian sage | Benzoic acid, phenylacetic acid, p-anisaldehyde, α-isophorone, 4-ketoisophorone, dehydrovomifoliol, 2,6,6-trimethyl-4-oxocyclohex-2-ene-1-carbaldehyde, 2,2,6-trimethylcyclohexane-1,4-dione, and coumaran | Jerkovic et al. 2006 |
| *Thymus capitatus* | 1,3-Diphenyl-2-propanone, (3-methylbutyl)benzene, 3,4,5-trimethoxybenzaldehyde, 3,4-dimethoxybenzaldehyde, vanilline, and thymol | Odeh et al. 2007 |
| *Thymelaea hirsuta* | Benzene propanol, benzyl alcohol, nonanol, hexanol, and 4-methoxyphenol | |
| *Tolpis virgata* | 3,5-Dihydroxytoluene and tridecane | |

There have been famous episodes of inebriation of humans from consuming toxic honey throughout history. For example, honey produced from the nectar of *Rhododendron ponticum* (also known as *Azalea pontica*) contains alkaloids that are poisonous to humans but do not harm bees (FDA 2001). Xenophon, Aristotle, Strabo, Pliny the Elder, and Columella all document the results of eating this "maddening" honey (Figure 17.2) (Kelhoffer 2005). Honey from these plants poisoned Roman troops in the first century BC under Pompey the Great when they were attacking the Heptakometes in Turkey. The soldiers were delirious and vomiting after eating the toxic honey. The Romans were easily defeated (Ambrose 1972; Georghiou 1980).

Honey produced from the nectar of *Andromeda* flowers contains grayanotoxins that can paralyze the limbs and eventually the diaphragm and result in death (Lensky 1997; Walderhaug 2001). Honey obtained from *Kalmia latifolia*, the calico bush, mountain laurel, or spoon-wood of the northern United States, and allied species such as sheep laurel (*Kalmia angustifolia*) can produce sickness or even death (Blanchan 1900; Walderhaug 2001). The nectar of the "wharangi bush," *Melicope ternata*, in New Zealand also produces toxic honey, and this has been fatal (Espina-Prez and Ordetx-Ros 1983). The dangers of toxic honey were also well known among the pre-Columbian residents of the Yucatán Peninsula, although this was honey produced by stingless bees, not by honeybees that are not native to the Americas (Ott 1998). Bee nectar collection from *Datura* plants in Mexico and Hungary, belladonna flowers, henbane (*Hyoscamus niger*) plants from Hungary, *Serjania lethalis* from Brazil, *Gelsemium sempervirens* from the American Southwest, and *Coriaria arborea* from New Zealand (Reid 2011) can all result in toxic honey (Crane 1975), as can honey made from other toxic plants such as oleander (Oleander 2009). Narcotic opium honey has also been reported from honey made in areas where opium poppy cultivation is widespread (McAlpine 2002).

**FIGURE 17.2**   *R. ponticum.*

Patterson's curse has been reported to produce toxins in Australian honeys. Deaths from the ingestion of large quantities of honey derived from *Rhododendron* species have been reported in the Caucasus region, and deaths were reported among Maori and settlers on the New Zealand North Island (Palmer-Jones 1965). Nonfatal poisoning was widespread in New Zealand until the 1920s, when honey production in the North Island was stopped. The source of poisoning was identified as being a combination of the plant *C. arborea* (tutu) and the production of honeydew from the passion vine hopper (*Scolypopa australis*). Some areas of the North Island have now been closed to honey production, and in other areas of New Zealand, beekeepers are advised not to collect honey at certain times of the year when vine hoppers are prevalent (Palmer-Jones 1965). Palmer-Jones also identified a range of other plants that have been implicated in the production of toxic honeys—*K. latifolia, Tripetalia paniculata*, and *Ledum palustre*. Symptoms of honey poisoning vary depending on the toxin but include dizziness, nausea, vomiting, convulsions, headache, palpitations, and death in some cases. Toxins identified include hyenanchin (in the case of the New Zealand poisonings), euphorbic acid, acetylandromedol, and ericolin (Palmer-Jones 1965). These toxins are apparently heat resistant. In the New Zealand poisoning cases, as little as 5 to 10 g could cause severe poisoning in humans.

## HYDROXYMETHYLFURFURAL

Hydroxymethylfurfural (HMF) is a cyclic aldehyde produced as a result of sugar degradation (Ramirez et al. 2000). It is said that the presence of simple sugars (glucose and fructose) and many acids in honey is a favorable condition for the production of this substance. It has been reported that HMF and its congener compounds are spontaneously formed in carbohydrate-containing foods by Maillard reactions (the nonenzymatic browning reaction) or by an acid-catalyzed dehydration of hexoses (Belitz and Grosch 1999).

Related to the production of furfural, HMF is produced from sugars. It arises via the dehydration of fructose (Román-Leshkov et al. 2006). Treatment of fructose with acids followed by liquid–liquid extraction into organic solvents such as methyl isobutyl ketone produces HMF. The conversion is affected by various additives such as DMSO, 2-butanol, and polyvinylpyrrolidone, which minimize the formation of side product. In an optimized system for fructose (but not raw biomass), conversion is 77%, with half the HMF ending up in the organic phase (Figure 17.3). Ionic liquids also facilitate the conversion of fructose to HMF (Ståhlberg et al. 2011).

Chromous chloride catalyzes the direct conversion of both fructose (yielding >90%) and glucose (yielding >70%) into an HMF (Zhao et al. 2007). Subsequently, cellulose has been directly converted into HMF (yielding 55%–96% purity) (Su et al. 2009). The chromium chloride catalyzes the conversion of glucose into fructose.

HMF is usually absent in fresh and untreated foods (Askar 1984), but its concentration is also reported to increase as a result of heating processes (Bath and Singh 1999; Fallico et al. 2004) or due to long-term storage. For this reason, HMF is a recognized parameter related to the freshness and quality of such foods. Several factors influence the formation of HMF in honey during storage conditions. These factors include the following: (1) the use of metallic containers (White 1978) and

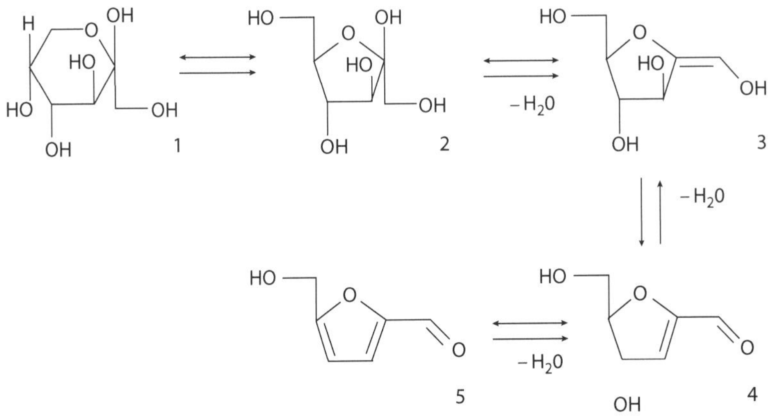

**FIGURE 17.3**  Chemical reactions for HMF formation: fructopyranose (1), fructofuranose (2), two intermediate stages of dehydration (not isolated) (3 and 4), and HMF (5).

(2) the physicochemical properties (the pH, total acidity, and mineral content) of honey itself, which are related to the floral source from which the honey has been extracted (Anam and Dart 1995), the humidity, and from thermal and/or photochemical stress (Spano et al. 2006).

The *Codex Alimentarius* (2000) has established that the HMF content of honey after processing and/or blending must not be higher than 80 mg/kg. The European Union (Council Directive of the European Union 2002), however, recommends a lower limit of 40 mg/kg with the following exceptions: 80 mg/kg is allowed for honey that originates from countries or regions with tropical temperatures, while a lower limit at only 15 mg/kg is allowed for honey with low enzymatic levels.

It has been shown that HMF at high concentrations is cytotoxic and irritating to eyes, upper respiratory tract, skin, and mucous membranes. Ulbricht et al. (1984) calculated an oral $LD_{50}$ of 3.1 g/kg body weight in rats. In another experiment in rats, U.S. EPA (1992) estimated an acute oral $LD_{50}$ of 2.5 g/kg and between 2.5 and 5.0 g/kg for males and females, respectively. Carcinogenic activity of HMF has been investigated in studies on rodents. HMF has been shown to induce and promote aberrant crypt foci (preneoplastic lesions) in rat colon (Archer et al. 1992; Bruce et al. 1993). Zhang et al. (1993) reported a significant dose-related increase in aberrant crypt foci in rats after oral administration of a single dose of 0 to 300 mg/kg body weight of HMF. Surh et al. (1994) described the induction of skin papillomas after topical administration of 10 to 25 mmol HMF to mice. Conversely, Miyakawa et al. (1991) reported no significant increase in the skin tumor rate after treatment with HMF. Finally, Schoental et al. (1971) reported the development of lipomatous tumors in rat kidney after subcutaneous administration of 200 mg/kg body weight of HMF. Recently, HMF has been reported to be a weak carcinogen in multiple intestinal neoplasia mice, increasing significantly the number of small intestine adenomas (Svendsen et al. 2009). HMF is converted in vitro and in vivo by sulfotransferases into sulfoxymethylfurfural (SMF), a compound that has been reported to be mutagenic in a conventional Ames test and to initiate tumors in mice skin (Surh et al.

1994; Lee et al. 1995). SMF, unlike HMF, forms DNA adducts in cell-free systems and was mutagenic to bacterial and mammalian cells without requiring an activating system (Surh et al. 1994; Glatt and Sommer 2006). In a recent study, SMF was found to be strongly nephrotoxic in the mouse in vivo (Bakhiya et al. 2009).

However, results obtained in short-term model studies for 5-HMF on the induction of neoplastic changes in the intestinal tract were negative or cannot be reliably interpreted as "carcinogenic." In the only long-term carcinogenicity study in rats and mice, no tumors or their precursory stages were induced by 5-HMF aside from liver adenomas in female mice, the relevance of which must be viewed as doubtful. Hence, no relevance for humans concerning carcinogenic and genotoxic effects can be derived. The remaining toxic potential is rather low.

Overall, the evidence of a carcinogenic potential is very limited. With regard to other toxic effects, no effects were observed at a daily dose in the range of 80 to 100 mg/kg body weight in animal experiments. There are no data on reproductive and developmental toxicity. Because of these uncertainties, it is not possible at the present time to establish a tolerable daily intake (Abraham et al. 2011).

## Fermented Honey

Honey that is not produced from the nectar of toxic plants can also ferment to produce ethanol, which is toxic. For example, Kettlewellh (1945) described finding an intoxicated bird, incapable of normal flight, that had been consuming honey that had fermented in the sun in Pretoria, Transvaal, South Africa (Kettlewellh 1945). Sometimes, honey is fermented intentionally to produce mead, a fermented alcoholic beverage made of honey, water, and yeast (called "meadhing"). Mead is also known as "honey wine" (Kerenyi 1976).

## Infantile Botulism

Infantile botulism is a rare form of food poisoning caused by ingestions of spores of the bacterium *Clostridium botulinum*. It occurs only in infants less than 12 months of age, with 95% of cases found in the first 6 months of life (Vanderbilt Medical Centre 1998). Honey appears to be one commonly implicated dietary source of *Clostridium* spores. The condition is rare and only affects infants. Because of the severity of the illness, Australian health authorities recommend that infants do not ingest honey.

## Honey Allergy

While apparently uncommon, allergies to honey have been reported and can involve reactions varying from cough to anaphylaxis (Kiistala et al. 1995). The incidence of honey allergy was reported (Bauer et al. 1996) to be 2.3% of a group of 173 patients with food allergies. Among patients with confirmed allergy to honey, 17% had suffered from anaphylaxis and 30% from asthma (Bauer et al. 1996). Bauer et al. (1996) report that the proteins responsible for honey allergy are derived from proteins secreted by bees and from proteins derived from plant pollens. Individuals with inhalational allergies to particular plants (e.g., members of the Compositae family) may sometimes demonstrate allergies to honeys produced by bees foraging on these plants (Bousquet et al. 1984; Helbling et al. 1992).

Kiistala et al. (1995) studied eight commercially available honeys in Finland and found, by RAST inhibition and immunospot methods, evidence of both pollen and insect allergens. They then challenged 46 pollen-allergic patients (mean age of 25 years) with 30 g honey and 32 patients with a placebo syrup. Minor symptoms were reported by 26% of those receiving honey and by 41% of those receiving the placebo. Symptoms reported included itching in the throat, nose, eyelid, and skin, feeling of edema in the throat or lips, runny nose, headache, and redness of the skin. The pattern of symptom distribution was similar between both groups. Although all participants had a history of atopy, including rhinitis and asthma, none had a history of bee or wasp venom allergies.

## REFERENCES

Abraham, K., Gürtler, R., Berg, K., Heinemeyer, G., Lampen, A., and Appel, K.E., 2011. Toxicology and risk assessment of 5-Hydroxymethylfurfural in food. *Molecular Nutrition and Food Research,* 55, 667–678.

Al-Waili, N.S., 2003. Effects of daily consumption of honey solution on hematological indices and blood levels of minerals and enzymes in normal individuals. *Journal of Medicinal Food,* 6, 135–140.

Alissandrakis, E., Kibaris, A.C., Tarantilis, P.A., Harizanis, P.C., and Polissiou, M., 2005a. Flavour compounds of Greek cotton honey. *Journal of the Science of Food and Agriculture,* 85, 1444–1452.

Alissandrakis, E., Tarantilis, P.A., Harizanis, P.C., and Polissiou, M., 2005b. Evaluation of four isolation techniques for honey aroma compounds. *Journal of the Science of Food and Agriculture,* 85, 91–97.

Alissandrakis, E., Tarantalis, P.A., Harizanis, P.C., and Polissiou, M., 2007. Aroma investigation of unifloral Greek citrus honey using solid-phase microextraction coupled to gas chromatographic-mass spectrometric analysis. *Food Chemistry,* 100, 396–404.

Aljadi, A.M. and Kamaruddin, M.Y., 2004. Evaluation of the phenolic contents and antioxidant capacities of two Malaysian floral honeys. *Food Chemistry,* 85, 513–518.

Aljadi, A.M. and Yusoff, K.M., 2003. Isolation and identification of phenolic acids in Malaysian honey with antibacterial properties. *Turkish Journal of Medicine and Sciences,* 33, 229–236.

Ambrose, J.T., 1972. *Bees and Warfare: Gleanings in Bee Culture*, 100, 343–346.

Anam, O.O. and Dart, R.K., 1995. Influence of metal ions on hydroxymethylfurfural formation in honey. *Analytical Proceedings Including Analytical Communications,* 32, 515–517.

Antony, S.M., Han, I.Y., Rieck, J.R., and Dawson, P.L., 2000. Antioxidative effect of Maillard reaction products formed from honey at different reaction times. *Journal of Agricultural and Food Chemistry,* 48, 3985–3989.

Anthony, S.M., Rieck, J.R., and Dawson, P.L., 2000. Effect of dry honey on oxidation in turkey breast meat. *Poultry Science,* 79, 1846–1850.

Archer, M.C., Bruce, W.R., Chan, C.C., Corpet, D.E., Medline, A., and Roncucci, L., 1992. Aberrant crypt foci and microadenoma as markers for colon cancer. *Environmental Health Perspectives,* 98, 195–197.

Askar, A., 1984. Flavour changes during production and storage of fruit juices. *Fluessiges Obst,* 51, 564–569.

Azeredo, L.C., Azeredo, M.A.A., Souza, S.R., and Dutra, V.M.L., 2003. Protein contents and physicochemical properties in honey samples of *Apis mellifera* of different floral origins. *Food Chemistry,* 80, 249–254.

Azzi, A., Davies, K.J.A., and Kelly, F., 2004. Free radical biology: Terminology and critical thinking. *FEBS Letters,* 558, 3–6.

Bakhiya, N., Monien, B., Frank, H., Seidel, A., and Glatt, G., 2009. Renal organic anion transporters OAT1 and OAT3 mediate the cellular accumulation of 5-sulfooxymethylfurfural, a reactive, nephrotoxic metabolite of the Maillard product 5-hydroxymethylfurfural. *Biochemical Pharmacology,* 78, 414–419.

Baltrušaitytė, V., Venskutonis, P.R., and Čeksterytė, V., 2007. Radical scavenging activity of different floral origin honey and beebread phenolic extracts. *Food Chemistry,* 101, 502–514.

Baroni, M.V., Nores, M.L., Díaz Mdel, P., Chiabrando, G.A., Fassano, J.P., Costa, C., and Wunderlin, D.A., 2006. Determination of volatile organic compound patterns characteristic of five unifloral honey by solid-phase microextraction–gas chromatography–mass spectrometry coupled to chemometrics. *Journal of Agricultural and Food Chemistry,* 54, 7235–7241.

Bath, P.K. and Singh, N., 1999. A comparison between Helianthus annuus and Eucalyptus lanceolatus honey. *Food Chemistry,* 67, 389–397.

Bauer, L., Kohlich, A., Hirschweir, R., Siemann, U., Ebner, H., Scheiner, O., Kraft, D., and Ebner, C., 1996. Food allergy to honey: Pollen or bee products? Characterisation of allergenic proteins in honey by means of immunoblotting. *The Journal of Allergy and Clinical Immunology,* 97, 65–73.

Belitz, H.D. and Grosch, W., 1999. *Food Chemistry.* Springer, New York. CAC, (2001, July). Draft Report of 24th Session, Geneva.

Bengsch, E., 1992. Connaissance du miel Des oligo-éléments pour la santé. *Review of France Apiculture,* 569, 383–386.

Bentivenga, G., D'auria, M., Fedeli, P., Mauriello, G., and Racioppi, R., 2004. SPME-GC-MS analysis of volatile organic compounds in honey from Basilicata. Evidence for the presence of pollutants from anthropogenic activities. *International Journal of Food Science and Technology,* 39, 1079–1086.

Beretta, G., Granata, P., Ferrero, M., Orioli, M., and Maffei Facino, R., 2005. Standardization of antioxidant properties of honey by a combination of spectrophotometric/fluorimetric assays and chemometrics. *Analytica Chimica Acta,* 533, 185–191.

Beretta, G., Orioli, M., and Facino, R.M., 2007. Antioxidant and radical scavenging activity of honey in endothelial cell cultures (EA.hy926). *Planta Medica,* 73, 1182–1189.

Blanchan, N., 1900. Consumption of the leaves of *Kalmia* can be fatal to cattle and grouse. *Nature's Garden: An Aid to Our Knowledge of Our Wild Flowers and Their Insect Visitors.* Garden City Pub. Co., New York.

Blasa, M., Candiracci, M., Accorsica, A., Piacentini, M.P., Albertini, M.C., and Piatt, E., 2006. Raw millefiorihoney is packed full of antioxidants. *Food Chemistry,* 97, 217–222.

Board, N.H., 2011. Food Technology Program. Available from: http://www.aaccnet.org/funcfood/content/releases/Honeyantioxidant.htm, cited July 10, 2011.

Bogdanov, S., Jurendic, T., and Sieber, R., 2008. Honey for nutrition and health: A review. *Journal of American College of Nutrition,* 27, 677–689.

Bonvehí, J.S. and Coll, F.V., 2003. Flavour index and aroma profiles of fresh and processed honeys. *Journal of the Science of Food and Agriculture,* 83, 275–282.

Bouseta, A., Collin, S., and Doufour, J., 1992. Characteristic aroma profiles of unifloral honeys obtained with a dynamic headspace GC-MS system. *Journal of Apicultural Research,* 32, 96–109.

Bousquet, J., Campos, J., and Michel, F.B., 1984. Food intolerance to honey. *Allergy,* 39, 73–75.

Bravo, L., 1998. Polyphenols: Chemistry, dietary sources, metabolism, and nutritional significance. *Nutrition Reviews,* 56, 317–333.

Bruce, W.R., Archer, M.C., Corpet, D.E., Medline, A., Minkin, S., and Stamp, D., 1993. Diet, aberrant crypt foci and colorectal cancer. *Mutation Research,* 290, 111–118.

Cabras, P., Angioni, A., Tuberoso, C., Froris, I., Reniero, F., Guillou, C., and Ghelli, S., 1999. *Journal of Agricultural and Food Chemistry,* 47, 4064–4067.

Catapano, A.L., 1997. Antioxidant effect of flavonoids. *Angiology,* 48, 39–44.

Chen, L., Mehta, A., Berenbaum, M., Zangerl, A.R., and Engeseth, N.J., 2000. Honeys from different floral sources as inhibitors of enzymatic browning in fruit and vegetable homogenates. *Journal of Agricultural and Food Chemistry,* 48, 4997–5000.

Cherbuliez, R. and Domerego, M., 2003. Medecine des Abeilles, editions Amyris, Collection Douce Alternative, 255.

Cherchi, A., Spanedda, L., Tuberoso, C., and Cabras, P., 1994. *Journal of Chromatography A,* 669, 59–64.

Chow, J., 2002. Probiotics and prebiotics: A brief overview. *Journal of Renal Nutrition,* 12, 76–86.

*Codex Alimentarius,* 2000. Draft revised standard for honey at step 8 of the Codex procedure.

Conti, M.E., 2000. Lazio region (central Italy) honeys: A survey of mineral content and typical quality parameters. *Food Control,* 11, 459–463.

Cook, N.C. and Sammon, S., 1996. Flavonoids—Chemistry, metabolism, cardio protective effects, and dietary sources. *Journal of Nutritional Biochemistry,* 7, 66–76.

Council Directive of the European Union, 2002. Council Directive 2001/110/EC of December 20, 2001 relating to honey. *Official Journal of the European Communities,* L10, 47–52.

Crane, E., 1975. *Honey: A Comprehensive Survey.* Bee Research Association, William Heinemann Ltd., London.

Deinzer, H.L., Thomson, P.A., Burgett, D.M., and Isaacson, D.L., 1977. Pyrrolizidine alkaloids: Their occurrence in honey from tansy ragwort. *Science,* 195, 497–499.

Dimitrova, B., Gevrenova, R., and Anklam, E., 2007. Analysis of phenolic acids in honeys of different floral origin by solid-phase extraction and high-performance liquid chromatography. *Phytochemical Analysis,* 18, 24–32.

Espina-Prez, D. and Ordetx-Ros, G.S., 1983. *Flora Apcola Tropical.* Editorial Tecnolgico de Costa Rica, Cartago, Costa Rica, 35.

Fahey, J.W. and Stephenson, K.K., 2002. Pinostrobin from honey and Thai ginger (Boesenbergia pandurata): A potent flavonoid inducer of mammalian phase 2 chemoprotective and antioxidant enzymes. *Journal of Agricultural Food Chemistry,* 50, 7472–7476.

Fallico, B., Zappal, M., Arena, E., and Verzera, A., 2004. Effects of heating process on chemical composition and HMF levels in Sicilian monofloral honeys. *Food Chemistry,* 85, 305.

FAO, 1996. *FOA Agricultural Services Bulletin.* FAO, Rome, Italy.

Ferreres, F., Andrade, P., and Tomás-Barberán, F.A., 1996. Natural occurrence of abscisic acid in heather honey and floral nectar. *Journal of Agricultural and Food Chemistry,* 44, 2053–2056.

Forman, H.J., Torres, M., and Fukuto, J., 2002. Redox signaling. *Molecular and Cellular Biochemistry,* 234–235, 49–62.

Frankel, S., Robinson, G.E., and Berenbaum, M.R., 1998. Antioxidant capacity and correlated characteristic of 14 unifloral honeys. *Journal of Apicultural Research,* 37, 27–31.

Georghiou, G.P., 1980. Ancient beekeeping. In Root, A.I. (Ed.), *The ABC and XYZ of Bee Culture.* A.I. Root Company, Medina, Ohio, 17–21.

Gheldof, N. and Engeseth, N.J., 2002. Antioxidant capacity of honeys from various floral sources based on the determination of oxygen radical absorbance capacity and inhibition of in vitro lipoprotein oxidation in human serum samples. *Journal of Agricultural and Food Chemistry,* 50, 3050–3055.

Gheldof, N., Wang, X.H., and Engeseth, N.J., 2003. Buckwheat honey increases serum antioxidant capacity in humans. *Journal of Agricultural and Food Chemistry,* 51, 1500–1505.

Gil, M.I., Ferreres, F., Ortiz, A., Subra, E., and Tomás-Barberán, F.A., 1995. Plant phenolic metabolites and floral origin of rose-mary honey. *Journal of Agricultural and Food Chemistry,* 43, 2833–2838.

Glatt, H.R. and Sommer, Y., 2006. Health risks by 5-hydroxymethylfurfural (HMF) and related compounds. In Skog, K. & Alexander, J. (Eds.), *Acrylamide and Other Health Hazardous Compounds in Heat-Treated Foods*. Woodhead Publishing, Cambridge, 328e357.

Goldberg, I., 1996. Fuctional Foods. *Designer Foods, Pharmafood, Nutraceuticals,* Chapman and Hall, London, U.K.

González-Paramás, A.M., Gómez-Bárez, J.A., and Cordón Marcos, C., 2006. HPLC-fluorimetric method for analysis of amino acids in products of the hive (honey and bee-pollen). *Food Chemistry,* 95, 148–156.

Graddon, A.D., Morrison, J.D., and Smith, J.S., 1979. Volatile constituents of some unifloral Australian honeys. *Journal of Agricultural and Food Chemistry,* 27, 832–837.

Guidotti, M. and Vitali, M., 1998. Identification of volatile organic compounds present in different honeys through SPME and GC/MS. *Industrie Alimentari,* XXXVII, 351–356.

Guyot, C., Bouseta, A., Scheirman, V., and Collin, S., 1998. Floral origin markers of chestnut and lime tree honeys. *Journal of Agricultural and Food Chemistry,* 46, 625–633.

Guyot, C., Scheirman, V., and Collin, S., 1999. Floral origin markers of heather honeys: *Calluna vulgaris* and *Erica arborea. Food Chemistry,* 64, 3–11.

Halliwell, B., Rafter, J., and Jenner, A., 2005. Health promotion by flavonoids, tocopherols, tocotrienols, and other phenols: Direct or indirect effects? Antioxidant or not? *The American Journal of Clinical Nutrition,* 81, 268S–276S.

Helbling, A., Peter, C., Berchtold, E., Bogdanov, S., and Muller, U., 1992. Allergy to honey: Relation to pollen and honey bee allergy. *Allergy,* 47, 41–49.

Hermosín, I., Chicón, R.M., and Cabezudo, M.D., 2003. Free amino acid composition and botanical origin of honey. *Food Chemistry,* 83, 263–268.

Herrero, M., Ibáñez, E., and Cifuentes, A., 2005. Analysis of natural antioxidants by capillary electromigration methods. *Journal of Separation Science,* 28, 883–897.

Iglesias, M.T., De Lorenzo, C., and Polo, M.C., 2004. Usefulness of amino acids composition to discriminate between honeydew and floral honeys. Application to honeys from a small geographic area. *Journal of Agricultural and Food Chemistry,* 52, 84–89.

Inoue, K., Murayarna, S., Seshimo, F., Takeba, K., Yoshimura, Y., and Nakazawa, H., 2005. Identification of phenolic compound in manuka honey as specific superoxide anion radical scavenger using electron spin resonance (ESR) and liquid chromatography with coulometric array detection. *Journal of the Science of Food and Agriculture,* 85, 872–878.

Jaganathan, S.K. and Mandal, M., 2009. Antiproliferative effects of honey and of its polyphenols: A review. *Journal of Biomedicine and Biotechnology,* 2009, 830616.

Jeffrey, A.E. and Echazarreta, C.M., 1996. Medical uses of honey. *Rev Biomedica Biochimica Acta,* 7, 43–49.

Jerkovic, I., Mastelic, J., and Marijanovic, Z., 2006. A variety of volatile compounds as markers in unifloral honey from dalmatian sage (*Salvia officinalis* L.). *Chemistry & Biodiversity,* 3, 1307–1316.

Kaškonienė, V., Maruška, A., and Kornyšova, O., 2009. Quantitative and qualitative determination of phenolic compounds in honey. *Cheminė Technologija,* Nr. 3, 74–80.

Kaškoniene, V., Venskutonis, P.R., and Ceksteryte, V., 2008. Composition of volatile compounds of honey of various floral origin and beebread collected in Lithuania. *Food Chemistry,* 111, 988–997.

Kassim, M., Achoui, M., Mustafa, M.R., Mohd, M.A., and Yusoff, K.M., 2010. Ellagic acid, phenolic acids, and flavonoids in Malaysian honey extracts demonstrate in vitro anti-inflammatory activity. *Nutrition Research,* 30, 650–659.

Keen, C.L., Holt, R.R., Oteiza, P.I., Fraga, C.G., and Schmitz, H.H., 2005. Cocoa antioxidants and cardiovascular health. *The American Journal of Clinical Nutrition,* 81, 298S–303S.

Kelhoffer, J.A. and James, A., 2005. John the Baptists "wild honey" and "honey" in antiquity. *Greek, Roman, and Byzantine Studies,* 45, 59–73.

Kerenyi, K., 1976. *Dionysus: Archetypal Image of Indestructible Life.* Princeton University Press.

Kettlewellh, B.D., 1945. A story of nature's debauch. *The Entomologist,* 88, 45–47.

Kiistala, R., Hannuksela, M., Makinen-Kiljunen, S., Niinimaki, A., and Haahtela, T., 1995. Honey allergy is rare in patients sensitive to pollens. *Allergy,* 50, 844–847.

Kücük, M., Kolayli, S., Karaoglu, S., Ulusoy, E., Baltaci, C., and Candan, F., 2007. Biological activities and chemical composition of three honeys of different types from Anatolia. *Food Chemistry,* 100, 526–534.

Lachman, J., Hejtmánková, A., Sýkora, J., Karban, J., Orsák, M., and Rygerová, B., 2010. Contents of Major Phenolic and Flavonoid Antioxidants in Selected Czech Honey. *Czech Journal of Food Sciences,* 28, 412–426.

Lambert, J.D., Hong, J., Yang, G., Liao, J., and Yang, C.S., 2005. Inhibition of carcinogenesis by polyphenols: Evidence from laboratory investigations. *The American Journal of Clinical Nutrition,* 81, 284S–291S.

Lee, Y.C., Shlyankevich, M., Jeong, H.K., Douglas, J.S., and Surh, Y.J., 1995. Bioactivation of 5-hydroxymethyl-2-furaldehyde to an electrophilic and mutagenic allylic sulfuric acid ester. *Biochemical and Biophysical Research Communications,* 209, 996–1002.

Lensky, Y., 1997. *Bee Products: Properties, Applications, and Apitherapy.* Springer, New York.

Leshkov, Y.R., Chheda, J.N., and A, D.J., 2006. Phase modifiers promote efficient production of hydroxymethylfurfural from fructose. *Science,* 312, 1933–1937.

Lesschaeve, I. and Noble, A.C., 2005. Polyphenols: Factors influencing their sensory properties and their effects on food and beverage preferences. *The American Journal of Clinical Nutrition,* 81, 330S–335S.

Madhavi, D.V., Despande, S.S., Salunkhe, D.K., Madhavi, D.L., Deshpande, S.S., and Salunkhe D.K. (Eds.), 1996. *Food Antioxidants.* Marcel Dekker, New York.

Manach, C., Scalbert, A., Morand, C., Rémésy, C., and Jimenez, L., 2004. Polyphenols: Food sources and bioavailability. *The American Journal of Clinical Nutrition,* 79, 727–747.

Marghitas, L.A., Dezmirean, D.S., Pocol, C.B., Ilea, M., Bobis, O., and Gergen, I., 2010. The development of a biochemical profile of acacia honey by identifying biochemical determinants of its quality. *Notulae Botanicae Horti Agrobotanici Cluj-Napoca,* 38, 84–90.

Martos, I., Cossentini, M., Ferreres, F., and Tomás-Barberán, F.A., 1997. Flavonoid composition of Tunisian honeys and propolis. *Journal of Agricultural and Food Chemistry,* 45, 2824–2829.

Martos, I., Ferreres, F., and Tomás-Barberán, F.A., 2000a. Identification of flavonoid markers for the botanical origin of *Eucalyptus* honey. *Journal of Agricultural and Food Chemistry,* 48, 1498–1502.

Martos, I., Ferreres, F., Yao, L., D'arcy, B., Caffin, N., and Tomás-Barberán, F.A., 2000b. Flavonoids in monospecific *Eucalyptus* honeys from Australia. *Journal of Agricultural and Food Chemistry,* 48, 4744–4748.

Mayor and Adrienne, 1995. Mad Honey [toxic honey in history]. *Archaeology,* 48.

Mcalpine, A., 2002. *Adventures of a Collector.* Allen & Unwin. ISBN 1865087866.

Mckibben, J. and Engeseth, N.J., 2002. Honey as a protective agent against lipid oxidation in ground turkey. *Journal of Agricultural and Food Chemistry,* 50, 592–595.

Mclellan, M.R., Kime, R.W., Lee, C.Y., and Long, T.M., 1995. *Journal of Food Processing and Preservation,* 19, 1–8.

Meda, C.E., Lamien, M., Romito, J., Millogo, O.G., and Nacoulma, O.G., 2005. Determination of the total phenolic, flavonoid and proline contents in Burkina Fasan honey, as well as their radical scavenging activity. *Food Chemistry,* 91, 571–577.

Mennen, L.I., Walker, R., Bennetau-Pelissero, C., and Scalbert, A., 2005. Risks and safety of polyphenol consumption. *The American Journal of Clinical Nutrition,* 81, 326S–329S.

Miyakawa, Y., Nishi, Y., Kato, K., Sato, H., Takahashi, M., and Hayashi, Y., 1991. Initiating activity of eight pyrolysates of carbohydrates in a two stage mouse skin tumorigenesis model. *Carcinogenesis,* 12, 1169–1173.

Moreira, R.F.A. and De Maria, C.A.B., 2001. Glícidos no mel. *Quim Nova,* 24, 516–525.

Moreira, R.F.A. and De Maria, C.A.B., 2005. Investigation of the aroma compounds from headspace and aqueous solution from the cambara (*Gochnatia Velutina*) honey. *Flavour and Fragrance Journal,* 20, 13–17.

Moreira, R.F.A., Trugo, L.C., Pietroluongo, M., and De Maria, C.A.B., 2002. Flavor composition of cashew (*Anacardium occidentale*) and marmeleiro (Croton Species) honeys. *Journal of Agricultural and Food Chemistry,* 50, 7616–7621.

Moskaug, J., Carlsen, H., Myhrstad, M.C.W., and Blomhoff, R., 2005. Polyphenols and glutathione synthesis regulation. *The American Journal of Clinical Nutrition,* 81, 277S–283S.

Nagy, G., Kim, J.H., Pang, Z.P., Matti, U., Rettig, J., Sudhof, T.C., and Sorensen, J.B., 2006. Different effects on fast exocytosis induced by synaptotagmin 1 and 2 isoforms and abundance but not by phosphorylation. *Journal of Neuroscience,* 26, 632–643.

Odeh, I., Lafi, S.A., Dewik, H., Al-Najjar, I., Imam, A., Dembitsky, V.M., and Hanuš, L.O., 2007. A variety of volatile compounds as markers in Palestinian honey from *Thymus capitatus, Thymelaea hirsuta,* and *Tolpis virgata. Food Chemistry,* 101, 1393–1397.

Oleander, 2009. MedLine Plus. Accessed on July 17, 2009.

Ott, J., 1998. The delphic bee: Bees and toxic honeys as pointers to psychoactive and other medicinal plants. *Economic Botany,* 52, 260–266.

Palmer-Jones, T., 1965. Poisonous honey overseas and in New Zealand. *The New Zealand Medical Journal,* 64, 631–637.

Pérez, A.R., Iglesias, M.T., and Pueyo, E., 2007. Amino acid composition and antioxidant capacity of Spanish honeys. *Journal of Agricultural and Food Chemistry,* 55, 360–365.

Pérez, R.A., 2002. Analysis of volatiles from Spanish honeys by solid-phase microextraction and gas chromatography-mass spectrometry. *Journal of Agricultural and Food Chemistry,* 50, 2633–2637.

Persano-Oddo, L. and Piro, R., 2004. Main European unifloral honeys: Descriptive sheets. *Apidologie,* 35, 38–81.

Piasenzotto, L., Gracco, L., and Conte, L., 2003. Solid phase microextraction (SPME) applied to honey quality control. *Journal of Science of Food and Agriculture,* 83, 1037–1044.

Pulcini, P., Allegrini, F., and Festuccia, N., 2006. Fast SPE extraction and LC-ESI-MS-MS analysis of flavonoids and phenolic acids in honey. *Apiacta,* 41, 21–27.

Radovic, B.S., Careri, M., Mangia, A., Musci, M., Gerboles, M., and Anklam, E., 2001. Contribution of dynamic headspace GC-MS analysis of aroma compounds to authenticity testing of honey. *Food Chemistry,* 72, 511–520.

Ramirez, C.M.A., Gonz Á Lez, N.S.A., and Sauri, D.E., 2000. Effect of the temporary thermic treatment of honey on variation of the quality of the same during storage. *Apiacta,* 35, 162–170.

Reid, M., background on toxic honey. *New Zealand Food Safety Authority,* Accessed on July 11, 2011.

Rowland, C.Y., Blackman, A.J., D'arcy, B.R., and Rintoul, G.B., 1995. Comparison of organic extractives found in leatherwood (*Eucryphia lucida*) honey and leatherwood flowers and leaves. *Journal of Agricultural and Food Chemistry,* 43, 753–763.

Salah, N., Millar, N.J., Paganda, G., Tijburg, L., Bolwell, G.P., and Rice-Evans, C., 1995. Polyphenolic flavanols as scavengers of aqueous phase radicals and as chain-breaking antioxidants. *Archives of Biochemistry and Biophysics,* 322, 339–346.

Scalbert, A., Manach, C., Morand, C., Rémésy, C., and Jiménez, L., 2005. Dietary polyphenols and the prevention of diseases. *Critical Reviews in Food Science and Nutrition,* 45, 287–306.

Scalbert, A. and Williamson, G., 2000. Dietary intake and bioavailability of polyphenols. *The Journal of Nutrition,* 130, 2073S–2085S.

Schoental, R., Hard, G.C., and Gibbard, S., 1971. Histopathology of renal lipomatous tumors in rats treated with the "natural" products, pyrrolizidine alkaloids and a, b-unsaturated aldehydes. *Journal of the National Cancer Institute,* 47, 1037–1044.

Schramm, D.D., Karim, M., Schrader, H.R., Holt, R.R., Cardetti, M., and Keen, C.L., 2003. Honey with high levels of antioxidants can provide protection to healthy human subjects. *Journal of Agricultural and Food Chemistry,* 51, 1732–1735.

Serra Bonvehı, J., Soliva Torrento, M., and Muntané Raich, J.I., 2000. Invertase activity in fresh and processed honeys. *Journal of the Science of Food and Agriculture,* 80, 507–512.

Serra Bonvehı, J. and Ventura Coll, F., 2000. Evaluation of purine alkaloids and diketopiperazines contents in processed cocoa powder. *European Food Research and Technology,* 210, 189–195.

Shimoda, M., Wu, Y., and Osajima, Y., 1996. Aroma compounds from aqueous solution of haze (Rhus sucedá nea) honey determined by absorptive column chromatography. *Journal of Agricultural and Food Chemistry,* 44, 3913–3918.

Sies, H., Schewe, T., Heiss, C., and Kelm, M., 2005. Cocoa polyphenols and inflammatory mediators. *The American Journal of Clinical Nutrition,* 81, 304S–312S.

Soria, A.C., Martínez-Castro, I., and Sanz, J., 2003. Analysis of volatile composition of honey by solid phase microextraction and gas chromatography-mass spectrometry. *Journal of Separation Science,* 26, 793–801.

Spano, N., Casula, L., Panzanelli, A., Pilo, M.I., Piu, P.C., Scanu, R., Tapparo, A., and Sanna, G., 2006. An RP-HPLC determination of 5-hydroxymethylfurfural in honey: The case of strawberry tree honey. *Talanta,* 68, 1390–1395.

Ståhlberg, T., Fu, W., Woodley, J.M., and Riisager, A., 2011. Synthesis of 5-(hydroxymethyl)furfural in ionic liquids: Paving the way to renewable chemicals. *Chemistry and Sustainability, Energy and Materials,* 4, 451–458.

Steeg, E. and Montag, A., 1987. Abbauprodukte des Phenylalanins als Aromakomponenten in Honig. *Dtsch Lebensm Rundsch,* 83, 103–107.

Steeg, E. and Montag, A., 1988. Quantitative Bestimmung aromatischer Carbosäuren in Honig. *Z Lebensm Untersuch Forsch,* 187, 115–120.

Stocker, A., Schramel, P., Kettrup, A., and Bengsch, E., 2005. Trace and mineral elements in royal jelly and homeostatic effects. *Journal of Trace Elements in Medicine and Biology,* 19, 183–189.

Su, Y., Brown, H.M., Huang, X., Zhou, X., Amonette, J.E., and Zhang, Z.C., 2009. Single-step conversion of cellulose to 5-hydroxymethylfurfural (HMF), a versatile platform chemical. *Applied Catalysis A General,* 361, 117.

Surh, Y.J., Liem, A., Miller, J.A., and Tannenbaum, S.R., 1994. 5-sulfooxymethylfurfural as a possible ultimate mutagenic and carcinogenic metabolite of the Maillard reaction-product, 5-hydroxymethylfurfural. *Carcinogenesis,* 15, 2375–2377.

Svendsen, C., Husøy, T., Glatt, H., Paulsen, J.E., and Alexander, J., 2009. 5-Hydroxymethylfurfural and 5-sulfooxymethylfurfural increase adenoma and flat ACF number in the intestine of Min/+ mice. *Anticancer Research,* 29, 1921–1926.

Tan, S.T., Wilkins, A.L., Holland, P.T., and Mcghie, T.K., 1989. Extractives from New Zealand unifloral honeys. 2. Degraded carotenoids and other substances from heather honey. *Journal of Agricultural and Food Chemistry,* 37, 1217–1221.

Terrab, A., Gonzále, M., and González, A., 2003. Characterisation of Moroccan unifloral honeys using multivariate analysis. *European Food Research and Technology,* 218, 88–95.

Tomás-Barberán, F.A., Martos, I., Ferreres, F., Radovic, B.S., and Anklam, E., 2001. HPLC flavonoid profiles as markers for the botanical origin of European unifloral honeys. *Journal of the Science of Food and Agriculture,* 81, 485–496.

Ulbricht, R.J., Northup, S.J., and Thomas, J.A., 1984. A review of 5-hydroxymethylfurfural (HMF) in parenteral solutions. *Fundamental and Applied Toxicology,* 4, 843–853.

U.S. EPA, 1992. Initial Submission: Acute Oral Ld50 Study with 5-Hydroxymethylfurfural in Rats. Cover letter dated 073192, EPA/OTS. Doc. #88-920005429, Chicago, Illinois.

Vanderbilt Medical Centre, 1998. Infant Botulism. http://www.mc.vanderbilt.edu/peds/pidl/neuro/botulism.htm.

Vela, L., De Lorenzo, C., and Pérez, R.A., 2007. Antioxidant capacity of Spanish honeys and its correlation with polyphenol content and other physicochemical properties. *Journal of the Science of Food and Agriculture,* 87, 1069–1075.

Verzera, A., Campisi, S., Zappala, M., and Bonaccorsi, I., 2001. SPME-GC-MS analysis of honey volatile components for the characterization of different floral origin. *American Laboratory,* 33, 18–21.

Vinson, J.A., Hao, Y., Su, X., and Zubik, L., 1998. Phenol antioxidant quantity and quality in foods: Vegetables. *Journal of Agricultural and Food Chemistry,* 46, 3630–3634.

Visser, F.R., Allen, J.M., and Shaw, G.J., 1988. The effect of heat on the volatile flavour fraction from a unifloral honey. *Journal of Apicultural Research,* 27, 175–181.

Vit, P., Soler, C., Tomás-Barberán, F.A., and Lebensm, Z., 1997. Profiles of phenolic compounds of *Apis mellifera* and *Melipona* spp. honeys from Venezuela. *Unters Forsch A,* 204, 43–47.

Vita, J.A., 2005. Polyphenols and cardiovascular disease: Effects on endothelial and platelet function. *The American Journal of Clinical Nutrition,* 81, 292S–297S.

Walderhaug, M., 2001. Grayanotoxin. *Foodborne Pathogenic Microorganisms and Natural Toxins Handbook.* U.S. FDA.

Weston, R.J., Mitchell, K.R., and Allen, K.L., 1999. Antibacterial phenolic components of New Zealand manuka honey. *Food Chemistry,* 64, 295–301.

White, J.W., 1975. Composition of honey. In Crane, E. (Ed.), *Honey, A Comprehensive Survey.* Crane, Russak: New York, 157–206.

White, J.W., 1978. Honey. *Advances in Food Research,* 24, 287–374.

Williamson, G. and Manach, C., 2005. Bioavailability and bioefficacy of polyphenols in humans. II. Review of 93 intervention studies. *The American Journal of Clinical Nutrition,* 81, 243S–255S.

Wollgast, J. and Anklam, E., 2002. Review on polyphenols in *Theobroma cacao*: Changes in composition during the manufacture of chocolate and methodology for identification and quantification. *Food Research International,* 33, 423–447.

Yao, L., Datta, N., Tomás-Barberán, F., Ferreres, F., Martos, I., and Singanusong, R., 2003. Flavonoids, phenolic acids and abscisic acid in Australian and New Zealand Leptospermum honeys. *Food Chemistry,* 81, 159–168.

Yao, L., Jiang, Y., D'arcy, B., Singanusong, R., Datta, N., Caffin, N., and Raymont, K., 2004a. *Journal of Agricultural and Food Chemistry,* 52, 210–214.

Yao, L., Jiang, Y., Singanusong, R., D'arcy, B., Datta, N., Caffin, N., and Raymont, K., 2004b. *Food Research International,* 37, 166–174.

Yao, L.H., Jiang, Y.M., Singanusong, R.T., Datta, N.D., and Raymont, K., 2004c. Phenolic acids and abscisic acid in Australian Eucalyptus honeys and their potential for floral authentication. *Food Chemistry,* 86, 169–177.

Zhang, X.M., Chan, C.C., Stamp, D., Minkin, S., Archer, M.C., and Bruce, W.R., 1993. Initiation and promotion of colonic aberrant crypt foci in rats by 5-hydroxymethyl-2-furaldehyde in thermolyzed sucrose. *Carcinogenesis,* 14, 773–775.

Zhao, H., Holladay, J.E., Brown, H., and Zhang, Z.C., 2007. Metal chlorides in ionic liquid solvents convert sugars to 5-hydroxymethylfurfural. *Science,* 316, 1597–1600.

# 18 Honey in the Food Industry

*Yuva Bellik and Mokrane Iguerouada*

## CONTENTS

## BACKGROUND

Recent years have seen growing interest on the part of consumers, the food industry, and researchers into food and the ways in which it may help maintain human health (Viuda-Martos et al. 2008).

The growing popularity toward natural products like honey is because of their potential to economic development, in terms of export earnings, value adding, and counteracting demand in the Northern Hemisphere. With regard to honey, although an ancient product, it is only recently that it went to commercialization and appeared on the market shelves and hence treated as a newcomer. Therefore, production of honey since it is a new crop carries large risks at this moment. Current and anticipated developments in the manufacture of corn and other syrups offer opportunity for falsification of honey that is increasingly difficult to detect.

There is an ever-increasing international honey trade, and this global market requires universal standards for the protection of consumers. To a certain extent, these global standards are provided by the *Codex Alimentarius* system. Interestingly, there has been a great deal of research into the development of new applications for existing analytical and techniques for food authentication since 2001 (Reid et al. 2006), particularly for honeys, and for detecting their adulteration (addition of cane or beet sugars and/or sugars obtained from starch hydrolysis) (Anklam 1998; Cordella et al. 2002), because guaranteeing honey quality is becoming increasingly important for consumers, producers, and regulatory authorities.

Thus, the purpose of this chapter is to introduce the honey industry and most applications of this natural product in food technology as a potential business opportunity and recovering recent and more interesting data on honey marketing all over the world. However, an effort will first be made to provide a brief update on the honey composition, properties, and adulteration, where a special emphasis is put on the promising parameters used to assess honey adulteration and which cover more recent developments on the subject.

## DEFINITION

Honey is the only food consumed by humans that is produced by insects. It is a very important energy food and is used as an ingredient in virtually hundreds of manufactured foods, mainly in cereal-based products, for sweetness, color, flavor, caramelization, pumpability, and viscosity (LaGrange and Sanders 1988).

The definition of honey in the *Codex Alimentarius* is as follows: "Honey is the natural sweet substance produced by honey bees from the nectar of plants or from secretions of living parts of plants or excretions of plant sucking insects on the living parts of plants, which the bees collect, transform by combining with specific substances of their own, deposit, dehydrate, store and leave in the honey comb to ripen and mature. Blossom honey or nectar honey is the honey that comes from nectars of plants. Honeydew honey is the honey which comes mainly from excretions of plant sucking insects (*Hemiptera*) on the living parts of plants or secretions of living parts of plants."

## CHEMICAL COMPOSITIONS OF HONEY

The composition of honey depends on the floral source as well as on some external factors, such as season, environmental factors, physiologic stage of the bees, and honey processing (Oddo et al. 1995).

Honey is one of the most complex mixtures of carbohydrates produced in nature (Swallow and Low 1990). The major carbohydrates in honey are fructose and glucose (Figure 18.1), which account for 65% to 75% of the total soluble solids in honey and 85% to 95% of honey carbohydrates (White 1979). The remaining carbohydrates are a mixture of at least 12 disaccharides, 7 trisaccharides (Low and Sporns 1988), and several higher oligosaccharides.

Water is the second most important component of honey. Its content is critical, since it affects the storage of honey (Cano et al. 2001). The final water content depends on numerous environmental factors during production, such as weather and humidity inside the hives, but also on nectar conditions and treatment of honey during extraction and storage (Olaitan et al. 2007).

Organic acids constitute 0.57% of honey and include gluconic acid, which is a byproduct of enzymatic digestion of glucose. The organic acids are responsible for the acidity of honey (Bogdanov et al. 2004) and contribute largely to its characteristic taste. Minerals are present in honey in very small quantities, with potassium as the most abundant. Others are calcium, copper, iron, manganese, and phosphorus storage (Olaitan et al. 2007). Nitrogenous compounds, among which the enzymes originate from salivary secretion of the worker honeybees, are also present. They

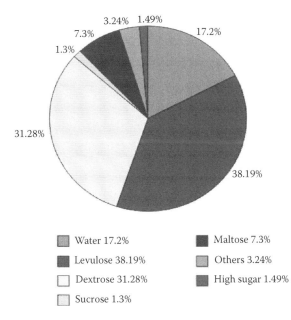

FIGURE 18.1 (See color insert.) Pie chart of honey composition indicating the percentage share of various sugars, water, and other minor constituents. (From Jaganathan, S.K. and Mandal, M., *Journal of Biomedicine and Biotechnology*, 1–13, 2009.)

**TABLE 18.1**

**Enzymes and Microorganisms in Honey**

| Nutrients | Average Amount per 100 g |
|---|---|
| Enzymes | Reaction on |
| Invertase | Saccharose |
| Glucose oxidase | Glucose |
| Diastase (amylase) | Starch |
| Catalase | Hydrogen peroxide |
| Microorganisms | Toxin produced |
| Clostridium botulinum | A, B, $C_1$, $C_2$, D, E, F, G |
| Clostridium butirycum | E |
| Clostridium barati | F |

*Source:*  Alvarez-Suarez, J.M. et al., *Mediterranean Journal of Nutrition and Metabolism*, 3, 15–23, 2010.

have an important role in the formation of honey. The main enzymes in honey are invertase (saccharase), diastase (amylase), and glucose oxidase (Oddo et al. 1999) (Table 18.1). Vitamin C, vitamin B (thiamine), and vitamin B2 complex such as riboflavin, nicotinic acid, and vitamin B6 (pantothenic acid) are also found.

Table 18.2 summarizes the composition of honey; however, these values can vary considerably according to the honey type and the geographical or botanical origin.

Honey is an entirely natural product. It contains neither additives nor preservatives. According to the definition of the *Codex Alimentarius* (*Codex Alimentarius* Commission 2001) and other international honey standards (2001/110/EC, EU Council 2002), honey shall not contain any food ingredient other than honey itself nor shall any particular constituent be removed from it. Honey shall not be tainted by any objectionable matter, flavor, aroma, or taint from foreign matter during harvesting, processing, and storage. Honey shall not have started to ferment or effervesce. No pollen or constituent particular to honey may be removed, except where this is unavoidable in the removal of foreign inorganic or organic matter. Honey shall not be heated or processed to such an extent that its essential composition is changed and/or its quality impaired (Bogdanov 2007).

## TYPES OF HONEY

Honey can be categorized according to its floral origin or source, mode of production, style of marketing, consistency and appearance, and color (Table 18.3).

### HONEY ADULTERATION

Economic adulteration has been practiced for centuries by unscrupulous farmers, manufacturers, or traders who seek benefits by extending or substituting expensive food products by low-cost materials (Prodolliet and Hischenhuber 1998). Nowadays, authentication of food products becomes of primary importance for both consumers and industries, at

## TABLE 18.2
## Composition of Honey (Data in g/100)

| Nutrients | Average Amount per 100 g |
|---|---|
| Water | 17.1 g |
| Total carbohydrates | 82.4 g |
| Fructose | 38.5 g |
| Glucose | 31.0 g |
| Maltose | 7.2 g |
| Sucrose | 1.5 g |
| Proteins and amino acids | 0.5 g |
| Vitamins | |
| Thiamine | <0.01 mg |
| Riboflavin | <0.3 mg |
| Niacin | <0.3 mg |
| Pantothenic acid | <0.25 mg |
| Vitamin B6 | <0.002 mg |
| Folate | <0.01 mg |
| Vitamin C | <0.5 mg |
| Minerals | |
| Calcium | 4.8 mg |
| Iron | 0.25 mg |
| Zinc | 0.15 mg |
| Potassium | 50 mg |
| Phosphorus | 5 mg |
| Magnesium | 2 mg |
| Selenium | 0.01 mg |
| Copper | 0.05 mg |
| Chromium | 0.02 mg |
| Manganese | 0.15 mg |
| Ash | 0.2 g |

*Source:* Arvanitoyannis, I.S. et al., *Critical Reviews in Food Science and Nutrition*, *45*, 193–203, 2005.

all levels of the production process, from raw materials to finished products. From the legislative point of view, quality standards have been established through the requirement of quality labels that specify the chemical composition of each product. From the economic point of view, product authentication is essential to avoid unfair competition that can create a destabilized market and disrupt the regional economy and even the national economy. All food products targeted for adulteration are high commercial value products and/or produced in high tonnage around the world (Cordella et al. 2002).

As honey is a product of limited supply and relatively high price, its quality assurance becomes extremely important. The adulteration of pure honey has become much more prevalent in recent years (Bertelli et al. 2010), which has a significant economic impact and undeniable nutritional and organoleptic consequences (Cordella et al. 2005), and more difficult to identify, particularly since the availability of cheap high-fructose corn syrup.

**TABLE 18.3**
**Types of Honey**

| Method of Characterization | Type of Honey |
|---|---|
| Origin | Blossom honey or nectar honey: Obtained from the nectar of plants. Consist mainly of the following: |
| | – Multifloral (plurifloral) honey: Contains pollen from the nectar of several plant species, with no domination by any single plant species. The properties of these honeys are much more variable, in relation to bee species, respective flowering, and climatic factors. |
| | – Unifloral (monofloral) honey: Originated from only one plant species (45% of the total pollen is from the same plant species). These honeys maintain always the same physicochemical and organoleptic characteristics (appearance, color, flavor, and taste) and are well appreciated for commerce. It is possible to determine their origin from flowers by the recognition of the dominant pollen grains. |
| | Honeydew honey: Made from excretions of plant-sucking insects on the living parts of plants or secretions of living parts of plants. |
| Mode of production | Squeezed honey: Honey obtained by pressing or squeezing of broodless combs with or without the application of moderate heat not exceeding 45°C. |
| | Filtered honey: Honey obtained by removing foreign inorganic or organic matter in such a way as to result in the significant removal of pollen. |
| | Drained honey: Honey obtained by draining decapped broodless combs. |
| | Extracted honey: Honey removed by centrifuging decapped honeycombs. This type of honey is produced by beekeepers who manage bees in moveable comb hives. |
| Style of marketing | Chunk honey: Obtained by cutting a piece of comb consisting only of sealed and undamaged honeycomb. This type of honey is generally packed in transparent glass vessels. |
| | Comb honey: Honey stored by bees in the cells of freshly built combs and sold in sealed whole combs or section of such combs. |
| | Comb honey in fluid honey: A cut comb inserted in fluid honey. |
| Consistency and appearance | Liquid: Honey free from visible crystals. |
| | Crystallized or granulated: Honey that is solidly crystallized irrespective of texture. |
| | Partially crystallized Honey: A mixture of liquid and crystallized. |
| Color | White honey: White or opaque. |
| | Dark brown (amber): Dark brown to amber. |
| | Golden honey: Golden. |

*Source:* Codex standard for honey. Codex Stan 12-1981. Adopted in 1981. Revisions 1987 and 2001, 1981; Barth, O.M., *Scientia Agricola, 61*, 342–350, 2004.

Honey fraud can involve the addition of industrial sugar syrups or its sale under a false name (Cotte et al. 2004). Corn syrups, invert syrups obtained by acidic or enzymatic hydrolysis from refined beet or cane sucrose, and high-fructose corn syrups, mainly produced by enzymatic hydrolysis and isomerization of corn starch, are the most common sweeteners that can be added directly to honey after harvest or fed to bees during harvest to improve yield (Swallow and Low 1994).

# DETECTION OF HONEY ADULTERATION

Many methods have been tested for adulteration proof, but most of them are not capable to detect unequivocal adulteration (Bogdanov and Martin 2002). Here, a special emphasis is put on the promising parameters used to assess honey adulteration and that cover more recent developments on the subject.

The major concern of honey quality is to inspect its adulteration with respect to the legislative requirements. Thus, a number of physical and physicochemical properties of honey are exploited to measure adulteration, which mostly include water content, enzyme activities (diastase and invertase), hydroxymethylfurfural (HMF), electrical conductivity, sugar profiles (glucose, fructose, and sucrose), proline, amino acid and proteins, and melissopalynological pattern (Schade et al. 1958; White 1979; Cotte et al. 2003, 2004; Ruoff et al. 2006). Several of these constituents are of great importance to the honey industry, as they influence the storage quality, granulation, texture, flavor, and the nutritional and medicinal qualities of the honey (Joshi et al. 2000).

## FERMENTATION

### Water Content

Water content is the most important measure related to honey quality, such as ripeness and shelf life, especially concerning the risk of spoilage due to fermentation processing (Ruoff et al. 2006). However, water content can be artificially altered during honey processing (Bogdanov 2007). The addition of water to honey showed remarkable changes in all of its physical properties, with major changes occurring in conductance and surface tension (Saif-ur-Rehman et al. 2008).

Good-quality honey essentially has low water content. Honey may be spoiled by fermentation and lose its freshness if its water content is greater than 18% (Ruoff et al. 2006). The reason is that all unpasteurized honey contains wild yeasts. Due to the high sugar concentration, these yeasts will pose little risk in low moisture honey because osmosis will draw sufficient water from the yeast to force them into dormancy. In honey that has a higher proportion of water, the yeast may survive and cause fermentation to begin in storage. This results in an increase in acidity, which then becomes an important quality criteria.

## HEAT DEFECTS

### Enzymes Activities

One of the characterization criteria of honey quality is the activity of enzymes. Enzymes as honey components have been the objects of much research over the years: the primary interest was a possible means of distinguishing between natural and artificial honey. Honey contains small amounts of different enzymes, the most important of which are diastase ($\alpha$-amylase), invertase ($\alpha$-glucosidase), glucose oxidase, catalase, and acid phosphatase (Oddo et al. 1999). Diastase and invertase are largely used in Europe as a measure of honey freshness, because their activity decreases in old or heated honeys (White et al. 1964; Sancho et al. 1992).

Invertase is the enzyme responsible for converting sucrose to fructose and glucose. It is more sensitive to heat than amylases and loses activity during storage faster compared with amylases. The use of diastase as a basis for measuring honey quality is questioned because of the extreme variation in beginning enzyme levels. Many honeys produced in the warm, dry areas of the world have fewer enzymes than cool-wet regions.

According to the Honey Quality and International Regulatory Standards, from the International Honey Commission, the diastase activity must not be less than or equal to 8, expressed as a diastase number (DN). DN in Schade scale, which corresponds to the Gothe scale number, is defined as gram starch hydrolyzed in 1 h at 40°C per 100 g honey. The activity of the diastase is closely related to its structure and can be modified by denaturation, brought about by heating (Tosi et al. 2008). Denaturation may be considered as a discontinuous phenomenon with various intermediate or transition states between the natural or native state and the completely denatured state. The *Codex Alimentarius* (1998) has established the minimum diastase activity value of 3 for honeys with natural low enzyme content. In honeys with a DN less than 8 and higher than or equal to 3, the HMF must not be higher than 15 mg/kg. If DN is equal to or higher than 8, the HMF limit is 60 mg/kg. Bogdanov (1993) pointed out that thermal treatment, which may destroy the diastase activity, should be as long as 31 days at 40°C, but they can be shortened to 1.2 h at 80°C. On the contrary, Cervantes et al. (2000) reported a 6.8 or 2.5 DN unit decrease in two honeys treated at 55°C for 15 min.

White (1992a, 1994) in a series of articles criticized severely the use of diastase content as a quality criterion. He suggested that diastase content is not useful in evaluating honey quality and especially heating, while HMF is more appropriate and can provide all the information needed to estimate the total heat exposure of any honey. In a previous study, White et al. (1964) demonstrated that invertase is preferable than diastase, as it is more sensitive to heating. Dustmann (1993) maintained that invertase, in combination with other analytical criteria, can detect damage by heating or overstorage and also HMF is rather inappropriate for the proof of heat damage, if taken into account as a sole criterion.

## Hydroxymethylfurfural

HMF, also 5-(hydroxymethyl)furfural, is a cyclic aldehyde compound derived from dehydration of sugars (Cervantes et al. 2000). This colorless solid is highly water-soluble. The molecule is a derivative of furan, containing both aldehyde and alcohol functional groups. HMF has been identified in a wide variety of heat-processed foods including milk, fruit juices, spirits, and honey.

The HMF measurement is used to evaluate the quality of honey; generally not present in fresh honey, its content increases during conditioning and storage (Ruoff et al. 2006). HMF levels increase gradually in the acidic honey solution through heat and storage; small amounts of HMF occur naturally in honey, resulting from the acid catalyzed dehydration of the hexoses, particularly fructose (Figure 18.2). It has been reported that HMF and its congener compounds are spontaneously formed in carbohydrate-containing foods by Maillard reactions (the nonenzymatic browning reaction) or by an acid-catalyzed dehydration of hexoses, and it is connected to the chemical properties of honey, such as pH, total acidity, and mineral content (Singh and Bath 1997).

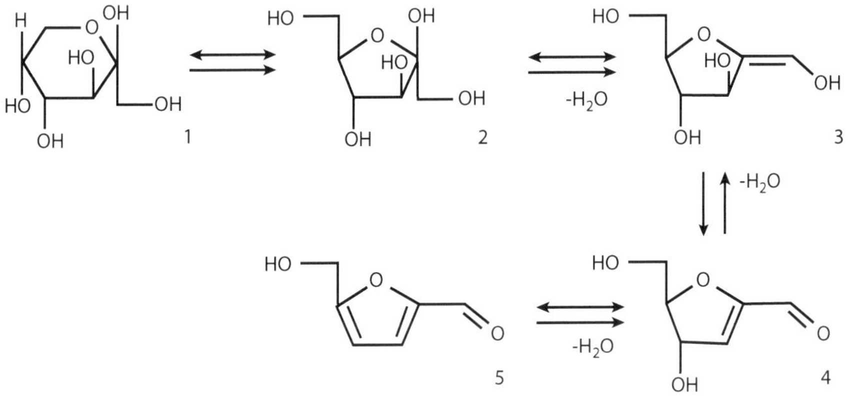

**FIGURE 18.2** Formation of HMF from fructose.

The *Codex Alimentarius* (Alinorm 01/25 2000) has established that the HMF content of honey after processing and/or blending must not be higher than 80 mg/kg. The European Union (EU Directive 110/2001), however, recommends a lower limit of 40 mg/kg with the following exceptions: 80 mg/kg is allowed for honey that originates from countries or regions with tropical temperatures, while a lower limit at only 15 mg/kg is allowed for honey with low enzymatic levels (Khalil et al. 2010).

## Aroma Compounds

The aroma profile is one of the most typical features of a food product both for its organoleptic quality and authenticity (Careri et al. 1994). Owing to the high number of volatile components, the aroma profile represents a "fingerprint" of the product, which could be used to determine its origin (Anklam and Radovic 2001). In the past decades, extensive research on aroma compounds has been carried out and more than 500 different volatile compounds have been identified in different types of honey (Alvarez-Suarez et al. 2010). Indeed, most aroma-building compounds vary in the different types of honey depending on its botanical origin (Bogdanov et al. 2004). Honey flavor is an important quality for its application in the food industry and is also a selection criterion for the consumer's choice. Aroma compounds are present in honey at very low concentrations as complex mixtures of volatile components of different functionality and relatively low molecular weight (Cuevas-Glory et al. 2007). An important number of organic compounds have been found as volatile components of different types of honeys. Thus, methyl anthranilate was identified as a compound characteristic of citrus honey. Other volatile compounds suggested as markers for citrus honey include lilac aldehyde (Piasenzotto et al. 2003; Alissandrakis et al. 2007), hotrienol (Piasenzotto et al. 2003), and 1-*p*-menthen-al (Alissandrakis et al. 2007). *Eucalyptus* honey was shown to be distinctive because of the content of the volatile compounds nonanol, nonanal, and nonanoic acid, and high levels of isophorone (3,5,5trimethylcyclohexen-2-enone) were found in heather honey (Piasenzotto et al. 2003; Cuevas-Glory et al. 2007).

## Honey Filtration

### Pollen Analysis (Melissopalyonology)

Pollen analysis of honey (or melissopalynology) is of great importance for its quality control (Von der Ohe et al. 2004). Honey always includes numerous pollen grains (mainly from the plant species foraged by honeybees) and honeydew elements (such as wax tubes, algae, and fungal spores) that altogether provide a good fingerprint of the environment where the honey comes from. Pollen analysis of honey considers the pollen grains, their morphologic characteristics, which lead to the indication of the species or the botanical taxa of their origin, as well as the quantity that may be indicative of the quality (Barth 2004). Thus, different types of pollens are used to indicate floral nectar sources utilized by bees to produce honey; relative pollen frequency is often used to verify and label a honey sample as to the major and minor nectar sources. This information has important commercial value because honey made from some plants commands a premium price. Moreover, pollen analysis provides some important information about honey extraction and filtration, fermentation, some kinds of adulteration (Kerkvliet et al. 1995), and hygienic aspects such as contamination with mineral dust, soot, or starch grains (Louveaux et al. 1970).

Microscopic analysis of the pollen in honey is the method used to determine its botanical origin. It is a simplified method for determining the total amount of pollen grains and the relative frequencies of pollen from various plant sources in honey. However, it requires highly specialized personnel, and pollen can be added fraudulently (Arvanitoyannis et al. 2005); in addition, there are difficulties in assuring a correct assignment of their origin (Serrano et al. 2004).

Quality-control methods, in conjunction with multivariate statistical analysis, have been found to be able to classify honey from different geographic regions, detect adulteration, and describe chemical characteristics (Cordella et al. 2002, 2003; Devillers et al. 2004). Multivariate analysis involves the use of mathematical and statistical techniques to extract information from complex data sets. It helps to look at the sample as a whole (holistically) and not just at a single component, allowing to untangle all the complicated interactions between the constituents and understand their combined effects on the whole matrix. Currently, the application of supervised pattern recognition and multivariate statistical techniques, such as principal component analysis, linear discriminant analysis, or discriminant analysis, provides the possibility to analyze the entire food sample matrix and to make a classification possible (Corbella and Cozzolino 2008).

## Sweeteners

### Carbohydrates

Regarding carbohydrate fraction, honey is probably the most complex mixture of oligosaccharides in nature, and the study of these compounds has been used to identify fraud. Honey can be adulterated by the use of invert syrup, corn syrup, or high-fructose corn syrup. Adulteration by addition of cane and corn sugar can be screened microscopically (Kerkvliet and Meijer 2000) and verified by measurement of $^{13}C/^{12}C$ isotopic ratio (White 1992b; White et al. 1998). Recently, infrared spectroscopic methods have been described for the detection of adulteration by adding beet and cane sugar to honey (Kelly et al. 2006).

It was reported that honey produced with corn syrup or high-fructose corn syrup is distinguished by a high amount of oligosaccharides (Anklam 1998). Cotte et al. (2003) investigated liquid sugar, invert sugar, and glucose syrups and determined that, in the liquid sugar syrup, the amount of sucrose was 40 times higher than in natural honey, while maltose and maltotriose were present in high quantities in the other syrups, 29.8 g/100 g and 6.5 g/100 g, respectively. It was suggested that these sugars could be used as markers, because they are not present in honey in so high concentrations. Usually, the amount of maltose in natural honey is less than 30 mg/g (Joshi et al. 2000; Cotte et al. 2003); however, in honeys from some plants, it can be up to 50 mg/g (Costa et al. 1999; Devillers et al. 2004). The ratios of fructose/glucose, maltose/isomaltose, sucrose/turanose, and maltose/turanose may be used to assess possible adulteration of honey with glucose or high-fructose syrups (Molnár-Perl and Horváth 1997).

## ELECTRICAL CONDUCTIVITY AND pH VALUE

Electrical conductivity and the pH value reflect the mineral and acid contents of honey. The electrical conductivity is used to distinguish between floral and honeydew honeys (Ruoff et al. 2006). It is also the most important physicochemical measure for the authentication of unifloral honeys (Vorwohl 1964; Mateo and Bosch-Reig 1998). The electrical conductivity value depends on the ash and acid content in honey: the higher their content, the higher the resulting conductivity (Bogdanov and Martin 2002). The electrical conductivity is a good criterion for the botanical origin of honey; therefore, it is very often used in routine honey control. The pH value can be used for the discrimination of floral and honeydew honey (Sanz et al. 2005) as well and is also helpful for the authentication of unifloral honeys (Oddo et al. 1995) and the differentiation of several honeydew honeys.

## FREE ACIDITY

The acid content in honey is characterized by the free acidity. The measure is useful for the evaluation of honey fermentation. A maximum of 40 mEq/kg is defined by the current standards. Furthermore, it is helpful for the authentication of unifloral honeys and especially for the differentiation between nectar and honeydew honeys. The reference method of equivalence point titration is relatively poor because of lactone hydrolysis during titration.

## AMINO ACID AND PROTEIN

Proteins are minor honey compounds; however, they are used in detecting adulteration. Honey proteins originate largely from honeybees as well as from the pollen and nectar of plants (White and Winters 1989; Bauer et al. 1996). At least 19 protein bands were detected by silver staining sodium dodecyl sulfate–polyacrylamide gel electrophoresis in honeys of different plant origin (Marshall and Williams 1987). Won et al. (2008) used the same technique and identified major proteins in honey produced by *Apis mellifera*. With regards to variations in the geographical and floral origin of honey, differences in the molecular weight of these major proteins in honey were noted as *Apis cerana* possessed a molecular weight of 56 kDa and *A. mellifera* had a molecular weight of 59 kDa (Lee et al. 1998).

Pirini et al. (1992) reported that the presence of amino acids, such as arginine, tryptophan, and cysteine, is characteristic for some honey types. Moreover, Conte et al. (1998) adapted the same methodology as Pirini et al. (1992) and analyzed honeys from various botanical sources and different areas. The quantities of free amino acids in the honey samples displayed high variation coefficients for honey types of the same botanical source.

## PROLINE

Proline, the main amino acid of honey, added to honey by the bee is a criterion of honey ripeness (Von der Ohe 1991) and is the second important quality parameter of the honey in *Codex Alimentarius*; the reference value for proline is given as 18 mg/100 g honey (Bogdanov et al. 1999). The proline content in honey is related to the degree of nectar processing by the bees (Ruoff et al. 2006). It is therefore used as an indicator of honey adulteration (White 1979). The proline content of honey changes according to its floral type (Oddo and Piro 2004) and also worker bees add proline to honey.

## TRACE ELEMENTS

Although the mineral content of honey can vary within a wide range (White 1975b), it usually amounts to 0.17%. The determination of heavy and transition metals in honey is of interest for quality control (Buldini et al. 2001).

The potential use of honey as an indicator in mineral prospecting and environmental contamination studies has been investigated (Tong et al. 1975). The bee can range over long distances, even up to 10 km$^2$ under exceptional circumstances: a hive can keep an area of 7 km$^2$ under its control (Crane 1984). Thus, since 1962, the bee has increasingly been employed as a biological indicator to monitor environmental pollution by heavy metals (Celli and Maccagnani 2003; Porrini et al. 2003). Elsewhere, Jones (1986) studied several samples of honeys from different regions of the United Kingdom and reported that samples from urban/industrial conurbations of Brigham and Liverpool contained elevated concentration of Ag, Cd, and Pb.

## HONEY IN THE FOOD INDUSTRY

Honey is most commonly consumed in its unprocessed state (i.e., liquid, crystallized, or in the comb). However, the antioxidant, antimicrobial, and antifungal properties of honey offer scope for applications in food technology. One special advantage is that, unlike some conventional preservatives, and as honey is an entirely natural product, it has a generally beneficial effect on human health. The following discussion describes most of the food industry applications in which honey is included.

Honey can be used for a variety of purposes, including mainly food, food ingredients, and as an ingredient in medicine-like products.

## Honey as Food

In the long human history, honey is a nutrient as well as a medicinal product. Today, with the increasing appreciation of more natural products in many countries, honey has been "rediscovered" as a valuable food. In addition, the nutritional and health-enhancing properties of honey are quite a new field of research.

Honey, a high-energy carbohydrate food, is one of the last untreated natural foods; most of the honey sold for food is used directly as table sweetener or spread. The amount of honey used for food far outweighs any of the miscellaneous nonfood uses that have been described in the technical and popular literature. For a long time in human history, it was the only known sweetener, until industrial sugar production began to replace it after 1800. Today, a large variety of packaging and semiprocessed and pure honey products are marketed.

## Honey as a Food Ingredient

Honey as a food ingredient deserves serious consideration, with its combination of interesting physical properties, fine flavor, and connotation of old-fashioned goodness (White). Honey is largely used on a small scale as well as at an industrial level: in bakery, confectionery, breakfast cereals, dairy, dressings and sauces, frozen foods, meats, snack bars (candy bars), spreads, ice creams, industrial nonalcoholic beverages, marmalades and jams, and many preserved products (Figure 18.3). In particular, the relatively new industry of "natural", health and biological products

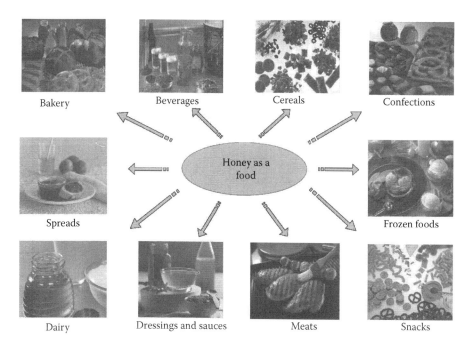

FIGURE 18.3 Main applications of honey as food ingredients.

uses honey abundantly as the sweetener of first choice, together with non-refined sugars in substitution of refined sucrose (cane and beet sugar).

An overview of the different applications of honey in food industry is given (Figure 18.3). The entire mentioned application showcases were commissioned by the American National Honey Board (www.honey.com).

The baking industry undoubtedly is the largest user of honey. Studies on the role of honey in commercial baking have shown that definite advantages in flavor, keeping quality, texture, and eating quality arise from judicious use of honey in many types of baked goods, including breads, yeast-raised sweet goods, cakes, fruitcakes, cookies, and pies of several types. Honey has long been used in confections. For the production of caramels, honey is only used in small quantities since its hygroscopicity presents a major disadvantage; it reduces the preservation time and softens the caramels at the surface, causing them to stick together. Several breakfast cereal products use honey (liquid, dried, or pulverized form) in their formulas for better flavor and increased consumer appeal. It can be mixed with cereal flakes and dried fruits or applied as a component in the sweetening and flavoring film, which covers the flakes. The dryness or hardness of the cereal can be adjusted with the honey content and the degree of drying. Candy bars often use honey as a binding and sweetening agent. The bar ingredients are chopped to various sizes and mixed with the hot honey and sugar. Depending on the composition and the degree of heating of the sugars (including the honey), a more or less solid product is obtained after cooling. In any case, all such products are fairly hygroscopic and need to be packed with material impermeable to moisture. Industrial nonalcoholic beverage industries use honey due to the wider distribution of "functional" drinks, such as health-oriented strengthening and replenishing isotonic drinks. Iced tea can also be flavored and clarified with the addition of honey. These beverages use a special ultrafiltration process to eliminate impurities. Ultrafiltered honey loses some of its flavor and color but is highly appreciated by food processors because it provides a more consistent product with lower production costs. Cake mixes, breads, and drink or energy health powders use dried or dehydrated honey.

The antibacterial effect of honey counteracts microbial spoilage of food (e.g., meat), while honey enhances the growth of dairy starter cultures in milk and milk products, especially species with weak growth rates in milk, such as bifidobacteria (Bf-1 and Bf-6). In addition, honey can be used as a prebiotic additive to probiotic milk products. Due to its antioxidant properties, honey prevents oxidation of food during storage and thus acts against lipid oxidation of meat (Nagai et al. 2006). The addition of honey to patties seems to prevent formation of heterocyclic aromatic amine and overall mutagenicity in fried ground-beef patties (Shin et al. 2003). Effects of honey against enzymatic browning of fruits and vegetables and soft drinks have been reported (Chen et al. 2000). Other physical and sensory properties make honey a good candidate for an additive to a wide variety of food: breads, cakes, spread yeast-raised sweet goods, cookies, and other foods stated above. All are improved by honey.

"Natural," health, and biological products use honey abundantly as a sweetener of first choice together with nonrefined sugars substituting for refined sucrose. In fact, honey can substitute for all or parts of the normal sugar in most products.

Limitations are presented on one side by costs and handling characteristics and on the other by the natural variations in honey characteristics, which change the end product, making it more variable and requiring more frequent adjustments in the industrial formulations.

Elsewhere, the U.S. National Honey Board provides useful information on the roles of honey characteristics on its application as a food additive (Table 18.4) in promotion and marketing to small and large industrial users of honey.

## HONEY AS AN INGREDIENT IN MEDICINE-LIKE PRODUCTS

The use of honey in medicine is a subject reported intermittently for the past 4000 years; its use as a therapeutic natural product has been reevaluated in a more scientific setting (Nasuti et al. 2006). The largest nonfood use of honey is in pharmaceuticals. Thus, several pharmaceutical preparations have honey as a useful adjunct. It has long been recommended in infant formulas. Luttinger (1922) highly recommended the use of honey in infant feeding because it does not produce acidosis, its rapid absorption prevents it from undergoing alcoholic fermentation, its free acids favor the absorption of fats, it complements the iron deficiency in human and cow's milk, it increases appetite and peristalsis, and it has a soothing effect that reduces fretfulness. These attributes of honey have been verified by subsequent investigators in tests using honey in feeding children of various ages and found special values for honey compared with other sugars. Included in the observed benefits were an increase in the hemoglobin content of the blood, better skin color, relief from constipation, a decrease in diarrhea and vomiting (Takuma 1955; Tropp 1957), more rapid increase in blood sugars than after sucrose administration (Müller 1956), better weight gains when honey was substituted for dextromaltose after faulty nutrition, and good honey tolerance with infants suffering from rickets, inflammation of the intestine, malnourishment, and prematurity. With this amount of definite evidence in the case of infants and children, there seems to be plenty of reasons for including honey, not only in the diets of infants and children, but also in the diets of adults as well, and particularly those who are undergoing vigorous exercise under exacting conditions (Kreider et al. 2002).

It is not hazardous when honey is used as an ingredient in various medicine-like products. At present, it is admitted that honey is a potential therapeutic agent and is efficacious against several diseases. The abundance of systematic scientific study and the proliferation of publications are now being offered on the diversity of its therapeutic properties. Both antibacterial and anti-inflammatory properties are useful in the stimulation of wound and burn healing (Molan 2001) and as a possible treatment of gastric ulcers and gastritis (Salem 1981).

## HONEY PRODUCTS MARKETED IN THE WORLD

Data about worldwide honey production are published every year by the Food and Agriculture Organization (FAO). At present, the annual world honey production is about 1.2 million tons, which is less than 1% of the total sugar production.

**TABLE 18.4**

### Bakery

| Honey's Characteristics | Functions |
| --- | --- |
| Antimicrobial properties | Delays spoilage |
| Color | Coloring agent |
| Flavor | Flavoring agent |
| Humectancy | Adds moisture |
| Miscibility | Water soluble |
| Nutrition | Healthy appeal |
| Preservation | Slows staling |
| Pumpable | Extrudable |
| Reducing sugars | Enhances browning |
| Spreadability | Improves reduced-fat products |
| Water activity | Extends shelf life |

### Beverages

| Honey's Characteristics | Functions |
| --- | --- |
| Carbohydrate composition | Flavor enhancement |
| Flavor | Flavoring agent |
| Hygroscopic | Retains moisture |
| Lower freezing point | Freezing point depression |
| Low glycemic index | Reduces rebound hypoglycemia |
| Nutrition | Healthy appeal |
| pH balance | Inhibits bacterial growth |
| Proteins | Clarification |
| Viscosity | Binding agent |

### Dressing and Sauces

| Honey's Characteristics | Functions |
| --- | --- |
| Antimicrobial properties | Delays spoilage |
| Carbohydrate composition | Flavor enhancement |
| Composition | Decrease burn perception |
| Flavor | Flavoring agent |
| Humectancy | Adds moisture |
| Hygroscopic | Retains moisture |
| Miscibility | Water |
| Pumpable | Extrudable |
| Reducing sugars | Enhances browning |
| Viscosity | Binding agent |
| Water activity | Extends shelf life |

### Cereals

| Honey's Characteristics | Functions |
| --- | --- |
| Flavor | Flavoring agent |
| Nutrition | Healthy appeal |
| Reducing sugars | Enhances browning |
| Spreadability | Improves reduced-fat products |
| Viscosity | Binding agent |
| Water activity | Extends shelf life |

### Confections

| Honey's Characteristics | Functions |
| --- | --- |
| Crystallization | Texture |
| Flavor | Flavoring agent |
| Nutrition | Healthy appeal |
| Pumpable | Extrudable |
| Spreadability | Improves reduced-fat products |

### Dairy

| Honey's Characteristics | Functions |
| --- | --- |
| Antimicrobial properties | Delays spoilage |
| Color | Coloring agent |
| Flavor | Flavoring agent |
| Lower freezing point | Freezing point depression |
| Miscibility | Water soluble |
| Nutrition | Healthy appeal |
| pH balance | Inhibits bacterial growth |
| Probiotic | Enhance bifidobacteria |

| Meats | | Snacks | | Spreads | |
|---|---|---|---|---|---|
| Honey's Characteristics | Functions | Honey's Characteristics | Functions | Honey's Characteristics | Functions |
| Antimicrobial properties | Delays spoilage | Antimicrobial properties | Delays spoilage | Antimicrobial properties | Delays spoilage |
| Carbohydrate composition | Flavor enhancement | Carbohydrate composition | Flavor enhancement | Carbohydrate composition | Flavor enhancement |
| Color | Coloring agent | Composition | Decreases burn perception | Crystallization | Texture |
| Composition | Decreases burn perception | Flavor | Flavoring agent | Flavor | Flavoring agent |
| Flavor | Flavoring agent | Low glycemic index | Reduces rebound hypoglycemia | Humectancy | Adds moisture |
| Maillard reaction precursor | Antioxidation | Nutrition | Healthy appeal | Miscibility | Water soluble |
| Reducing sugars | Enhances browning | Reducing sugars | Enhances browning | Nutrition | Healthy appeal |
| Spreadability | Improves reduced-fat products | Viscosity | Binding agent | Pumpable | Extrudable |
| Viscosity | Binding agent | | | Spreadability | Improves reduced-fat products |

*Note:* Roles of honey characteristics on its application as a food additive (interpolated from the National Honey Board [http://www.nhb.org]).

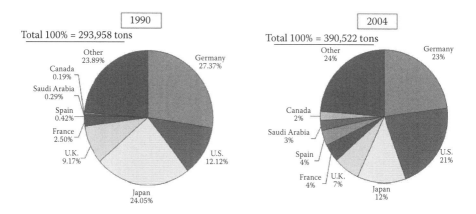

**FIGURE 18.4**  (**See color insert.**) Change of share of honey importation from 1993 to 2004. (Courtesy of Food and Agriculture Organization of the United Nations, www.fao.org.)

## Honey Importation and Consumption in the World

World honey imports are mainly concentrated in Europe; North America and Asia are in the next place (Figure 18.4). It is stated that Germany and the United States were the biggest importing countries in 2004 with more than 20% market share. However, the import growth rate for Germany was lower than the world average. U.S. growth was much higher, as was the growth of emerging potential countries such as Saudi Arabia, Spain, and Canada.

Honey consumption is higher in developed (industrialized) countries, where domestic production does not always meet the market demand; only 25% to 28% of production was exported between 1990 and 2004.

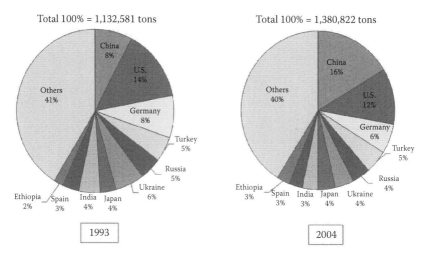

**FIGURE 18.5**  (**See color insert.**) Change of share of honey consumption from 1993 to 2004. (Courtesy of Food and Agriculture Organization of the United Nations, www.fao.org.)

China has become the world's biggest honey consumer, significantly increasing its share of the global market from 8% in 1993 to 16% in 2004 (Figure 18.5). Chinese honey consumption was nearly two times higher in 2004 than in 1990, and its additional consumption was slightly above its additional production (109,000 tons consumption compared with 103,000 tons) partly because of its economic growth leading to honey becoming an affordable product. The same situation appeared in Russia, Spain, and Ethiopia. Even in the United States, production decreased while its consumption increased. In the European Union, which is both a major honey importer and producer, the annual consumption per capita varies from medium (0.3–0.4 kg) in Italy, France, Great Britain, Denmark, and Portugal to high (1–1.8 kg) in Germany, Austria, Switzerland, Portugal, Hungary, and Greece, while in countries such as the United States, Canada, and Australia, the average per capita consumption is 0.6 to 0.8 kg/year (Bogdanov et al. 2008).

## HONEY PRODUCTION AND EXPORTATION IN THE WORLD

The major producers are China, United States, Argentina, Turkey, Mexico, Spain, and Canada. World honey production increased 1% compounded annual growth rate (CAGR) from 1990 to 2004 (Figure 18.6). However, production of honey decreased and fluctuated after peaking at 1991 until 1996 when production increased regularly.

China is the biggest honey-producing country and dominates honey production in the world, accounting for 22% of production and growing nearly three times faster than the world average, while colony collapse disorder and a decrease in beekeeper numbers are the main reasons for the reduction in growth in the United States and Germany.

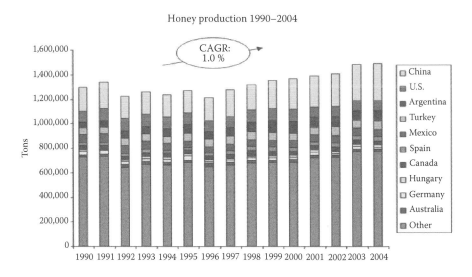

**FIGURE 18.6** World honey production increased at a CAGR of 1% from 1990 to 2004. (Courtesy of Food and Agriculture Organization of the United Nations, www.fao.org.)

Honey exports 1990–2004

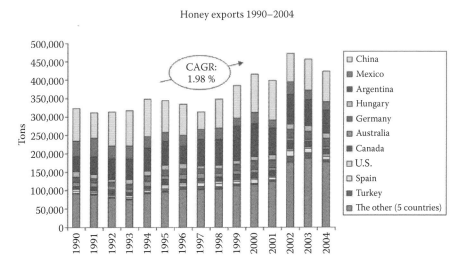

**FIGURE 18.7** World honey exports increased nearly 2% per annum from 1990 to 2004. (Courtesy of Food and Agriculture Organization of the United Nations, www.fao.org.)

The five biggest exporters in the world are China, Argentina, Mexico, Germany, and Brazil and accounted for more than 65% of world honey exports in 2004 (Figure 18.7). However, the highest colony yields are recorded in Australia and Canada, which have a favorable environment as well as highly developed colony management.

Although the market for honey in China has fluctuated in recent years, the country is currently by far the leading honey-producing nation in the world, with approximately 40% slice of the market. The next biggest producers are the United States and Argentina. According to the American Honey Producers Association, China and Argentina have been adversely affecting America's domestic honey industry with cheap imports, although there is a counterargument that both China and Argentina have been helping to counterbalance the falling production in the United States.

China imports of honey have grown steadily since 1995. By 1999, total honey imports reached 1400 net tons, a growth of 775% from 1995. In part, this growth in imports has been due to the rise in demand for specific types of honey from the food manufacturing industry. Although domestic demand promises to be strong, the outlook for China's honey exports does not look so promising. According to the China National Chamber of Food, Native Produce and Animal By-products Importers and Exporters, exports of Chinese honey will decrease in the long term due to rising global competitiveness.

## REFERENCES

Alissandrakis, E., Tarantilis, P.A., Harizanis, P.C., and Polissiou, M. 2007. Aroma investigation of unifloral Greek citrus honey using solid-phase microextraction coupled to gas chromatographic–mass spectrometric analysis. *Food Chemistry* 100: 396–404.

Alvarez-Suarez, J.M., Tulipani, S., Romandini, S., Bertoli, E., and Battino, M. 2010. Contribution of honey in nutrition and human health: A review. *Mediterranean Journal of Nutrition and Metabolism* 3: 15–23.

Anklam, E. 1998. A review of the analytical methods to determine the geographical and botanical origin of honey. *Food Chemistry* 63: 549–562.

Anklam, E. and Radovic, B. 2001. Suitable analytical methods for determining the origin of European honey. *International Scientific Communications* 33: 60–64.

Arvanitoyannis, I.S., Chalhoub, C., Gotsiou, P., Lydakis-Simantiris, N., and Kefalas, P. 2005. Novel quality control methods in conjunction with chemometrics (Multivariate analysis) for detecting honey authenticity. *Critical Reviews in Food Science and Nutrition* 45: 193–203.

Barth, O.M. 2004. Melissopalynology in Brazil: A review of pollen analysis of honeys, propolis and pollen loads of bees. *Scientia Agricola* 61: 342–350.

Bauer, L., Kohlich, A., Hirschwehr, R., Siemann, U., Ebner, H., Scheiner, O., Kraft, D., and Ebner, C. 1996. Food allergy to honey: Pollen or bee products? Characterization of allergenic proteins in honey by means of immunoblotting. *Journal of Allergy and Clinical Immunology* 97: 65–73.

Bertelli, D., Lolli, M., Papotti, G., Bortolotti, L., Serra, G., and Plessi, M. 2010. Detection of honey adulteration by sugar syrups using one-dimensional and two-dimensional high-resolution nuclear magnetic resonance. *Journal of Agricultural and Food Chemistry* 58: 8495–8501.

Bogdanov, S. 1993. Liquefaction of honey. *Apiacta* XXVIII, 4–10.

Bogdanov, S. and 21 other members of the International Commission. 1999. Honey quality methods of analysis and international regulatory standards: Review of the work of the International Honey Commission. *Mitteilungenaus Lebensmitteluntersuchung und Hygiene* 90: 108–125.

Bogdanov, S. and Martin, P. 2002. Honey authenticity. Mitteilungen aus dem Gebiete der. *Lebensmitteluntersuchung und Hygiene* 93: 232–254.

Bogdanov, S., Ruoff, K., and Oddo, P.L. 2004. Physico-chemical methods for the characterisation of unifloral honeys: A review. *Apidologie* 35: S4–S17.

Bogdanov, S. 2007. Authenticity of honey and other bee products: State of the art. *Bulletin USAMV-CN* 63–64.

Bogdanov, S., Jurendic, T., Sieber, R., and Gallmann, P. 2008. Honey for nutrition and health: A review. *American Journal of the College of Nutrition* 27: 677–689.

Buldini, P.L., Cavalli, S., Mevoli, A., and Sharma, J.L. 2001. Ion chromatographic and voltammetric determination of heavy and transition metals in honey. *Food Chemistry* 73: 487–495.

Cano, C.B., Felsner, M.L., Matos, J.R., Bruns, R.E., Whatanabe, H.M., and Almeida-Muradian, L.B. 2001. Comparison of methods for determining moisture content of citrus and eucalyptus Brazilian honeys by refractometry. *Journal of Food Composition and Analysis* 14: 101–109.

Careri, M., Mangia, A., Barbieri, G., Bouoni, L., Virgili, R., and Parolari, G. 1993. Sensory property relationships to chemical data of Italian-type dry-cured ham. *Journal of Food Science* 58: 968–972.

Celli, G. and Maccagnani, B. 2003. Honey bees as bioindicators of environmental pollution. *Bulletin of Insectology* 56: 137–139.

Cervantes, M.A.R., Novelo, S.A.G., and Duch, E.S. 2000. Effect of the temporary thermic treatment of honey on variation of the quality of the same during storage. *Apiacta* 35: 162–170.

Chen, L., Mehta, A., Berenbaum, M., Zangerl, A.R., and Engeseth, N.J. 2000. Honeys from different floral sources as inhibitors of enzymatic browning in fruit and vegetable homogenates. *Journal of Agricultural and Food Chemistry* 48: 4997–5000.

*Codex Alimentarius* Comission. Revised codex standard for honey, Codex STAN 12-1981, Rev. 1 (1987), Rev. 2 (2001).

Conte, L.S., Mirioni, M., Giomo, A., Bertacco, G., and Zironi, R. 1998. Evaluation of some fixed components for unifloral honey. *Journal of Agricultural and Food Chemistry* 46: 1844–1849.

Corbella, E. and Cozzolino, D. 2008. Combining multivariate analysis and pollen count to classify honey samples accordingly to different botanical origins. *Chilean Journal of Agricultural Research* 68: 102–107.

Cordella, C., Moussa, I., Martel, A.-C., Sbirrazzuoli, N., and Lizzani-Cuvelier, L. 2002. Recent developments in food characterization and adulteration detection: Techniques oriented perspectives. *Journal of Agricultural and Food Chemistry* 50: 1751–1764.

Cordella, C.B.Y., Militao, J.S.L.T., Clement, M.-C., and Cabrol-Bass, D. 2003. Honey characterization and adulteration detection by pattern recognition applied on HPAEC-PAD profiles. 1. Honey floral species characterization. *Journal of Agricultural and Food Chemistry* 5: 3234–3242.

Cordella, C., Militao, J.S.L.T., Clement, M.-C., Drajnudel, P., and Cabrol-Bass, D. 2005. Detection and quantification of honey adulteration via direct incorporation of sugar syrups or bee-feeding: Preliminary study using high-performance anion exchange chromatography with pulsed amperometric detection (HPAEC-PAD) and chemometrics. *Analytica Chimica Acta* 531: 239–248.

Costa, L.S.M., Albuquerque, M.L., Trugo, L.C., Quinteiro, L.M.C., Barth, O.M., Ribeiro, M., and De Maria, C.A.B. 1999. Determination of non-volatile compounds of different botanical origin Brazilian honeys. *Food Chemistry* 65: 347–352.

Cotte, J.F., Casabianca, H., Chardon, S., Lheritier, J., and Grenier-Loustalot, M.F. 2003. Application of carbohydrate analysis to verify honey authenticity. *Journal of Chromatography A* 1021: 145–155.

Cotte, J.F., Casabianca, H., Giroud, B., Albert, M., Lheritier, J., and Grenier-Loustalot, M.F. 2004. Characterization of honey amino acid profiles using high-pressure liquid chromatography to control authenticity. *Analytical and Bioanalytical Chemistry* 378: 1342–1350.

Crane, E. 1984. Bees, honey and pollen as indicators of metals in the environment. *Bee World* 55: 47–49.

Cuevas-Glory, L.F., Pino, J.A., Santiago, L.S., and Sauri-Duch, E. 2007. A review of volatile analytical methods for determining the botanical origin of honey. *Food Chemistry* 103: 1032–1043.

Devillers, J., Morlot, M., Pham-Delègue, M.H., and Doré, J.C. 2004. Classification of monofloral honeys based on their quality control data. *Food Chemistry* 86: 305–312.

Dustmann, J.H. 1993. Honey quality and its control. *American Bee Journal* 133: 648–651.

Food and Agriculture Organization of the United Nations. Available at www.fao.org.

Jaganathan, S.K. and Mandal, M. 2009. Antiproliferative effects of honey and of its polyphenols: A review. *Journal of Biomedicine and Biotechnology* 2009: 1–13.

Jones, K.C. 1986. Honey as an indicator of heavy metal contamination. *Water, Air, and Soil Pollution* 33: 179–189.

Joshi, S.R., Pechhacker, H., Willam, A., and Von der Ohe, W. 2000. Physico-chemical characteristics of *Apis dorsata, A. cerana* and *A. mellifera* honey from Chitwan district, central Nepal. *Apidologie* 31: 367–375.

Kelly, J.D., Petisco, C., and Downey, G. 2006. Potential of near infrared transflectance spectroscopy to detect adulteration of Irish honey by beet invert syrup and high fructose corn syrup. *Journal of Near Infrared Spectroscopy* 14: 139–146.

Kerkvliet, J.D., Shrestha, M., Tuladhar, K., and Manandhar, H. 1995. Microscopic detection of adulteration of honey with cane sugar and cane sugar products. *Apidologie* 26: 131–139.

Kerkvliet, J.D. and Meijer, H.A.J. 2000. Adulteration of honey: Relation between microscopic analysis and δ13C measurements. *Apidologie* 31: 717–726.

Khalil, M.I., Sulaiman, S.A., and Gan, S.H. 2010. High 5-hydroxymethylfurfural concentrations are found in Malaysian honey samples stored for more than one year. *Food and Chemical Toxicology* 48: 2388–2392.

Kreider, R.B., Rasmussen, C.J., Lancaster, S.L., Kerksick, C., and Greenwood, M. 2002. Honey: An alternative sports gel. *National Strength and Conditioning Association* 24: 50–51.

LaGrange, V. and Sanders, S.W. 1988. Honey in cereal-based new food products. *American Association of Cereal Chemists* 33: 833–838.

Lee, D.C., Lee, S.Y., Cha, S.H., Choi, Y.S., and Rhee, H.I. 1998. Discrimination of native bee-honey and foreign bee-honey by SDS–PAGE. *Korean Journal of Food Science* 30: 1–5.

Louveaux, J., Maurizio, A., and Vorwohl, G. 1978. Methods of melissopalynology. *Bee World* 59: 139–157.

Low, N.I.H. and Sporns, P.E. 1988. Analysis and quantitation of minor di- and trisaccharides in honey, using capillary gas chromatography. *Journal of Food Science* 53: 558–561.

Luttinger, P. 1922. Bees' honey. *New York Medicine Journal* 116: 153.

Marshall, T. and Williams, K.M. 1987. Electrophoresis of honey: Characterization of trace proteins from a complex biological matrix by silver staining. *Analytical Biochemistry* 167: 301–303.

Mateo, R. and Bosch-Reig, F. 1998. Classification of Spanish unifloral honeys by discriminant analysis of electrical conductivity, color, water content, sugars, and pH. *Journal of Agricultural and Food Chemistry* 46: 393–400.

Molan, P.C. 2001. Potential of honey in the treatment of wounds and burns. *American Journal of Clinical Dermatology* 2: 9–13.

Molnár-Perl, I. and Horváth, K. 1997. Simultaneous quantitation of mono-, di- and trisaccharides as their TMS ether oxime derivatives by GC-MS: I. In model solutions. *Chromatographia* 45: 321–327.

Müller, L. 1956. Der Bienenhonig in der Säuglingsernährung bei Berücksichtigung einer neuen Fertignahrung. *Medizinische Monatsschrift* 10: 729–732.

Nagai, T., Inoue, R., Kanamori, N., Suzuki, N., and Nagashima, T. 2006. Characterization of honey from different floral sources. Its functional properties and effects of honey species on storage of meat. *Food Chemistry* 97: 256–262.

Nasuti, C., Gabbianelli, R., Falcioni, G., and Cantalamessa, F. 2006. Antioxidative and gastro-protective activities of anti-inflammatory formulations derived from chestnut honey in rats. *Nutrition Research* 26: 130–137.

National Honey Board (NHB). Available at http://www.nhb. org/foodtech/index.html.

Oddo, L.P., Piazza, M.G., Sabatini, A.G., and Accorti, M. 1995. Characterization of unifloral honeys. *Apidologie* 26: 453–465.

Oddo, L.P., Piazaa, M.G., and Pulcini, P. 1999. Invertase activity in honey. *Apidologie* 30: 57–65.

Oddo, L.P. and Piro, R. 2004. Main European unifloral honeys: Descriptive sheet 1. *Apidologie* 35: S38–S81.

Olaitan, P.B., Adeleke, O.E., and Ola, I.O. 2007. Honey: A reservoir for microorganisms and an inhibitory agent for microbes. *African Health Sciences* 7: 159–165.

Piasenzotto, L., Gracco, L., and Conte, L. 2003. Solid phase microextraction (SPME) applied to honey quality control. *Journal of the Science of Food and Agriculture* 83: 1037–1044.

Pirini, A., Conte, L.S., Francioso, O., and Lercker, G. 1992. Capillary gas chromatographic determination of free amino acids in honey as a means discrimination between different botanical sources. *Journal of High Resolution Chromatography* 15: 165–170.

Porrini, C., Sabatini, A.G., Girotti, S., Ghini, S., Medrzycki, P., Grillenzoni, F., Bortolotti, L., Gattavecchia, E., and Celli, G. 2003. Honey bees and bee products as monitors of the environmental contamination. *APIACTA* 28: 63–70.

Prodolliet, J. and Hischenhuber, C. 1998. Food authentication by carbohydrate chromatography. *Zeitschrift für Lebensmitteluntersuchung und Forschung A* 207: 1–12.

Reid, L.M., O'Donnell, C.P., and Downey, G. 2006. Recent technological advances for the determination of food authenticity. *Trends in Food Science & Technology* 17: 344–353.

Ruoff, K., Iglesias, M.T., Luginbuhl, W., Bosset, J.-O., Bogdanov, S., and Amado, R. 2006. Quantitative analysis of physical and chemical measurands in honey by mid-infrared spectrometry. *European Food Research and Technology* 223: 22–29.

Saif-ur-Rehman, Khan, Z.F., and Maqbool, T. 2008. Physical and spectroscopic characterization of Pakistani honey. *Ciencia Investigacion Agraria* 35: 199–204.

Salem, S.N. 1981. Treatment of gastroenteritis by the use of honey. *Islam Medicine* 1: 358–362.

Sancho, M.T., Muniategui, S., Huidobro, J.F., and Lozano, J.S. 1992. Aging of honey. *Journal of Agricultural and Food Chemistry* 40: 134–138.

Sanz, M.L., Gonzalez, M., de Lorenzo, C., Sanz, J., and Martinez-Castro, I. 2005. A contribution to the differentiation between nectar honey and honeydew honey. *Food Chemistry* 91: 313–317.

Schade, J.E., Marsh, G.L., and Eckert, J.E. 1958. Diastase activity and hydroxy-methyl-furfural in honey and their usefulness in detecting heat alteration. *Journal of Food Science* 23: 446–463.

Serrano, S., Villarejo, M., Espejo, R., and Jodral, M. 2004. Chemical and physical parameters of Andalusian honey: Classification of Citrus and Eucalyptus honeys by discriminant analysis. *Food Chemistry* 87: 619–625.

Shin, H.S., Strasburg, G.M., and Ustunol, Z. 2003. Influence of different unifloral honeys on heterocyclic aromatic amine formation and overall mutagenicity in fried ground-beef patties. *Journal of Food Science* 68: 810–815.

Singh, N. and Bath, P.K. 1997. Quality evaluation of different types of Indian honey. *Food Chemistry* 58: 2129–2133.

Swallow, K.W., and Low, N.H. 1990. Analysis and quantification of carbohydrates in honey using high performance liquid chromatography. *Journal of Agricultural and Food Chemistry* 38: 1828–1832.

Swallow, K.W., and Low, N.H. 1994. Determination of honey authenticity by anion-exchange liquid chromatography. *Journal of Association of Official Analytical Chemists International* 77: 695–702.

Takuma, D.T. 1955. Honig bei der Aufzucht von Säuglingen. *Monatsschrift für Kinderheilkunde* 103: 160–161.

Tropp, C. 1957. Der Honig und seine Bedeutung in der Säuglings und Kinderernährung. *Der Landarzt* 33: 250–252.

Tong, S.C., Morse, R.A., Bache, C.A., and Lisk, D.J. 1975. Elemental analysis of honey as an indicator of pollution. Forty-seven elements in honeys produced near highway, industrial, and mining areas. *Archives of Environmental Health* 30: 329–332.

Tosi, E., Martinet, R., Ortega, M., Lucero, H., and Re, E. 2008. Honey diastase activity modified by heating. *Food Chemistry* 106: 883–887.

Viuda-Martos, M., Ruiz-Navajas, Y., Fernandez-Lopez, J., and Perez-Alvarez, J.A. 2008. Functional properties of honey, propolis, and royal jelly. *Journal of Food Science* 73: R117–R124.

Von der Ohe, W., Dustmann, J.H., and Von der Ohe, K. 1991. Prolin als kriterium der reife des honigs. *Deutsche Lebensmittel-Rundschau* 87: 383–386.

Von der Ohe, W., Oddo, L.P., Piana, M.L., Morlot, M., and Martin, P. 2004. Harmonized methods of melissopalynology. *Apidologie* 35: S18–S25.

Vorwohl, G. 1964. Die Beziehungen zwischen der elektrischen Leitfähigkeit der Honige und ihrer trachtmässigen Herkunft. *Annales de l'Abeille* 7: 301–309.

White, J.W., Kushnir, I., and Subers, M.H. 1964. Effect of Storage and processing temperatures on honey quality. *Food Technology* 18: 153–156.

White, J.W. 1979. Methods for determinging carbohydrates, hydroxymethylfurfural, and proline in honey: Collaborative study. *Journal of Association of Official Analytical Chemists* 62: 515–526.

White, J.W. and Winters, K. 1989. Honey protein as internal standard for stable carbon isotope ratio detection of adulteration of honey. *Journal of Association of Official Analytical Chemists* 72: 907–911.

White, J.W. 1992a. Quality evaluation of honey: Role of HMF and diastase. *American Bee Journal* 737–743.

White, J.W. 1992b. Internal standard stable carbon isotope ratio method for determination of C4-plants sugars in honey: Collaborative study, and evaluation of improved protein preparation procedure. *Journal of Association of Official Analytical Chemists International* 75: 543–548.

White, J.W. 1994. The role of HMF and diastase assays in honey quality evaluation. *Bee World* 75: 104–117.

White, J.W., Winters, K., Martin, P., and Rossmann, A. 1998. Stable carbon isotope ratio analysis of honey: Validation of internal standard procedure for worldwide application. *Journal of Association of Official Analytical Chemists International* 81: 610–619.

Won, S.-R., Lee, D.-C., Ko, S.H., Kim, J.-W., and Rhee, H.-I. 2008. Honey major protein characterization and its application to adulteration detection. *Food Research International* 41: 952–956.

# 19 Culinary Uses of Honey

*Oktay Yildiz, Hüseyin Sahin, and Sevgi Kolayli*

## CONTENTS

## INTRODUCTION

Honey is a sweet, viscous, aromatic, and valuable food made by bees (the genus *Apis*) using nectar from flowers. Sugar is the main constituent of honey and is comprised of about 95% of honey dry weight (Bogdanov et al. 2004). Due to the high sugar content, honey and fruit molasses (pekmez) had been used as sweeteners until the production of industrial sugar began to replace them after 1800 (Crane 1975). In the long human history, honey has not only been used as a nutrient but also as an apiterapic product for the treatment of various kinds of disease, such as ulcer, sore throat, asthma, and chronic wounds (Bogdanov 2011; Jones 2001).

Honey is consumed in its unprocessed state (i.e., liquid [filtered], crystallized, or in the comb [Krell 1996]) or used for food preparation and culinary as a sweetener, seasoning, and flavoring. Some products as fruit juices, sausages, marmalades, jams, baked products, cakes, confectionaries, gums, candy, spreads, breakfast cereals, beverages, milk and milk products, fruit leathers, energy drinks, energy bars, chocolates, alcoholic beverages, and many traditional foods can be enhanced by honey in the culinary, small-, medium-, and large-scale food industries. In addition, it is also utilized for protective properties, which are antioxidant and antimicrobial in some foods as fruits, fruit juices, vegetables, meat, some meals, and coffee beans (Crane 1990; Nahmias 1981; Chen et al. 2000; Mckibben and Engeseth 2002; Nagai et al. 2006; Bogdanov 2011).

## SOME FOOD APPLICATIONS WITH HONEY

### Fruit/Vegetable Juices, Nectars, and Still Drinks with Honey

Juice is a popular beverage made from fruit or vegetables. It is naturally contained in fruit or vegetable tissue. Juice is prepared by mechanically squeezing or macerating fruit or vegetable flesh without the solvents in factories or kitchens. Fruit juices have the essential physical, chemical, organoleptical, and nutritional characteristics of the fruit(s) from which the juice was extracted and honey (Codex Stan 247-2005). Properly obtained juices are quite similar to the source fruit and vegetable; they contain most substances that are found in the original ripe and sound fruit/vegetable from which the juice was extracted.

Juices can be made by adding sugar (glucose, fructose, sucrose, etc.) or by not adding sugar. Especially, honey is a very good sweetener used for this purpose only, except grape and pear juices (Turkish Food Codex 2002).

Fruit or vegetable nectars have 25% to 99% juice content and added sugar. Still drinks contain 0% to 24% juice content in fruit, vegetable, or other flavors. All these ingredients can be derivatized with sugar or artificial sweetener (Turkish Food Codex 2002; Food and Drug Administration 2003; Fruit Juices and Fruit Nectars [England] Regulations 2003; Fruit Juices and Fruit Nectars [Scotland] Regulations 2003).

Honey can be used instead of all or part of the sugars used in prescription. A fruit juice is made from the whole fruit (edible parts) and it does not contain more sugar than the corresponding fruit. The use of honey in juices has been increasing as a result of the need to turn to nature.

Different brands and kinds of juices with honey are found in research (Lee and Kime 1984; Wakayama and Lee 1987; Krell 1996; Radojčić et al. 2008; de la Rosa et al. 2011). The usage of honey in juices is more common on a small scale as culinary rather than in an industrial scale. In culinary, many kinds of juices with honey are produced and consumed without being stored. Also, cocktails with honey are non-alcoholic versions of drinks produced in the kitchen. Prescriptions may vary according to consumer preferences and the cook's skills.

### Alcoholic Beverages with Honey

Alcoholic beverages are drinks that contain ethyl alcohol, commonly known as alcohol. General alcoholic beverage classes are beers, wines, spirits, and liquor. Many countries have laws for regulating their production, sale, and consumption (International Center for Alcohol Policies 2011).

Honey is usually used as a source of sugar and/or flavoring in these drinks and is added to distilled alcoholic beverages after distillation, for example, Benedictine in France, Drambuie in Scotland, Irish Mist in Ireland, Grappa al Miele in Italy, Krupnik in Poland, Barenfang in Germany, and many others (Krell 1996). Besides, mead and honey beer are commonly produced in some countries by fermentation. In traditional and culinary production, using filtered liquid honey and occasionally adding sugars, aromatic plants such as clove are used and naturally fermented. After fermentation, they are rested and filtered.

## Honey Vinegar

Honey vinegar is sour liquid obtained by the acetic acid fermentation of ethanol that is produced in honey syrup, usually without any additional flavoring. The main ingredient of vinegar is acetic acid, which is responsible for its acidic taste.

Honey vinegar is made by two consecutive biological processes. The first one is called alcoholic fermentation and occurs when yeasts change natural sugars to alcohol under controlled conditions. The other one occurs with acetic acid fermentation, whereby a group of bacteria (called *Acetobacter*) converts the alcohol portion to acid.

Honey vinegar contains many vitamins, phenolic compounds, and other compounds such as riboflavin, thiamine, and mineral salts from the honey that imparts vinegar with its distinct flavor.

Honey vinegar is used for cooking and preparation of foods such as salad dressings (vinaigrettes) and pickles, and is also used for production of mayonnaise.

## Jams, Jellies, and Marmalades with Honey

In the preparation of jams, jellies, and marmalades, commercial or natural pectin is used as a gelling agent, although sugar or honey may be used as well. Jams contain both fruit juice and whole fruit, cut into pieces, or crushed, but marmalades contain fruit pulp/purée instead of fruit parts and are homogeneous, of viscous structure, and arable.

The EU Council Directive (2001) and Turkish Food Codex (2002), relating to jams, jellies, marmalades, and sweet chestnut purée intended for human consumption, specify both definitions and labels of jams and related products (Fügel et al. 2005). According to these directives, marmalade is a mixture brought to a suitable gelled consistency of sugars, the pulp, and/or purée of one or more kinds of fruit and water. The quantity of pulp and/or purée used for the manufacture of 1000 g of finished traditional marmalade products shall not be less than 450 g as a general rule. Traditional marmalades must have a soluble dry matter content of 55% or more determined by a refractometer (Yildiz and Alpaslan 2012).

Jelly is an appropriately gelled mixture of sugars and/or honey with the juice and/or aqueous extracts of fruits.

For jam production, the fruits are washed and heated with water and sugar/honey to activate the pectin that is in fruits. In marmalades, fruit pulp or purée is first obtained by washing, heating, breaking, and separating. Then, pulp or purée is processed to marmalade by a sugar and/or honey supplement up to 55 oBrix with 45% pulp (12 oBrix) and maximum 30% sugar content (within Turkish Food Codex 2002). The mixture is pasteurized and put into packages or consumed.

## Bakery Products with Honey

Bakery products are foods that are cooked by dry heat with some kind of oven. These products, which include bread, rolls, cookies, pies, pastries, and muffins, are usually prepared from flour or meal derived from some form of grain.

Fresh bakery products such as breads, cakes, and pies have a relatively short shelf life. Some microbiological and especially physical and chemical changes, which lead to the staling process, occur in these types of foods (Selomulyo and Zhou 2007; Minervini et al. 2011).

In bakery products, different sugars (glucose, fructose, sucrose, maltose, etc.), artificial sweeteners, and some concentrated fruit syrups such as pekmez (molasses) are used for sweetener and color changes, especially browning. Honey is used instead of all or part of these ingredients/additives in culinary and industry.

Nonenzymatic browning reactions, Maillard reactions, and caramelization are the main chemical reactions occurring in bakery products. Maillard reactions are (partly) responsible for the flavor of bread, cookies, cakes, and cooked rice. In many cases, the flavor is a combination of Maillard reactions and caramelization. However, caramelization only takes place above 120°C to 150°C, whereas Maillard reactions already occur at room temperature (Maillard 1912; Tyl and Crump 2003; Karagöz 2009).

Honey and other sweeteners are added in the mixture (desserts as pies, cakes, baklava, shortbread cookies, muffins, etc.) or sprayed on the surface (not sweet as bread, Turkish bagels, pastries, flatbread, etc.) of the products.

Honey can be successfully incorporated at levels as high as 12% to 15% in dry products such as breakfast cereals (Neumann and Chambers 1993), potato chips (Demetriades et al. 1995), and extruded snacks (National Honey Board 1995).

## CONFECTIONARIES WITH HONEY

Confectionary is foodstuff produced by cooking sugars (sucrose, glucose, fructose, etc.) and some food additives such as citric acid, tartaric acid, or potassium bitartarate in all types and milk, milk powder, gelatine, oil, nuts, aroma, etc., in different types. Some types of confectionary are hard candy, soft candy, jelly confectionery, candy-coated tablets, tablet sugars, candied fruits, almond paste, torrone, fund, krokan, and bars filled with confectionary products. In gelatinous or gum products, chocolate honey can be used. One Swiss chocolate, a few Turkish chocolates in particular, in which honey is included in the form of broken nougat, can be found worldwide. Some caramels made with special machinery have a liquid honey core (Krell 1996).

In culinary, honey is used in these products as flavoring and this product's name is sugar with honey. Finally, honey can be used as an ingredient in the production of confectionary instead of sugar.

## MILK AND MILK PRODUCTS WITH HONEY

There have been some different production areas such as milk and milk products since ancient times used against colds and infections of the throat by using honey. Honey can be mixed with milk, yoghurt, butter, ice cream, cream cheese, milk pudding, etc., namely, milk productions.

Honey enhances the growth of dairy starter cultures in milk and milk products. The growth rate of two bifidobacteria Bf-1 and Bf-6 in milk can be stimulated by the addition of honey to milk. Thus, honey can be used as a prebiotic additive to probiotic milk products (Bogdanov 2011; Ustunol and Gandhi 2001).

## TRADITIONAL FOODS WITH HONEY

Honey is still included in many traditional products that are consumed locally in considerable quantities, such as Turkish delight with honey (Batu 2006), torrone from Italy, nougat from France (Krell 1996), pestil and köme from Turkey (Yildiz et al. 2011), and halvah from Turkey and Greece. In culinary, and in these productions, honey can be used for different aims: flavor, sweetener, viscosity, softness, and color.

## OTHER USES OF HONEY

Honey is an important ingredient and additive in the preparation of some meals because of some of its features such as biological activity, sugar content, and viscosity. It is used in the production of meat dishes for the antibacterial and antioxidant effects of honey (Mckibben and Engeseth 2002; Nagai et al. 2006). Honey acts against enzymatic browning and is thus an efficient meal additive for fruit and vegetable meals (Chen et al. 2000; Bogdanov 2011). Honey can also be used in culinary as an additive to flour bagels, cereals, chicken marinades, French fries, breads, pasta, extruded snacks, corn chips, potato chips, salsas, and culinary sauces (Bogdanov 2011).

## REFERENCES

Batu, A. 2006. Türk Lokumu Üretim Tekniği Ve Kalitesi. *Gida Teknolojileri Elektronik Dergisi*, 2006(1): 35–46.

Bogdanov, S. 2011. *Honey as Nutrient and Functional Food: A Review*. Bee Product Science, available at www.bee-hexagon.net, July 9, 2011.

Bogdanov, S., Ruoff, K., and Persano Oddo, L. 2004. Physico-chemical methods for the characterisation of unifloral honeys: A review. *Apidologie*, 35: 4–17.

Chen, L., Mehta, A., Berenbaum, M., Zangerl, A.R., and Engeseth, N.J. 2000. Honeys from different floral sources as inhibitors of enzymatic browning in fruit and vegetable homogenates. *J Agric Food Chem*, 48(10): 4997–5000.

Codex general standard for fruit juices and nectars (CODEX STAN 247-2005).

Crane, E. 1990. *Bees and Beekeeping: Science, Practice and World Resources*. Cornstock Publ., Ithaca, NY, 593 pp.

Crane, E. 1975. History of honey. In Crane, E. (ed), *Honey, A Comprehensive Survey*. William Heinemann, London, pp. 439–488.

de la Rosa, L.A., Alvarez-Parrilla, E., Moyers-Montoya, E., Villegas-Ochoa, M., Ayala-Zavala, F., Hernández, J., Ruiz-Cruz, S., and González-Aguilar, G.A. 2011. Mechanism for the inhibition of apple juice enzymatic browning by palo fierro (desert ironweed) honey extract and other natural compounds. *LWT-Food Sci Technol*, 44: 269–276.

Demetriades, K., Guffey, C., and Khalil, M.H. 1995. Evaluating the role of honey in fat-free potato chips. *Food Technol*, 49(10): 66–67.

FDA Juice HACCP Regulation: Questions & Answers, September 4, 2003.

Fruit Juices & Fruit Nectars (Scotland) Regulations,–3 February–SSI15842003.

Fruit Juices and Fruit Nectars (England) Regulations, 11th Draft, LEG 06/325;3.2.03, 2003.

Fügel, R., Carle, R., and Schieber, A. 2005. Quality and authenticity control of fruit purées, fruit preparations and jams—A review. *Trends Food Sci Technol*, 16: 433–441.

International Center for Alcohol Policies, Minimum Age Limits Worldwide. http://icap.org/table/MinimumAgeLimitsWorldwide. Retrieved November 3, 2011.

Jones, R. 2001. Honey and healing through the ages. In Munn, P. and Jones, R. (eds), *Honey and Healing*. International Bee Research Association IBRA, Cardiff, GB, 1–4.

Karagöz, A. 2009. Akrilamid ve Gidalarda Bulunuşu. *TAF Prev Med Bull*, 8(2).

Krell, R. 1996. *Value-Added Products from Beekeeping*. FAO Agricultural Services Bulletin No. 124 Food and Agriculture Organization of the United Nations, Rome.

Lee, C.Y. and Kime, R.W. 1984. The use of honey for clarifying apple juice. *J Apic Res*, 23(1): 45–49.

Maillard, L.C. 1912. Réaction générale des acides aminés sur les sucres: Ses conséquences biologiques. Compte-rendu Société de Biologie, tome 72 (LXXII), http://www.lc-maillard.org/.

Mckibben, J. and Engeseth, N.J. 2002. Honey as a protective agent against lipid oxidation in ground turkey. *J Agric Food Chem*, 50(3): 592–595.

Minervini, F., Pinto, D., Cagno, R.D., Angelis, M.D., and Gobbetti, M. 2011. Scouting the application of sourdough to frozen dough bread technology. *J Cereal Sci*, 54(3): 296–304.

Nagai, T., Inoue, R., Kanamori, N., Suzuki, N., and Nagashima, T. 2006. Characterization of honey from different floral sources. Its functional properties and effects of honey species on storage of meat. *Food Chem*, 15(2): 256–262.

Nahmias, F. 1981. *[Cure Yourself with Honey.] Curatevi con il miele*. De Vecchi Editore, Milano, 125 pp.

Neumann, P.E. and Chambers, E. 1993. Effects of honey type and level on the sensory and physical properties of an extruded honey-graham formula breakfast cereal. *Cereal Food World*, 38(6): 418.

NHB. 1995. *Honey in Extruded Snacks*. National Honey Board, Longmont, CO.

Radojčić, I., Berković, K., Kovač, S., and Vorkapić-Furač, J. 2008. Natural honey and black radish juice as tin corrosion inhibitors. *Corros Sci*, 50: 1498–1504.

Selomulyo, V.O. and Zhou, W. 2007. Frozen bread dough: effects of freezing storage and dough improvers. *J Cereal Sci*, 45: 1–17.

The EU Council Directive (2001/113/EC of December 20, 2001).

Turkish Food Codex (Communiqué on Jam, Jelly, Marmalade and Sweetened Chestnut Puree) (2002/10).

Tyl, R. and Crump, K. 2003. Acrylamide in food. *Food Standards Agency*, 5: 215–222.

Ustunol, Z. and Gandhi, H. 2001. Growth and viability of commercial Bifidobacterium spp. in honeysweetened skim milk. *J Food Prot*, 64(11): 1775–1779.

Wakayama, T. and Lee, C.Y. 1987. Factors influencing the clarification of apple juice with honey. *Food Chem*, 25: 111–116.

Yildiz, O., Aliyazicioglu, R., Sahin, H., Aydin, Ö., and Kolayli, S. 2011. Ak dut (Morus alba) pekmezi pestili ve kömesinin üretim metotlari. *Gümüşhane University Journal of Science and Technology Institute*, 1(1): 47–56.

Yildiz, O. and Alpaslan, M. 2012. Properties of rosehip marmalades, food technology and biotechnology. *Food Technol Biotechnol*, 50(1): 98–106.

Yildiz, O., Sahin, H., Kara, M., Aliyazicioglu, R., Tarhan, Ö., and Kolayli, S. 2010. Maillard Reaksiyonlari ve Reaksiyon Ürünlerinin Gidalardaki Önemi. *Academic Food Journal*, 8(6): 44–51.

# Index

Page numbers followed by f and t indicate figures and tables, respectively.